Refinement in Z and Object-Z

John Derrick • Eerke A. Boiten

Refinement in Z and Object-Z

Foundations and Advanced Applications

Second Edition

John Derrick
Department of Computer Science
University of Sheffield
Sheffield, UK

Eerke A. Boiten
School of Computing
University of Kent
Canterbury, UK

ISBN 978-1-4471-6167-7
ISBN 978-1-4471-5355-9 (eBook)
DOI 10.1007/978-1-4471-5355-9
Springer London Heidelberg New York Dordrecht

Printed on acid-free paper

Springer is part of Springer Science+Business Media (www.springer.com)

To our families:
Michelle and Bea;
Gwen, Sara, Ivor, Rowan and Jasmyn

Preface

What Is Refinement?

Refinement is one of the cornerstones of a formal approach to software engineering. The traditional purpose of refinement is to show that an implementation meets the requirements set out in an initial specification. Armed with such a methodology, program development can then begin with an abstract specification and proceed via a number of steps, each step producing a slightly more detailed design which is shown to be a refinement of the previous specification.

In order to help verify refinements, different notations use different approaches, and in Z, Object-Z and other state-based languages, the techniques of downward and upward simulations emerged as the standard method of verifying refinements. However, modern specification languages and practices offer challenges, and variations to the model of computation, such as partial specifications, informal specifications, and combinations of state-based and behavioural (e.g. process-algebraic) styles. These required generalisations of the existing formal basis, and have led to new applications. The purpose of this book is to bring together that work in one volume.

Z and Object-Z

Refinement has been studied in many notations and contexts, however in this book we will concentrate on Z and Object-Z.

Z is a formal specification language, which uses mathematical notation (set theory and logic) to describe the system under consideration. Z, along with Object-Z, (Event-)B, ASM, Alloy, and VDM, is often called a state-based language because it specifies a system by describing the states it can be in, together with a description of how the state evolves over time. In effect a specification builds a model of the system.

Object-Z is an object-oriented extension of Z, which adds the notion of class as an additional structuring device, enabling an Object-Z specification to represent a number of communicating objects.

The importance of Z for refinement is that Z specifications are abstract and often loose, allowing refinement to many different implementations. In this book we discuss some of the techniques for doing so.

Structure of the Book

The book is structured into a series of parts, each focusing on a particular theme.

In Part I we provide an introduction to data refinement, and look at its applicability in Z. This leads to the formalisation of both downward and upward simulations in the Z schema calculus. We then discuss how refinements can be calculated and look at the application of this work to the derivation of tests from formal specifications. We also show how we can derive a single complete simulation rule by using possibility mappings.

Part II looks at generalisations of refinement which allow refinement of the interface and atomicity of operations. The standard presentation of refinement requires that the interface (i.e., input/output) of an operation remains unchanged, however for a variety of applications this is unsuitable and there is a need for more general refinements which this part discusses. The refinement of an interface is intimately tied up with the problem of non-atomic refinement, i.e., where we decompose a single abstract operation into a number of concrete operations upon refinement. This part discusses this problem in some detail, presenting refinement rules for specifications containing internal operations, as well as the general approach to non-atomic refinement in Z.

In Part III we look at how the theory of refinement can be used in an object oriented context. To do so we introduce the Object-Z specification language as a canonical example, and show how refinement between classes (and collections of classes) can be verified. We also discuss the relationship between refinement and inheritance and look at the formalisation of interface refinement and non-atomic refinement in Object-Z.

Finally, in Part IV we turn our attention to combining notions of state and behaviour when specifying and refining systems. The use of Object-Z provides a notion of state, which we combine with the notion of behaviour that is provided by the process algebra CSP.

In a concluding chapter we also briefly discuss a number of related formalisms, like ASM, B and Event-B, the refinement calculus, and VDM.

The dependencies between the chapters of the book are shown in Fig. 1.

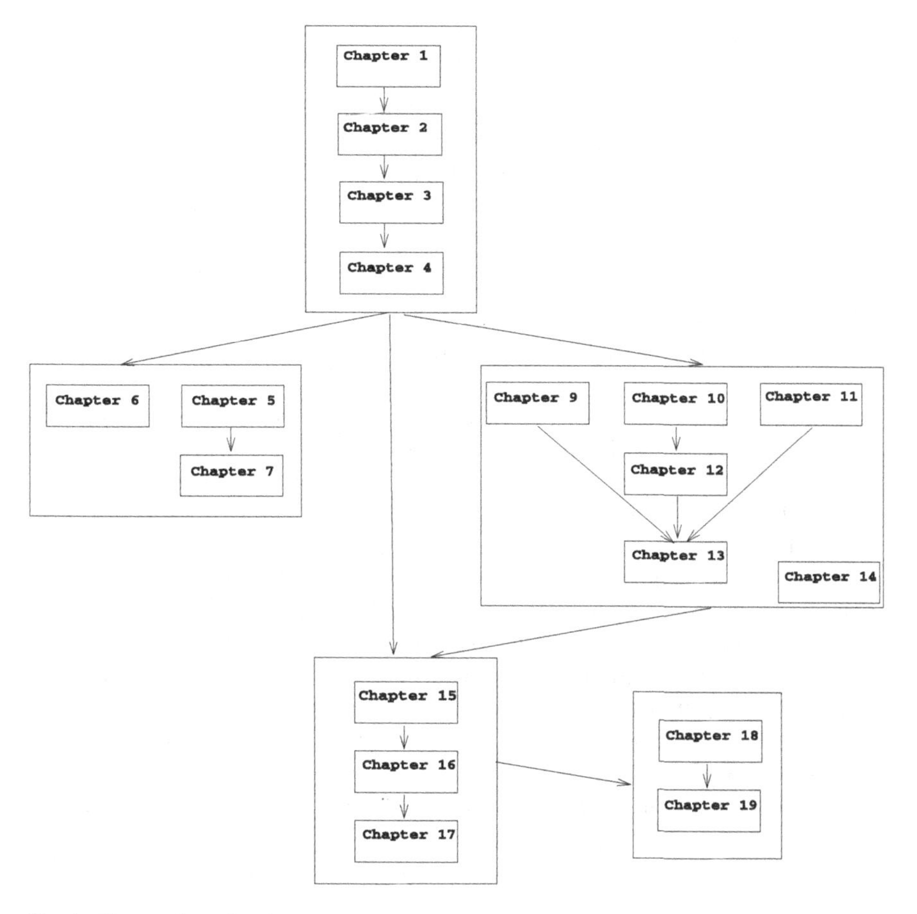

Fig. 1 Chapter dependencies

Readership

The book is intended primarily as a monograph aimed at researchers in the field of formal methods, academics, industrialists using formal methods, and postgraduate researchers. By including an introduction to the languages and notations used, the book is self contained, and should appeal to all those with an interest in formal methods and formal development.

The book is also relevant to a number of courses, particularly at postgraduate level, on software engineering, formal methods, object-orientation and distributed systems.

There is an associated web page to accompany this book:

http://www.refinenet.org.uk/refinezoz/

Second Edition

The first edition of this book incorporated most of our research into Z and Object-Z refinement up to that point. The major changes to this second edition are based on new research insights developed since then, particularly from our research programme on "relational concurrent refinement" as well as on action refinement and other topics. In addition to numerous small changes throughout the text, this second edition differs from the first edition in the following aspects.

- The fundamental Chaps. 3 and 4 have been extended to also include the trace refinement relation, based directly on partial relations rather than through totalisation.
- In Part II, Chap. 11 contains an updated discussion on divergence, and Chap. 12 a discussion on coupled simulations as a means to verify non-atomic refinements. In addition, Chap. 14 now contains material on approximate refinement.
- In Part III on Object-Oriented Refinement we have strengthened several specifications involving object aggregates with framing conditions.
- In Part IV on modelling state and behaviour we have added to Chap. 18 some material on differing semantics of operations and outputs and how they affect the abstraction of models written using Object-Z and CSP. In addition, Chap. 19 has been significantly revised, based on research into combining state-based and behavioural specification methods in the last ten years. In particular, we give a fuller account of the relationship between relational refinement and various models of refinement in CSP.
- Bibliographic notes at the end of each chapter have been extended with more recent research.
- The concluding chapter has been extended with discussion of ASM and Event-B.

Acknowledgements

Fragments of Chap. 3 and some later chapters were reprinted from: *Proceedings of Mathematics of Program Construction 2000*, R.C. Backhouse and J.N. Oliveira (Eds.), Lecture Notes in Computer Science 1837, E.A. Boiten and J. Derrick, "Liberating data refinement", pp. 144–166, copyright (2000), with kind permission of Springer Science and Business Media.

Fragments of Chap. 5 were reprinted from: *Information and Software Technology*, Volume 41, J. Derrick and E.A. Boiten, "Calculating upward and downward simulations of state-based specifications", pp. 917–923, copyright (1999), with permission from Elsevier Science.

Fragments of Chap. 7 are reproduced with permission from: *Software Testing, Verification and Reliability*, Volume 9, J. Derrick and E.A. Boiten, "Testing refinements of state-based formal specifications", pp. 27–50, 1999 © John Wiley & Sons Limited.

Fragments of Chap. 8 were reprinted from: *The Journal of Logic and Computation*, Volume 10, Number 5, J. Derrick, "A single complete simulation rule in Z", pp. 663–675, copyright (2000), by permission from Oxford University Press.

Fragments of Chap. 9 were reprinted from: *International Refinement Workshop & Formal Methods Pacific '98*, J. Grundy, M. Schwenke and T. Vickers (Eds.), Series in Discrete Mathematics and Theoretical Computer Science, E.A. Boiten and J. Derrick, "Grey box data refinement", pp. 45–60, copyright (1998), with kind permission of Springer Science and Business Media.

Fragments of Chap. 11 were reprinted from: *Formal Aspects of Computing*, Volume 10, J. Derrick, E.A. Boiten, H. Bowman and M.W.A. Steen, "Specifying and refining internal operations in Z", pp. 125–159, copyright (1998), with kind permission of Springer Science and Business Media.

Fragments of Chap. 11 were reprinted from: *Formal Aspects of Computing*, Volume 21, E.A. Boiten, J. Derrick, and G. Schellhorn, "Relational concurrent refinement Part II: internal operations and outputs", pp. 65–102, copyright (2009), with kind permission of Springer Science and Business Media.

Fragments of Chap. 12 were reprinted from: *FM'99 World Congress on Formal Methods in the Development of Computing Systems*, J.M. Wing and J.C.P. Woodcock and J. Davies (Eds.), Lecture Notes in Computer Science 1708, J. Derrick and E.A. Boiten, "Non-atomic refinement in Z", pp. 1477–1496, copyright (1999), with kind permission of Springer Science and Business Media.

Fragments of Chap. 12 were reprinted from: *ZB 2003: Formal Specification and Development in Z and B, 3rd International Conference of B and Z Users*, D. Bert, J.P. Bowen, S. King and M.A. Waldén (Eds.), Lecture Notes in Computer Science 2651, J. Derrick and H. Wehrheim, "Using coupled simulations with non-atomic refinement", pp. 127–147, copyright (2003), with kind permission of Springer Science and Business Media.

Fragments of Chap. 14 were reprinted from: *ZB 2005: Formal Specification and Development in Z and B, 4th International Conference of B and Z Users*, H. Treharne and S.A. Schneider (Eds.), Lecture Notes in Computer Science 3455, E.A. Boiten and J. Derrick, "Formal program development with approximations", pp. 374–392, copyright (2005), with kind permission of Springer Science and Business Media.

Fragments of Chap. 18 were reprinted from: *ICFEM 2002: Formal Methods and Software Engineering, 4th International Conference on Formal Engineering Methods*, C. George and H. Miao (Eds.), Lecture Notes in Computer Science 2495, G. Smith and J. Derrick, "Abstract specification in Object-Z and CSP", pp. 108–119, copyright (2002), with kind permission of Springer Science and Business Media.

Fragments of Chap. 19 were reprinted from: *Formal Aspects of Computing*, Volume 15, J. Derrick and E.A. Boiten, "Relational concurrent refinement", pp. 182–214, copyright (2003), with kind permission of Springer Science and Business Media.

Many people have contributed to this book in a variety of ways. We have had numerous discussions on aspects of specification and refinement and benefited from related work in this area with other researchers including: Ralph Back, Richard Banach, Howard Bowman, Michael Butler, David Cooper, Jim Davies, Brijesh Dongol, Steve Dunne, Clemens Fischer, Lindsay Groves, Ian Hayes, Martin Henson, Karl Lermer, Peter Linington, Christie Marr (née Bolton), Carroll Morgan, Ernst-Rüdiger Olderog, Steve Reeves, Gerhard Schellhorn, Steve Schneider, Kaisa Sere, Maarten Steen, Susan Stepney, Ketil Stølen, David Streader, Ramsay Taylor, Marina Waldén, and Jim Woodcock.

Thanks are especially due to Behzad Bordbar, Rob Hierons, Ralph Miarka, Graeme Smith, Chris Taylor, Ian Toyn, Helen Treharne, and Heike Wehrheim, for their careful reading of earlier drafts of the first edition and helpful suggestions. Thanks also to the reviewers of the first edition, in particular Perdita Stevens, for their useful comments.

We also wish to thank staff at Springer UK for their help and guidance in preparing the first and second editions of this book: Rosie Kemp, Steve Schuman, Beverley Ford, and Ben Bishop.

Finally, thanks to our friends and families who have provided love and support throughout this project. We would not have completed this book without them.

Contents

Part I
Refining Z Specifications

This part of the book provides an introduction to data refinement, and considers its application to Z. This leads to the formalisation of both downwards and upwards simulations in the Z schema calculus. We also consider how refinements can be calculated and the application of refinement in the specific contexts of promotion and testing. Finally, we present a single complete refinement rule.

Chapter 1 introduces the Z specification notation. It presents the notations for logic, sets and relations, the schema notation and the schema calculus, leading to the definition of an abstract data type in the "states-and-operations" style.

Chapter 2 introduces the notion of refinement informally, and discusses the refinement of operations on a given data type.

In Chap. 3 the formal definitions of data refinement and upward and downward simulation are presented, initially for total relations. Various ways of applying these definitions to abstract data types with partial relations are then considered. This chapter is phrased in terms of relations, rather than Z schemas. Those readers who are mostly interested in the application of refinement to Z may consider skipping Chaps. 3 and 4 up to the definitions of upward and downward simulation at the end of Sect. 4.4.

Chapter 4 translates the relational refinement rules for upward and downward simulation into rules for specifications consisting of Z schemas. Particular attention is given to the *rôle* of inputs and outputs in Z.

Chapter 5 shows how, given a simulation relation, refinements can be calculated. This is first expressed relationally, and then applied to Z.

Chapter 6 discusses a technique for structuring specifications known as *promotion*, and in which circumstances this allows component-wise refinement in Z.

Chapter 7 explores the relationship between testing and refinement. In particular, it looks at how tests for a refinement can be derived from tests for the abstract system.

The final chapter in this part, Chap. 8, presents a single complete refinement rule for Z, formulated using possibility mappings.

Chapter 1
An Introduction to Z

1.1 Z: A Language for Specifying Systems

Z is a formal specification language. It is a language in that it provides a notation for describing the behaviour of a system, and it is formal in that it uses mathematics to do so. The mathematics is simple, consisting of first order predicate logic and set theory. However, based upon this, it offers a very elegant way of *structuring* the mathematics to provide a specification of the system under consideration.

This elegance was not the result of a happy accident but the product of a language that had been developed alongside its use in an industrial environment. In particular, Z grew out of work at Oxford University's Programming Research Group, and benefited from being used at IBM Hursley for the re-specification of their CICS system. Since then, Z has been applied to numerous projects involving both software and hardware in a number of application areas (financial, security, safety critical, and so on), e.g., see [27, 33, 37, 39, 58] for a representative selection.

Two texts helped establish Z and stabilise the notation. Hayes [28] edited a collection of case studies, and then Spivey produced a reference manual [51] which became the *de facto* language definition for many years. There is now a substantial body of literature devoted to the language and related techniques, some of which we mention in Sect. 1.10, including its own conference series, now merged into the ABZ series. In 2002 it underwent standardisation by ISO [35].

One of the advantages of Z is that it can be used in a number of different ways according to the style appropriate to the application area. For example, its use to specify the functionality of a database will probably be radically different from when it is used to specify a security policy. One of the disadvantages, however, is that although it is often called a formal method, Z provides little in the way of methodology.

The word methodology can mean two things: a management methodology describing how and to what Z can be applied, and how to control the process of developing a system from a Z specification; or a formal methodology which describes a set of rules for transforming a Z specification into another (perhaps more detailed) description.

J. Derrick, E.A. Boiten, *Refinement in Z and Object-Z*,
DOI 10.1007/978-1-4471-5355-9_1, © Springer-Verlag London 2014

The former is of considerable importance, and many books touch on the issue, in particular the text by Barden, Stepney and Cooper [3] provides some useful advice written from an industrial perspective. The purpose of *this* book, however, is to explore the latter.

This idea of transformation is known as refinement. Refinement in Z has, like the language, been developed over a number of years and contrasts with, for example, the B notation which has a particular notion of refinement embedded as part of the language definition. A number of techniques and methods of refinement have been developed for Z, and it has reached a sufficient level of maturity to be applied in an industrial context (see, for example, [53, 58]).

In this book we shall be concerned with the refinement of one Z specification into another more detailed Z specification where further design decisions have been made. The process of turning a detailed Z specification into program code is a separate area of study, and is addressed, for example, by the refinement calculus (see Chap. 20).

The Z specifications we consider will be written in the (usual) "states-and-operations" style. In this style a system is described by specifying operations which describe changes to the state of the system. The state of the system and the operations acting on it are written in Z using *schemas* which structure the specification into convenient components. A *calculus* is provided to combine the schemas in appropriate ways, and this schema calculus helps to structure the specification.

Used in this way, Z is very similar to other state-based languages such as (Event-)B, ASM, VDM and Object-Z, and many of the comments we make in this book are equally applicable to these other notations. Part III of this book is devoted to Object-Z, an object oriented extension of Z, whilst refinement in ASM, (Event-)B and VDM is discussed briefly in Chap. 20.

In this chapter we provide a brief introduction to the Z notation. Section 1.2 discusses the use of logic and set theory in Z, this leads to Sect. 1.3 covering types, declarations and abbreviations. Relations, functions, sequences and bags are then discussed in Sect. 1.4.

Section 1.5 introduces the idea of a schema as a means to structure a specification, and illustrates the use of the schema calculus. Section 1.6 contains a first example of refinement in Z. Section 1.7 discusses the issue of semantics versus conventional interpretation in Z. Finally, Sects. 1.8 and 1.9 discuss standardisation and tool support respectively. To those familiar with the language there are few surprises here, and this chapter could be safely skipped. The discussion of refinement in Z begins in earnest in Chap. 2, although we do provide a motivating example here in Sect. 1.6.

1.2 Predicate Logic and Set Theory

Z uses propositional and predicate logic to express relationships between the components of a system, and to constrain the behaviour accordingly. For example, the specification of an operation will usually consist of inputs, outputs and changes to the state of the system. The relationship between input, output and before- and after-state will be described by a predicate relating and constraining these values.

The propositional logic used contains a number of connectives with their usual meanings, given as follows in descending order of operator precedence.

$$\begin{array}{ll} \neg & \text{negation} \\ \wedge & \text{conjunction} \\ \vee & \text{disjunction} \\ \Rightarrow & \text{implication} \\ \Leftrightarrow & \text{equivalence} \end{array}$$

A standard set of introduction and elimination rules is usually used, see for example [46] or [57]. However, although we shall be concerned with verification of refinements, we adopt the usual informality of mathematics in many of our derivations. Detailed step by step proofs can be supported by tools built specifically or tailored for Z, in particular those allowing meta-theorems, see Sect. 1.9 for an overview. For example, the refinement theory we developed in [6] was mechanically checked by Schellhorn using the KIV tool.

Predicate logic is provided by the introduction of quantifiers into the language, together with free and bound variables. Z is a typed language which means that every variable ranges over a fixed set of values, and all quantifications are similarly typed.

Universal quantification (for all) has the form

$$\forall x : T \mid P \bullet Q$$

and is interpreted as meaning that for all x in T satisfying P, Q holds.

Existential quantification (there exists) has the form

$$\exists x : T \mid P \bullet Q$$

and is interpreted as meaning that there exists an x in T satisfying P such that Q holds.

The constraint P is optional (omission is equivalent to a constraint of *true*), and is in fact an abbreviation defined as follows.

$$\begin{array}{lcl} \forall x : T \mid P \bullet Q & \Leftrightarrow & \forall x : T \bullet P \Rightarrow Q \\ \exists x : T \mid P \bullet Q & \Leftrightarrow & \exists x : T \bullet P \wedge Q \end{array}$$

Quantifiers introduce bound variables, and the usual scoping laws apply. For example, in

$$\forall x : \mathbb{N} \bullet x \leq 7$$

the scope of x is the whole expression, whereas in

$$(\forall y : \mathbb{N} \bullet \exists z : \mathbb{Z} \bullet y + z = y) \land (\exists y : \mathbb{N} \bullet y = 4)$$

the scopes of the two declarations of y are each constrained to one of the bracketed subexpressions.

Variables that are not bound in predicates are said to be free. Substitution of free variables allows occurrences of one variable to be substituted for existing occurrences. A substitution $P[y/x]$ denotes the predicate that results from substituting y for each free occurrence of x in P. We will normally only use this as a meta-notation.

Using substitution, we can express a proof rule that occurs very frequently: one might call it $\exists$-elimination, but we call it the *one point rule*. For any predicate P, expression e, and set T:

$$\exists x : T \bullet x = e \land P \quad \equiv \quad P[e/x] \land e \in T$$

That is, if we are required to demonstrate the existence of a variable ($\exists x : T$) and a candidate value is named ($x = e$), then we can eliminate the existential quantifier under an appropriate substitution. We shall see an example of its use in Sect. 4.3. The need for the second conjunct in the rule should be clear from a case like $\exists x : \{1\} \bullet x = 0$.

Equality ($=$) between expressions is provided and allows the definition of unique existential quantification, which has the form

$$\exists_1 x : T \mid P \bullet Q$$

and asserts that there exists one, and only one, element of T satisfying P such that Q holds.

Z makes extensive use of set theory (hence the name, from Zermelo-Fraenkel), both as a descriptive medium and also to define further data structures such as relations, functions and sequences. Membership ($\in$), the empty set ($\varnothing$), equality ($=$) and subset ($\subseteq$) are all defined as usual. Sets can be defined by extension (i.e., listing their elements), as in

$$\{Dennis, Gnasher, Bea\}$$

There are also two forms of set comprehension. The set of all elements k in S that satisfy the predicate P is written

$$\{k : S \mid P\}$$

For example, $\{x : \mathbb{Z} \mid (x \text{ is even}) \land (1 \leq x \leq 4)\}$ describes a set consisting of 2 and 4. There is also a form of comprehension that returns an expression constructed from values satisfying a predicate. Thus

$$\{k : S \mid P \bullet e\}$$

denotes the set consisting of all expressions e such that $k \in S$ satisfies P. For example, $\{n, m : \mathbb{N} \mid \{m, n\} \subseteq 1..2 \bullet (n^2, m^3)\}$ describes a finite set consisting of the elements $(1, 1)$, $(1, 8)$, $(4, 1)$ and $(4, 8)$.

We have occasionally preferred the untyped notation of traditional set theory, but have used it only in ways that could be typed.

The power set constructor, Cartesian products, union, intersection and difference are all available within the language. Thus $\mathbb{P}\, S$ represents the set of all subsets of S. Hence

$$\mathbb{P}\{1, 2\} = \big\{\varnothing, \{1\}, \{2\}, \{1, 2\}\big\}$$

The set of all *finite* subsets of a set S is constructed using $\mathbb{F}$. The size of a finite set S is determined by its cardinality (#), thus

$$\#\{5, 8, 7\} = 3$$

Union ($\cup$), intersection ($\cap$) and difference ($\setminus$) all have their usual meaning.

Cartesian products allow the construction of sets of ordered pairs. For any sets S, T the product $S \times T$ denotes the set of ordered pairs whose first element is in S and whose second element is in T. Elements of this set are written (s, t) where $s \in S$ and $t \in T$. Hence $(Dennis, 4)$ and $(Gnasher, 2)$ are members of $\{Dennis, Gnasher, Bea\} \times \mathbb{N}$.

1.3 Types, Declarations and Abbreviations

So far we have introduced some background notation without describing what a specification contains, and how it is defined. We begin this discussion now. In general our specifications will contain the following information in roughly the same order.

- The types and global constants for the specification.
- A description of the abstract state, usually written using schemas.
- A description of the initial state of the system.
- One or more operations, which define the behaviour of the system. Again these will be written using schemas and the schema calculus.

However, Z does not just consist of mathematics, but allows informal commentary to be interleaved with formal definitions and declarations throughout a specification. Indeed, such informal commentary, providing a natural language description of the intended meaning of the specification, is at least as important as its formal counterpart.

1.3.1 Types

There are a number of ways in which a specifier can define the types needed for a specification. Z provides a single built-in type, namely the type of integers $\mathbb{Z}$. Every value in Z has a *unique* type; when x is declared as $x : S$ then the type of x is the largest set containing S. Thus, every type is a set, and every set[1] is contained in exactly one type. The sets $\mathbb{N}$ and $\mathbb{N}_1$ are defined as subsets of $\mathbb{Z}$, being the sets $\{0, 1, 2, \ldots\}$ and $\{1, 2, \ldots\}$ respectively. Thus, the type of 7 is $\mathbb{Z}$, not $\mathbb{N}$.

For illustrative purposes, we sometimes use the type of real numbers $\mathbb{R}$. The Z standard uses *true* and *false* as the possible values of Boolean expressions; we go beyond the standard in identifying them as the only members of a type of Booleans $\mathbb{B}$, and we use elements of $\mathbb{B}$ interchangeably with predicates and will write e.g.

$$b' = (x > 7)$$

rather than

$$\textbf{if } x > 7 \textbf{ then } b' = \mathit{true} \textbf{ else } b' = \mathit{false}$$

and $\neg b$ instead of $b = \mathit{false}$, etc.

Although Z only provides a single pre-existing type, there are a number of ways for a specifier to define types relevant to the specification under construction.

One way to build further types is to simply declare them. A *given set* is a declaration of the form

$$[\mathit{TYPE}]$$

which introduces a new type *TYPE*. Thus

$$[\mathit{PERSON}]$$

defines a set *PERSON*, but at this stage no information is given about its values and the relationships between them.

Types can also be constructed from existing types in a variety of ways. These include Cartesian products, for example, $\mathit{PERSON} \times \mathbb{Z}$ is a type consisting of ordered pairs. Arbitrary n-ary Cartesian products are defined and used analogously. Hence if *Dennis* is of type *PERSON* then $(\mathit{Dennis}, 4) \in \mathit{PERSON} \times \mathbb{Z}$.

Various collection types can also be defined. The power set constructor $\mathbb{P}$ is the only one of these which is elementary. Finite sets, sequences and bags can also be used but actually abbreviate sets with restrictions or a particular structure, see below for details.

Another important type constructor is the *free type*. Formally it is defined as a given type with additional constraints, see the next section. Free types are declared

[1] The symbol $\varnothing$ is overloaded, denoting the empty sets of all possible types.

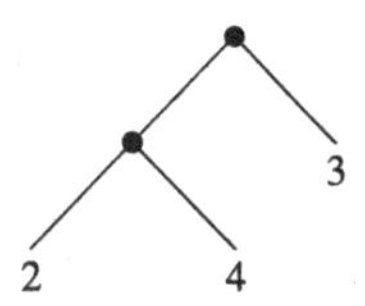

Fig. 1.1 The tree *branch*(*branch*(*leaf*(2), *leaf*(4)), *leaf*(3))

in a BNF-like notation, e.g.,

$$PRIMARY ::= red \mid yellow \mid blue$$

denotes a type *PRIMARY* containing exactly three different constants *red*, *yellow* and *blue*. Similarly the Booleans might be defined by $\mathbb{B} ::= true \mid false$.

Other types may also occur on the right-hand side of a free type definition. An expression like

$$anint\langle\langle \mathbb{Z} \rangle\rangle$$

denotes an integer value labelled with the constructor function *anint*. Thus we can represent a type of values which are either Booleans or integers by

$$INTORBOOL ::= anint\langle\langle \mathbb{Z} \rangle\rangle \mid abool\langle\langle \mathbb{B} \rangle\rangle$$

values of which are denoted, for example, by *anint*(7) and *abool*(*true*). Even the newly defined type may occur on the right-hand side of its definition, e.g.,

$$NATTREE ::= leaf\langle\langle \mathbb{N} \rangle\rangle \mid branch\langle\langle NATTREE \times NATTREE \rangle\rangle$$

is a definition of binary trees with natural number values at the leaves only, see, for example, Fig. 1.1. There are restrictions on recursive occurrences of the newly defined type on the right-hand side which are detailed in [2] and meticulously observed in this book.

A further kind of type is defined in Z, the *schema type*, which we will introduce in Sect. 1.5.

1.3.2 Axiomatic Definitions

To introduce variables of any type T we use the standard form of declaration $x : T$, which is known as a signature. Z also allows subsets of types to be used in a declaration, and these can be expanded into a signature plus constraint if necessary. Axiomatic definitions introduce new objects into a specification which are subject to constraints. An axiomatic definition is written

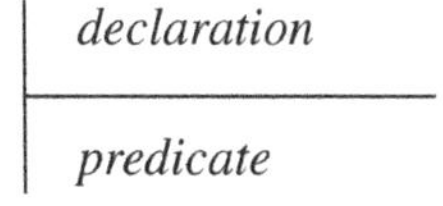

and the purpose of the predicate is to constrain the values introduced in the declaration. The predicate is optional, thus

$$PLAYER : \mathbb{P}\, PERSON$$

introduces a global constant *PLAYER* as a subset of *PERSON*. Predicates can uniquely determine value introduced, e.g., we can introduce a constant representing the number of associated football clubs in the English premier league as

$$leaguesize : \mathbb{N}$$

$$leaguesize = 20$$

However, axiomatic definitions can also be loosely specified, e.g.,

$$squadsize : \mathbb{N}$$

$$squadsize \geq 11$$

Free types are formally defined in terms of a given type with an axiomatic definition, e.g., the earlier definition of *PRIMARY* represents

$$[PRIMARY]$$

$$red, yellow, blue : PRIMARY$$

$$PRIMARY = \{red, yellow, blue\}$$
$$red \neq yellow \land blue \neq red \land yellow \neq blue$$

1.3.3 Abbreviations

Abbreviations are vital to the readability of a specification. In Z they are written as follows

$$name == expression$$

Thus, we could have defined

$$leaguesize == 20$$

and

$$SMALLNAT == \{0, 1, 2\}$$

introduces a set consisting of the three elements listed. An abbreviation does not introduce new values, so we cannot write

$$POSITION == \{goal, back, mid, forward\}$$

until the names *goal* etc. have been introduced into the specification. One way to do this would be to use an axiomatic definition, but a neater way would be to use a free type. So, the definition

$$POSITION ::= goal \mid back \mid mid \mid forward$$

would introduce *POSITION* as a type consisting exactly of these four, different, values, which had not been previously defined.

1.4 Relations, Functions, Sequences and Bags

We can now write down types and constants for a specification. Before we go on to describe how state and operations are structured using schemas, it is useful to define some additional data structures for use in their description, and Z makes extensive use of relations, functions and sequences.

1.4.1 Relations

A relation is a set of ordered pairs, and for sets X and Y, the set of all relations between X and Y is denoted $X \leftrightarrow Y$. Thus $X \leftrightarrow Y$ is just an abbreviation for $\mathbb{P}(X \times Y)$. When defining relations the so-called maplet notation $x \mapsto y$ is often used as an alternative to (x, y).

The simplest relation is the *identity relation*, defined (using set comprehension) by

$$id\, X == \{x : X \bullet x \mapsto x\}$$

This is sometimes also denoted by id_X.

Relations can be defined by abbreviation or by axiomatic definition. If our set of football players contains the following elements

$$\{Gullit, Lineker, Carlos, Schmeichel\} \subset PLAYER$$

we can describe the position that these players are capable of playing in as a relation *position*. This could be defined by writing

$$position == \big\{(Gullit, back), (Gullit, forward), (Lineker, forward), (Carlos, back), (Schmeichel, goal)\big\}$$

or equivalently as

$$\begin{array}{|l} position : PLAYER \leftrightarrow POSITION \\ \hline position = \{(Gullit, back), (Gullit, forward), (Lineker, forward), \\ \qquad (Carlos, back), (Schmeichel, goal)\} \end{array}$$

Then $(Schmeichel, goal) \in position$ formalises the statement that *Schmeichel* plays in goal. First and second *component projections* are provided. For any ordered pair p, *first* p and *second* p are the first and second components of the pair respectively, e.g., $first(Schmeichel, goal) = Schmeichel$. As an alternative, numeric selectors can also be used, as in $(Lineker, forward).2$ and $(Gullit, back).1$. These generalise to allow an arbitrary selection from an n-tuple.

The *domain* of a relation $R : X \leftrightarrow Y$ is the set of elements in X related to something in Y. Similarly the *range* of R is the set of elements of Y to which some element of X is related.

$$\mathrm{dom}\, R = \{p : R \bullet first\ p\}$$
$$\mathrm{ran}\, R = \{p : R \bullet second\ p\}$$

For example,

$$\mathrm{dom}\, position = \{Gullit, Lineker, Carlos, Schmeichel\}$$
$$\mathrm{ran}\, position = \{goal, back, forward\}$$

Instead of considering the whole of the domain and range we will frequently need to consider restrictions of these sets. The *domain restriction* of a relation $R : X \leftrightarrow Y$ by a set $S \subseteq X$ is the subset of R whose first components are in S. Similarly the *range restriction* of R by a set $T \subseteq Y$ is the set of pairs whose second components are in T.

$$S \lhd R = \{p : R \mid first\ p \in S\}$$
$$R \rhd T = \{p : R \mid second\ p \in T\}$$

For example,

$$\{Gullit\} \lhd position = \{Gullit \mapsto back, Gullit \mapsto forward\}$$
$$position \rhd \{forward\} = \{Gullit \mapsto forward, Lineker \mapsto forward\}$$

In a similar fashion we also often need subtractions from the domain and range. The *domain subtraction* of a relation $R : X \leftrightarrow Y$ by a set $S \subseteq X$ is the subset of R whose first components are not in S. Similarly the *range subtraction* of R by a set $T \subseteq Y$ is the set of pairs whose second components are not in T.

$$S \ntriangleleft R = \{p : R \mid first\ p \notin S\}$$
$$R \ntriangleright T = \{p : R \mid second\ p \notin T\}$$

For example,

$$\{Gullit, Lineker\} \ndres position = \{Carlos \mapsto back, Schmeichel \mapsto goal\}$$
$$position \nrres \{forward, goal\} = \{Gullit \mapsto back, Carlos \mapsto back\}$$

Sometimes we will need to determine the effect of a relation on a particular subset of elements, and for this we use *relational image*. The relational image of a set $S \subseteq X$ through $R : X \leftrightarrow Y$ is the subset of Y whose elements are related by R to S:

$$R(\!| S |\!) = \{p : R \mid first\ p \in S \bullet second\ p\} = \mathrm{ran}(S \lhd R)$$

So, for example, $position(\!| \{Gullit\} |\!) = \{back, forward\}$.

In a state-based description we shall often want to specify that a new state is the same as the old except for a small alteration. For a relation this will often involve changing some of the pairs in a relation to new ones, and to do this we use *overriding*. If R and Q are relations between X and Y the relational overriding, $Q \oplus R$, of Q by R is Q outside the domain of R but is R wherever R is defined:

$$Q \oplus R = \big((\mathrm{dom}\, R) \ndres Q\big) \cup R$$

Thus

$$\begin{aligned} &position \oplus \{Carlos \mapsto forward\} \\ &\quad = \big\{(Gullit, back), (Gullit, forward), (Lineker, forward), \\ &\qquad (Carlos, forward), (Schmeichel, goal)\big\} \end{aligned}$$

Notice that this is different from simply adding the pair $(Carlos, forward)$ to the relation, i.e., forming $position \cup \{Carlos \mapsto forward\}$. However, the two operators do coincide when the intersection between the domains of the relations is empty.

Suppose that the relation[2] $playedfor : PLAYER \leftrightarrow CLUBS$ is defined by

$$\begin{aligned} playedfor == \{&Ronaldo \mapsto madrid, Lineker \mapsto barcelona, \\ &Ronaldo \mapsto man\_utd, Carlos \mapsto madrid, Lineker \mapsto tottenham\} \end{aligned}$$

for an appropriate definition of $CLUBS$, then

$$playedfor \oplus \{Gullit \mapsto milan\} = playedfor \cup \{Gullit \mapsto milan\}$$

The *inverse* of a relation R, denoted $R^{\sim}$ or R^{-1}, is the relation obtained by reversing every ordered pair in the relation. Thus

$$\begin{aligned} playedfor^{-1} = \{&madrid \mapsto Ronaldo, barcelona \mapsto Lineker, \\ &man\_utd \mapsto Ronaldo, madrid \mapsto Carlos, \\ &tottenham \mapsto Lineker\} \end{aligned}$$

---

[2]Representing only very partial information about which player has ever played for which club.

Perhaps the most important operator on relations is *relational composition*. If $R_1 : X \leftrightarrow Y$ and $R_2 : Y \leftrightarrow Z$ then $R_1 \mathbin{;} R_2$ denotes their composition, and $(x, z) \in R_1 \mathbin{;} R_2$ whenever there exists y with $(x, y) \in R_1$ and $(y, z) \in R_2$:

$$R_1 \mathbin{;} R_2 = \{q : R_1, r : R_2 \mid (second\ q = first\ r) \bullet (first\ q) \mapsto (second\ r)\}$$

For example,

$$playedfor^{-1} \mathbin{;} position = \{barcelona \mapsto forward, madrid \mapsto back, \\ tottenham \mapsto forward\}$$

To determine which players played for the same team we could define

$$playedforsameteam == (playedfor \mathbin{;} playedfor^{-1}) \setminus id$$

which would be the relation

$$\{(Ronaldo, Carlos), (Carlos, Ronaldo)\}$$

The *transitive closure* R^+ of a relation R is the union of all finite compositions of R with itself:

$$R^+ = R \cup (R \mathbin{;} R) \cup (R \mathbin{;} R \mathbin{;} R) \cup \cdots$$

The *reflexive and transitive closure* R^* of a relation R of type $T \leftrightarrow T$ adds the identity relation on T to the transitive closure:

$$R^* = R^+ \cup id_T$$

Finally, a pair of relations can be combined using parallel composition $\|$. If we have $R : X \leftrightarrow Y$ and $S : U \leftrightarrow V$ then

$$R \| S : X \times U \leftrightarrow Y \times V$$

$$\forall x : X; y : Y; u : U; v : V \bullet \\ (x, u) \mapsto (y, v) \in R \| S \Leftrightarrow (x, y) \in R \land (u, v) \in S$$

Relations, functions and sequences play a dual *rôle* in this book. Not only are they used within specifications, but we also use them extensively to reason *about* specifications (cf. Chap. 3). That is, we use them as a convenient meta-notation when discussing Z and its refinement rules. To highlight and distinguish the use of Z as a meta-language, it will be typeset in sans serif.

For example, in deriving some of the refinement rules we shall use the standard notation for the weakest post- and pre-specification. These are defined by

$$X/R == \overline{(R^{-1} \mathbin{;} \overline{X})}$$

and

$$L \setminus X == \overline{(\overline{X} \mathbin{;} L^{-1})}$$

respectively where $\overline{X}$ represents complement. These operators are useful because they are the approximate left and right inverses for composition (i.e., $R \mathbin{;} T \subseteq X$ iff $T \subseteq X/R$ and $T \mathbin{;} L \subseteq X$ iff $T \subseteq L \setminus X$). Their pointwise characterisations are

$$(a, b) \in X/R \quad \Leftrightarrow \quad \forall c \bullet (c, a) \in R \Rightarrow (c, b) \in X$$
$$(a, b) \in L \setminus X \quad \Leftrightarrow \quad \forall c \bullet (b, c) \in L \Rightarrow (a, c) \in X$$

1.4.2 Functions

A function is, of course, a relation such that each element in the domain is mapped to at most one element in the range. Because functions are, in fact, just relations with particular properties, the operators on relations discussed above are all defined on functions, as in $\mathrm{dom} f$.

This means that relational composition is defined on functions, however we usually use a different notation, that of *functional composition*, to conform with normal mathematical usage. This is denoted $f \circ g$ where

$$f \circ g = g \mathbin{;} f$$

Thus $(f \circ g)(t) = f(g(t))$ for all $t \in \mathrm{dom}\, g$, where, as usual $g(t)$ denotes function application of g to t. Although function application in Z can be denoted using juxtaposition ($g\ t$ rather than $g(t)$), for clarity we will often use a bracketed notation.

In addition, Z provides a collection of symbols to denote the sets of functions with differing sets of properties.

The set of all *partial functions* from X to Y is denoted $X \nrightarrow Y$, and is the set of all relations between X and Y such that each $x \in X$ is related to at most one $y \in Y$. Somewhat confusingly, but in keeping with standard terminology, the concepts of a "function" and a "partial function" coincide. The set $X \rightarrow Y$ denotes all *total functions*, i.e., functions where $\mathrm{dom} f = X$. Thus we might write

$$shirt : PLAYER \rightarrow \mathbb{N}$$
$$player : \mathbb{N} \nrightarrow PLAYER$$

for functions *shirt* and *player* where *shirt*(p) is the number on the football jersey worn by p and *player*(n) is the player wearing jersey number n. Every player has a shirt number, so *shirt* is total, but not every number is used, so *player* is partial.

Functions can be *injective*, *surjective* or *bijective*. A function from X to Y is injective iff each $y \in Y$ is related to no more than one $x \in X$. A function from X to Y is surjective iff the range of it is equal to Y, and it is bijective iff it is both injective and surjective. A function is *finite* iff it consists of a finite set of pairs. Different arrows are used to denote different combinations as follows

$X \nrightarrow Y$	partial functions
$X \rightarrow Y$	total functions
$X \rightarrowtail Y$	total injective functions
$X \twoheadrightarrow Y$	total surjective functions
$X \rightarrowtail\!\!\!\!\!+ Y$	partial injective functions
$X \twoheadrightarrow\!\!\!\!\!+ Y$	partial surjective functions
$X \nrightarrow\!\!\!\!\!+ Y$	finite functions
$X \rightarrowtail\!\!\!\!\!++ Y$	finite injective functions
$X \rightarrowtail\!\!\!\!\twoheadrightarrow Y$	total bijections

Thus we could have written $shirt : PLAYER \rightarrowtail \mathbb{N}$ and $player : \mathbb{N} \twoheadrightarrow\!\!\!\!\!+ PLAYER$.

Global functions can be introduced by axiomatic definition, e.g., given four-tuples of natural numbers,

$$ROW == \mathbb{N} \times \mathbb{N} \times \mathbb{N} \times \mathbb{N}$$

the fourth of which is used to denote "points" (*pts*), we could define a "symbolic" accessor function

$$\begin{array}{|l}
pts : ROW \rightarrow \mathbb{N} \\
\hline
\forall r : ROW \bullet pts(r) = r.4
\end{array}$$

The usual lambda notation can also be used to define functions, e.g., the equivalent

$$pts == \lambda\, r : ROW \bullet r.4$$

1.4.3 A Pitfall: Undefined Expressions

Most of the historic discussion about Z semantics and logics for Z has concentrated on an issue closely related to the use of functions, namely that of ill-defined expressions. These include the application of a function to a value outside its domain, e.g division by zero, head of an empty sequence, and the "function application" of a relation which actually is not a function, i.e. relates more than one value from its range to a given value from its domain. Different approaches to addressing this problem have led to convoluted explanations, fiery discussions about the use of three-valued logic, and inconsistent proposals of logic for Z—in the end, the Z standard (see Sect. 1.8) ended up without a logic, partly due to this issue. As the emphasis in this book is on refinement, we aim to avoid this particular dark corner of Z usage, and guard all applications of partial functions with context information ensuring these applications are well-defined.

1.4.4 Sequences

Sequences are heavily used in Z to model collections of elements where order is important. Sequences are written listing elements between angle brackets:

$$\langle\rangle$$
$$\langle tottenham, arsenal, man\_utd\rangle$$

The first of these denotes the *empty sequence* and the second the sequence containing three elements in the order given.

In Z the elements in a given sequence have to have the same type, and $\mathrm{seq}\, X$ denotes the set of all finite sequences with elements taken from X. In fact we model sequences over X as functions from $\mathbb{N}$ to X where the number gives the position in the sequence, with indexing starting from 1. Thus

$$\mathrm{seq}\ X == \{f : \mathbb{N} \twoheadrightarrow X \mid (\exists n : \mathbb{N} \bullet \mathrm{dom} f = 1..n)\}$$

For example, we might represent the clubs in a football league as a sequence which represents their position in the league.

$$clubs : \mathrm{seq}\, CLUBS$$
$$clubs = \langle tottenham, arsenal, man\_utd\rangle$$

If we want to consider non-empty sequences we use the notation $\mathrm{seq}_1\, X$. If we want to assume that no element in a sequence appears more than once then we use injective sequences, written $\mathrm{iseq}\, X$. Thus

$$\langle man\_utd, arsenal, chelsea\rangle \in \mathrm{iseq}\, CLUBS$$
$$\langle man\_utd, man\_utd, arsenal\rangle \notin \mathrm{iseq}\, CLUBS$$

The elements in a sequence can have structure, thus $\mathrm{seq}(CLUBS \times \mathbb{N})$ consists of sequences of pairs so that the points that a club has gained so far in a season might be incorporated into a league table by defining:

$$clubs : \mathrm{seq}(CLUBS \times \mathbb{N})$$
$$\forall i, j : \mathrm{dom}\, clubs \mid i \leq j \bullet second(clubs(i)) \geq second(clubs(j))$$

the invariant saying that the order in the sequence is in descending point order. Because sequences are in fact special kinds of functions, writing $i : \mathrm{dom}\, clubs$ and $clubs(i)$ makes perfect sense since $\mathrm{dom}\, clubs$ is the set $1..n$, where n is the length of $clubs$ and $clubs(i)$ is the ith element in the sequence. Note that the length of a sequence s is given by $\#s$. A common Z idiom, abstracting away from the ordering and frequency of the elements in a sequence s, is $\mathrm{ran}\, s$: by forgetting the indices in the (index, value) pairs one indeed ends up with just the set of its elements.

A basic operation on sequences is *concatenation* which adds together two sequences preserving their order, thus the concatenation of s and t is a sequence which begins with all the elements of s and then continues with all the elements of t. Thus

$$\langle man\_utd, arsenal, man\_utd\rangle \frown \langle arsenal, chelsea, chelsea\rangle$$
$$= \langle man\_utd, arsenal, man\_utd, arsenal, chelsea, chelsea\rangle$$

Four other operations are also useful. The *head* of a sequence is the first element in the sequence, and the *tail* is the remainder of the sequence apart from the head. In a similar fashion functions *last* and *front* return the last element and all but the final element respectively. Hence

$$head\, s = s(1)$$
$$tail\, s = \lambda\, n : 1..(\#s - 1) \bullet s(n+1)$$
$$front\, s = \{\#s\} \lhd s$$
$$last\, s = s(\#s)$$

Note that all four operations are only defined for non-empty sequences.

We can also construct sequences of sequences. For example, we might represent the four English football tables as a sequence with four elements, each element being an individual league table.

$$leagues : \mathrm{seq}\, \mathrm{seq}\, CLUBS$$
$$\#leagues = 4$$
$$\#(head\, leagues) = 20$$
$$\forall\, l : tail\, leagues \bullet \#l = 24$$

The invariant states that there are four elements in *leagues*; the first element is a sequence with 20 elements, and the remainder each have 24 elements.

Distributed concatenation ($\frown/$) flattens a sequence of sequences to a single sequence, so that

$$\frown/\langle\langle leeds, arsenal, tottenham\rangle, \langle\rangle, \langle liverpool, man\_city\rangle\rangle$$
$$= \langle leeds, arsenal, tottenham, liverpool, man\_city\rangle$$

In general, for any associative operation $\otimes$ of type $X \times X \to X$, we can define distributed application of $\otimes$ over a sequence (of type $\mathrm{seq}\, X$) by

$$\otimes/\langle x_1, \ldots, x_n\rangle = x_1 \otimes \cdots \otimes x_n$$

We shall most often have cause to use distributed composition ($⨾/$) and distributed addition ($+/$).

Another operation that we frequently apply to sequences is the *map* operator $*$ which applies the same function to every element in a sequence. For a function

$f : X \rightarrow Y$ and sequence of type X we have

$$f * \langle x_1, \ldots, x_n \rangle = \langle f(x_1), \ldots, f(x_n) \rangle$$

In fact, it is also defined for relations. Thus for $R : X \leftrightarrow Y$

$$\begin{array}{|l} R* : \operatorname{seq} X \leftrightarrow \operatorname{seq} Y \\ \hline \forall s : \operatorname{seq} X;\ t : \operatorname{seq} Y \bullet \\ \qquad (s, t) \in R* \Leftrightarrow \# s = \# t \wedge \forall i : \operatorname{dom} s \bullet (s(i), t(i)) \in R \end{array}$$

For example, given $squad : \operatorname{seq} PLAYER$, $shirt * squad$ gives the sequence of the shirt numbers of all players in the squad.

1.4.5 Bags

Sequences represent both ordering and number of occurrences, while sets just represent occurrence. *Bags*, on the other hand, represent the number of occurrences without consideration of order. Bags are written as follows.

$$\begin{array}{l} [\![\ ]\!] \\ [\![a, b, a, b, c]\!] \\ [\![b, b, c, a, a]\!] \end{array}$$

The first is the empty bag, and the second and third denote the same bag, namely one where a and b both occur twice and c occurs once. Bags of type X are modelled as a function $X \rightarrow\!\!\!\!\!\!\!+ \;\; \mathbb{N} \setminus \{0\}$ where the number associated with an element represents its occurrence.

A collection of functions are defined on bags. Bag membership is denoted $x \operatorname{in} B$, $items\, s$ turns a sequence into the appropriate bag and $count\, B\, x$ denotes the number of times x occurs in B. Bag union and difference are also used. $B \uplus C$ contains as many copies of each element as B and C put together. For bag difference, suppose that $count\, B\ x = m$, $count\, C\ x = n$, then $B \uminus C$ contains $m - n$ occurrences of x if $m \geq n$ and 0 otherwise.

1.5 Schemas

The language constructs introduced so far have been as one would expect in any typed specification language using logic and set theory. Schemas are probably what Z is best known for, if only for their representation on paper by boxes. Schemas form a new layer on top of the constructs defined in the previous sections, and quite often *the* top layer of a Z specification.

Essentially, a schema denotes a *labelled product*, much like simple record types in programming languages, or tuples in relational databases. For example, the following schema

```
┌─ FBT ──────────────────────────────
│ manager, coach : PERSON
│ squad : ℙ PLAYER
└────────────────────────────────────
```

describes (unordered) tuples of three components, two of which (with labels *manager* and *coach*) are *PERSON*s, and one of which (with label *squad*) is a set of *PLAYER*s. *FBT* is the name of the type of all such tuples.

In a *specification* language like Z we can do more than just describe record types: we can also impose restrictions on the allowable records. This is done by adding a predicate to the schema, for example

```
┌─ PlayerManagerFBT ─────────────────
│ manager, coach : PERSON
│ squad : ℙ PLAYER
├──────────────
│ manager ∈ squad
└────────────────────────────────────
```

(which indirectly also forces $squad \neq \varnothing$, and is type-correct because $PLAYER \subseteq PERSON$), or (recall that *leaguesize* was defined as a global constant)

```
┌─ League ───────────────────────────
│ clubs : iseq CLUBS
├──────────────
│ #clubs = leaguesize
└────────────────────────────────────
```

Omitting the predicate, as in the schema *FBT*, is equivalent to having the predicate *true*.

The declarations of the components are *local* to the schema, so we would need to have the schema's declarations in context in order to use the component names. This can be done using the traditional dot-notation for records, e.g., when we have a declaration $Leeds : FBT$ in context, then *Leeds.manager* is a well-defined expression. This is not the most common way of introducing schema components in context, however, see Sect. 1.5.6.

1.5.1 Schema Syntax

Declarations may be split across lines; they may also be put on the same line, separated by semicolons. A predicate split across several lines denotes a *conjunction*, unless a line is broken with another operator, e.g.,

$$(x = 7) \lor$$
$$(x = 9)$$

means $(x = 7) \lor (x = 9)$. We also use indentation to indicate scopes, e.g.,

$$\begin{array}{l} \exists x : squad \bullet \\ \qquad x \in \{manager, coach\} \lor \\ \qquad x \notin \operatorname{ran} shirt \\ coach \notin \operatorname{ran} shirt \end{array}$$

means (there is a squad member who is manager or coach, or does not have a shirt number; the coach does not have a shirt number)

$$\big(\exists x : squad \bullet x \in \{manager, coach\} \lor x \notin \operatorname{ran} shirt\big) \land coach \notin \operatorname{ran} shirt$$

Schemas can also be written in "horizontal form", e.g.,

$$League \mathrel{\widehat{=}} [clubs : \operatorname{iseq} Clubs \mid \#clubs = leaguesize]$$

Here, the naming of the schema is more explicit (the name is optional in either form); the actual schema text is enclosed in square brackets, and a vertical bar separates declarations and predicates.

1.5.2 Schema Inclusion

A schema can be *included* in another schema, for example to define a schema with extra restrictions, e.g.,

$$\begin{array}{|l} \multicolumn{1}{l}{\textit{PremierLeague}} \\ League \\ \hline \{arsenal, everton\} \subseteq \operatorname{ran} clubs \\ \hline \end{array}$$

Such a schema is equivalent to the one obtained by expanding all declarations and conjoining all predicates, i.e.,

$$\begin{array}{|l} \multicolumn{1}{l}{\textit{PremierLeague}} \\ clubs : \operatorname{iseq} CLUBS \\ \hline \#clubs = leaguesize \\ \{arsenal, everton\} \subseteq \operatorname{ran} clubs \\ \hline \end{array}$$

denotes the same schema. We can also create a schema with more components using schema inclusion, e.g.,

```
┌─ LeagueC ──────────────────────────────
│ League
│ champs : CLUBS
├────────────
│ champs ∈ ran clubs
└────────────────────────────────────────
```

which abbreviates

```
┌─ LeagueC ──────────────────────────────
│ clubs : iseq CLUBS
│ champs : CLUBS
├────────────
│ #clubs = leaguesize
│ champs ∈ ran clubs
└────────────────────────────────────────
```

1.5.3 Decorations and Conventions

The example schemas given so far all used sequences of lower case letters for their component identifiers. A much richer syntax for identifiers exists, however in this book we will use conventions which relate the *rôle* of a schema to the kinds of identifiers used in it. The conventions used are permitted but not enforced by the Z standard; most of them have been commonly used (in the "states-and-operations" specification style) since the early days of Z.

Convention 1.1 (Component names) An identifier (component name) ending in ? denotes an *input*. An identifier ending in ! denotes an *output*. An identifier ending in $'$ denotes an *after-state* component. All other component names (usually consisting of just lower case letters, digits, and underscores) denote (before-) state components, or global constants. □

(What a (before- or after-) "state" is will be made clear later.)

Convention 1.2 (Subscripts) We use subscripts on schema and component names to indicate "version" information: A for "abstract", which usually precedes a version with C for "concrete"; or numbers to indicate multiple versions. □

Subscripts and characters like ?, ! and $'$ at the end of names are called *decorations*. Decorations are significant, in the sense that $x?$, $x!$ and x are different identifiers, and (using Convention 1.1) they carry an interpretation, e.g., $x?$ will be viewed as an input variable for an operation schema (see below). The identifier without all the decorations is called the *base name*. Decorations occur so commonly in Z that decoration has been defined as an operation on schemas, too.

Definition 1.1 (Decoration) For any decoration character $\maltese$, the decoration of schema S with $\maltese$ is denoted $S\maltese$, and contains:

- all declarations (explicit and included) of S, with $\maltese$ added to the end of every component name;
- all predicates (explicit and included) of S, with $\maltese$ added to the end of every free occurrence of a name declared in S (i.e., excluding global constants and bound names) in those predicates. □

For example, given *League* defined above, the schema *League′* is defined as

$$
\begin{array}{|l}
\textit{League}' \\
\hline
\textit{clubs}' : \mathrm{iseq}\, \textit{CLUBS} \\
\hline
\#\textit{clubs}' = \textit{leaguesize} \\
\hline
\end{array}
$$

Note that *leaguesize*, being a global constant, does *not* get decorated.

1.5.4 States, Operations and ADTs

Schemas whose component names are all undecorated are often interpreted as *state schemas*, i.e., schemas describing (snapshots of) the state of the system under consideration. The components of a state schema represent all variables of interest; the predicate represents the invariant of the state. *Operations* on such a state are defined in terms of two copies of the state: an undecorated copy which represents the *before*-state, and a primed copy representing the *after*-state. The definition of decoration implies that the primed version of the state invariant is automatically imposed on the after-state. The combination of before- and after-state is so common that a special convention is used to represent it.

Convention 1.3 (Delta) The schema ΔS, if not given another definition, represents the schema

$$
\begin{array}{|l}
\Delta S \\
\hline
S \\
S' \\
\hline
\end{array}
$$

□

For example, ΔFBT is

$$\begin{array}{|l}\hline \Delta FBT \\ manager, coach, manager', coach' : PERSON \\ squad, squad' : \mathbb{P}\,PLAYER \\ \hline\end{array}$$

Note that, very occasionally, we (and others) define ΔS to include extra restrictions that apply to every operation, for example, that a particular state component never changes.

An *operation schema* on a state (schema) S is thus a schema whose components include those of S and S'. In addition, it may declare and use inputs (names ending in ?) and outputs (names ending in !). For example, an operation (buying the player *Ronaldo*, assuming $Ronaldo : PLAYER$) is described by

$$\begin{array}{|l}\hline BuyRonaldo \\ \Delta FBT \\ \hline manager' = manager \land coach' = coach \\ squad' = squad \cup \{Ronaldo\} \\ \hline\end{array}$$

The operation to buy a player specified by input $p?$ is given by

$$\begin{array}{|l}\hline BuyAny \\ \Delta FBT \\ p? : PLAYER \\ \hline manager' = manager \land coach' = coach \\ squad' = squad \cup \{p?\} \\ \hline\end{array}$$

If we wish state components *not* to change in an operation, we have to state so explicitly. This is inconvenient, and can often be avoided by use of the following convention.

Convention 1.4 (Xi) The schema ΞS represents the schema $[S;\ S' \mid p]$ with the predicate p which denotes the equality between all components in S and their primed versions. □

In Sect. 1.5.8 we will introduce the notation required to express such a predicate p.

If we define a schema *Coach_Manager* by

$$\begin{array}{|l}\hline Coach\_Manager \\ manager, coach : PERSON \\ \hline\end{array}$$

then we could give the following alternative definition of *BuyAny*:

$$\begin{array}{|l}
\textit{BuyAny} \\
\hline
\Delta FBT \\
\Xi Coach_Manager \\
p? : PLAYER \\
\hline
squad' = squad \cup \{p?\}
\end{array}$$

It is acceptable to effectively declare components twice in a schema (e.g. *manager* and *coach* are introduced in *BuyAny* both through *FBT* and through *CoachManager*) provided the declarations are *compatible*, in the sense that all declarations of the same name have the same type. (They do not need to be from the same *set*, e.g., $x : \mathbb{N}$ and $x : \{-1, 1\}$ are compatible since both occurrences of x have type $\mathbb{Z}$.)

Apart from a state and operations on that state (possibly with IO: input and output variables), for a full state-based description we also need an *initial state*. The collection of possible initial states is described by an *initialisation schema*. Syntactically this is represented as a degenerate operation which has no before-state (and no IO), i.e., a schema on S' when the state is given by S.

The initialisation schema and operation schemas do not necessarily define a *single* after-state (and collection of outputs) for a given before-state, i.e., they can be *non-deterministic*.

A state schema with an initialisation schema and a collection of operations on that state schema, forms an *abstract data type (ADT)*. These are not defined in the Z standard, and not even usually defined explicitly in the states-and-operations style, but the encapsulation of these schemas will prove useful in the refinement theory in the rest of this book.

Definition 1.2 (Standard Z ADT) A standard Z ADT is a 3-tuple $(State, Init, \{Op_i\}_{i \in I})$ such that *State* is a state schema, *Init* is a schema on *State*$'$, and Op_i are operations on $\Delta State$. The index set I is the "alphabet" of the ADT. □

The ADT provides an interface in the form of a list of visible operations. Some other schemas may also appear in the specification, but only for structuring purposes, e.g., the schemas *FBT* and *Coach_Manager* in the example below.

Example 1.1 An ADT representing the management of a football squad, slightly expanding our earlier examples, is given by

$$(FBTA, Init, \{BuyPlayer, SellPlayer, NewManager\})$$

where

$$\begin{array}{|l} \textit{FBT} \\ \hline manager, coach : PERSON \\ squad : \mathbb{P}\,PLAYER \\ \hline \end{array}$$

$$\begin{array}{|l} \textit{FBTA} \\ \hline FBT \\ available : \mathbb{P}\,PLAYER \\ \hline available \subseteq squad \\ \hline \end{array}$$

$$\begin{array}{|l} \textit{Init} \\ \hline FBTA' \\ \hline squad' = \varnothing \\ \hline \end{array}$$

$$\begin{array}{|l} \textit{Coach\_Manager} \\ \hline manager, coach : PERSON \\ \hline \end{array}$$

$$\begin{array}{|l} \textit{SellPlayer} \\ \hline \Delta FBTA \\ \Xi Coach\_Manager \\ p? : PLAYER \\ \hline p? \neq manager \\ p? \in squad \\ squad' = squad \setminus \{p?\} \\ available' = available \setminus \{p?\} \\ \hline \end{array}$$

$$\begin{array}{|l} \textit{BuyPlayer} \\ \hline \Delta FBTA \\ \Xi Coach\_Manager \\ p? : PLAYER \\ \hline p? \notin squad \cup \{manager, coach\} \\ squad' = squad \cup \{p?\} \\ available' \setminus \{p?\} = available \\ \hline \end{array}$$

$$\begin{array}{|l} \textit{NewManager} \\ \hline \Delta FBTA \\ p? : PERSON \\ \hline manager' = p? \\ coach' = coach \\ squad' = squad \setminus \{manager\} \\ available' = available \setminus \{manager\} \\ \hline \end{array}$$

Some of the subtleties of this specification are worth pointing out. The predicate for $available'$ in *BuyPlayer* indicates that a newly bought player may or may not be immediately available (due to suspension or injury etc.), but the availability of the other players is not affected. A predicate like $squad' = squad \cup \{p?\}$ in *BuyPlayer* represents a deterministic assignment, indicating how the after-value can be constructed. However, more general predicates, not necessarily deterministic, are also allowed—e.g. the predicate $available' \setminus \{p?\} = available$ leaves it open whether the new player is available immediately (i.e., $p? \in available'$ may or may not hold). The specification also allows the possibility of a player-manager ($manager \in squad$, formally). Player-managers do not usually get sold (thus $p? \neq manager$ in *SellPlayer*), rather they get sacked and replaced, thus the removal of the old manager from the squad in *NewManager*. *Init* does not explicitly define $available'$, but from the state invariant $available \subseteq squad$ and $squad' = \varnothing$ in *Init* it follows that $available'$ can only be $\varnothing$. Finally, the schema *FBT* is included in the specification as a building block for *FBTA*, but it is not itself a part of the ADT. Building blocks for operations, e.g. constructed using the schema calculus (Sect. 1.5.5) may be in a similar position. □

It is ADTs like the one above which will form the main focus of our theory of refinement. However, before we consider refinement, we will look at the various ways in which schemas can be used and combined.

1.5.5 The Schema Calculus

Much of the power of Z as a specification notation derives from the operators that are provided for combining schemas. A precondition for combining two schemas is that their declarations are *compatible*, in the sense described previously.

1.5.5.1 Renaming

The components of a schema can be renamed. We will apply this operator only when the new name is not already used in the schema. For schema S, $S[p/q]$ denotes S with q renamed to p (except for any occurrences of q bound inside predicates of S).

For example, we have[3]

```
__FBT[gaffer/coach]_______________________
| manager, gaffer : PERSON
| squad : ℙ PLAYER
|__________________________________________
```

1.5.5.2 Schema Conjunction

The *schema conjunction* operator is closely related to schema inclusion as defined above. The conjunction of two schemas S and T is identical to a schema which includes both S and T (and nothing else).

Thus, an alternative (and probably more common) notation for the schema $[\,S;\ S'\,]$ is $S \wedge S'$.

Schema conjunction is useful both on states and on operations. On states it can be used as a sort of product (when there are no overlaps in components), or as a restriction (when all components of one occur in the other), or a combination of the two (closely related to the natural join in relational databases). For example, the schema conjunction of *LeagueC* and *PremierLeague* (p. 21) is

[3]In graphical presentations of expressions returning schemas, we often put the expression in place of the name. This is done purely for presentational purposes, as expressions denote unnamed schemas.

$$\begin{array}{|l}
\multicolumn{1}{l}{\underline{\quad LeagueC \land PremierLeague \quad}} \\
League \\
champs : CLUBS \\
\hline
champs \in \operatorname{ran} clubs \\
\{arsenal, everton\} \subseteq \operatorname{ran} clubs \\
\hline
\end{array}$$

Conjunction on operation schemas can often be viewed as a way of composing restrictions, in particular when the schemas have components in common. For example, the conjunction of *BuyAny* and *BuyRonaldo* (p. 24) is

$$\begin{array}{|l}
\multicolumn{1}{l}{\underline{\quad BuyAny \land BuyRonaldo \quad}} \\
\Delta FBT \\
p? : PLAYER \\
\hline
manager' = manager \land coach' = coach \\
squad' = squad \cup \{p?\} \\
squad' = squad \cup \{Ronaldo\} \\
\hline
\end{array}$$

(This is nearly equivalent to *BuyRonaldo* with a redundant input $p?$ which necessarily equals *Ronaldo*—except when $p?$ and *Ronaldo* are both in the *squad* already.)

Schema conjunction between operations on the same state can rarely be used to achieve the effects of both operations (unlike in Object-Z, see Chap. 15). For example, $BuyPlayer \land NewManager$ would *not* describe the appointment of a new player-manager, as it would contain predicates $p? \notin \{manager, coach\}$, $manager' = manager$ (from $\Xi Coach\_Manager$) and $manager' = p?$ which are contradictory. Schema *composition* (see below) should be used for this instead.

Schema conjunction is only well-defined when components have compatible types. If they are taken from different sets, then the intersection of those sets should be taken. For example, $[x : \mathbb{N}] \land [x : \{-1, 1\}] = [x : \mathbb{N} \cap \{-1, 1\}] = [x : \{1\}]$. Although this may appear complicated and arbitrary, it is consistent with the approach which first normalises the schemas (see Definition 1.3 below), ensuring multiple occurrences of the same name have the same type, and then takes the conjunction of the respective restrictions.

1.5.5.3 Schema Disjunction

The other logical schema operators are defined similarly, as one might expect—combining all the declarations, and then applying the logical operator to the predicates. However, for all but conjunction and equivalence, the distinction between *sets* and *types* becomes relevant when components are taken from different but compatible sets.

When there are no predicates present, we can take the union of such sets, i.e., $[x:\mathbb{N}] \vee [x:\{-1,1\}] = [x:\mathbb{N}\cup\{-1,1\}] = [x:\mathbb{N}\cup\{-1\}]$. However, the disjunction of the following schemas cannot be computed in a similar way:

$S1$

$x:\mathbb{N}$

$even(x)$

$S2$

$x:\{-1,1\}$

The disjunction of these is *not*

$x:\mathbb{N}\cup\{-1\}$

$even(x) \vee true$

whose predicate would simplify to *true*, and thus allow values of x allowed by neither $S1$ nor $S2$. The solution here lies in normalisation, which will produce an equivalent schema where all components are declared to be members of their *types* (rather than of sets contained in those types).

Definition 1.3 (Normalisation) Consider a schema S with components $x_1 : X_1, \ldots, x_n : X_n$, such that the type of x_i is T_i. The normalisation of S is obtained from S by replacing all declarations $x_i : X_i$ by $x_i : T_i$, and adding predicates $x_i \in X_i$. □

Thus, the normalisations of $S1$ and $S2$ would be the following (equivalent) schemas:

$S1$

$x:\mathbb{Z}$

$x\in\mathbb{N}$
$even(x)$

$S2$

$x:\mathbb{Z}$

$x\in\{-1,1\}$

The schema disjunction is constructed from *normalised* schemas, which are guaranteed to have *identical* types for common component names. It would be in this case

$S1 \vee S2$

$x:\mathbb{Z}$

$(x\in\mathbb{N} \wedge even(x)) \vee x\in\{-1,1\}$

which allows exactly all values of x allowed by either $S1$ or $S2$.

Schema disjunction is rarely used on state schemas (it makes little sense to satisfy one state invariant *or* the other), but very frequently on operation schemas, often in combination with conjunction.

The operation *BuyPlayer* is only applicable for certain inputs (namely when $p? \notin squad \cup \{manager, coach\}$). For robustness, we might like to specify a *total* operation which returns an appropriate error message when the original *BuyPlayer* is not applicable. For this purpose, we define an enumerated type

$$REPORT ::= ok \mid already\_employed$$

and a trivial operation (recall that an identifier decorated with ! denotes an output)

ReportOK

$r! : REPORT$

$r! = ok$

Then, the normal situation is represented by $BuyPlayer \wedge ReportOK$. The error case leaves the state unchanged and reports that the input $p?$ is already employed, giving

AlreadyEmployed

$\Xi FBTA$
$p? : PLAYER$
$r! : REPORT$

$r! = already\_employed$
$p? \in squad \cup \{manager, coach\}$

Then, a total operation covering all possible inputs is given by

$$(BuyPlayer \wedge ReportOK) \vee AlreadyEmployed$$

1.5.5.4 Schema Negation

Schema negation also requires normalisation. Thus, for specifiers who are not acutely aware which sets are types and which sets are not, it is likely to have surprising results. For example, the negation of

S1

$x : \mathbb{N}$

$even(x)$

is

$$\begin{array}{|l} \multicolumn{1}{l}{\neg S1} \\ \hline x : \mathbb{Z} \\ \hline x \notin \mathbb{N} \lor \neg even(x) \\ \hline \end{array}$$

For this reason, we will not use schema negation in this book, with the following exception.

By taking the disjunction with the negation of a schema, we obtain a schema on the same components whose predicate is universally true—i.e., the normalised schema with all predicates removed. We will call such a schema the signature of a schema, because it only retains information about the included components, and not about their relations and restrictions.

Definition 1.4 (Signature of a schema) The *signature* of a schema S is defined by

$$\Sigma S = S \lor \neg S$$ □

This definition is not part of standard Z, although it occurs in at least one textbook [3].

The signature of *SellPlayer* is

$$\begin{array}{|l} \multicolumn{1}{l}{\Sigma SellPlayer} \\ \hline manager, coach, manager', coach', p? : PERSON \\ squad, squad' : \mathbb{P}\, PERSON \\ \hline \end{array}$$

(note that all occurrences of *PLAYER* have been replaced by the containing type *PERSON*). This is also the signature of the other operations, *BuyPlayer* and *NewManager*. The signature of *League* is

$$\begin{array}{|l} \multicolumn{1}{l}{\Sigma League} \\ \hline clubs : \mathbb{P}(\mathbb{Z} \times CLUBS) \\ \hline \end{array}$$

This clearly demonstrates how the fact that seq is not a basic type constructor (but represented by functions, which in turn are sets of pairs) sometimes obscures specifications.

1.5.5.5 Schema Implication and Equivalence

Schema implication $\Rightarrow$ and schema equivalence $\Leftrightarrow$ have the usual meaning but, as implication is defined in terms of negation, they are defined in terms of *normalised* schemas. This makes these schema operators somewhat unpredictable and, as a consequence, they are rarely used to construct new stand-alone schemas. Their most common use is as "predicates", e.g., in refinement conditions, when all their components are universally quantified. (See below for a full explanation of universal quantification over schemas.) For example, for two operations Op_1 and Op_2 on the same state whose only component is $x : T$, the predicate

$$\forall x, x' : T \bullet Op_1 \Rightarrow Op_2$$

states that the effect of Op_1 is always consistent with Op_2, and

$$\forall x, x' : T \bullet Op_1 \Leftrightarrow Op_2$$

states that their effects are identical. More examples of this kind will be given in Sect. 1.5.6.

1.5.5.6 Projections

A number of projection operations on schemas exist. What they have in common is that, from a given schema, they produce a schema containing a subset of its components, with a predicate that is obtained by quantifying over the removed components.

One way of projecting a schema to some of its components is by *existential quantification*. Assume a schema

```
┌─ S ──────────────────────────────
│ x : T
│ Declarations
├──────────
│ pred
└──────────────────────────────────
```

where *Declarations* consists of declarations and schema references not (re-)declaring x. Then existential quantification over x in S yields:

```
┌─ ∃x : T • S ─────────────────────
│ Declarations
├──────────
│ ∃x : T • pred
└──────────────────────────────────
```

Thus, $\exists x : T \bullet S$ is a schema on all components of S except for x, which allows all component values such that a matching x in T to satisfy S would have existed.

For example,

$$\exists p? : PLAYER \bullet BuyAny$$

$$=$$

$$[\Delta FBT \mid \exists p? : PLAYER \bullet manager' = manager \wedge coach' = coach \wedge squad' = squad \cup \{p?\}]$$

If we existentially quantify this schema over (all the components of) ΔFBT, this simplifies to $[\,|\, PLAYER \neq \varnothing]$—i.e., a set of empty tuples with at most one element, depending on the truth of the predicate, and indeed a schema representation of that predicate, cf. Sect. 1.5.7.

Schema *hiding* is identical to existential quantification but has a different syntax: $S \setminus (x)$ denotes the schema S existentially quantified over x. In general, a bracketed list of components separated by commas may be hidden this way.

Schema *projection* is the "positive" version of hiding. Schema projection of schema S on another schema T is denoted $S \upharpoonright T$. Its meaning is that of the conjunction of S and T, existentially quantifying over all components *not* contained in T.

Finally, *universal quantification* over schemas is also possible, but used much less frequently, as it is a very strong requirement. If we have a schema

```
┌─ S ──────────────────────────────
│ x : T
│ Declarations
├──────────────
│ pred
└──────────────────────────────────
```

where *Declarations* consists of declarations and schema references not (re-)declaring x, then universal quantification over x in S yields:

```
┌─ ∀x : T • S ─────────────────────
│ Declarations
├──────────────
│ ∀x : T • pred
└──────────────────────────────────
```

Thus, $\forall x : T \bullet S$ is a schema on all components of S except x. The most common use of universal quantification is to turn a schema equivalence or implication into an empty schema (whose predicate is either true or false). For example, the formalisation that *BuyAny* is in some sense a generalisation of *BuyRonaldo*,

$$\begin{array}{l} \forall\, manager, manager', coach, coach' : PERSON; \\ \quad squad, squad' : \mathbb{P}\, PLAYER \bullet \\ \qquad BuyRonaldo \Rightarrow (\exists\, p? : PLAYER \bullet BuyAny) \end{array}$$

can be simplified to the empty schema with predicate *true*. (The implication expands to:

$$\begin{array}{l} (manager' = manager \wedge coach' = coach \wedge squad' = squad \cup \{Ronaldo\}) \\ \Rightarrow \\ (\exists\, p? : PLAYER \bullet manager' = manager \wedge coach' = coach \wedge \\ \qquad\qquad\qquad\qquad squad' = squad \cup \{p?\}) \end{array}$$

which holds by instantiation of $p?$ to *Ronaldo*.)

1.5.5.7 Schema Composition

This is a crucial operation which is only meaningful when applied to operation schemas on the same state. The schema composition of Op_1 and Op_2 is denoted $Op_1 \fatsemi Op_2$, and intuitively has the effect of first performing Op_1, and then Op_2 on the resulting state.

It can be constructed as follows. Assume Op_1 and Op_2 operate on *State*. Decorate all components of *State'* in Op_1 with an extra prime, and all components of *State* in Op_2 with two primes. Thus, the intermediate state is now *State''* in both operations. Then take the conjunction of these modified operations and hide (existentially quantify—like in relational composition) the intermediate state *State''*. This results in a schema on $State \wedge State'$, i.e., a single operation. The only complication arises from inputs and outputs, which are not renamed or hidden and therefore have to have compatible types if they occur in both Op_1 and Op_2.

Illustrating this for the composition $BuyPlayer \fatsemi SellPlayer$, which intuitively should result in no change whenever this composition is applicable, the modified operations are ($\Xi Coach\_Manager$ needs to be expanded out):

$$
\begin{array}{|l}
\Delta FBTA \\
p? : PLAYER \\
\hline
coach'' = coach \\
manager'' = manager \\
p? \notin squad \cup \{manager, coach\} \\
squad'' = squad \cup \{p?\} \\
available'' \setminus \{p?\} = available
\end{array}
$$

$$
\begin{array}{|l}
\Delta FBTA \\
p? : PLAYER \\
\hline
coach' = coach'' \\
manager' = manager'' \\
p? \neq manager'' \\
p? \in squad'' \\
squad' = squad'' \setminus \{p?\} \\
available' = available'' \setminus \{p?\}
\end{array}
$$

From $p? \notin squad$ it follows that $(squad \cup \{p?\}) \setminus \{p?\} = squad$, and thus the conjunction of these schemas, hiding the components of $FBTA''$, results in

$$
\begin{array}{|l}
\Xi FBTA \\
p? : PLAYER \\
\hline
p? \notin squad \cup \{manager, coach\}
\end{array}
$$

It can also be checked that $BuyPlayer \fatsemi NewManager$ describes the appointment of a player-manager (who was not previously employed by the club). The identification of the inputs $p?$ in the two operations is essential for this.

1.5.5.8 Schema Piping

As was clear in the previous example, schema composition is analogous to relational composition as far as the state is concerned, but leaves input and output untouched. Schema piping, denoted $\gg$, is the complementary operation on inputs and outputs. The schema $S \gg T$ normally represents the conjunction of S and T, equating outputs of S with inputs of T that have the same base name, and hiding those matching inputs and outputs. However, when this would result in names of inputs of S or outputs of T being captured, the matching inputs and outputs need to be renamed first. The application of schema piping is most relevant when only one of the two schemas involved refers to the state.

For example, when we have a schema

```
┌─ Renamepq ─────────────────────
│ q?, p! : PLAYER
├───────────
│ q? = p!
└────────────────────────────────
```

then $Renamepq \gg BuyPlayer = [Renamepq;\ BuyPlayer \mid p! = p?] \setminus (p?, p!) = BuyPlayer[q?/p?]$.

1.5.6 Schemas as Declarations

The syntax of the quantified variable(s) in existential and universal quantification, as described in Sect. 1.2, is already very close to the horizontal schema notation: a collection of typed variables, and a predicate on those, separated by $\mid$. Z does in fact take this analogy to its logical conclusion, and allows any schema to occur in the place of a declaration.

Using this feature, we can express many of our previous examples much more succinctly. For example, we considered

$$\exists FBT';\ p? : PLAYER \bullet BuyAny$$

and

$$\forall \Delta FBT \bullet BuyRonaldo \Rightarrow (\exists p? : PLAYER \bullet BuyAny)$$

Example 1.2 (Satisfiable initialisation) A sensible condition to impose on ADTs (see Chap. 3) is that they allow at least one initial state. When the initialisation of an ADT is given by $Init == [State' \mid init]$, then the existence of an initial state can be expressed as

$$\exists State' \bullet init$$

or even as $\exists\, Init$. The initialisation of the ADT in Example 1.1 is indeed satisfiable:

$$\exists\, manager', coach' : PERSON;\ squad', available' : \mathbb{P}\, PLAYER \mid$$
$$available' \subseteq squad' \bullet squad' = available'$$

which is clearly *true*. □

A central operator on operation schemas is defined in terms of existential quantification: the precondition operator "pre".

Definition 1.5 (Precondition) For an operation *Op* on state *State*, with inputs *Inps* and outputs *Outs*, its precondition is defined by

$$\text{pre}\, Op = \exists\, State';\ Outs \bullet Op$$

Thus, pre *Op* will be a schema on *State* and *Inps* indicating for which before-states and inputs *Op* provides a possible after-state and output. □

For example

$$\text{pre}\, SellPlayer = [FBTA;\ p? : PLAYER \mid p? \neq manager \wedge p? \in squad]$$

which describes that we can only sell players that we own. We can always replace the manager, as

$$\text{pre}\, NewManager = [FBTA;\ p? : PLAYER]$$

The precondition of *AlreadyEmployed* (p. 30) not only hides the after-state, but also the output, i.e., it is

$$\begin{array}{|l} \hline \text{pre}\, AlreadyEmployed \\ FBTA \\ p? : PLAYER \\ \hline p? \in squad \cup \{manager, coach\} \\ \hline \end{array}$$

The following operations do not occur in the standard Z literature but are used in this book to avoid having to say (except in the definition of pre) "suppose we have inputs $x_i? : X_i$" etc.

Definition 1.6 (Input and output signature) The *input signature* of an operation schema *Op* on state *State* is its projection on input components only, i.e.,

$$?Op = \Sigma(\exists\, State \bullet \text{pre}\, Op)$$

The *output signature* of an operation schema *Op* on state *State* is its projection on output components only, i.e.,

$$!Op = \Sigma(\exists \operatorname{pre} Op;\ State' \bullet Op) \qquad \square$$

For example, the output signature of *AlreadyEmployed* (abbreviated here to *AE*) is given as follows:

$$!AE$$

$$\equiv \{\text{definition}\}$$

$$\Sigma(\exists \operatorname{pre} AE;\ FBTA' \bullet AE)$$

$$\equiv \{\text{definition of} \operatorname{pre} AE\text{, see above}\}$$

$$\Sigma\big(\exists[FBTA;\ p? : PLAYER \bullet AE];\ FBTA' \bullet AE\big)$$

$$\equiv \{\text{calculus, } \Delta FBTA = FBTA \land FBTA'\}$$

$$\Sigma(\exists\, \Delta FBTA;\ p? : PLAYER \bullet AE)$$

$$\equiv \{\text{only remaining component of } AE\text{; predicate removed by } \Sigma\}$$

$$[r! : REPORT]$$

1.5.7 Schemas as Predicates

It should be clear from the predicate logic operators defined on schemas above that schemas are closely related to predicates, with the components of the schema playing the *rôle* of the free variables. This analogy holds fully in Z, because when the components of a schema are bound by the context, we are always allowed to use a schema in the *rôle* of a predicate. (From the definitions of schema quantification it was clear they could occur in the body of a quantification, but in fact they can occur in set comprehensions etc. as well.) Moreover, a predicate without any free variables represents a truth value—similarly, a schema with no components (but only a predicate which can be either *true* or false) effectively represents a truth value, too.

Example 1.3 (Satisfiable initialisation) Using schemas as declarations *and* as predicates is a more convenient way of expressing the existence of an initial state (see Example 1.2). If the initialisation is *Init* and the state is *State*, the satisfiability of initialisation is expressed by

$$\exists\, State' \bullet Init \qquad \square$$

1.5.8 Schemas as Types

Thus far, we have avoided saying what exactly the meaning of a schema is. From the fact that a declaration $Leeds : FBT$ is possible, it follows that a schema like FBT must denote a set which is contained in some type. The elements of the set FBT are called *bindings*. The type of these bindings is the signature of FBT, which (when viewed as a set) is the largest set of bindings containing all elements of FBT.

Bindings can be denoted explicitly. For example, if we have $VanGaal : PERSON$ then

$$\langle\!| \, manager == VanGaal, manager' == VanGaal, coach == VanGaal, \\ coach' == VanGaal, squad == \varnothing, squad' == \{Ronaldo\} \, |\!\rangle$$

is one of the many bindings of the schema *BuyRonaldo*.

A special operator exists to construct bindings (elements of a schema type) in a context where all the component names (possibly all with the same decoration) are declared. This is the θ operator. For example,

$$\theta FBT = \langle\!| \, manager == manager, coach == coach, squad == squad \, |\!\rangle$$

This looks distinctly odd, but the two occurrences of each of the names have rather different *rôles*: the first is local to the binding, just the name of a field in the tuple; the second must refer to a value, namely the value of the variable of that name which must be in context. So, the following predicates are equivalent:

$$Leeds \in FBT \\ \exists \, FBT \bullet Leeds = \theta FBT$$

When θ is applied to decorated schemas, it is taken into account that the labels of the binding are local names, and therefore not subject to the decoration. Thus,

$$\theta FBT' = \langle\!| \, manager == manager', coach == coach', squad == squad' \, |\!\rangle$$

Of course this expression is only meaningful when FBT' is in context, for example, the fact that PSG (of type FBT) have bought $Ronaldo$ can be expressed as

$$\exists \, BuyRonaldo \bullet PSG = \theta FBT' \tag{1.1}$$

A common use of the θ notation in this book is to turn an operation into a relation between states. If we have an operation Op on $\Delta State$, its relational form is given by the set comprehension

$$\{Op \bullet (\theta State \mapsto \theta State')\}$$

(for each possible binding of Op, include a pair consisting of the included bindings for before-state $State$ and those for the after-state $State'$). For example, when $State == [x : \mathbb{N}]$ and

```
┌─ Op ──────────────────────────────
│ ΔState
├──────────────
│ (x = 5 ∧ x' = 7) ∨ (x = 6 ∧ x' = 0)
└───────────────────────────────────
```

its relational interpretation is

$$\{(\langle\!| x == 5 |\!\rangle, \langle\!| x == 7 |\!\rangle), (\langle\!| x == 6 |\!\rangle, \langle\!| x == 0 |\!\rangle)\}$$

Finally, the general definition of ΞS can now be expressed as:

$$\Xi S == [\, S;\ S' \mid \theta S = \theta S' \,]$$

The θ notation can be avoided if there are no variables of *schema* type around, and it can also be circumvented by using the dot notation when the names of the relevant components are known, for example,

$$Ronaldo \in PSG.squad$$

expresses the same constraint as (1.1) above. However, the θ notation is near unavoidable in our reasoning at the meta-level (using relations over schema types) in this book, and for describing promotion: see Chap. 6.

1.5.9 Schema Equality

In our reasoning about Z, we often have to consider whether schemas are *equal*. The equality operator $=$ is not normally defined between sets of a different *type*, so it only applies to schemas of the same signature (in which case it is just the comparison between the sets of allowed bindings of S and T). Universal quantification over schemas might appear to be a solution, defining S and T to be equal whenever

$$\forall\, \Sigma S;\ \Sigma T \bullet S \Leftrightarrow T$$

However, this predicate also holds for $[x : \mathbb{Z}]$ and $[x, y : \mathbb{Z}]$ which are different.

Thus, we have to assume that equality tests on schemas sometimes occur on the meta-level.

Convention 1.5 (Schema equality) For every equality test $S = T$ between schemas S and T in this book, the result is false whenever $S = T$ is a type error in Z, and the Z meaning of $S = T$ otherwise. □

In proofs, we will mostly use $\equiv$ for equivalence, even when it is strictly speaking between schemas rather than between predicates.

1.6 Example Refinements

As we have seen, Z can be used to describe the behaviour of a system by writing a specification in the states-and-operations style, and this in effect describes an ADT. Given such an ADT we will probably want to develop it further, for several reasons:

- the operations specified might be partial;
- the operations might be non-deterministic; and
- the data structures used might be inappropriate for use in an implementation.

Our final aim is to produce a development in which the behaviour is fully specified (i.e., operations are now total), all appropriate design decisions have been made (operations are now deterministic), and the data structures used are more readily implemented (e.g., using sequences instead of functions). However, whatever development is produced it must certainly be consistent with the requirements made in the initial specification. Refinement in Z tackles all of these issues.

Chapters 2–4 describe in some detail which refinements are appropriate.

To motivate that discussion we present two examples: a small one involving numbers and simple data structures, followed by a larger one involving more complex types.

Example 1.4 Consider the ADT $(A, AInit, \{AAdd, ARemove\})$ with

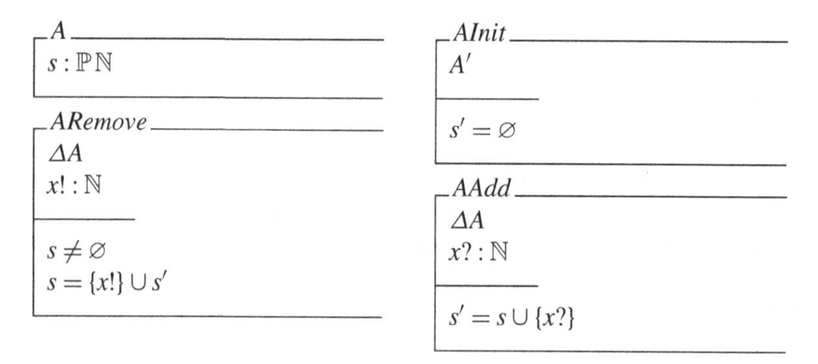

This generic "collection" interface leaves a few things undefined: it does not define the output value $x!$ for *ARemove* when the set s is empty and it does not specify how the output $x!$ is selected from s in *ARemove*. Additionally, it uses the mathematically convenient type $\mathbb{P}\,\mathbb{N}$ of possibly infinite sets for s—however, this type is not available in most programming languages, and for good reasons.

Refinement can address all these issues. It can produce a return value of -1 for removing an element from an empty set,[4] it can maintain the elements of s in a finite sequence, and by particular choices for adding and removing from that

[4] Note that this is a refinement example, not an example of good programming practice!

Fig. 1.2 A football league table

CLUBS	p	f	a	pts
leeds	14	28	16	27
man_utd	15	27	16	27
...				
arsenal (← A row)	15	21	19	22
...				

sequence decide a selection discipline for removing elements—for example FIFO (queue) behaviour in the ADT $(C, CInit, \{CAdd, CRemove\})$ below:

C

$q : \mathrm{seq}\,\mathbb{N}$

CInit

C'

$q' = \langle\rangle$

CRemove

ΔC

$x! : \mathbb{Z}$

$q = \langle\rangle \land q' = \langle\rangle \land x! = -1 \lor$

$q = \langle x! \rangle \frown q'$

CAdd

ΔC

$x? : \mathbb{N}$

$q' = q \frown \langle x? \rangle$

□

Example 1.5 Continuing with the football theme, we wish to specify a league table which we can display and update as matches are played (see Fig. 1.2). We describe the table as a function from $CLUBS$ to ROWs, where a ROW records matches played (p), goals scored (f), goals conceded (a) and total points (pts). We assume throughout an appropriate definition of $CLUBS$ containing 20 elements.

$$ROW == \mathbb{N} \times \mathbb{N} \times \mathbb{N} \times \mathbb{N}$$

$$p, f, a, pts : ROW \to \mathbb{N}$$

$$\forall r : ROW \bullet p(r) = r.1 \land f(r) = r.2 \land a(r) = r.3 \land pts(r) = r.4$$

LeagueTable

$$table : CLUBS \to ROW$$

Init

$$LeagueTable'$$

$$\forall c : CLUBS \bullet table'(c) = (0, 0, 0, 0)$$

A number of operations are specified. *Play* updates the table after a match on the basis of two (different) teams playing and producing a *score*?. The predicate updates the necessary rows in the table in terms of the current table, using an auxiliary function *update*.

Points can also be deducted from a team for bad behaviour (failing to turn up for a match, fighting on the pitch etc.). This is described by *Deduct*. The *Champion* is the club with the most points at the end of a season (i.e., when all clubs have played 38 matches). Finally, we might have an operation to *Display* the league in table form, presenting the clubs together with their points. The use of $\exists$ in the bodies of axiomatic definitions and schemas below is a common idiom, effectively creating local definitions.

$$update : \mathbb{N} \times \mathbb{N} \to ROW \to ROW$$

$$\begin{aligned}
&\forall for, aga : \mathbb{N};\ r : ROW \bullet \exists sc : \mathbb{N} \bullet \\
&\quad (for < aga \Rightarrow sc = 0) \land (for > aga \Rightarrow sc = 3) \\
&\quad (for = aga \Rightarrow sc = 1) \\
&\quad update\ (for, aga)\ r = (p(r) + 1, f(r) + for, a(r) + aga, pts(r) + sc)
\end{aligned}$$

Play

$$\begin{aligned}
&\Delta LeagueTable \\
&team? : CLUBS \times CLUBS \\
&score? : \mathbb{N} \times \mathbb{N}
\end{aligned}$$

$$\begin{aligned}
&first\ team? \neq second\ team? \\
&\exists new1, new2 : ROW \bullet \\
&\qquad new1 = update\ score?\ table(first\ team?) \\
&\qquad new2 = update\ (score?.2, score?.1)\ table(second\ team?) \\
&\qquad table' = table \oplus \{first\ team? \mapsto new1, second\ team? \mapsto new2\}
\end{aligned}$$

Deduct

$$
\begin{array}{|l}
\Delta LeagueTable \\
team? : CLUBS \\
points? : \mathbb{N} \\
\hline
table' = table \oplus \{team? \mapsto (p(team?), f(team?), \\
\qquad\qquad a(team?), pts(team?) - points?)\}
\end{array}
$$

Champion

$$
\begin{array}{|l}
\Xi LeagueTable \\
team! : CLUBS \\
\hline
\forall c : CLUBS \bullet p(table(c)) = 38 \\
\qquad pts(table(c)) \leq pts(table(team!))
\end{array}
$$

Display

$$
\begin{array}{|l}
\Xi LeagueTable \\
display! : \mathrm{seq}(CLUBS \times \mathbb{N}) \\
\hline
\#display! = 20 \\
\#\{t.1 \mid t \in \mathrm{ran}\, display!\} = 20 \\
\forall i, j : 1..20 \mid i \leq j \bullet display!(i).2 \geq display!(j).2 \\
\forall i : 1..20 \bullet \exists c : CLUBS \bullet display!(i) = (c, pts(table(c)))
\end{array}
$$

Having written this specification there are a number of issues to consider. For example, *Play* is a partial operation, what should happen if we apply it outside its precondition? In addition, the specification is non-deterministic in that we have left open the ordering of clubs that have the same number of points; what choices are appropriate in an implementation? Finally, we have described the table as a function. This might not be convenient in an implementation and to what extent will a specification be acceptable if it uses a different representation of the table?

We can resolve some of these issues by refining this specification to another one. The new specification will use a different state space, the table now being stored as an ordered collection of clubs (i.e., a sequence). We shall also make some specific design decisions. In particular, we will define what will happen to operations outside their precondition, and we will resolve some of the non-determinism by deciding to take the goals scored into account when ordering the league table.

The predicate in $LeagueTable_C$ ensures that the league table contains all the clubs, with no duplicates, and that it is an ordered sequence with the leading club first. A subscript C denotes the equivalent state or operation to the one named above. We have elided the predicate in *Deduct* and the ... in *Play* represents updating the table in a manner similar to above. We have omitted the obvious definitions of symbolic accessor functions p_C, f_C, a_C and pts_C (operating on 5-tuples).

LeagueTable$_C$

$table_C : \mathrm{seq}(\mathbb{N} \times \mathbb{N} \times \mathbb{N} \times \mathbb{N} \times CLUBS)$

$\#table_C = 20$
$\#\{s.5 \mid s \in \mathrm{ran}\, table_C\} = 20$
$\forall i, j : 1..20 \mid i \leq j \bullet pts_C(table_C(i)) \geq pts_C(table_C(j))$
$\forall i, j : 1..20 \mid i \leq j \bullet pts_C(table_C(i)) = pts_C(table_C(j))$
$\quad \Rightarrow f_C(table_C(i)) \geq f_C(table_C(j))$

Init$_C$

$LeagueTable'_C$

$\forall i : 1..20 \bullet \exists c : CLUBS \bullet table'_C(i) = (0, 0, 0, 0, c)$

Play$_C$

$\Delta LeagueTable_C$
$team? : CLUBS \times CLUBS$
$score? : \mathbb{N} \times \mathbb{N}$

$(\mathit{first}\ team? \neq second\ team? \Rightarrow \cdots)$
$(\mathit{first}\ team? = second\ team? \Rightarrow \Xi LeagueTable_C)$

Deduct$_C$

$\Delta LeagueTable_C$
$team? : CLUBS$
$points? : \mathbb{N}$

$\ldots$

Champion$_C$

$\Xi LeagueTable_C$
$team! : CLUBS$

$\forall i : 1..20 \bullet p_C(table_C(i)) = 38$
$team! = head(table_C).5$

Display$_C$

$\Xi LeagueTable_C$
$display! : \mathrm{seq}(CLUBS \times \mathbb{N})$

$display! = \{i : \mathrm{dom}\, table_C \bullet i \mapsto (table_C(i).5, pts_C(table_C(i)))\}$

□

We claim that this specification is an acceptable refinement of the one above. This claim is based upon the following observations:

- the internal representation (*table* vs. $table_C$) does not matter, all that matters is the observable behaviour, e.g., the effect of the outputs in the operations;
- if options have been left open, we are free to make a choice.

The first observation leads to the idea of *data refinement* where state spaces can be altered between developments. The second, i.e., the reduction of non-determinism, allows us both to weaken the precondition of an operation and to resolve the non-determinism in any outputs.

Subsequent chapters explore these questions in detail. Chapter 2 looks at refinements where the state schema does not change. This restriction is relaxed in Chap. 3 to discuss *data* refinements and simulations (which are a means to verify a refinement). The notion of observation plays an important part in this discussion, as does the treatment of partial operations. Chapter 3 looks at these issues from the point of view of simple data types described as relations, however, ultimately we want to derive refinement rules for use in Z and Chap. 4 describes the link between the relational theory and the expression of the rules in the Z schema calculus.

1.7 What Does a Z Specification Mean?

The meaning of Z is fixed in the Z standard, in terms of Zermelo-Fraenkel set theory and logic, defining what is a mostly non-controversial intuitive mathematics. However, in this book we add a (mostly independent) layer of interpretation to this, namely the most commonly adopted one of "states-and-operations" (included in the non-prescriptive part of the standard). But if this is an interpretation, what does a Z specification mean, and how does refinement relate to this question?

In particular, the use of Z to describe the functionality of a system depends on a number of conventions. Thus the use of ? and ! to represent input and output, and the use of $'$ to represent an after-state, are not logical characterisations, but rather intuition to help us understand the description.

Indeed logically a variable v' has no relation to its unprimed counterpart v; they are distinct declarations. Furthermore, Z does not have an operational semantics, that is, no formal representation as a transition system by which to formalise our intuitive ideas of behaviour.[5]

However, our interpretation of Z specifications does have meaning and, in particular, we can think of refinement as a formalisation of this intuition. This is because refinement is concerned with producing collections of acceptable implementations. Since the refinement rules will encode our conventions and interpretations, what they effectively do is formalise this interpretation in terms of acceptable implementations or developments (albeit written in the same notation).

---

[5]However, Object-Z *does* have such a semantics.

We will see this in action, since by changing the interpretation we adapt the rules accordingly. The most obvious example we shall come across is the interpretation of the precondition of an operation, where different refinement rules encode different possibilities and this leads to different collections of acceptable developments. However, this is but one example and many of the generalisations of refinement we discuss in later parts of this book are motivated by interpreting Z specifications in particular ways.

1.8 The Z Standard

After living for many years with [52] acting as a *de facto* standard for Z specifications, Z underwent full standardisation by ISO [34]. Standardisation has cleaned up some of the irregularities of the original definition and has addressed issues that have been resolved in different ways by different users. It provides a formal definition of Z, defining its syntax and semantics, however no logic is provided. Here we briefly list some of the major issues addressed in the standard, and explain to what extent we conform to standard Z.

To help structure a large specification, standard Z introduces a section notation. A Z section contains a sequence of paragraphs with a header (giving the section name) and lists the other sections that are its parents. For backwards compatibility, a specification with no section header is accepted as a single section with the toolkit as defined in the standard as its only parent.

Standard Z also introduces a notation to introduce new operators, called an operator template paragraph. This permits a wider range of operators to be introduced than in [52] (see [34, 54] for details). A standardised form for conjectures has also been defined using the symbol $\models ?$. The use of free types has been expanded to allow mutually-recursive free types to be written within a single paragraph.

Schemas, as always, are defined as sets of bindings, and standard Z additionally allows bindings to be constructed explicitly using a binding extension expression, as in $\langle\!|\, d == 7, m == 3 \,|\!\rangle$. Selectors from tuples are also provided. Thus if $(x, y, z) : X \times Y \times Z$ then $(x, y, z).3$ selects the third component.

The categories of schema expressions and expressions have been combined, so now a schema expression can appear as a predicate, an operand to θ or as an expression. This means that $==$ can be used instead of $\widehat{=}$ when defining schemas. Empty schemas (i.e., schemas with no declaration) are also now allowed.

A set "arithmos", denoted $\mathbb{A}$, representing an unrestricted concept of number is now the basis for numeric operations. Properties of its subsets, $\mathbb{Z}$, $\mathbb{N}$ and $\mathbb{N}_1$, are defined in the toolkit.

Apart from a couple of notable exceptions, we have attempted to conform to the syntax of the standard. However, we depart from it in the following respects:

- We use $\mathbb{Z}$ as a built-in type provided by Z, rather than as a subset of $\mathbb{A}$. The only consequence of this is in the normalisation of a schema needed for schema negation (which we do not use other than for the definition of signatures—see Sect. 4.9 for a pitfall associated with this).

- We use bags using the syntax of [52], and some additional operations on relations.
- We make occasional use of the type of real numbers $\mathbb{R}$ and assume it and its operators are defined appropriately.
- We make more extensive use of the Booleans $\mathbb{B}$, which are assumed to comprise the values *true* and *false* which are the existing possible values for predicates. This avoids coercions between values of predicates and values of boolean variables, and allows us to write predicates like $\neg b'$ rather than $b' = false$.
- Occasionally we rely on indentation to determine the meaning of multi-line formulas inside schemas, more than the strict Z standard would allow.

However, the main difference is in our use of decorated references to schemas. Standard Z requires that any decoration on a reference to a schema must be separated from the schema name (e.g., by white space or parentheses around the name). The necessity for this comes from merging schema expressions and expressions, therefore $S' == [x : \mathbb{N}]$ is a valid schema definition according to the standard. To decorate a schema we must then write either $S\ '$ or $(S)'$.

We have avoided this problem by disallowing definitions to have names using $'$, i.e., we do not allow the definition $S' == [x : \mathbb{N}]$. Thus all occurrences of S' in this book for any name S refer to the traditional interpretation as a reference to the schema S with its components decorated.

Clearly it is necessary to be able to verify properties of our Z specifications, and to do so an appropriate logic is needed. Earlier Committee Drafts of the Z standard contained such a logic, but this was omitted from later Committee Drafts (see the discussion by Henson [29] on the logic contained in an earlier draft of the standard, and the comments by King [38]). The final position on this was for the ISO standard to contain a semantics against which any logic could be verified, and that incorporation of a particular set of inference rules will be left to future revisions of the standard.

However, work on a logic for Z continued, for example by Henson and Reeves. Henson and Reeves [31, 32] describe their work which provides an interpretation of Z within a specification logic Z_C, which comprises a logic and a semantics within which soundness of the logic is proved. A further development of this is the νZ language [30].

1.9 Tool Support

There are a number of tools to support Z. These offer varying degrees of functionality covering aspects of typesetting and pretty printing; syntax and type checking; theorem proving and animation. Many tools were commercial products at one stage, but most or all are now available free of charge or even as open source. At the time of writing, *Z/EVES* and *Z Formaliser* no longer appear to be available. *FuZZ* by Spivey and *ZTC* by Xiaoping Jia are older tools consistent with [52]. *CADiZ* by Toyn is a set of integrated tools for preparing, type-checking and analysing Z specifications which has followed the standard. *ProofPower-Z* by Lemma One is designed to support specification and proof in Higher Order Logic (HOL) and Z, and is now open source software.

The *ProB* animator and model checker by Leuschel and others [40, 41], originally designed for B, also supports Z notation, ProB uses *FuZZ* as the front-end for Z. The ProB approach to animation and model checking in Z is described in [44], and [45] illustrates the approach in Z and other languages as well as describing the LTL model checking. At a more technical level [42] describes the symmetry reduction employed by ProB, that is also applicable to Z. ProB supports some limited refinement checking, specifically trace refinement [41].

More recently Derrick, North and Simons have developed model-checking support [19, 20, 23] for Z via translation of Z specifications into the SAL toolkit [16]. This has recently included a prototype refinement checker [22], based upon some earlier foundational work [24, 48, 49].

The largest concerted effort in developing Z tools in the last ten years has been the Community Z Tools (CZT) project, based at czt.sourceforge.net. It is an open source framework using Java and an XML markup for Z, and includes contributions from across the world. At the time of writing, it includes parsers, typecheckers, and printers for Z, Object-Z and Circus, an animator for a Z subset, and IDE support for developing Z specifications.

1.10 Bibliographical Notes

A large collection of books (and articles) are available which provide an introduction to Z and its use. *Specification Case Studies* edited by Hayes [28] served as an early definition of the language and contains many examples of particular aspects of the notation. Spivey's *The Z Notation: A Reference Manual* [51, 52] came to be the *de facto* definition of the language, and until the standard appeared, [50] was the main resource for Z semantics.

Z soon became commonplace in the CS undergraduate curriculum, and this led to the publication of many books providing an introduction to Z. These include *An Introduction to Formal Specification and Z* by Potter et al. [46] and books by Ratcliff [47], Bottaci and Jones [8], Bowen [9] and Jacky [36].

Two more advanced books are also worth mentioning. *Z in Practice* [3] by Barden, Stepney and Cooper aimed at those who already understood the basics of Z, and *Using Z: Specification, Refinement and Proof* [57] by Woodcock and Davies looks at proof and refinement in Z in some detail. *Using Z* acted as a precursor to this book in many ways, and there is an overlap in some of the material covered in Chaps. 1–4 of this book with that in [57]. Many of the later refinement examples in [57] are particularly interesting and motivated some of the generalisations to refinement discussed below.

The regular series of conferences, *ZUM: The Z Formal Specification Notation* ran throughout the 1990s [11–15, 43] and has since developed into joint conferences with users of the B notation [4, 5, 10, 55], and most recently also with ASM, Alloy and VDM in the ABZ series [7, 18, 26]. The *FME* (since 2005: FM) conferences also regularly contain papers concerning aspects of the notation.

More details of publications, tools etc. can be found on the extensive Z wiki and archive maintained at:

http://formalmethods.wikia.com/wiki/Z_notation

References

1. Araki, K., Gnesi, S., & Mandrioli, D. (Eds.) (2003). *Formal Methods Europe (FME 2003). Lecture Notes in Computer Science: Vol. 2805*. Berlin: Springer.
2. Arthan, R. D. (1991). On free type definitions in Z. In J. E. Nicholls (Ed.), *Sixth Annual Z User Workshop* (pp. 40–58). Berlin: Springer.
3. Barden, R., Stepney, S., & Cooper, D. (1994). *Z in Practice. BCS Practitioner Series*. New York: Prentice Hall.
4. Bert, D., Bowen, J. P., Henson, M. C., & Robinson, K. (Eds.) (2002). *ZB 2002: Formal Specification and Development in Z and B, 2nd International Conference of B and Z Users. Lecture Notes in Computer Science: Vol. 2272*. Berlin: Springer.
5. Bert, D., Bowen, J. P., King, S., & Waldén, M. A. (Eds.) (2003). *ZB 2003: Formal Specification and Development in Z and B, Third International Conference of B and Z Users. Lecture Notes in Computer Science: Vol. 2651*. Berlin: Springer.
6. Boiten, E. A., Derrick, J., & Schellhorn, G. (2009). Relational concurrent refinement II: internal operations and outputs. *Formal Aspects of Computing*, *21*(1–2), 65–102.
7. Börger, E., Butler, M. J., Bowen, J. P., & Boca, P. (Eds.) (2008). *ABZ 2008. Lecture Notes in Computer Science: Vol. 5238*. Berlin: Springer.
8. Bottaci, L., & Jones, J. (1995). *Formal Specification Using Z: A Modelling Approach*. International Thomson Publishing.
9. Bowen, J. P. (1996). *Formal Specification and Documentation Using Z: A Case Study Approach*. International Thomson Computer Press.
10. Bowen, J. P., Dunne, S., Galloway, A., & King, S. (Eds.) (2000). *ZB2000: Formal Specification and Development in Z and B. Lecture Notes in Computer Science: Vol. 1878*. Berlin: Springer.
11. Bowen, J. P., Fett, A., & Hinchey, M. G. (Eds.) (1998). *ZUM'98: The Z Formal Specification Notation. Lecture Notes in Computer Science: Vol. 1493*. Berlin: Springer.
12. Bowen, J. P., & Hall, J. A. (Eds.) (1994). *ZUM'94, Z User Workshop. Workshops in Computing*. Cambridge: Springer.
13. Bowen, J. P. & Hinchey, M. G. (Eds.) (1995). *ZUM'95: The Z Formal Specification Notation. Lecture Notes in Computer Science: Vol. 967*. Limerick: Springer.
14. Bowen, J. P., Hinchey, M. G., & Till, D. (Eds.) (1997). *ZUM'97: The Z Formal Specification Notation. Lecture Notes in Computer Science: Vol. 1212*. Berlin: Springer.
15. Bowen, J. P. & Nicholls, J. E. (Eds.) (1992). *Seventh Annual Z User Workshop*. London: Springer.
16. de Moura, L., Owre, S., Rueß, H., Rushby, J., Shankar, N., Sorea, M., & Tiwari, A. (2004). SAL 2. In R. Alur & D. Peled (Eds.), *International Conference on Computer Aided Verification (CAV 2004). Lecture Notes in Computer Science: Vol. 3114* (pp. 496–500). Berlin: Springer.
17. Davies, J. & Gibbons, J. (Eds.) (2007). *Integrated Formal Methods, 6th International Conference, IFM 2007. Lecture Notes in Computer Science: Vol. 4591*. Berlin: Springer.
18. Derrick, J., Fitzgerald, J. A., Gnesi, S., Khurshid, S., Leuschel, M., Reeves, S., & Riccobene, E. (Eds.) (2012). *Abstract State Machines, Alloy, B, VDM, and Z—Third International Conference, ABZ 2012. Lecture Notes in Computer Science: Vol. 7316*. Berlin: Springer.

19. Derrick, J., North, S., & Simons, A. J. H. (2006). Issues in implementing a model checker for Z. In Z. Liu & J. He (Eds.), *ICFEM*. *Lecture Notes in Computer Science: Vol. 4260* (pp. 678–696). Berlin: Springer.
20. Derrick, J., North, S., & Simons, A. J. H. (2008) Z2SAL—building a model checker for Z. In Börger et al. [7] (pp. 280–293).
21. Derrick, J., Boiten, E. A., & Reeves, S. (Eds.) (2011). *Proceedings of the 15th International Refinement Workshop*. *Electronic Proceedings in Theoretical Computer Science: Vol. 55*. Open Publishing Association.
22. Derrick, J., North, S., & Simons, A. J. H. (2011) Building a refinement checker for Z. In Derrick et al. [21] (pp. 37–52).
23. Derrick, J., North, S. D., & Simons, A. J. H. (2011). Z2SAL: a translation-based model checker for Z. *Formal Aspects of Computing*, *23*(1), 43–71.
24. Derrick, J., & Smith, G. (2008). Using model checking to automatically find retrieve relations. *Electronic Notes in Theoretical Computer Science*, *201*, 155–175.
25. Fitzgerald, J. A., Jones, C. B., & Lucas, P. (Eds.) (1997). *FME'97: Industrial Application and Strengthened Foundations of Formal Methods*. *Lecture Notes in Computer Science: Vol. 1313*. Berlin: Springer.
26. Frappier, M., Glässer, U., Khurshid, S., Laleau, R., & Reeves, S. (Eds.) (2010). *ABZ 2010*. *Lecture Notes in Computer Science: Vol. 5977*. Berlin: Springer.
27. Grimm, K. (1998) Industrial requirements for the efficient development of reliable embedded systems. In Bowen et al. [11] (pp. 1–4).
28. Hayes, I. J. (Ed.) (1987). *Specification Case Studies*. *International Series in Computer Science*. New York: Prentice Hall. 2nd ed., 1993.
29. Henson, M. C. (1998). The standard logic of Z is inconsistent. *Formal Aspects of Computing*, *10*(3), 243–247.
30. Henson, M. C., Deutsch, M., & Kajtazi, B. (2006). The specification logic nuZ. *Formal Aspects of Computing*, *18*(3), 364–395.
31. Henson, M. C., & Reeves, S. (1999). Revising Z: Part I—Logic and semantics. *Formal Aspects of Computing*, *11*(4), 359–380.
32. Henson, M. C., & Reeves, S. (1999). Revising Z: Part II—Logical development. *Formal Aspects of Computing*, *11*(4), 381–401.
33. Hinchey, M. G., & Bowen, J. P. (1995). *Applications of Formal Methods*. *International Series in Computer Science*. New York: Prentice Hall.
34. ISO/IEC (1999). Z notation: final committee draft. International Standard CD 13568.2. International Standards Organization.
35. ISO/IEC (2002). *Information Technology—Z Formal Specification Notation—Syntax, Type System and Semantics*. International Standard 13568, International Standards Organization. http://standards.iso.org/ittf/PubliclyAvailableStandards/c021573_ISO_IEC_13568_2002(E).zip.
36. Jacky, J. (1997). *The Way of Z: Practical Programming with Formal Methods*. Cambridge: Cambridge University Press.
37. Jacky, J., Unger, J., Patrick, M., Reid, R., & Risler, R. (1997) Experience with Z developing a control program for a radiation therapy machine. In Bowen et al. [14] (pp. 317–328).
38. King, S. (1999). 'The standard logic for Z': a clarification. *Formal Aspects of Computing*, *11*(4), 472–473.
39. King, S., Hammond, J., Chapman, R., & Prior, A. (1999) The value of verification: positive experience of industrial proof. In Wing et al. [56] (pp. 1527–1545).
40. Leuschel, M., & Butler, M. (2003) ProB: A model checker for B. In Araki et al. [1] (pp. 855–874).
41. Leuschel, M., & Butler, M. (2005). Automatic refinement checking for B. In K. Lau & R. Banach (Eds.), *International Conference on Formal Engineering Methods, ICFEM 2005*. *Lecture Notes in Computer Science: Vol. 3785* (pp. 345–359). Berlin: Springer.
42. Leuschel, M., & Massart, T. (2010). Efficient approximate verification of B and Z models via symmetry markers. *Annals of Mathematics and Artificial Intelligence*, *59*(1), 81–106.

43. Nicholls, J. E. (Ed.) (1990). *Z User Workshop*, Oxford. *Workshops in Computing*. Berlin: Springer.
44. Plagge, D., & Leuschel, M. (2007) Validating Z specifications using the ProB animator and model checker. In Davies and Gibbons [17] (pp. 480–500).
45. Plagge, D., & Leuschel, M. (2010). Seven at one stroke: LTL model checking for high-level specifications in B, Z, CSP, and more. *International Journal on Software Tools for Technology Transfer*, *12*(1), 9–21.
46. Potter, B., Sinclair, J., & Till, D. (1991). *An Introduction to Formal Specification and Z. International Series in Computer Science*. New York: Prentice Hall. 2nd ed., 1996.
47. Ratcliff, B. (1994). *Introducing Specification Using Z: A Practical Case Study Approach*. New York: McGraw-Hill.
48. Smith, G., & Derrick, J. (2005). Model checking downward simulations. *Electronic Notes in Theoretical Computer Science*, *137*(2), 205–224.
49. Smith, G., & Derrick, J. (2006). Verifying data refinements using a model checker. *Formal Aspects of Computing*, *18*(3), 264–287.
50. Spivey, J. M. (1988). *Understanding Z: A Specification Language and Its Formal Semantics*. Cambridge: Cambridge University Press.
51. Spivey, J. M. (1989). *The Z Notation: A Reference Manual. International Series in Computer Science*. New York: Prentice Hall.
52. Spivey, J. M. (1992). *The Z Notation: A Reference Manual* (2nd ed.). *International Series in Computer Science*. New York: Prentice Hall.
53. Stringer-Calvert, D. W. J., Stepney, S., & Wand, I. (1997) Using PVS to prove a Z refinement: a case study. In Fitzgerald et al. [25] (pp. 573–588).
54. Toyn, I. (1998) Innovations in standard Z notation. In Bowen et al. [11] (pp. 193–213).
55. Treharne, H., King, S., Henson, M. C., & Schneider, S. A. (Eds.) (2005). *ZB 2005: Formal Specification and Development in Z and B, 4th International Conference of B and Z Users. Lecture Notes in Computer Science: Vol. 3455*. Berlin: Springer.
56. Wing, J. M., Woodcock, J. C. P., & Davies, J. (Eds.) (1999). *FM'99 World Congress on Formal Methods in the Development of Computing Systems. Lecture Notes in Computer Science: Vol. 1708*. Berlin: Springer.
57. Woodcock, J. C. P., & Davies, J. (1996). *Using Z: Specification, Refinement, and Proof*. New York: Prentice Hall.
58. Woodcock, J. C. P., Stepney, S., Cooper, D., Clark, J. A., & Jacob, J. (2008). The certification of the Mondex electronic purse to ITSEC Level E6. *Formal Aspects of Computing*, *20*(1), 5–19.

Chapter 2
Simple Refinement

We use the term *simple refinement* to describe refinements of operations (and collections of operations) where the state schema does not change. These rules apply to ADTs (as introduced in Chap. 1), and some even apply to "concrete" data types, which we will call *repertoires*.

We first present a semi-formal motivation of the simplest refinement relation of all, *viz. operation refinement*. Operation refinement can be applied to individual operations without reference to the other operations present in the ADT, and will be presented in detail in Sect. 2.2. The other simple refinement rules, *establishing* and *imposing invariants*, which we will discuss in Sect. 2.4, only apply in the context of a full ADT, and require abstraction, in the sense of the state not being observable.

2.1 What Is Refinement?

Through this book we hope to convey to our readers that refinement in a sufficiently flexible and expressive notation (like Z or Object-Z) forms a rich and interesting subject of study. Research into this area has certainly unearthed a variety of complex issues. The first definition of refinement in Z (Definition 2.5) contains some 10 different symbols, and quantifies (implicitly or explicitly) over some 5 variables, and these numbers are only increased by later definitions. All this seems to suggest that refinement is an inherently complex notion.

However, that is not really the case. The basic idea of refinement is fortunately a simple one—fortunately because any theory that starts from a set of complex definitions is likely to have a very tenuous intuitive foundation. The intuition behind refinement is just the following:

> **Principle of Substitutivity:** it is acceptable to replace one program by another, *provided* it is impossible for a user of the programs to observe that the substitution has taken place. If a program can be acceptably substituted by another, then the second program is said to be a *refinement* of the first.

For example, replacing the current scheduling of trains in South-East England with one in which all trains run according to the timetable would be a valid refine-

J. Derrick, E.A. Boiten, *Refinement in Z and Object-Z*,
DOI 10.1007/978-1-4471-5355-9_2, © Springer-Verlag London 2014

ment. Of course some suspicion about a substitution having taken place would still arise, but a casual train traveller could never be entirely sure of that. This is due to the fact that all trains running on time is also an imaginable behaviour of the current system.[1]

Any definition of refinement presented in this book should be validated. This tends to involve a proof that previous definition(s) are a special case of the new definition. Ideally, however, it should also include a demonstration that the refinement relation is based on a particular principle of substitutivity.

Note that, although it is as yet informal, the principle of substitutivity strongly suggests some properties of refinement. One way of guaranteeing that no difference can be observed is by substituting a program by itself. This implies that refinement will be *reflexive*. Two substitutions in a row, both of which are impossible to observe, would constitute another example of an unobservable substitution. So refinement will be *transitive* as well, allowing for "step-wise" refinement. It is not likely to be symmetric or anti-symmetric, though. Often we will present *refinement-equivalent* specifications, which may not be equal but which can be safely substituted for each other in both directions. Finally, it is desirable that refinement is a pre-congruence[2] or "compositional": if a component on its own can be substituted by another, this substitution should also be acceptable when the component is part of a larger system. This allows for "piece-wise" refinement.

(So, if refinement is a simple notion in principle, where does the complexity of the refinement rules come from? In our view, there is a trade-off between abstract relational "point-free" characterisations, and notations like Z which tend to name every object being related. Relational characterisations can sometimes be very short and are then easier to manipulate, but certainly once they encode formulas containing multiple occurrences of the same variable, they quickly become less comprehensible. Conversely, the explicit naming and quantification in Z provide much "formal noise" for simple formulas, but make it easier to comprehend more complicated formulas piece by piece. For these reasons, Chap. 3 will define the central notions of data refinement in a relational setting, which will only get interpreted and derived in Z in Chap. 4.)

Of course the informal definition of refinement above is incomplete. It does not define programs, it does not define what a "user" is allowed to do, or what constitutes an "observation". These are all questions with different possible answers, leading to different formal definitions of refinement later on. For the moment, we remain semi-formal:

- a *program* is a finite sequence of *operations*, interpreted as their *sequential composition*;

---

[1]On the other hand, replacing the machine that performs the weekly lottery draw with one which returns the same set of numbers every week is probably not an acceptable refinement. This requires a more sophisticated explanation, however, concerning the difference between non-deterministic and probabilistic choice, which is outside the scope of this book—see [5] for a comprehensive treatment.

[2]A partial order $\leq$ is a pre-congruence if for all contexts $C[.]$, whenever $x \leq y$ also $C[x] \leq C[y]$.

- an *operation* is a binary relation over the state space of interest, taken from a fixed collection (indexed set) of such operations. If that state space is represented by the Z schema *State*, then operations are represented by schemas over $\Delta State$;
- an *observation* is a pair of states: the state before execution of a program, and a state after.

In this interpretation, the notion of *repertoire* (or *interface*, *signature*, or *alphabet*) of a system is evident.

Definition 2.1 (Repertoire) A repertoire is an indexed collection of operations $(Op_i)_{i \in I}$ over the same state. □

Because the observations for these repertoires are pairs of states, we could also view repertoires as *concrete* data types. We did not introduce repertoires in the previous chapter, because their function in this book is mostly as an intermediate stage for the introduction and explanation of *abstract* data types.

Given two repertoires, we are not normally interested in replacing a program using the one with an *arbitrary* program using the other. Rather, we are interested in the relation between two programs of the *same structure*, but using the corresponding operations from each of the two repertoires in corresponding places. This is formalised in the following definitions, which presuppose a definition of refinement between programs, to be given later.

Definition 2.2 (Conformal repertoires) Two repertoires are conformal iff their operations are indexed by the same index set, and defined over the same state. □

Definition 2.3 (Repertoire refinement) For conformal repertoires $A = (AOp_i)_{i \in I}$ and $C = (COp_i)_{i \in I}$, C is a repertoire refinement of A iff for every finite sequence $\langle i_1, i_2, \ldots, i_n \rangle$ the program $\langle COp_{i_1}, COp_{i_2}, \ldots, COp_{i_n} \rangle$ is a refinement of $\langle AOp_{i_1}, AOp_{i_2}, \ldots, AOp_{i_n} \rangle$. □

However, this is not a very practical definition, as it involves quantification over all finite sequences. The only plausible way of proving such a refinement would be by structural induction, which however requires an extra condition for it to work. The base case[3] of such an induction could be found in the following definition.

Definition 2.4 (Operation-wise refinement) Given two repertoires $A = (AOp_i)_{i \in I}$ and $C = (COp_i)_{i \in I}$, both operating over the same state. C is an operation-wise refinement of A iff for all $i \in I$, the program $\langle COp_i \rangle$ is a refinement of $\langle AOp_i \rangle$. □

The validity of the induction step depends on the condition in the following theorem.

---

[3]Minus the obvious case for the empty sequence.

Theorem 2.1 *If sequential composition is monotonic with respect to refinement, i.e., (using* $\mathbin{;}$ *for composition and* $\sqsubseteq$ *for refinement)*

$$S_1 \sqsubseteq S_2 \land T_1 \sqsubseteq T_2 \Rightarrow S_1 \mathbin{;} T_1 \sqsubseteq S_2 \mathbin{;} T_2$$

then repertoire refinement and operation-wise refinement coincide.

Proof

- Repertoire refinement obviously implies operation-wise refinement, which considers only the programs of length 1.
- Operation-wise refinement implies repertoire refinement, by induction over the program sequences. Monotonicity allows the induction step. □

Monotonicity of program constructors with respect to refinement has been listed before as a desirable property: it makes refinement a pre-congruence, and thus allows "piecewise" refinement.

However, the definition of refinement between programs is at this stage still missing, as we have not made precise what the possible observations are.

2.2 Operation Refinement

In this section we motivate the traditional definition of operation refinement in Z by making precise what we mean by "observing" a program—in particular, a program consisting of a single operation.

Recall that operations and programs are binary relations over a state space, and observations are, for the moment, pairs of concrete states. The most obvious connection between those is to interpret the set of observations as a relation itself, which should be equal to the operation. This is perfectly acceptable when the operation is total. When it is not, however, it would be impossible to discuss observations starting from a state which is outside the operation's domain—whereas the traditional Z approach *does* allow such observations.

An operation allows two kinds of possible observations (which are illustrated in Fig. 2.1):

- a before-state and an after-state which are related by the operation;
- a before-state which is not in the domain of the operation, with an *arbitrary* after-state. For technical reasons, this after-state may also be the distinguished state $\perp$, representing a non-terminating computation; $\perp$ as a before-state is also included in this case.

As a consequence, observing the original *AOp* and the substituting operation *COp* from a particular before-state s:

- if s was in the domain of *AOp*, then the after-state for *COp* should always be one of those possible according to *AOp*. This in turn means that s should be in the

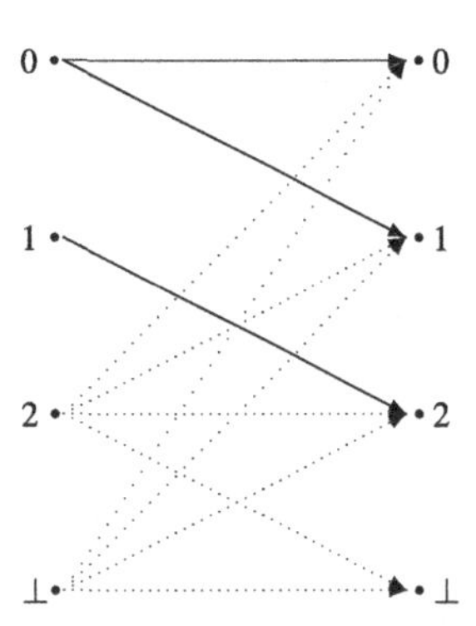

Fig. 2.1 Observations of $Op = \{(0,0), (0,1), (1,2)\}$ over the state $\{0,1,2\}$. The *dotted lines* indicate the observations from before-states outside the domain

domain of COp (or else $\perp$, which AOp does not allow, would be observable in COp);

- if s was not in the domain of AOp, any collection of possible observations for COp would be acceptable[4]—so s may be in the domain of COp, but it may not be.

For the definitions of refinement rules which follow now, we will assume that operations are represented by schemas in Z. When we take a more formal approach (starting from the next chapter), we will introduce a clear separation between operations as binary relations and operations as schemas, and indicate (in Chap. 4) precisely how they are related.

Definition 2.5 (Operation refinement) An operation COp is an operation refinement of an operation AOp over the same state space $State$ iff

Correctness

$$\forall State; State' \bullet \text{pre}\, AOp \wedge COp \Rightarrow AOp$$

Applicability

$$\forall State \bullet \text{pre}\, AOp \Rightarrow \text{pre}\, COp$$

□

At the level of binary relations, operation refinement coincides with inclusion of relations only when their domains are equal:

Lemma 2.1 *If* $\text{pre}\, AOp = \text{pre}\, COp$, *then* COp *is an operation refinement of* AOp *iff*

$$\forall State; State' \bullet COp \Rightarrow AOp$$

Proof Applicability is clearly satisfied under the given condition. The antecedent of correctness can be simplified using the tautology $Op \wedge \text{pre}\, Op = Op$. □

[4]Note, however, that given this definition of observations, it is impossible to define COp such that $(x, \perp)$ is a possible observation but not (x, y) for some y.

Lemma 2.2 *If* pre $COp = State$*, i.e., COp is total on State, then AOp is an operation refinement of COp if*

$$\forall State; State' \bullet \text{pre}\, AOp \wedge COp \Rightarrow AOp$$

Proof Applicability holds trivially as its conclusion is true, so only correctness remains. □

In some applications (cf. Chaps. 15 and 18) equality of domains is always required in operation refinement, due to a different interpretation of partial operations.

Theorem 2.2 *Operation refinement is a partial order on operations over the same state, i.e. it is transitive and reflexive.*

Proof Reflexivity is obvious from the conditions. Assume BOp is an operation refinement of AOp, and COp is an operation refinement of BOp. Now we need to prove that COp is an operation refinement of AOp. Applicability is trivial, as it follows from transitivity of $\Rightarrow$. For correctness we have:

$$\text{pre}\, AOp \wedge COp$$

$$\Rightarrow \{\text{applicability } AOp \text{ to } BOp\}$$

$$\text{pre}\, AOp \wedge \text{pre}\, BOp \wedge COp$$

$$\Rightarrow \{\text{correctness } BOp \text{ to } COp\}$$

$$\text{pre}\, AOp \wedge BOp$$

$$\Rightarrow \{\text{correctness } AOp \text{ to } BOp\}$$

$$AOp$$

□

Operation refinements can be obtained through schema conjunction. This automatically guarantees correctness, and applicability amounts to the domain of the operation being preserved. This is expressed in the following lemma.

Lemma 2.3 *The operation* $AOp \wedge X$*, for X and AOp operations over* $\Delta State$*, is an operation refinement of AOp iff*

$$\forall State \bullet \text{pre}\, AOp \Rightarrow \text{pre}(AOp \wedge X)$$

□

So far we have only discussed operation refinement from the perspectives of repertoires, which form a sort of *concrete* data types (as their state is observable). Shortly, we will be focusing on ADTs, whose state space is *not* observable. Consequently, in order to make observations of ADT programs, we will need to introduce *inputs* and *outputs*. Operation refinement is also meaningful in the presence of inputs and outputs, which must form part of the observation. As a consequence of that, they cannot change in operation refinement, and must be simply added to the quantification.

Definition 2.6 (Operation refinement with inputs and outputs) An operation *COp* is an operation refinement of an operation *AOp* over the same state space *State* and with the same inputs *?AOp* and the same outputs *!AOp* iff

Correctness

$$\forall State; State'; ?AOp; !AOp \bullet \text{pre}\, AOp \land COp \Rightarrow AOp$$

Applicability

$$\forall State; ?AOp \bullet \text{pre}\, AOp \Rightarrow \text{pre}\, COp$$

□

If we look back at Example 1.5 we can see that these definitions of operation refinement tackle two out of the three issues we highlighted, namely allowing preconditions to be weakened and non-determinism to be reduced. In particular, applicability requires a concrete operation to be defined everywhere the abstract operation was defined, however it also allows the concrete operation to be defined in states for which the precondition of the abstract operation was false. That is, the precondition of the operation can be weakened.

On the other hand, correctness requires that a concrete operation is consistent with the abstract whenever it is applied in a state where the abstract operation is defined. However, the outcome of the concrete operation only has to be consistent with the abstract, and not identical. Thus if the abstract allowed a number of options, the concrete operation is free to use any subset of these choices. In other words non-determinism can be resolved.

Example 2.1 In the league table example (Example 1.5) we can make a number of valid refinements even without changing the state space. For example, we might weaken the precondition of *Play* to describe the effect of this operation when two identical teams were used as parameters. Clearly a team cannot play against itself (despite the best efforts of some players) and we sensibly choose that *Play* should not change the state in these circumstances:

$Play_D$

$\Delta LeagueTable$
$team? : CLUBS \times CLUBS$
$score? : \mathbb{N} \times \mathbb{N}$

$Play \lor (first\ team? = second\ team? \land \Xi LeagueTable)$

We might also choose to reduce the non-determinism in the operation *Champion*, which currently outputs any team with maximum points, to one that also considers goal difference.

$$\begin{array}{|l}
\hline \textit{Champion} \\
\Xi LeagueTable \\
team! : CLUBS \\
\hline
\forall c : CLUBS \bullet p(table(c)) = 38 \\
\quad pts(table(c)) \leq pts(table(team!)) \\
\quad (f(table(c)) - a(table(c))) \leq (f(table(team!)) - a(table(team!))) \\
\hline
\end{array}$$

Then

$$\begin{array}{l}
\text{pre}\, Play = [LeagueTable;\ team? : CLUBS \times CLUBS;\ score? : \mathbb{N} \times \mathbb{N} \mid \\
\qquad\qquad first\ team? \neq second\ team?] \\
\text{pre}\, Play_D = [LeagueTable;\ team? : CLUBS \times CLUBS;\ score? : \mathbb{N} \times \mathbb{N}]
\end{array}$$

Applicability and correctness for *Play* easily follow. Similarly $\text{pre}\, Champion = \text{pre}\, Champion_D$ and $\forall\, \Delta LeagueTable;\ team! : CLUBS \bullet Champion_D \Rightarrow Champion$. The operations $Play_D$ and $Champion_D$ are thus operation refinements of their abstract counterparts. □

Example 2.2 We model the payment of a given amount by means of a sequence of coins. The set of all different coins is modelled by a set of their values.

$$\mid\ coins : \mathbb{F}\, \mathbb{N}_1$$

For example, in the UK the normal collection of coins (amounts in pence) is

$$coins = \{1, 2, 5, 10, 20, 50, 100, 200\}$$

An operation to return a required amount in any combination of coins is then

$$\begin{array}{|l}
\hline \textit{GiveAmt} \\
amt? : \mathbb{N} \\
c! : \text{seq}\, coins \\
\hline
amt? = +/c! \\
\hline
\end{array}$$

Note that there is no state in this operation. If required, this can be taken as an abbreviation for using a trivial state space which has only one possible value.

The operation *GiveAmt* is clearly operation-refined by

$$\begin{array}{|l}
\hline \textit{GiveFewCoins} \\
amt? : \mathbb{N} \\
c! : \text{seq}\, coins \\
\hline
amt? = +/c! \\
\forall c : \text{seq}\, coins \bullet +/c = amt? \Rightarrow \#c \geq \#c! \\
\hline
\end{array}$$

as, if a sequence of coins exists, so does a shortest sequence. Correctness follows from Lemma 2.3.

In many Eurozone countries, prices are expressed in cents, but supermarkets actually round amounts to the nearest 5 cents, as if the 1 and 2 cent coins did not exist and the smallest coin was worth 5 cents. In such a situation, *GiveAmt* is partial, and we might need to refine it to

```
┌─ GiveRounded ──────────────────────────────
│ amt? : ℕ
│ c! : seq coins
├──────────────
│ ∀ c : seq coins • |+/c − amt?| ≥ |+/c! − amt?|
└────────────────────────────────────────────
```

GiveRounded implies *GiveAmt* in those cases where the exact amount can be paid out. For applicability, observe that *GiveRounded* is total. (It is even defined when $coins = \varnothing$.) □

2.3 From Concrete to Abstract Data Types

In the setting of repertoires, operation refinement is the only *simple* refinement possible, due to the actual states being observable. The canonical interpretation of an *abstract* data type in Z has as its observations only the outputs generated by a particular sequence of inputs. In addition, through a construction similar to the one for state-observations in the previous section, partiality of operations may also be "observed".

Not only has the notion of observation to change, but also the notion of program. This is due to ADTs (unlike repertoires) having an initialisation. Programs are still characterised by sequences of indices, and the evolution of the state is governed by sequential composition of (initialisation and) operations. Observations are now pairs of sequences: the sequence of outputs produced by a particular sequence of inputs. However, the formalisation of this is postponed to Chap. 3, which gives a relational interpretation of Z ADTs.

We will, however, present more possibilities for "simple" refinement, which do not change the state space, although they do require consideration of the full ADT for their verification.

2.4 Establishing and Imposing Invariants

In Chap. 1 we discussed how invariant relations between the various components would be described in a state schema. However, in first drafts of specifications quite often you will not have thought of all the invariants that are required to hold in the system so, as a consequence, intuitively correct properties cannot be verified. Sometimes this means a return to the drawing board, but sometimes refinement will actually allow the introduction of invariants. How this can be done will be described in this section.

Formally, one could view the introduction of invariants either as a change to the state schema, or as a change to all operations. In the former view, introduction of invariants is a special (degenerate) case of data refinement, which we will describe in Chap. 4. In this chapter, however, we will concentrate on the latter approach, with the formal justification deriving from the data refinement conditions to be presented later.

We describe two ways of introducing invariants: *establishing* invariants is a refinement step in both directions, and *imposing* invariants is, in general, a refinement in only one direction. What the two have in common, is the paradoxical effect that preconditions can be *strengthened* in refinement—though only if the strengthening is an acceptable invariant of the original system already.

2.4.1 Establishing Invariants

Establishing invariants concerns the elimination of unreachable parts of the state space. This may be a useful and necessary step if certain safety properties of the system appear hard to prove from the specification.

Example 2.3 Consider the following (artificial) example.

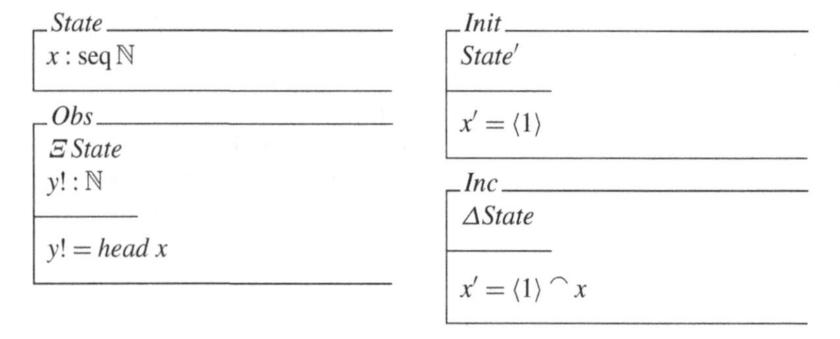

The operation *Obs* is always going to be applicable, as it is "obvious" from the rest of the specification that x will never be empty. □

Another example is that which occurs in low-level refinement steps, where one often requires that the data structures to be implemented are finite. This clashes with the common habit of Z specifiers using $\mathbb{P}$ where $\mathbb{F}$ would have worked equally well.

Such desirable invariants can in many cases be added to an ADT at the cost of a relatively simple proof. If the invariant holds for the ADT as given, adding it results in a refinement-equivalent specification.

Theorem 2.3 *Given an ADT* $(State, Init, \{Op_i\}_{i \in I})$, *the schema Inv on the state State is an* invariant *of the ADT iff*

1. *it is guaranteed to be established by the initialisation*:

$$\forall Init \bullet Inv'$$

2. *it is preserved by every operation*:

$$\forall i : I \bullet \forall \Delta State \bullet Inv \wedge Op_i \Rightarrow Inv'$$

In that case both $(State \wedge Inv, Init, \{Op_i\}_{i \in I})$ *and* $(State, Init, \{Op_i \wedge Inv \wedge Inv'\}_{i \in I})$ *are ADTs which are refinement equivalent to the given ADT.*

Proof The original ADT's being a refinement of either of the modified ones follows from operation refinement on each individual operation (widening the preconditions to include the possibility of $\neg Inv$; condition 2 ensures that the postcondition is not weakened).

The modified ADTs' being a refinement of the original follows from Theorem 2.4. □

The strongest invariant that can be established is by definition the strongest property satisfying the above conditions, which can easily be seen to be *reachability*: the property that characterises that the state can be reached from some initial state through a sequence of operations.

Example 2.4 Continuing from Example 2.3, the property that $x \neq \langle\rangle$ is an invariant; the proof conditions reduce to $\langle 1 \rangle \neq \langle\rangle$ and $x \neq \langle\rangle \Rightarrow x \frown \langle 1 \rangle \neq \langle\rangle$, both of which obviously hold. Thus, this invariant can safely be added to the state (or to every operation). □

Similarly, finiteness can be proved invariant for any set in a state schema which is initialised to a finite set and changed using only the "standard" binary set operators with finite arguments. In that case, $\mathbb{F}$ can be used rather than $\mathbb{P}$.

For example, *squad*, in Example 1.1, is declared as having the type $\mathbb{P}\,PLAYER$. However, it is initially empty and always remains finite. So states with infinite squads are unreachable and could be eliminated from the state space.

2.4.2 Imposing Invariants

Even when the required invariant does not follow directly from the specification, it may well be the case that it can be added by resolving non-determinism in the specification in a suitable way.

Theorem 2.4 *Given an ADT* $(State, Init, \{Op_i\}_{i \in I})$, *the schema Inv on the state State can be* imposed *as an invariant on the ADT iff*

1. *it is allowed by the initialisation*:

$$\exists Init \bullet Inv'$$

2. *it can be imposed on every operation*:

$$\forall i : I \bullet \forall State \bullet Inv \wedge \mathrm{pre}\, Op_i \Rightarrow \mathrm{pre}(Op_i \wedge Inv')$$

In that case both $(State \wedge Inv, Init, \{Op_i\}_{i \in I})$ *and* $(State, Init \wedge Inv', \{Op_i \wedge Inv \wedge Inv'\}_{i \in I})$ *are refinements of the original ADT.*

Proof This follows from a degenerate case of data refinement, a full proof will be given after Definition 4.3. The concrete state is a subset of the abstract state, with the invariant as the retrieve relation. The correctness condition follows from Lemma 2.3; applicability simplifies to condition 2. □

2.5 Example: London Underground

This example describes people travelling on the London Underground in a very abstract sense—just how abstract will become clear when we manage to prove refinements that display rather unexpected (if not unacceptable!) behaviour.

We assume types of stations and lines:

$$[STATION, LINE]$$

A network consists of some (names of) lines with the lists of the stations that are on them. There is always at least 1 station on every line (i.e., even the shortest possible line has a starting and ending station).

$$NET == LINE \nrightarrow \mathrm{seq}_1\, STATION$$

We will need to describe graph-like aspects of such a network. The relation *sameline* describes a graph linking every two stations on the same line; *connected* then represents the reachability relation on that graph. It is a transitive relation because it is a transitive closure, and reflexive and symmetric because *sameline* is reflexive and symmetric (this is obvious from the form of its definition). We also introduce *lines* and *stations* to make the specifications that follow more readable. (Understanding Z should not be about deciphering expressions involving multiple instances of "ran" and "dom".)

$$
\begin{array}{|l}
connected : NET \to (STATION \leftrightarrow STATION) \\
sameline : NET \to (STATION \leftrightarrow STATION) \\
lines : NET \to \mathbb{P}\, LINE \\
stations : NET \to \mathbb{P}\, STATION \\
\hline
\forall x, y : STATION;\ n : NET \bullet \\
\quad (x, y) \in sameline\ n \Leftrightarrow \exists\, l : lines\ n \bullet x \in \operatorname{ran}(n\ l) \wedge y \in \operatorname{ran}(n\ l) \\
\quad connected\ n = (sameline\ n)^{+} \\
\quad lines\ n = \operatorname{dom} n \\
\quad stations\ n = \operatorname{ran}(\bigcup \operatorname{ran} n)
\end{array}
$$

Rather than describing the entire London Underground map, we will record only some crucial properties that will be used later.

$$
\begin{array}{|l}
lu_0 : NET \\
victoria, mc : STATION \\
\hline
victoria \in \operatorname{dom}((connected\ lu_0) \setminus \operatorname{id}) \\
mc \notin stations\ lu_0
\end{array}
$$

The map represented by lu_0 connects at least two *different* stations, Victoria being one of these. Mornington Crescent is also special—for not being part of the map. Historically, this was the case from 1992 to 1998. Together these properties imply the existence of at least three stations.

Finally, we present our rather abstract ADT.

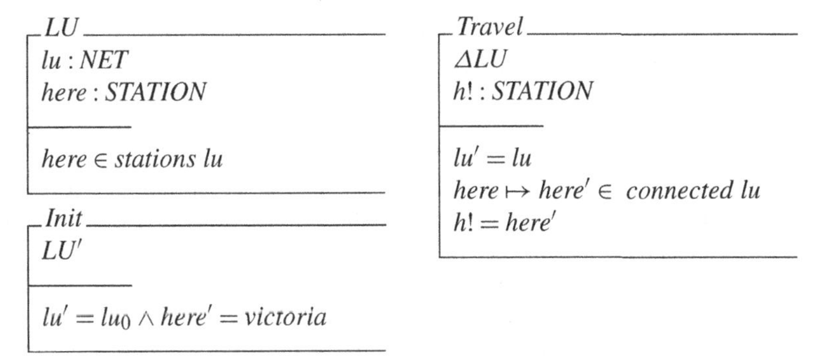

The *Travel* operation represents a trip between two stations, possibly with changes.

We can now illustrate a number of simple refinements, of the various types described above. First, an operation refinement of *Travel* is given by

$$
\begin{array}{|l}
\multicolumn{1}{l}{DontTravel} \\
\Xi LU \\
h! : STATION \\
\hline
h! = here
\end{array}
$$

Its precondition is equal to the precondition of *Travel*, which is *true* in any possible state. The postcondition can be strengthened to $here' = here$ because we know $here \in stations\ lu$ and *connected lu* is reflexive.

The rules for operation refinement do not allow us to refine *Travel* to

$$TrueTravel == Travel \land \neg DontTravel$$

This operation, which requires the starting station and end station to be *different*, is a realistic mode of travel for anyone but kids, trainspotters, and the terminally forgetful. Moreover, we know that in the above specification, starting from Victoria, we will always end up in a station connected to some different station, which could be Victoria itself when we are not there already. (In graph terms: we always remain in the component containing Victoria, and this component has at least 2 nodes due to the restriction on lu_0.) So this condition should always be satisfiable.

However, imagine a station in *stations* lu_0 which only occurs as the single station on a single line—let's call it *mdome*. (In graph theoretic terms, *mdome* is an isolated node in *connected* lu_0.) The precondition of *Travel* holds when $here = mdome$, but not the precondition of *TrueTravel*. Thus, operation refinement fails (for some possible values of lu_0).

The solution to this is to establish the knowledge that we will always be in the component of the graph to which Victoria belongs as an invariant. This also requires the information that the map is as it was initially.

```
InVicComp
LU
---
(here, victoria) ∈ connected lu
lu = lu₀
```

The conditions for establishing an invariant are proved using the properties of *connected* lu_0. Reflexivity of *connected* ensures that the invariant is established initially ($(victoria, victoria) \in connected\ lu_0$). Transitivity and symmetry of *connected* ensure that the invariant is preserved by *Travel*.

After this invariant has been added to *Travel* (or to the state—technically this is a data refinement step as argued above) *TrueTravel* is provably an operation refinement of *Travel*.

Another invariant that could be established is $here \neq mc$. This would also require $lu = lu_0$ or some weaker condition which would prevent Mornington Crescent being reconnected.

Finally, we will look at imposing an invariant. Our travellers will like this even less than the previous refinements.[5]

[5] In other words, by having the single operation *Travel* as its interface, our ADT does not express the level of external choice that would be necessary from the customer's point of view.

Consider the operation

$$\begin{array}{|l}
\multicolumn{1}{l}{_NearVic___} \\
\Delta LU \\
h! : STATION \\
\hline
lu' = lu \\
here \mapsto here' \in connected\ lu \\
(here', victoria) \in sameline\ lu \\
h! = here' \\
\hline
\end{array}$$

That is, you can travel anywhere provided there is a direct train back to Victoria. This is not an operation refinement of the initial *Travel*, because the postcondition is not satisfiable from any station *here* which is at least 2 changes away from Victoria[6] (nor is it satisfiable from any station in a different component of the graph such as *mdome*, of course). Moreover, we cannot establish the invariant $(here, victoria) \in sameline\ lu$, because it is *not* guaranteed to be preserved by *Travel*.

However, this invariant *can be imposed* on the specification. Initially it is certainly true. Then, in *Travel*, whenever we start out in a station on one line with Victoria, we can always find a connected station that is also on one line with Victoria – for example, by staying where we are, or travelling to Victoria itself! In fact, even the invariant $here = victoria$ can be imposed, but that (in combination with $lu = lu_0$) should certainly be the strongest possible one.

2.6 Bibliographical Notes

The principle of substitutivity is so intuitive that it has often been proposed under various other names. It is called "semantic implementation correctness" in [3], attributing the term to Gardiner and Morgan [4]. Nipkow [6] gives a similar intuitive definition of "implementation". Related fundamental definitions follow from formalising different methods of testing a specification and comparing their observations, see the work by De Nicola [2] and the extensive hierarchy of refinement relations explored by Van Glabbeek [9, 10], and also more recently work by Reeves and Streader [7]. For other references to early work on refinement we refer to the monograph by de Roever and Engelhardt [3].

Typically, Z textbooks do cover operation refinement separately from data refinement, as we do here. However, the other forms of simple refinement are not normally covered independently, although the Z community has been aware of them [8]. Some textbooks even require the abstract and concrete state to be always different in data refinement, to avoid name capture.

[6] At the time of writing, there were actually very few such stations.

References

1. de Bakker, J. W., de Roever, W.-P., & Rozenberg, G. (Eds.) (1990). *REX Workshop on Stepwise Refinement of Distributed Systems*, Nijmegen, 1989. *Lecture Notes in Computer Science: Vol. 430*. Berlin: Springer.
2. de Nicola, R. (1987). Extensional equivalences for transition systems. *Acta Informatica*, *24*(2), 211–237.
3. de Roever, W.-P., & Engelhardt, K. (1998). *Data Refinement: Model-Oriented Proof Methods and Their Comparison*. Cambridge: Cambridge University Press.
4. Gardiner, P. H. B., & Morgan, C. C. (1993). A single complete rule for data refinement. *Formal Aspects of Computing*, *5*, 367–382.
5. McIver, A., & Morgan, C. C. (2004). *Abstraction, Refinement and Proof for Probabilistic Systems*. Berlin: Springer.
6. Nipkow, T. (1990) Formal verification of data type refinement—theory and practice. In de Bakker et al. [1] (pp. 561–591).
7. Reeves, S., & Streader, D. (2011). Contexts, refinement and determinism. *Science of Computer Programming*, *76*(9), 774–791.
8. Strulo, B. (1995). Email communication.
9. van Glabbeek, R. J. (1993). The linear time–branching time spectrum II. The semantics of sequential systems with silent moves (extended abstract). In E. Best (Ed.), *CONCUR'93, 4th International Conference on Concurrency Theory*. *Lecture Notes in Computer Science: Vol. 715* (pp. 66–81). Berlin: Springer.
10. van Glabbeek, R. J. (2001). The linear time–branching time spectrum I. The semantics of concrete sequential processes. In J. A. Bergstra, A. Ponse, & S. A. Smolka (Eds.), *Handbook of Process Algebra* (pp. 3–99). Amsterdam: North-Holland.

Chapter 3
Data Refinement and Simulations

In the previous chapter, a semi-formal motivation was given for the operation refinement rule, and it was stated that this rule also applies in the presence of inputs and outputs. Furthermore, other "simple refinement" rules were given whose justification was postponed to the general data refinement theorem, to be presented in this chapter and formulated in Z in the next. Apart from that, the following issues were not fully exposed:

- What exactly are the observations of an *abstract* data type, and what is their relation to ADT *programs*?
- How do inputs and outputs fit in with this?
- What is the effect of using operations specified by relations which are not necessarily total?
- What is the function of initialisation?

This chapter answers most of these questions by presenting the standard refinement theory for ADTs in a relational setting. The notion of observation will be fixed by introducing an explicit concrete ("global") state which is used to initialise (and finalise) ADTs. How inputs and outputs are embedded in this will be postponed until the next chapter, where we transfer the results of this chapter to Z.

The central part of this chapter is the standard definition of data refinement for relational data types, the definitions of upward and downward simulations, and the statement of their soundness and joint completeness. This theory was initially formulated for ADTs whose operations are *total* relations.

In order to apply this theory to a specification language we look at how operations in a specification are modelled as partial relations. The application of the simulation rules to specifications with partial operations leads to the simulation rules as they are normally presented.

Much of this book is concerned with *different* refinement relations for (notations like) Z, and this is where it starts. Not only are there different ways of interpreting

---

J. Derrick, E.A. Boiten, *Refinement in Z and Object-Z*,
DOI 10.1007/978-1-4471-5355-9_3, © Springer-Verlag London 2014

partial relations as total relations on an extended domain; the standard theory is equally valid for *partial* relations, and thus can be applied directly as well—leading to a total of at least three different notions of refinement.

3.1 Programs and Observations for ADTs

Recall that observations of concrete data types ("repertoires") were pairs of concrete states. (Some extra complications arose from the operations' not being total, but for the moment we will only consider ADTs with *total* relations as their operations.) For abstract data types, it is not immediately obvious what the observations should be. They could not be pairs of states of the ADT, as these are supposed to be abstract and thus not observable. One possible approach would be just to observe that a certain sequence of operations, or branching structure of choices between operations, is possible. This sort of behavioural interpretation is commonly used for algebraic specifications and process algebras (about which more later), but it is rather artificial for a data-oriented notation like Z (e.g., a stack containing the single value 17 is behaviourally characterised by the fact that it will allow *pop* exactly once, resulting in a value which will allow *predecessor* exactly seventeen times, rather than giving the output 17). Recall also that in the eventual Z presentation of the rules, inputs and outputs will form part of the observation, which would make for a very convoluted behavioural description.

A solution to this issue is to introduce a notion of *global* state, which plays a *rôle* comparable to the state of a concrete data type in the previous chapter. One might say that we now consider every ADT *program* as an *operation* on the global state, for which operation-wise refinement on the global level should hold. The state as included in the ADT will be referred to as the *local* state. It is the local state which is "abstract", in the sense of not being considered observable to the user of the ADT.

For the repertoires in the previous chapter, a program was characterised by a sequence of indices, and denoted the sequential composition of the sequence of operations characterised by those indices. This would not be sufficient for ADTs, as this sequential composition would still be a relation on the *local* state. For that reason, we will prepend an *initialisation* operation to each such sequence, which will take every global state into some local (starting) state, and append a *finalisation*, which will translate back from local to global state, and call the result a *complete* program. This will lead to a complete program being interpreted as a relation on the global state as required. The initialisation and finalisation will form part of the abstract data type.

In order to distinguish between relational formulations (which use Z as a meta-language) and expressions in terms of Z schemas etc., we introduce the convention that expressions and identifiers in the world of relational data types are typeset in a sans serif font.

In the following definition we present the traditional definition of a (abstract relational) data type. This differs from the standard Z ADT introduced in Chap. 1

in three ways: it is expressed in terms of relations rather than schemas; besides an initialisation it also contains a finalisation; initialisation and finalisation are not represented by sets but by relations between a (fixed) global state G and the ADT state.

Definition 3.1 (Data type) A *data type* is a quadruple $(\mathsf{State}, \mathsf{Init}, \{\mathsf{Op_i}\}_{i \in I}, \mathsf{Fin})$. The operations $\{\mathsf{Op_i}\}$, indexed by $i \in I$, are relations on the set State; Init is a total relation from G to State; Fin is a total relation from State to G. □

Insisting that Init and Fin be total merely records the facts that we can always start a program sequence (the extension to partial initialisations is trivial) and that we can always make an observation afterwards.

Particular classes of data types are those which have only total or functional operations.

Definition 3.2 (Total and canonical data type) A data type $(\mathsf{State}, \mathsf{Init}, \{\mathsf{Op_i}\}_{i \in I}, \mathsf{Fin})$ is *total* if $\mathsf{Op_i}$ are all *total*, and *canonical* if Init, and $\mathsf{Op_i}$ are all *functions*. □

Just like for repertoires, we only consider refinement between conformal data types.

Definition 3.3 (Conformal data types) Two data types are *conformal* if their global data space G and the indexing sets of their operations are equal. □

For the rest of this chapter, assume that all data types considered are conformal, using some fixed index set I. In later chapters we will consider more liberal notions of conformity.

Definition 3.4 (Complete program) A *complete program* over a data type $D = (\mathsf{State}, \mathsf{Init}, \{\mathsf{Op_i}\}_{i \in I}, \mathsf{Fin})$ is an expression of the form $\mathsf{Init} \fatsemi P \fatsemi \mathsf{Fin}$, where P, a relation over State, is a *program* over $\{\mathsf{Op_i}\}_{i \in I}$.

In particular, for a sequence p over I, and data type D, p_D denotes the complete program over D characterised by p (i.e., for D as above, if $p = \langle p_1, \ldots, p_n \rangle$ then $p_D = \mathsf{Init} \fatsemi \mathsf{Op}_{p_1} \fatsemi \cdots \fatsemi \mathsf{Op}_{p_n} \fatsemi \mathsf{Fin}$). □

Example 3.1 We apologise for the artificial nature of all the relational examples in this chapter—"realistic" relational specifications just do not allow the clear separation between presentational structure and mathematical content that is possible with Z schemas.

Let $G = \mathbb{N}$, $D = (\mathbb{N}, \mathsf{Init}, \{\mathsf{Op_1}, \mathsf{Op_2}\}, \mathsf{Fin})$ where $\mathsf{Init} = \{x : \mathbb{N} \bullet (x, 0)\}$, $\mathsf{Op_1} = \{x : \mathbb{N} \bullet (x, x+1)\}$, $\mathsf{Op_2} = \{x, y : \mathbb{N} \mid x = y \lor x + 2 = y \bullet (x, y)\}$ and $\mathsf{Fin} = \mathsf{id}_{\mathbb{N}}$, then the program $\langle 1, 2, 1 \rangle_D$ denotes $\mathsf{Init} \fatsemi \mathsf{Op_1} \fatsemi \mathsf{Op_2} \fatsemi \mathsf{Op_1} \fatsemi \mathsf{Fin}$ which is $\{x : \mathbb{N}; y : \{2, 4\} \bullet (x, y)\}$, see Fig. 3.1. □

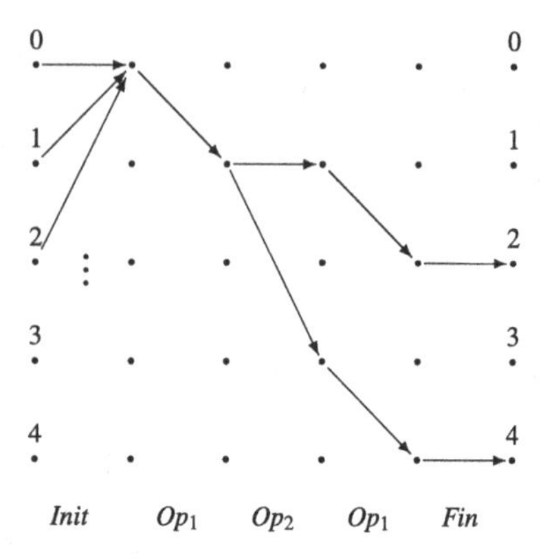

Fig. 3.1 The program $[1,2,1]_D$

Choices can be made as to which constructors can be used for programs, however sequential composition is normally included. In the literature, one often finds Dijkstra's guarded command language as the programming language for ADTs. Here we will stick to programs as finite sequences of operations only, as this will be sufficient for the required notions of refinement in Z. Any program in Dijkstra's language would represent a collection of such straight line programs. However, the combination of recursion (or while-loops) and unbounded non-determinism that is offered in such a language presents an extra level of complication, which we will not go into here. Analogous to the definition of refinement for repertoires, we have the standard definition of refinement between data types, by quantification over all programs using them:

Definition 3.5 (Data refinement) For data types A and C, C *refines* A (also denoted $\mathsf{A} \sqsubseteq \mathsf{C}$) iff for each finite sequence p over I, $\mathsf{p}_\mathsf{C} \subseteq \mathsf{p}_\mathsf{A}$. □

Observe that data refinement is defined here in terms of inclusion of *total* relations, and is thus *only* concerned with "reduction of non-determinism".

Because set inclusion satisfies many useful properties, the following is immediate.

Theorem 3.1 (Data refinement is a preorder) *Data refinement is transitive and reflexive, i.e., it is a preorder.* □

However, data refinement does not inherit all properties of $\subseteq$, in particular it is not anti-symmetric: $\mathsf{p}_\mathsf{C} = \mathsf{p}_\mathsf{A}$ may hold for all p even when C and A are different. Other lattice-like properties may hold but are non-obvious from this definition.

Example 3.2 Let $\mathsf{G} = \mathbb{N}$, $\mathsf{A} = (\mathbb{N}, \mathsf{id}_\mathbb{N}, \{\mathsf{AOp}\}, \mathsf{id}_\mathbb{N})$, $\mathsf{C} = (\mathbb{N}, \mathsf{id}_\mathbb{N}, \{\mathsf{COp}\}, \mathsf{id}_\mathbb{N})$ where $\mathsf{AOp} = \{\mathsf{x}, \mathsf{y} : \mathbb{N} \mid \mathsf{x} < \mathsf{y} \bullet (\mathsf{x}, \mathsf{y})\}$, $\mathsf{COp} = \{\mathsf{x} : \mathbb{N} \bullet (\mathsf{x}, \mathsf{x}+1)\}$.

It can be proved that for any sequence p (as there is only one operation, only its length is relevant): $\mathsf{p}_\mathsf{A} = \{\mathsf{x}, \mathsf{y} : \mathbb{N} \mid \mathsf{x} + \#\mathsf{p} \leq \mathsf{y} \bullet (\mathsf{x}, \mathsf{y})\}$, whereas $\mathsf{p}_\mathsf{C} = \{\mathsf{x} : \mathbb{N} \bullet (\mathsf{x}, \mathsf{x} + \#\mathsf{p})\}$. Clearly $\mathsf{p}_\mathsf{C} \subseteq \mathsf{p}_\mathsf{A}$, and therefore C refines A. □

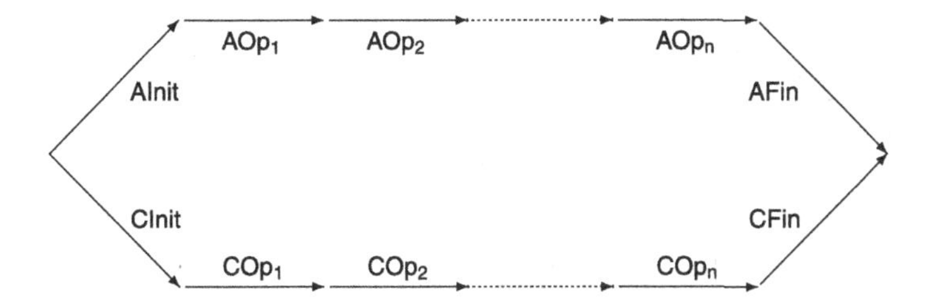

Fig. 3.2 Data refinement

For all but the most trivial examples, however, Definition 3.5 is very difficult to verify directly as it quantifies over all programs. As before, we need to find a definition of refinement which quantifies over all operations instead.

The solution to this is well-known and described below, but not immediate. There are two simpler candidate solutions, drawn from the analogy with the previous chapter, to be discounted first.

First, consider operation-wise refinement between conformal data types $\mathsf{A} = (\mathsf{AState}, \mathsf{AInit}, \{\mathsf{AOp}_i\}_{i\in I}, \mathsf{AFin})$ and $\mathsf{C} = (\mathsf{CState}, \mathsf{CInit}, \{\mathsf{COp}_i\}_{i\in I}, \mathsf{CFin})$. Comparing AOp_i with COp_i directly is impossible, in general, for reasons of typing: one is a relation on AState, the other on CState.

However, looking more closely at the definition of operation-wise refinement, we can see that it is not defined between operations, but between *programs* consisting of a single operation each. Taking this approach for data types as well solves the typing problem: the programs $\langle i \rangle_A$ and $\langle i \rangle_C$ *are* comparable. They are both relations on the global state, which must be the same for A and C due to conformity.

Unfortunately, knowing that $\langle i \rangle_C \subseteq \langle i \rangle_A$ (which is: $\mathsf{CInit} \mathbin{⨾} \mathsf{COp}_i \mathbin{⨾} \mathsf{CFin} \subseteq \mathsf{AInit} \mathbin{⨾} \mathsf{AOp}_i \mathbin{⨾} \mathsf{AFin}$) for all i does not help us very much. In general, unlike for repertoires, it is not the case that for any data type A and all sequences p and q, the program represented by their concatenation can be obtained from relational composition of the individual programs for p and q. This would require the unrealistically strong assumption that $\mathsf{AFin} \mathbin{⨾} \mathsf{AInit} = \mathsf{id}_{\mathsf{AState}}$. (Try composing two instances of Fig. 3.2 to get another instance—the extra condition would remove the X shape in the middle.) In other words, composition of programs is unrelated to composition of complete programs. As a consequence, we could not prove data refinement from this sort of operation-wise refinement using induction.

3.2 Upward and Downward Simulations

Figure 3.2 represents our target: we want to be able to compare any two structurally equivalent programs. In particular, we want to do it inductively, by considering a limited number of base cases, and then by a finite number of compositions of such results obtain the result for any finite sequence of operations. Above we have argued that two diagrams like Fig. 3.2 cannot be concatenated to form another diagram of

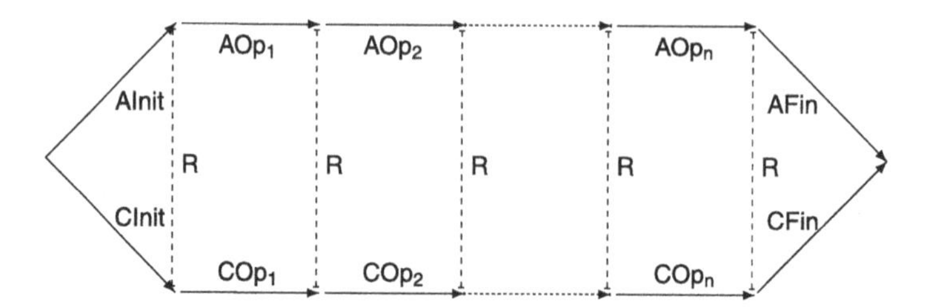

Fig. 3.3 Data refinement with simulations

Fig. 3.4 Data refinement: a single operation

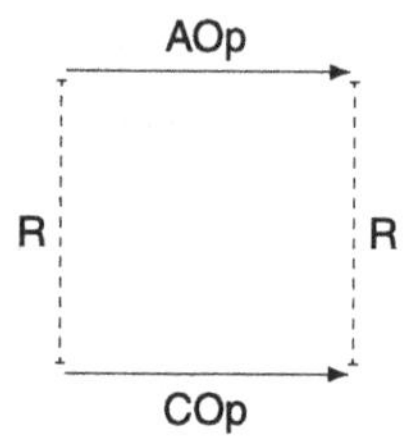

the same kind. If we did want to form a diagram like Fig. 3.2 from simpler diagrams, what shape should they take? The answer to that question will still appear as a bit of a *eureka*—except that it is a common and natural answer in the real world as well.

Example 3.3 Assume we are given the problem of cycling from Dover to Whitstable, and we are provided with a detailed road map of East Kent which clearly indicates Dover and Whitstable.

On the map we would pick out a route—one which is fairly direct, probably without using dual carriageways. Now how would we remember our route? Most likely not as a sequence of left/right turns only! Rather, we would use the map and its legend to identify intermediate points on the route with places we know in East Kent, and remember our journey as going through (say) Patrixbourne and Canterbury.

The map gives an abstract representation of our concrete real world. We construct our concrete cycle trip from an abstract curve on the map not just by inferring left/right instructions from it; we validate it by using the relation between the abstract world (the map) and the concrete world (East Kent), at the endpoints but also at intermediate points. This relation is a *simulation*. □

Simulations are also known as *retrieve relations*, *abstraction relations*, or *coupling invariants*. Certainly in the context of model-based specification languages with an *explicit* notion of state it is only natural to base the comparison of two data types on the relation between their states. Adding a linking relation R between concrete and abstract states in Fig. 3.2 leads to Fig. 3.3. Apart from triangles at the beginning and at the end, we get instances of the commuting square that is usually required for data type implementation, see Fig. 3.4. In a setting where all operations and the abstraction are specified by *functions*, the directions of the functions

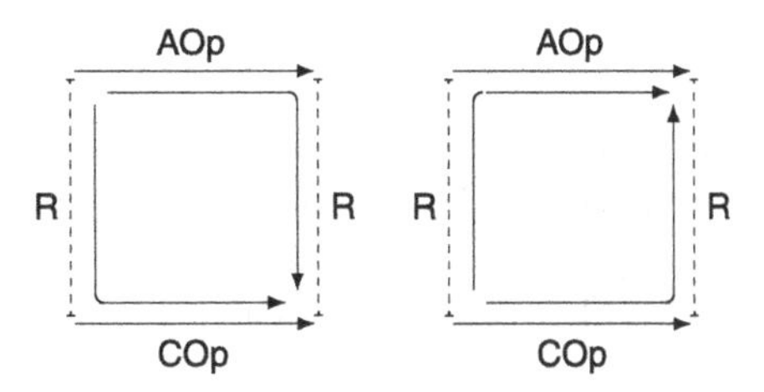

Fig. 3.5 L-simulation and L^{-1}-simulation

determine precisely how this square could commute, and both paths are necessarily equal. However, in our setting, where operations and simulations are relations, there are four ways of "going around the square". De Roever and Engelhardt [12] name them after the shapes of the paths. U-simulation (the top line versus the rest) and U^{-1}-simulation (the bottom line versus the rest) are not useful for us, as they do not guarantee that squares can be concatenated. However, L-simulation (the two paths from top-left to bottom-right) and L^{-1}-simulation (bottom-left to top-right, see Fig. 3.5) do have that property, cf. Theorem 3.4.

Additionally, since we do not want to establish equality in Fig. 3.3 but inclusion of the "concrete" path in the "abstract" path, we will need "semi-commutativity" of the squares, with the path involving COp *included* in, rather than equal to, the other path.

There are also two possible ways for the initialisation/finalisation triangles to semi-commute, the particular choices for L- and L^{-1}-simulation are dictated by the need for the diagrams to be concatenated, and again to have the concrete component on the smaller side of the inclusion.

In the rest of this book we will use the terms *downward* and *upward* simulation for L- and L^{-1}-simulation, respectively. They are defined as follows.

Definition 3.6 (Downward simulation) Let $A = (AState, AInit, \{AOp_i\}_{i \in I}, AFin)$ and $C = (CState, CInit, \{COp_i\}_{i \in I}, CFin)$ be data types. A *downward* simulation is a relation R from AState to CState satisfying

$$CInit \subseteq AInit \mathbin{⨾} R \tag{3.1}$$

$$R \mathbin{⨾} CFin \subseteq AFin \tag{3.2}$$

$$\forall i : I \bullet R \mathbin{⨾} COp_I \subseteq AOp_i \mathbin{⨾} R \tag{3.3}$$

If such a simulation exists, we also say that C is a downward simulation of A, also denoted $A \sqsubseteq_{DS} C$, and similarly for corresponding operations of A and C. □

Besides "L-simulation", another term for this is *forward* simulation.

Definition 3.7 (Upward simulation) For data types A and C as above, an *upward* simulation is a relation T from CState to AState such that

$$CInit \mathbin{⨾} T \subseteq AInit \tag{3.4}$$

$$CFin \subseteq T \mathbin{⨾} AFin \tag{3.5}$$

$$\forall i : I \bullet COp_i \,⨾\, T \subseteq T \,⨾\, AOp_i \tag{3.6}$$

If such a simulation exists, we also say that C is an upward simulation of A, also denoted $A \sqsubseteq_{US} C$, and similarly for corresponding operations of A and C. □

Besides "L^{-1}-simulation", another term for this is *backward* simulation.

Theorem 3.2 (Vertical composition) *The composition of two upward simulations is an upward simulation; the composition of two downward simulations is a downward simulation.*

Proof Immediate from the definitions, using associativity and monotonicity of ⨾ and transitivity of $\subseteq$. □

Theorem 3.3 (Functional simulations) *For data types* A *and* C *as above, and a relation* T *between* CState *and* AState *such that* T *is total and functional, then* T *is an upward simulation between* A *and* C *if and only if* T^{-1} *is a downward simulation between* A *and* C.

Proof Totality of T is encoded relationally as $T \,⨾\, T^{-1} \supseteq id_{CState}$, and functionality as $T^{-1} \,⨾\, T \subseteq id_{AState}$. Together these allow the proof of equivalence for the corresponding upward and downward simulation conditions. For example, the following proves the equivalence of the conditions for operations by mutual implication.

$$COp_i \,⨾\, T \subseteq T \,⨾\, AOp_i$$
$$\Rightarrow$$
$$T^{-1} \,⨾\, COp_i \,⨾\, T \,⨾\, T^{-1} \subseteq T^{-1} \,⨾\, T \,⨾\, AOp_i \,⨾\, T^{-1}$$
$$\Rightarrow \{ \text{ totality on lhs, functionality on rhs } \}$$
$$T^{-1} \,⨾\, COp_i \subseteq AOp_i \,⨾\, T^{-1}$$
$$\Rightarrow$$
$$T \,⨾\, T^{-1} \,⨾\, COp_i \,⨾\, T \subseteq T \,⨾\, AOp_i \,⨾\, T^{-1} \,⨾\, T$$
$$\Rightarrow \{ \text{ totality on lhs, functionality on rhs } \}$$
$$COp_i \,⨾\, T \subseteq T \,⨾\, AOp_i$$

(The middle line is the condition for T^{-1} to be a downward simulation, which both implies and is implied by the top/bottom line.) □

In order to prove that the existence of an upward or downward simulation guarantees refinement, i.e., that they are *sound*, we need to establish that the sub-diagrams of Fig. 3.3 can be combined in the appropriate way. If we put two diagrams together which semi-commute according to upward/downward simulation rules, the outer paths of the combined diagram should semi-commute in the same way. This is expressed for all combinations of diagrams in the following theorem.

Theorem 3.4 (Horizontal composition) *For conformal data types* A *and* C *as above, and appropriately typed relations* R *and* T,

(a) *if* (3.1) *and* (3.3) *hold, then* (3.1) *holds for* $\mathsf{AInit} := \mathsf{AInit} \mathbin{⨾} \mathsf{AOp_i}$ *and* $\mathsf{CInit} := \mathsf{CInit} \mathbin{⨾} \mathsf{COp_i}$
(b) *if* (3.3) *holds, then it also holds for any sequence of two operations, i.e.,*

$$\forall \mathsf{i}, \mathsf{j} : \mathsf{I} \bullet \mathsf{R} \mathbin{⨾} \mathsf{COp_i} \mathbin{⨾} \mathsf{COp_j} \subseteq \mathsf{AOp_i} \mathbin{⨾} \mathsf{AOp_j} \mathbin{⨾} \mathsf{R}$$

(c) *if* (3.3) *and* (3.2) *hold, then* (3.2) *holds for* $\mathsf{AFin} := \mathsf{AOp_i} \mathbin{⨾} \mathsf{AFin}$ *and* $\mathsf{CFin} := \mathsf{COp_i} \mathbin{⨾} \mathsf{CFin}$
(d) *if* (3.4) *and* (3.6) *hold, then* (3.4) *holds for* $\mathsf{AInit} := \mathsf{AInit} \mathbin{⨾} \mathsf{AOp_i}$ *and* $\mathsf{CInit} := \mathsf{CInit} \mathbin{⨾} \mathsf{COp_i}$
(e) *if* (3.6) *holds, then it also holds for any sequence of two operations, i.e.,*

$$\forall \mathsf{i}, \mathsf{j} : \mathsf{I} \bullet \mathsf{COp_i} \mathbin{⨾} \mathsf{COp_j} \mathbin{⨾} \mathsf{T} \subseteq \mathsf{T} \mathbin{⨾} \mathsf{AOp_i} \mathbin{⨾} \mathsf{AOp_j}$$

(f) *if* (3.6) *and* (3.5) *hold, then* (3.5) *holds for* $\mathsf{AFin} := \mathsf{AOp_i} \mathbin{⨾} \mathsf{AFin}$ *and* $\mathsf{CFin} := \mathsf{COp_i} \mathbin{⨾} \mathsf{CFin}$.

Proof The proofs are all elementary, with the proofs for (d)–(f) (which we have omitted) dual to those for (a)–(c).

(a)

$$\begin{array}{l} \mathsf{CInit} \mathbin{⨾} \mathsf{COp_i} \\ \quad \subseteq \{\text{by } (3.1)\} \\ \mathsf{AInit} \mathbin{⨾} \mathsf{R} \mathbin{⨾} \mathsf{COp_i} \\ \quad \subseteq \{\text{by } (3.3)\} \\ \mathsf{AInit} \mathbin{⨾} \mathsf{AOp_i} \mathbin{⨾} \mathsf{R} \end{array}$$

(b)

$$\begin{array}{l} \mathsf{R} \mathbin{⨾} \mathsf{COp_i} \mathbin{⨾} \mathsf{COp_j} \\ \quad \subseteq \{\text{by } (3.3)\} \\ \mathsf{AOp_i} \mathbin{⨾} \mathsf{R} \mathbin{⨾} \mathsf{COp_j} \\ \quad \subseteq \{\text{by } (3.3)\} \\ \mathsf{AOp_i} \mathbin{⨾} \mathsf{COp_j} \mathbin{⨾} \mathsf{R} \end{array}$$

(c)

$$\begin{array}{l} \mathsf{R} \mathbin{⨾} \mathsf{COp_i} \mathbin{⨾} \mathsf{CFin} \\ \quad \subseteq \{\text{by } (3.3)\} \\ \mathsf{AOp_i} \mathbin{⨾} \mathsf{R} \mathbin{⨾} \mathsf{CFin} \\ \quad \subseteq \{\text{by } (3.2)\} \\ \mathsf{AOp_i} \mathbin{⨾} \mathsf{AFin} \end{array}$$

□

The base case of the soundness proofs is the empty program.

Theorem 3.5 (Simulations for the empty program) *If an upward simulation* T *or downward simulation* R *exists between data types* A *and* C, *then data refinement (Definition* 3.5) *holds when we consider empty programs only.*

Proof The proof required is that

$$\mathsf{CInit} \fatsemi \mathsf{CFin} \subseteq \mathsf{AInit} \fatsemi \mathsf{AFin}$$

For downward simulation:

$$\begin{array}{l} \mathsf{CInit} \fatsemi \mathsf{CFin} \\ \qquad \subseteq \{\text{by } (3.1)\} \\ \mathsf{AInit} \fatsemi \mathsf{R} \fatsemi \mathsf{CFin} \\ \qquad \subseteq \{\text{by } (3.2)\} \\ \mathsf{AInit} \fatsemi \mathsf{AFin} \end{array}$$

For upward simulation:

$$\begin{array}{l} \mathsf{CInit} \fatsemi \mathsf{CFin} \\ \qquad \subseteq \{\text{by } (3.5)\} \\ \mathsf{CInit} \fatsemi \mathsf{T} \fatsemi \mathsf{CFin} \\ \qquad \subseteq \{\text{by } (3.4)\} \\ \mathsf{AInit} \fatsemi \mathsf{AFin} \end{array}$$

□

Theorem 3.6 (Soundness of simulations) *If an upward or downward simulation exists between conformal data types* A *and* C, *then* C *is a data refinement of* A.

Proof By induction on the (complete) programs, using Theorem 3.5 as the base case, and Theorem 3.4 (a)/(d) as the induction step. □

However, neither downward simulation nor upward simulation is *complete*, see the following example. In fact, refinements are possible which require *both* upward and downward simulation for their proof.

Example 3.4 ADTs which are related by data refinement where no downward simulation exists are as follows. (The essence of this counter-example, as represented in Fig. 3.6, can be found in nearly all the books and papers on simulations. It can be summarised as "downward simulation does not allow postponement of non-deterministic choice".)

Let $\mathsf{G} = 0..4$, which acts as the global state and one of the local states; the other local state is $\mathsf{YS} = \{0, 1, 3, 4\}$. Define $\mathsf{X} = (\mathsf{G}, \mathsf{Init}, \{\mathsf{XOp}_1, \mathsf{XOp}_2\}, \mathsf{id}_\mathsf{G})$,

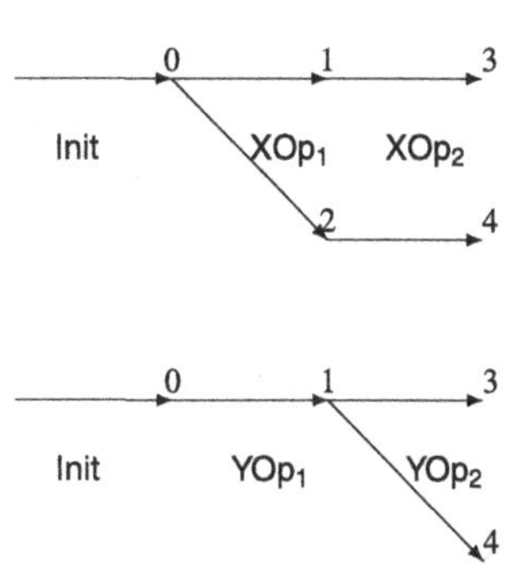

Fig. 3.6 Incompleteness of simulations: infamous counter-example

$\mathsf{Y} = (\mathsf{YS}, \mathsf{Init}, \{\mathsf{YOp}_1, \mathsf{YOp}_2, \}, \mathsf{id}_{\mathsf{YS}})$, where

$$
\begin{aligned}
&\mathsf{Init} = \{\mathsf{x} : \mathsf{G} \bullet (\mathsf{x}, 0)\}\\
&\mathsf{XOp}_1 = \{(0,1), (0,2), (1,1), (2,2), (3,3), (4,4)\}\\
&\mathsf{XOp}_2 = \{(0,0), (1,3), (2,4), (3,3), (4,4)\}\\
&\mathsf{YOp}_1 = \{(0,1), (1,1), (3,3), (4,4)\}\\
&\mathsf{YOp}_2 = \{(0,0), (1,3), (1,4), (3,3), (4,4)\}
\end{aligned}
$$

Y is a data refinement of X. A program here is by definition a finite sequence over the numbers 1 and 2. The finalisation is the identity, so the final state is the only observable outcome of a program.

- Any program which does not include operation 1 (including the empty program) will end in state 0.
- Any program containing operation 1 but not 2 after it will end in state 1 or 2 for X, and in state 1 for Y.
- Any other program (i.e., containing at least operation 1 followed by 2 sometime later) could end in either state 3 or 4 for both X and Y.

As the outcomes for Y are always included in those for X, data refinement holds.

Now assume we have a downward simulation R from G to YS. From the finalisation condition (3.2) we get $\mathsf{R} \subseteq \mathsf{id}_{\mathsf{YS}}$. From initialisation and refinement for the first operation we then get that $(1,1) \in \mathsf{R}$. The refinement condition for the second operation then gives us:

$$
\begin{aligned}
&\mathsf{R} \mathbin{⨾} \mathsf{YOp}_2 \subseteq \mathsf{XOp}_2 \mathbin{⨾} \mathsf{R}\\
&\qquad \Rightarrow \{(1,1) \in \mathsf{R}, \text{transitivity of } \subseteq\}\\
&\{(1,3), (1,4)\} \subseteq \mathsf{XOp}_2 \mathbin{⨾} \mathsf{R}\\
&\qquad \equiv\\
&\{(1,3), (1,4)\} \subseteq \{(1,3)\} \mathbin{⨾} \mathsf{R}\\
&\qquad \Rightarrow \{⨾\}\\
&(3,4) \in \mathsf{R}
\end{aligned}
$$

which contradicts that R is contained in the identity relation. Thus, no such R can exist.

However, it can be proved that $\{(0,0),(1,1),(1,2),(3,3),(4,4)\}$ does constitute an upward simulation in this case. A similar example[1] (swapping abstract and concrete data type) can be used to show a data refinement where no upward simulation exists. A modification where we identify the abstract states 0 and 2 would even be a data refinement where neither simulation exists. □

The good news is, however, that for our definition of data refinement, downward and upward simulation are jointly complete.

Theorem 3.7 *Upward and downward simulation are jointly complete, i.e., any data refinement can be proved by a combination of simulations.*

Proof In fact, any data refinement can be proved by an upward simulation, followed by a downward simulation. The upward simulation constructs a canonical ADT, i.e., a deterministic equivalent of the "abstract" type, using the familiar powerset construction or using a construction where states are characterised as traces (sequences of operations). A downward simulation can then always be found to the "concrete" type. For details of the proof, see the literature, e.g., [12, 18, 21]. The construction involved will reappear in Chap. 8, where we define a single complete simulation rule. A variant of the completeness proof applies the normal form construction, which is a refinement equivalence, to both ADTs, and then uses an auxiliary lemma that downward simulation is complete for canonical ADTs to complete the proof. □

3.3 Dealing with (Partial) Relations

The standard theory of relational data types using simulations as described above was originally developed for *total* data types, e.g., in the most commonly referenced paper by He, Hoare, and Sanders [18]. However, the restriction to total operations turned out to be unnecessary and disappeared in subsequent publications by He and Hoare, e.g. [17], without the authors emphasising this relaxation. The literature on refinement in Z, including the previous edition of this book, did not recognise this. Thus, refinement notions for Z (which allows partiality) were based only on embeddings of relations into total relations and then using the He/Hoare/Sanders simulation theory, when there is the alternative of basing them directly on simulations for (not necessarily total) relations.[2] Below, we first discuss the approaches based on embeddings, including the one which leads us to the traditional refinement relation for Z, in Sects. 3.3.1 and 3.3.2 and then the approach of using the relational theory directly in Sect. 3.3.3.

[1]It would need $\{(1,2),(2,1)\}$ added to both finalisations, cf. Example 3.5.

[2]In this section of the book, we use "partial" (for relations and operations) to mean "not necessarily total". Mathematically this is uninformative as this characterises *all* relations, but methodologically it serves to remind us that totality may not be assumed, which has its consequences.

3.3.1 Partial Relations and Totalisations

In Chap. 2 (particularly Fig. 2.1 on p. 57) we gave a first indication of how partial relations could be interpreted for concrete data types, by defining particular collections of observations for before-states outside an operation's domain, involving a fictitious value $\perp$. The construction given there embedded a partial operation on some state *State* into a total relation on an enhanced state $State_\perp$. There are at least two fundamentally different approaches to such embeddings, each representing a different semantic view of the world of state based systems.

In the "*contract*" or "non-blocking" approach to ADTs, the domain (precondition) of an operation describes the area within which the operation should be guaranteed to deliver a well-defined result, as specified by the relation. Outside that domain, the operation *may* be applied, but "anything may happen", which is modelled by the operation returning any value, even an undefined one (modelling, e.g., non-termination). This represents a situation in which the customer employs a black box data type, with its precondition representing the situations in which they expect to be using it. This is also the approach taken in Chap. 2, as evident in the definition of an "observation".

In the "*behavioural*" or "blocking" approach to ADTs, operations may not be applied outside their precondition; doing so anyway leads to an "error" result. The precondition is often called the *guard* in this context. This represents the situation where the customer employs an independently operating machine, the operations of the ADT corresponding to possible interactions between the system and its environment. This approach is close to the one used in Object-Z, see Part III of this book, which links to the notion of "refusal" or "blocking" in process algebra. As a consequence, it is also the normal interpretation in integrations of state based systems with process algebras, e.g., in Part IV.

Traditionally, the first approach is the more common one in Z, although it really should depend on the area of application as to which of the two interpretations is the "correct" one. In fact, combinations of the two perspectives can be constructed, as can be seen in [4, 7, 12].

In both approaches the simulation rules for partial operations are derived from those for total operations. This is done analogous to the definition of observations in Chap. 2: partial relations on a set S are modelled as total relations on a set $S_\perp$, which is S extended with a distinguished value $\perp$ not in S.[3] Rather than discussing observations (which are now elements of our visible global state), we will now talk about "totalisation" of relations, in the first place on the invisible local state. In the contract approach, the totalisation of an operation is as already illustrated in

[3] In order to define $S_\perp$ smoothly, we have to take a small liberty with our use of Z as a "meta-language". Z has a strong decidable type system, where every type is itself a set. For a set S which is a type, it would not be possible to find a distinguished element $\perp$ not in S such that $S \cup \{\perp\}$ is an expression that type-checks. However, in this chapter we will pretend that it is always possible to define $S_\perp$ in such a way, as the alternative would be to define it as a free type (disjoint sum), which would result in an abundance of injection functions and their inverses in all our expressions.

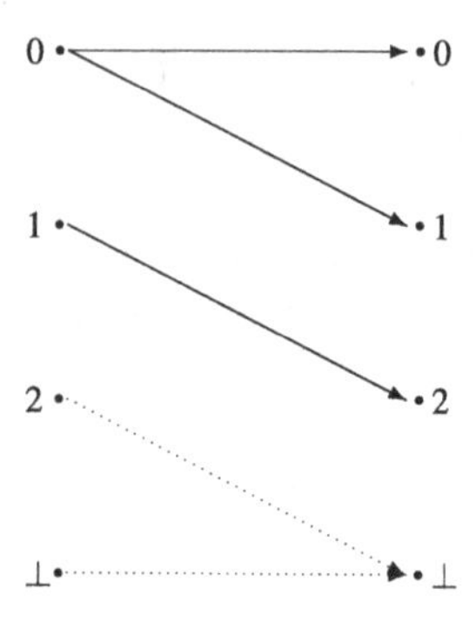

Fig. 3.7 Totalisation of $\{(0,0),(0,1),(1,2)\}$ over the state $\{0,1,2\}$ in the behavioural approach

Fig. 2.1 (p. 57): values outside the domain are linked to all values including $\perp$. In the behavioural approach, it is illustrated for the same example in Fig. 3.7: values outside the domain are linked to $\perp$ only.

Definition 3.8 (Totalisation) For a partial relation Op on State, its totalisation is a total relation $\widehat{\mathsf{Op}}$ on $\mathsf{State}_\perp$, defined in the "contract" approach to ADTs by

$$\widehat{\mathsf{Op}} == \mathsf{Op} \cup \{\mathsf{x}, \mathsf{y} : \mathsf{State}_\perp \mid \mathsf{x} \notin \operatorname{dom} \mathsf{Op} \bullet (\mathsf{x}, \mathsf{y})\}$$

or in the behavioural approach to ADTs by

$$\widehat{\mathsf{Op}} == \mathsf{Op} \cup \{\mathsf{x} : \mathsf{State}_\perp \mid \mathsf{x} \notin \operatorname{dom} \mathsf{Op} \bullet (\mathsf{x}, \perp)\}$$

Totalisations of initialisation and finalisation are defined analogously. □

We will not consider partial initialisations. This means that we "always" allow the ADT and its program to be run by the environment. As we do not (yet) make many assumptions about the global state G, this could often be redefined in such a way that initialisation *is* total anyway. Furthermore, allowing it to be partial would introduce the issue of what partiality *means* once again, now at the global level. Not requiring finalisation to be total will be illustrative at this stage, although the implicit finalisation for Z ADTs will turn out to be total, cf. Chap. 4.

To preserve undefinedness, simulation relations of particular forms only are considered, *viz.* those which relate $\perp$ to every other value, or $\perp$ to $\perp$, depending on the ADT interpretation.

Definition 3.9 (Extension) A relation R between AState and CState is extended to a relation $\widetilde{\mathsf{R}}$ between $\mathsf{AState}_\perp$ and $\mathsf{CState}_\perp$, defined in the contract approach by

$$\widetilde{\mathsf{R}} == \mathsf{R} \cup (\{\perp_{\mathsf{AState}}\} \times \mathsf{CState}_\perp)$$

and in the behavioural approach by

$$\widetilde{\mathsf{R}} == \mathsf{R} \cup \{(\perp_{\mathsf{AState}}, \perp_{\mathsf{CState}})\}$$ □

The resulting refinement rules for totalised relations can then be simplified to remove any reference to $\perp$. This simplification ("relaxing") is given in detail for the contract approach for downward simulation in [25]. As an illustration, we include the corresponding derivations for the behavioural approach here.

Downward Simulation for Partial Relations in the Behavioural Approach For simplicity, we will consider ADTs with a single operation $(\mathsf{AState}, \mathsf{AInit}, \{\mathsf{AOp}\}, \mathsf{AFin})$ and $(\mathsf{CState}, \mathsf{CInit}, \{\mathsf{COp}\}, \mathsf{CFin})$. Assume initialisations are total. Our goal is to apply the simulation rules to the totalised versions of these data types, and then to remove all occurrences of $\widetilde{\ }$, $\widehat{\ }$ and $\perp$ in the rules.

For initialisation we have:

$$
\begin{array}{l}
\widehat{\mathsf{CInit}} \subseteq \widehat{\mathsf{AInit}} \fatsemi \widetilde{\mathsf{R}} \\
\qquad\qquad \equiv \{ \text{ definition of } \widetilde{\ }, \widehat{\ }; \text{ totality of } \mathsf{CInit}, \mathsf{AInit} \ \} \\
\mathsf{CInit} \cup \{(\perp_{\mathsf{G}}, \perp_{\mathsf{CState}})\} \\
\quad \subseteq \\
(\mathsf{AInit} \cup \{(\perp_{\mathsf{G}}, \perp_{\mathsf{AState}})\}) \fatsemi (\mathsf{R} \cup \{(\perp_{\mathsf{AState}}, \perp_{\mathsf{CState}})\}) \\
\qquad\qquad \equiv \{ \text{ distribution of } \fatsemi \text{ over } \cup; \perp \notin \mathrm{dom}\, \mathsf{R}; \perp \notin \mathrm{ran}\, \mathsf{AInit} \ \} \\
\mathsf{CInit} \cup \{(\perp_{\mathsf{G}}, \perp_{\mathsf{CState}})\} \subseteq \mathsf{AInit} \fatsemi \mathsf{R} \cup \{(\perp_{\mathsf{G}}, \perp_{\mathsf{CState}})\} \\
\qquad\qquad \equiv \{ \ \perp_{\mathsf{G}} \notin \mathrm{dom}\, \mathsf{CInit} \ \} \\
\mathsf{CInit} \subseteq \mathsf{AInit} \fatsemi \mathsf{R}
\end{array}
$$

We use the following abbreviation ($\overline{\mathsf{A}}$ denotes A's complement):

$$
\mathsf{X}\ntriangleleft == (\overline{\mathrm{dom}\, \mathsf{X}})_{\perp}
$$

For the operations we get the following:

$$
\begin{array}{l}
\widetilde{\mathsf{R}} \fatsemi \widehat{\mathsf{COp}} \subseteq \widehat{\mathsf{AOp}} \fatsemi \widetilde{\mathsf{R}} \\
\qquad\qquad \equiv \{ \text{ definition of } \widetilde{\ }, \widehat{\ } \ \} \\
(\mathsf{R} \cup \{(\perp_{\mathsf{AState}}, \perp_{\mathsf{CState}})\}) \fatsemi (\mathsf{COp} \cup \mathsf{COp}\ntriangleleft \times \{\perp_{\mathsf{CState}}\}) \\
\quad \subseteq \\
(\mathsf{AOp} \cup \mathsf{AOp}\ntriangleleft \times \{\perp_{\mathsf{AState}}\}) \fatsemi (\mathsf{R} \cup \{(\perp_{\mathsf{AState}}, \perp_{\mathsf{CState}})\}) \\
\qquad\qquad \equiv \{ \text{ distribution of } \fatsemi \text{ over } \cup; \text{ simplifications } \} \\
(\mathsf{R} \fatsemi \mathsf{COp}) \cup (\mathsf{R} \fatsemi (\mathsf{COp}\ntriangleleft \times \{\perp_{\mathsf{CState}}\})) \cup \{(\perp_{\mathsf{AState}}, \perp_{\mathsf{CState}})\} \\
\quad \subseteq \\
(\mathsf{AOp} \fatsemi \mathsf{R}) \cup (\mathsf{AOp}\ntriangleleft \times \{\perp_{\mathsf{CState}}\}) \cup \{(\perp_{\mathsf{AState}}, \perp_{\mathsf{CState}})\} \\
\qquad\qquad \equiv \{ \ \perp_{\mathsf{AState}} \in \mathsf{AOp}\ntriangleleft \ \} \\
(\mathsf{R} \fatsemi \mathsf{COp}) \cup (\mathsf{R} \fatsemi (\mathsf{COp}\ntriangleleft \times \{\perp_{\mathsf{CState}}\})) \\
\quad \subseteq \\
(\mathsf{AOp} \fatsemi \mathsf{R}) \cup (\mathsf{AOp}\ntriangleleft \times \{\perp_{\mathsf{CState}}\}) \\
\qquad\qquad \equiv \{ \text{ disjoint ranges } \}
\end{array}
$$

$$\mathsf{R} \mathbin{;} \mathsf{COp} \subseteq \mathsf{AOp} \mathbin{;} \mathsf{R}$$
$$\wedge$$
$$(\mathsf{R} \mathbin{;} (\mathsf{COp} \not\lhd \times \{\bot_{\mathsf{CState}}\})) \subseteq (\mathsf{AOp} \not\lhd \times \{\bot_{\mathsf{CState}}\})$$

The second conjunct can be further simplified. As the ranges of both relations are $\{\bot_{\mathsf{CState}}\}$, only their domains are relevant. Removing also the complements, we get

$$\mathrm{ran}(\mathrm{dom}\,\mathsf{AOp} \lhd \mathsf{R}) \subseteq \mathrm{dom}\,\mathsf{COp}$$

For finalisation, using analogous steps:

$$\widetilde{\mathsf{R}} \mathbin{;} \widehat{\mathsf{CFin}} \subseteq \widehat{\mathsf{AFin}}$$
$$\equiv$$
$$(\mathsf{R} \cup \{(\bot_{\mathsf{AState}}, \bot_{\mathsf{CState}})\}) \mathbin{;} (\mathsf{CFin} \cup \mathsf{CFin} \not\lhd \times \{\bot_{\mathsf{G}}\})$$
$$\subseteq$$
$$\mathsf{AFin} \cup \mathsf{AFin} \not\lhd \times \{\bot_{\mathsf{G}}\}$$
$$\equiv$$
$$\mathsf{R} \mathbin{;} \mathsf{CFin} \subseteq \mathsf{AFin}$$
$$\wedge$$
$$\mathrm{ran}(\mathrm{dom}\,\mathsf{AFin} \lhd \mathsf{R}) \subseteq \mathrm{dom}\,\mathsf{CFin}$$

This calculation, and a similar one for the contract interpretation [25], lead to the following definition.

Definition 3.10 (Downward simulation for partial relations) Let $\mathsf{A} = (\mathsf{AState}, \mathsf{AInit}, \{\mathsf{AOp_i}\}_{\mathsf{i} \in \mathsf{I}}, \mathsf{AFin})$ and $\mathsf{C} = (\mathsf{CState}, \mathsf{CInit}, \{\mathsf{COp_i}\}_{\mathsf{i} \in \mathsf{I}}, \mathsf{CFin})$ be data types where the operations may be partial.

A *downward* simulation is a relation R from AState to CState satisfying, in the contract interpretation

$$\mathsf{CInit} \subseteq \mathsf{AInit} \mathbin{;} \mathsf{R}$$
$$\mathsf{R} \mathbin{;} \mathsf{CFin} \subseteq \mathsf{AFin}$$
$$\mathrm{ran}(\mathrm{dom}\,\mathsf{AFin} \lhd \mathsf{R}) \subseteq \mathrm{dom}\,\mathsf{CFin}$$
$$\forall\, \mathsf{i} : \mathsf{I} \bullet \mathrm{ran}(\mathrm{dom}\,\mathsf{AOp_i} \lhd \mathsf{R}) \subseteq \mathrm{dom}\,\mathsf{COp_i}$$
$$\forall\, \mathsf{i} : \mathsf{I} \bullet (\mathrm{dom}\,\mathsf{AOp_i} \lhd \mathsf{R}) \mathbin{;} \mathsf{COp_i} \subseteq \mathsf{AOp_i} \mathbin{;} \mathsf{R}$$

The five conditions are commonly referred to as "initialisation", "finalisation", "finalisation applicability", "applicability" and "correctness". In the behavioural interpretation, correctness is strengthened to:

$$\forall\, \mathsf{i} : \mathsf{I} \bullet \mathsf{R} \mathbin{;} \mathsf{COp_i} \subseteq \mathsf{AOp_i} \mathbin{;} \mathsf{R} \qquad \square$$

Some more discussion of the behavioural approach is in order here. Note that in the case where R is the identity relation on a state space (i.e., for "operation refinement"), correctness reduces to $\mathsf{COp} \subseteq \mathsf{AOp}$ and applicability to $\mathrm{dom}\,\mathsf{AOp} \subseteq \mathrm{dom}\,\mathsf{COp}$. As a consequence, we obtain $\mathrm{dom}\,\mathsf{AOp} = \mathrm{dom}\,\mathsf{COp}$ as expected: the precondition may not be widened in refinement. However, due to the generality of our setup (arbitrary relations as simulations), widening of preconditions *is* possible in data refinement—in other words, the inclusion in the applicability condition can *not* be strengthened to an equality. This is more a formal problem than anything else: the only concrete states which *may* be added to a precondition are states which are not in the range of the simulation. The initialisation and correctness conditions together ensure that such states are unreachable anyway.

Also in the contract approach, the conditions of Definition 3.10 reduce to those for operation refinement (Definition 2.5) when R is taken to be the identity relation. As the rules in Chap. 2 are expressed in terms of Z schemas rather than in this chapter's relational setup, we cannot verify the rules concerning establishment and imposition of invariants yet. However, the following lemma can be proved.

Lemma 3.1 (Establishing relational invariants) *Let* $\mathsf{A} = (\mathsf{AState}, \mathsf{AInit}, \{\mathsf{AOp_i}\}_{\mathsf{i} \in \mathsf{I}}, \mathsf{AFin})$ *and* Inv *satisfy*

$$\mathsf{AInit} \subseteq \mathsf{Inv}$$
$$\forall\, \mathsf{i} : \mathsf{I} \bullet \mathrm{ran}(\mathsf{Inv} \lhd \mathsf{AOp_i}) \subseteq \mathsf{Inv}$$

then $(\mathsf{Inv}, \mathsf{AInit}, \{\mathsf{Inv} \lhd \mathsf{AOp_i}\}_{\mathsf{i} \in \mathsf{I}}, \mathsf{Inv} \lhd \mathsf{AFin})$ *is a downward simulation of* A *and vice versa, in either interpretation.*

Proof Using $\mathsf{R} = \mathrm{id}_{\mathsf{Inv}}$, the applicability and finalisation conditions are immediate in either direction. Initialisation uses the first condition on Inv, and correctness (in either interpretation) the second, in both directions. □

Upward Simulation for Partial Relations in the Behavioural Approach The derivation of rules for initialisation in this case is analogous to above: $\widehat{\mathsf{CInit}} \operatorname{⨾} \widetilde{\mathsf{T}} \subseteq \widehat{\mathsf{AInit}}$ simplifies to $\mathsf{CInit} \operatorname{⨾} \mathsf{T} \subseteq \mathsf{AInit}$. For the operations we get:

$$
\begin{array}{l}
\widehat{\mathsf{COp}} \operatorname{⨾} \widetilde{\mathsf{T}} \subseteq \widetilde{\mathsf{T}} \operatorname{⨾} \widehat{\mathsf{AOp}} \\
\qquad \equiv \{\ \text{definitions}\ \} \\
(\mathsf{COp} \cup \overline{\mathrm{dom}\,\mathsf{COp}} \times \{\bot_{\mathsf{CState}}\}) \operatorname{⨾} (\mathsf{T} \cup \{(\bot_{\mathsf{CState}}, \bot_{\mathsf{AState}})\}) \\
\quad \subseteq \\
(\mathsf{T} \cup \{(\bot_{\mathsf{CState}}, \bot_{\mathsf{AState}})\}) \operatorname{⨾} (\mathsf{AOp} \cup \overline{\mathrm{dom}\,\mathsf{AOp}} \times \{\bot_{\mathsf{AState}}\}) \\
\qquad \equiv \{\ \text{distribution, simplifications}\ \} \\
\mathsf{COp} \operatorname{⨾} \mathsf{T} \cup (\overline{\mathrm{dom}\,\mathsf{COp}} \times \{\bot_{\mathsf{AState}}\}) \\
\quad \subseteq \\
\mathsf{T} \operatorname{⨾} \mathsf{AOp} \cup (\mathrm{dom}(\mathsf{T} \rhd \mathrm{dom}\,\mathsf{AOp}) \times \{\bot_{\mathsf{AState}}\}) \cup \{(\bot_{\mathsf{CState}}, \bot_{\mathsf{AState}})\} \\
\qquad \equiv \{\ \text{disjoint ranges, simplifications}\ \}
\end{array}
$$

$$\mathsf{COp} \fatsemi \mathsf{T} \subseteq \mathsf{T} \fatsemi \mathsf{AOp}$$
$$\wedge$$
$$\overline{\overline{\mathrm{dom}\,\mathsf{COp}}} \subseteq \mathrm{dom}(\mathsf{T} \rhd \mathrm{dom}\,\mathsf{AOp})$$

The second conjunct is somewhat unwieldy due to the double negative. Using explicit quantification and contraposition we could rephrase it as:

$$\forall \mathsf{c} : \mathsf{CState} \bullet \mathsf{T}(\!|\, \{\mathsf{c}\} \,|\!) \subseteq \mathrm{dom}\,\mathsf{AOp} \Rightarrow \mathsf{c} \in \mathrm{dom}\,\mathsf{COp}$$

In words, if an abstract operation is enabled in all of the image of a concrete state c under the simulation T, then its concrete counterpart must be enabled in c.

For finalisation, as with downward simulation, we obtain an applicability condition comparable to that for operations

$$\forall \mathsf{c} : \mathsf{CState} \bullet \mathsf{T}(\!|\, \{\mathsf{c}\} \,|\!) \subseteq \mathrm{dom}\,\mathsf{AFin} \Rightarrow \mathsf{c} \in \mathrm{dom}\,\mathsf{CFin}$$

plus the "undecorated" version of the original condition

$$\mathsf{CFin} \subseteq \mathsf{T} \fatsemi \mathsf{AFin}$$

Definition 3.11 (Upward simulation for partial relations) Assume data types $\mathsf{A} = (\mathsf{AState}, \mathsf{AInit}, \{\mathsf{AOp_i}\}_{\mathsf{i} \in \mathsf{I}}, \mathsf{AFin})$ and $\mathsf{C} = (\mathsf{CState}, \mathsf{CInit}, \{\mathsf{COp_i}\}_{\mathsf{i} \in \mathsf{I}}, \mathsf{CFin})$ where the operations may be partial.

An *upward* simulation is a relation T from CState to AState satisfying, in the contract interpretation

$$\mathsf{CInit} \fatsemi \mathsf{T} \subseteq \mathsf{AInit}$$
$$\mathsf{CFin} \subseteq \mathsf{T} \fatsemi \mathsf{AFin}$$
$$\forall \mathsf{c} : \mathsf{CState} \bullet \mathsf{T}(\!|\, \{\mathsf{c}\} \,|\!) \subseteq \mathrm{dom}\,\mathsf{AFin} \Rightarrow \mathsf{c} \in \mathrm{dom}\,\mathsf{CFin}$$
$$\forall \mathsf{i} : \mathsf{I} \bullet \overline{\mathrm{dom}\,\mathsf{COp_i}} \subseteq \mathrm{dom}(\mathsf{T} \rhd \mathrm{dom}\,\mathsf{AOp_i})$$
$$\forall \mathsf{i} : \mathsf{I} \bullet \mathrm{dom}(\mathsf{T} \rhd \mathrm{dom}\,\mathsf{AOp_i}) \lhd \mathsf{COp_i} \fatsemi \mathsf{T} \subseteq \mathsf{T} \fatsemi \mathsf{AOp_i}$$

The five conditions are commonly referred to as "initialisation", "finalisation", "finalisation applicability", "applicability" and "correctness". In the behavioural interpretation, correctness is strengthened to:

$$\forall \mathsf{i} : \mathsf{I} \bullet \mathsf{COp_i} \fatsemi \mathsf{T} \subseteq \mathsf{T} \fatsemi \mathsf{AOp_i} \qquad \square$$

An important consequence of this definition (which has no obvious counterpart for downward simulations) is the following.

Theorem 3.8 (Upward simulations are total) *When the concrete finalisation is total, the upward simulation* T *from* CState *to* AState *is* total *on* CState.

Proof From the finalisation condition:

$$\begin{aligned}
&\mathsf{CFin} \subseteq \mathsf{T} \mathbin{;} \mathsf{AFin} \\
&\qquad \Rightarrow \{\text{taking domains on both sides}\} \\
&\mathrm{dom}\,\mathsf{CFin} \subseteq \mathrm{dom}(\mathsf{T} \mathbin{;} \mathsf{AFin}) \\
&\qquad \equiv \{\mathsf{CFin} \text{ is total}\} \\
&\mathsf{CState} \subseteq \mathrm{dom}(\mathsf{T} \mathbin{;} \mathsf{AFin}) \\
&\qquad \equiv \{\mathsf{T} \text{ is a relation on } \mathsf{CState}\} \\
&\mathrm{dom}(\mathsf{T} \mathbin{;} \mathsf{AFin}) = \mathsf{CState} \\
&\qquad \Rightarrow \{\mathrm{dom}(\mathsf{P} \mathbin{;} \mathsf{Q}) \subseteq \mathrm{dom}\,\mathsf{P}\} \\
&\mathrm{dom}\,\mathsf{T} = \mathsf{CState}
\end{aligned}$$

□

Using Lemma 3.1 one might redefine upward simulation such that it allows the introduction of unreachable states, in which case the simulation relation only needs to be total on the reachable ones.

Example 3.5 We return to the traditional example as represented in Fig. 3.6. The previous presentation of it, necessarily for total relations, incidentally used yet another totalisation: outside its domain, an operation has a null effect.

For partial relations, the example goes as follows. Let $\mathsf{G} = 0..4$, which acts as the global state and as one of the local states. The other local state is $\mathsf{YS} = \{0, 1, 3, 4\}$. Define $\mathsf{X} = (\mathsf{G}, \mathsf{Init}, \{\mathsf{XOp}_1, \mathsf{XOp}_2\}, \mathsf{XFin})$, $\mathsf{Y} = (\mathsf{YS}, \mathsf{Init}, \{\mathsf{YOp}_1, \mathsf{YOp}_2, \}, \mathsf{YFin})$, where

$$\begin{aligned}
&\mathsf{Init} = \{x : \mathsf{G} \bullet (x, 0)\} \\
&\mathsf{XOp}_1 = \{(0, 1), (0, 2)\} \\
&\mathsf{XOp}_2 = \{(1, 3), (2, 4)\} \\
&\mathsf{XFin} = \mathsf{id}_\mathsf{G} \cup \{(1, 2), (2, 1)\} \\
&\mathsf{YOp}_1 = \{(0, 1)\} \\
&\mathsf{YOp}_2 = \{(1, 3), (1, 4)\} \\
&\mathsf{YFin} = \mathsf{id}_\mathsf{YS} \cup \{(1, 2))\}
\end{aligned}$$

Y is a data refinement of X. However it can be shown, analogous to Example 3.4, that no downward simulation exists to verify this: we need an upward simulation.

The required upward simulation from Y to X is $\mathsf{T} = \{(0, 0), (1, 1), (1, 2), (3, 3), (4, 4)\}$. This can be verified as follows:

- $\mathsf{Init} \mathbin{;} \mathsf{T} = \mathsf{Init} \subseteq \mathsf{Init}$;
- $\mathsf{YFin} = \mathsf{T} \subseteq \mathsf{T} \mathbin{;} \mathsf{XFin}$;
- both finalisations are total;
- $\mathrm{dom}\,\mathsf{YOp}_1 = \{1, 3, 4\}$; $\mathrm{dom}(\mathsf{T} \rhd \mathrm{dom}\,\mathsf{XOp}_1) = \{1, 3, 4\}$;
- $\{1, 3, 4\} \lhd \mathsf{YOp}_1 = \mathsf{YOp}_1$; $\mathsf{YOp}_1 \mathbin{;} \mathsf{T} = \{(0, 1), (0, 2)\} = \mathsf{XOp}_1 \subseteq \mathsf{T} \mathbin{;} \mathsf{XOp}_1$;
- $\mathrm{dom}\,\mathsf{YOp}_2 = \{0, 3, 4\}$; $\mathrm{dom}(\mathsf{T} \rhd \mathrm{dom}\,\mathsf{XOp}_2) = \{0, 3, 4\}$;
- $\{0, 3, 4\} \lhd \mathsf{YOp}_2 = \mathsf{YOp}_2$; $\mathsf{YOp}_2 \mathbin{;} \mathsf{T} = \{(1, 3), (1, 4)\}$;
  $\mathsf{T} \mathbin{;} \mathsf{XOp}_2 = \{(1, 2), (1, 1)\} \mathbin{;} \{(1, 3), (2, 4)\} = \{(1, 3), (1, 4)\}$.

In addition, X is a data refinement of Y. This can only be proved using a *downward* simulation—which turns out to be the same T. This can be verified as follows:

- $\mathsf{Init} = \mathsf{Init} \mathbin{⨾} \{(0,0)\} \subseteq \mathsf{Init} \mathbin{⨾} \mathsf{T}$;
- $\mathsf{T} \mathbin{⨾} \mathsf{XFin} = \mathsf{T} \subseteq \mathsf{YFin}$;
- both finalisations are total;
- $\text{ran}(\text{dom}\,\mathsf{YOp_1} \lhd \mathsf{T}) = \text{ran}(\{0\} \lhd \mathsf{T}) = \{0\}$; $\text{dom}\,\mathsf{XOp_1} = \{0\}$;
- $\text{ran}(\text{dom}\,\mathsf{YOp_2} \lhd \mathsf{T}) = \text{ran}(\{1\} \lhd \mathsf{T}) = \{1,2\}$; $\text{dom}\,\mathsf{XOp_2} = \{1,2\}$;
- $(\{0\} \lhd \mathsf{T}) \mathbin{⨾} \mathsf{XOp_1} = \{(0,1),(0,2)\} = \mathsf{YOp_1} \mathbin{⨾} \mathsf{T}$;
- $(\{1\} \lhd \mathsf{T}) \mathbin{⨾} \mathsf{XOp_2} = \{(1,3),(1,4)\} = \mathsf{YOp_2} \mathbin{⨾} \mathsf{T}$.

It may appear unsatisfactory to the reader that examples only verify that a particular simulation has the required properties, rather than constructing such a simulation. The construction of a simulation is definitely possible, but like the search for (bi)simulations in process algebra, it is not normally a linear process, e.g., here we can derive immediately from the condition on initialisation that $(0,0) \in \mathsf{T}$ and that $(0,\mathsf{x}) \notin \mathsf{T}$ for $\mathsf{x} \neq 0$. This however tells us nothing for values of $\mathsf{YS} \neq 0$. From the finalisation condition it follows that $\mathsf{id_{YS}} \subseteq \mathsf{T}$. Further determination of T is more problematic, as the conditions for operations contain occurrences of T on either side of $\subseteq$. Of course Fig. 3.6 strongly hints at the crucial observation, that a link from 1 below to 2 above may be necessary. □

3.3.2 Soundness and Completeness for Embeddings

Theorem 3.7 states the joint completeness of upward and downward simulations in the basic theory. Soundness for the full domain (all relations, Theorem 3.6) implies soundness of the same rules for a subdomain (relations constructed as embeddings of partial relations), and thus the derived refinement rules are sound for the general relational refinement notion (Definition 3.5) as applied to the subdomain. However, completeness on the full domain does *not* carry over to completeness on the subdomain. We briefly discuss the known results in this area.

For the contract approach, upward and downward simulations are jointly complete. We are not aware of a published proof of exactly this—however, De Roever and Engelhardt [12] published the corresponding result and proof in a slightly larger domain. This does not imply the result we are looking for, but the same proof carries over in the required setting.

For the behavioural approach, we have shown that completeness does *not* hold [6], contradicting previously published results. However, very similar simulation rules for the closely related notion of *failures refinement* have been shown to be complete by Josephs [21], and this may be an additional reason for using this CSP inspired behavioural refinement relation for Z as well, instead of the behavioural notion explored in this chapter. Indeed, in Chap. 16 we will be using Josephs' notion of refinement for Object-Z.

3.3.3 Partial Relations, Directly

From the fact that the simulation theory also applies to not necessarily total relations, it follows that the simulation definitions (Definitions 3.6 and 3.7) could also be applied directly to Z, providing it with a sound and complete simulation set up. However, this would encode a different notion of refinement, also of a more "behavioural" nature. Rather than returning an error value when an operation is applied outside its guard (when it "blocks"), it returns no value in such a case. As a consequence, the semantics records *certain* blocking (a particular program has the empty relation as its semantics), but not *possible* blocking (where the behavioural approach above would have recorded $\perp$ as well as other results). This semantics is the familiar "trace refinement" relation, which is about preservation of safety: if something bad does not happen in the abstract ADT, it also will not happen in the concrete ADT. In the extreme case, the empty program (and the operation that is never applicable) is *always* a refinement in this semantics. That makes it rather weak for our purposes, and we will not consider it any further in the next chapter when deriving Z refinement conditions from the relational characterisations in this chapter. However, in [7, 13] we have shown how by including richer observations in the finalisation we can still derive interesting refinement relations for Z from this partial relation semantics. Some of these results are highlighted in Chap. 19.

3.4 Bibliographical Notes

The theory of data refinement presented here was first set out in a seminal paper by He, Hoare and Sanders in ESOP'86 [18] and a companion paper in IPL [20], following on from an earlier paper by Hoare [19]. A more extensive account of the theory is in [17], which also dropped the restriction to total operations.

Using only finite sequences of operations as programs, as was done by Woodcock and Davies [25], avoids some of the complexity of the general theory. Characterisation of recursion through domains fixes $\perp$ as the semantics of non-terminating recursion. This, in combination with unbounded non-determinism may cause L^{-1}-simulation to become unsound in a setting of total correctness. A lucid explanation of this issue is in Dijkstra's book [15]: using Dijkstra's language it is provably impossible to define the program "set *n* to *any* natural number" such that it is guaranteed to terminate.

The most comprehensive account of the theory of simulations is given in the book by De Roever and Engelhardt [12]. This also contains theorems stating the circumstances in which upward and downward simulation are equivalent or sufficient. In addition, it contains an extensive history and bibliography of the area.

The "contract" approach to partial operations in Z is the traditional one; Josephs [22] used the behavioural approach in specifying reactive systems in Z. Strulo [24] describes the "firing condition" approach, which differs from the behavioural approach described here by viewing an unconstrained operation as divergent. A systematic study of the two kinds of totalisations was carried out by Henson and

Deutsch [14]. The relevance of $\perp$ and possibility of multiple error elements are explored in some detail in a note by Boiten and De Roever [4]. Relational refinement rules for such formalisms are derived in detail by Boiten, Derrick and Schellhorn [7], see also further discussion of these results in Chaps. 11 and 19. The language B [1] has both explicit guards and preconditions, but is embedded in the richer semantic domain of predicate transformers. (Its successor Event-B [2] on the other hand uses the partial relations framework described above, with small extensions inspired by action systems—see Chap. 20 for further discussion.) Versions of Z with explicit guards added have been explored by Fischer [16], and by us with Miarka [23].

The totalisation of partial relations and subsequent derivation of simulation rules for partial relations is strongly inspired by the explanation by Woodcock and Davies [25]. Their book only contains the totalisation necessary for the "contract" approach to ADTs. In a paper with Bolton [8] they also defined totalisation and derived simulations in the behavioural approach, but note the incompleteness for these rules shown in [6]. Our work with Schellhorn [7] on using multiple error values generalises both approaches.

A dual approach to totalisation is taken by De Roever and Engelhardt who consider the consequences of restricting refinement to total relations of a particular shape, *viz.* those which (we would say) represent totalised relations. However, their semantic domain is richer than that because it also includes partial variants of such relations.

References

1. Abrial, J.-R. (1996). *The B-Book: Assigning Programs to Meanings*. Cambridge: Cambridge University Press.
2. Abrial, J.-R. (2010). *Modelling in Event-B*. Cambridge: Cambridge University Press.
3. Araki, K., Galloway, A., & Taguchi, K. (Eds.) (1999). *International Conference on Integrated Formal Methods 1999 (IFM'99)*. York: Springer.
4. Boiten, E. A., & de Roever, W.-P. (2003). Getting to the bottom of relational refinement: relations and correctness, partial and total. In R. Berghammer & B. Möller (Eds.), *7th International Seminar on Relational Methods in Computer Science (RelMiCS 7)*, University of Kiel, May 2003 (pp. 82–88).
5. Boiten, E. A., & Derrick, J. (2000). Liberating data refinement. In R. C. Backhouse & J. N. Oliveira (Eds.), *Mathematics of Program Construction. Lecture Notes in Computer Science: Vol. 1837* (pp. 144–166). Berlin: Springer.
6. Boiten, E. A., & Derrick, J. (2010). Incompleteness of relational simulations in the blocking paradigm. *Science of Computer Programming*, *75*(12), 1262–1269.
7. Boiten, E. A., Derrick, J., & Schellhorn, G. (2009). Relational concurrent refinement II: internal operations and outputs. *Formal Aspects of Computing*, *21*(1–2), 65–102.
8. Bolton, C., Davies, J., & Woodcock, J. C. P. (1999) On the refinement and simulation of data types and processes. In Araki et al. [3] (pp. 273–292).
9. Bowen, J. P., Dunne, S., Galloway, A., & King, S. (Eds.) (2000). *ZB2000: Formal Specification and Development in Z and B. Lecture Notes in Computer Science: Vol. 1878*. Berlin: Springer.
10. Bowen, J. P., Fett, A., & Hinchey, M. G. (Eds.) (1998). *ZUM'98: the Z Formal Specification Notation. Lecture Notes in Computer Science: Vol. 1493*. Berlin: Springer.

11. Bowen, J. P. & Hinchey, M. G. (Eds.) (1995). *ZUM'95: the Z Formal Specification Notation. Lecture Notes in Computer Science: Vol. 967*. Limerick: Springer.
12. de Roever, W.-P., & Engelhardt, K. (1998). *Data Refinement: Model-Oriented Proof Methods and Their Comparison*. Cambridge: Cambridge University Press.
13. Derrick, J., & Boiten, E. A. (2003). Relational concurrent refinement. *Formal Aspects of Computing*, *15*(2–3), 182–214.
14. Deutsch, M., & Henson, M. C. (2006). An analysis of refinement in an abortive paradigm. *Formal Aspects of Computing*, *18*(3), 329–363.
15. Dijkstra, E. W. (1976). *A Discipline of Programming*. Englewood Cliffs: Prentice-Hall.
16. Fischer, C. (1998) How to combine Z with a process algebra. In Bowen et al. [10] (pp. 5–23).
17. He, J., & Hoare, C. A. R. (1990). Prespecification and data refinement. In *Data Refinement in a Categorical Setting*. *Technical Monograph PRG-90*. Oxford: Oxford University Computing Laboratory.
18. He, J., Hoare, C. A. R., & Sanders, J. W. (1986). Data refinement refined. In B. Robinet & R. Wilhelm (Eds.), *Proc. ESOP'86*. *Lecture Notes in Computer Science: Vol. 213* (pp. 187–196). Berlin: Springer.
19. Hoare, C. A. R. (1972). Proof of correctness of data representations. *Acta Informatica*, *1*, 271–281.
20. Hoare, C. A. R., He, J., & Sanders, J. W. (1987). Prespecification in data refinement. *Information Processing Letters*, *25*(2), 71–76.
21. Josephs, M. B. (1988). A state-based approach to communicating processes. *Distributed Computing*, *3*, 9–18.
22. Josephs, M. B. (1991). *Specifying reactive systems in Z* (Technical Report PRG-19). Programming Research Group, Oxford University Computing Laboratory.
23. Miarka, R., Boiten, E. A., & Derrick, J. (2000) Guards, preconditions and refinement in Z. In Bowen et al. [9] (pp. 286–303).
24. Strulo, B. (1995) How firing conditions help inheritance. In Bowen and Hinchey [11] (pp. 264–275).
25. Woodcock, J. C. P., & Davies, J. (1996). *Using Z: Specification, Refinement, and Proof*. New York: Prentice Hall.

Chapter 4
Refinement in Z

The previous chapter presented the underlying theory of data refinement for ADTs—its definition for total relations, its verification using upward and downward simulations and, finally, how simulation rules for partial operations are derived from these. All of the questions listed at the beginning of Chap. 3 have now been answered, with the exception of the one concerning input and output.

In this chapter, we will formulate the theory of data refinement for Z. This will be done systematically: a relational interpretation will be given for the standard Z ADT as defined in Chap. 1. We apply the simulation based refinement rules from Chap. 3 to these interpretations, and then reformulate them directly for Z schemas (i.e., without using the relational meta-language). In order to separate the issues involved with this derivation, we first derive rules for Z ADTs without inputs and outputs, and then show the more complicated derivation in the presence of inputs and outputs (IO). Finally, this chapter presents a collection of examples of data refinement in Z.

4.1 The Relational Basis of a Z Specification

The refinement theory described in Chap. 3 was based on data types of the following form:

$$(\mathsf{State}, \mathsf{Init}, \{\mathsf{Op_i}\}_{\mathsf{i} \in \mathsf{I}}, \mathsf{Fin})$$

(for partial relations $\mathsf{Op_i}$) whereas the definition of a standard Z ADT was given in Chap. 1 as a tuple

$$(State, Init, \{Op_i\}_{i \in I})$$

What is the "obvious" or "correct" relation between these tuples?

Let us first answer this question in a (too) simple context, where the Z operations have no input or output. The combination of having no outputs and the Z data type's

J. Derrick, E.A. Boiten, *Refinement in Z and Object-Z*,
DOI 10.1007/978-1-4471-5355-9_4, © Springer-Verlag London 2014

being *abstract* means that all we could really observe in this context is whether a sequence of operations would be "possible", in the sense of having a possible outcome different from $\perp$ in its embedded form.

Starting with the easiest correspondence, the index sets I in both tuples should clearly be identical. This leaves us with the task of finding correspondences between $\mathsf{Op_i}$ and Op_i. Before doing so, we first relate the types that they range over.

It seems peculiar that there is no finalisation in the Z ADT. This means that no information is conveyed at the point of finalisation. Conceptually this corresponds to an operation from the local state (which will correspond to *State*) to a global state which contains no information, i.e., a one point domain, which we define as follows:

$$\mathsf{G} == \{*\}$$

The state in the relational data type needs to be a set. Clearly any Z schema can be interpreted as a set, *viz.* the set of all its possible bindings. It does not matter for the use of a relational interpretation that this is a set of bindings, i.e., labelled tuples, with a fixed set of labels.

$$\mathsf{State} == State$$

The initialisation in the Z ADT is represented by a schema with the same signature as the state schema. As a consequence, it could also be interpreted as a set of bindings, and due to its syntactic form it is guaranteed to be a subset of the state space. The relational formalisation requires a total relation from the global state to the local state, however. Since we have established (when we considered finalisation) that the global state G is a singleton domain $\{*\}$, the relation taking $*$ to any value in the set of initial states is the required relation. So we have

$$\mathsf{Init} == \{Init \bullet * \mapsto \theta State'\}$$

(Recall that $\theta State'$ produces the binding which is represented by the components in $State'$, but removing the primes on the labels. Thus, the target type of Init is indeed State.)

From the fact that the relational initialisation needs to be *total*, we must conclude that the initialisation schema in a Z ADT needs to be *satisfiable*, i.e., describe at least one possible state.

Definition 4.1 (Satisfiable initialisation) A Z ADT $(State, Init, \{Op_i\}_{i \in I})$ has a *satisfiable initialisation* iff

$$\exists State' \bullet Init \qquad \square$$

In the rest of this book we will generally assume that the ADTs dealt with have satisfiable initialisations. Occasionally, when explicitly constructing a new ADT, we will make this requirement explicit and indeed may need to check that it is indeed satisfied.

An ADT's not having a satisfiable initialisation does not make it (entirely) meaningless, nor excluded from consideration in a theory of refinement. However, such ADTs are not part of this theory of refinement, as they are (the only) ADTs which cannot be embedded in the relational data types from which we inherit our refinement theory.

Finally, we define an embedding for the operations. The Z operations are also represented by schemas, but interpreting these as sets does not help, as they are not sets of pairs. A binding of a Z operation (with no inputs or outputs) can be viewed as a pair, though, by separating the primed (after-state) labels from the unprimed (before-state) ones. In the context where operation *Op* on $\Delta State$ has been declared, this separation can be achieved using the "schema binding" operator θ as follows:

$$\mathsf{Op} == \{Op \bullet \theta State \mapsto \theta State'\}$$

I.e., for each binding of *Op*, we include a pair to the relation, its first element consisting of all the before-state components with their values, and its second element containing all the after-state components.

This completes the representation of a standard Z ADT without inputs and outputs by a relational data type, as summarised in the definition below.

Definition 4.2 (Relational interpretation of ADT) The Z ADT $(State, Init, \{Op_i\}_{i \in I})$ is interpreted by the relational data type $(\mathsf{State}, \mathsf{Init}, \{\mathsf{Op_i}\}_{\mathsf{i} \in \mathsf{I}}, \mathsf{Fin})$, where

$$\begin{aligned}
&\mathsf{State} == State \\
&\mathsf{Init} == \{Init \bullet * \mapsto \theta State'\} \\
&\mathsf{Op_i} == \{Op_i \bullet \theta State \mapsto \theta State'\} \\
&\mathsf{Fin} == \{State \bullet \theta State \mapsto *\}
\end{aligned}$$

□

Observe that in this interpretation, pre corresponds to dom, in the sense that

$$\mathrm{pre}\,\mathsf{Op} == \mathrm{dom}\,\mathsf{Op} = \{Op \bullet \theta State\}$$

(This will no longer be the case once we take inputs and outputs into account.)

Example 4.1 The standard Z ADT $(B, Init, \{On, Off\})$ where

```
┌─ B ──────────────
│ b : 𝔹
└──────────────────

┌─ Init ───────────
│ B'
├──────
│ b'
└──────────────────

┌─ On ─────────────
│ ΔB
├──────
│ ¬b ∧ b'
└──────────────────

┌─ Off ────────────
│ ΔB
├──────
│ b ∧ ¬b'
└──────────────────
```

is interpreted as the relational data type (B, Init, {On, Off}, Fin) where

$$\begin{aligned}
&\mathsf{B} == \{\lparen\!| b == true |\!\rparen, \lparen\!| b == false |\!\rparen\} \\
&\mathsf{Init} == \{* \mapsto \lparen\!| b == true |\!\rparen\} \\
&\mathsf{On} == \{\lparen\!| b == false |\!\rparen \mapsto \lparen\!| b == true |\!\rparen\} \\
&\mathsf{Off} == \{\lparen\!| b == true |\!\rparen \mapsto \lparen\!| b == false |\!\rparen\} \\
&\mathsf{Fin} == \{\lparen\!| b == true |\!\rparen \mapsto *, \lparen\!| b == false |\!\rparen \mapsto *\}
\end{aligned}$$ □

A retrieve relation R between $AState$ and $CState$ must be embedded in the relational setting similarly to how we embedded operations:

$$\mathsf{R} == \{R \bullet \theta AState \mapsto \theta CState\}$$

Example 4.2 The retrieve relation R between B in Example 4.1 and $N == [x : \{0, 1\}]$ given by

```
┌─R──────────────────────
│ B; N
├──────────
│ b = (x = 1)
└────────────────────────
```

is represented by the relation

$$\mathsf{R} == \{\lparen\!| b == true |\!\rparen \mapsto \lparen\!| x == 1 |\!\rparen, \lparen\!| b == false |\!\rparen \mapsto \lparen\!| x == 0 |\!\rparen\}$$ □

Refinement conditions will contain occurrences of both R and R' for retrieve relation R. Both of these will be represented by R, which could equivalently be defined by

$$\mathsf{R} == \{R' \bullet \theta AState' \mapsto \theta CState'\}$$

so in translating relational formulations back to Z schemas we need to consider carefully whether we are applying the retrieve relation to the before- or after-states.

4.2 Deriving Downward Simulation in Z

Given the embedding above, we can translate the relational refinement conditions of Definitions 3.6 and 3.7 into refinement conditions for Z ADTs. Assuming the Z data types $(AState, AInit, \{AOp_i\}_{i \in I})$ and $(CState, CInit, \{COp_i\}_{i \in I})$, and retrieve relation R, with their relational interpretations in sans serif, we derive the conditions for downward simulation below. The earlier derivations have more detailed steps, giving the full θ-laden expressions in the interpretations. Later derivations have larger, but mostly analogous, steps. Variable names are chosen suggestively: g for global state (i.e., at this point always a trivial quantification over the singleton set $\{*\}$), a for abstract, c for concrete; primes for after-states.

For initialisation:

$$
\begin{array}{l}
\mathsf{CInit} \subseteq \mathsf{AInit} \mathbin{;} \mathsf{R} \\
\qquad \equiv \{ \text{ definition of } \subseteq \} \\
\forall g, c' \bullet (g, c') \in \mathsf{CInit} \Rightarrow (g, c') \in \mathsf{AInit} \mathbin{;} \mathsf{R} \\
\qquad \equiv \{ \text{ definition of } \mathbin{;} \} \\
\forall g, c' \bullet (g, c') \in \mathsf{CInit} \Rightarrow \exists a' \bullet (g, a') \in \mathsf{AInit} \wedge (a', c') \in \mathsf{R} \\
\qquad \equiv \{ \text{ interpretation of } \mathit{Init} \text{ and } R';\ g = * \} \\
\forall c' \bullet c' \in \{\mathit{CInit} \bullet \theta \mathit{CState}'\} \Rightarrow \\
\qquad \exists a' \bullet a' \in \{\mathit{AInit} \bullet \theta \mathit{AState}'\} \\
\qquad\qquad \wedge\ (a', c') \in \{R' \bullet \theta \mathit{AState}' \mapsto \theta \mathit{CState}'\} \\
\qquad \equiv \{ \text{ introduce schema quantification to remove } \theta \} \\
\forall \mathit{CState}' \bullet \mathit{CInit} \Rightarrow \exists \mathit{AState}' \bullet \mathit{AInit} \wedge R'
\end{array}
$$

For finalisation:

$$
\begin{array}{l}
\mathsf{R} \mathbin{;} \mathsf{CFin} \subseteq \mathsf{AFin} \\
\qquad \equiv \{ \text{ definition of } \subseteq \} \\
\forall g, a \bullet (a, g) \in \mathsf{R} \mathbin{;} \mathsf{CFin} \Rightarrow (a, g) \in \mathsf{AFin} \\
\qquad \equiv \{ \text{ definition of } \mathsf{AFin} \} \\
\forall g, a \bullet (a, g) \in \mathsf{R} \mathbin{;} \mathsf{CFin} \Rightarrow (a \in \mathsf{AState} \wedge g = *) \\
\qquad \equiv \{ \operatorname{ran} \mathsf{CFin} = \{*\},\ \operatorname{dom} \mathsf{R} \subseteq \mathsf{AState} \} \\
\mathit{true}
\end{array}
$$

and

$$
\begin{array}{l}
\operatorname{ran}(\operatorname{dom} \mathsf{AFin} \lhd \mathsf{R}) \subseteq \operatorname{dom} \mathsf{CFin} \\
\qquad \equiv \{ \operatorname{dom} \mathsf{CFin} = \mathsf{CState} \} \\
\mathit{true}
\end{array}
$$

Thus, for downward simulation, having a trivial finalisation does not incur any obligations in refinement. We will see that this does *not* hold for upward simulation.

For applicability:

$$
\begin{array}{l}
\operatorname{ran}(\operatorname{dom} \mathsf{AOp_i} \lhd \mathsf{R}) \subseteq \operatorname{dom} \mathsf{COp_i} \\
\qquad \equiv \{ \text{ definition of } \subseteq \} \\
\forall c \bullet (c \in \operatorname{ran}(\operatorname{dom} \mathsf{AOp_i} \lhd \mathsf{R}) \Rightarrow c \in \operatorname{dom} \mathsf{COp_i}) \\
\qquad \equiv \{ \text{ definition of ran } \} \\
\forall c \bullet ((\exists a \bullet (a, c) \in \operatorname{dom} \mathsf{AOp_i} \lhd \mathsf{R}) \Rightarrow c \in \operatorname{dom} \mathsf{COp_i}) \\
\qquad \equiv \{ \text{ definition of } \lhd \} \\
\forall c \bullet ((\exists a \bullet (a, c) \in \mathsf{R} \wedge a \in \operatorname{dom} \mathsf{AOp_i}) \Rightarrow c \in \operatorname{dom} \mathsf{COp_i}) \\
\qquad \equiv \{ \text{ predicate calculus } \}
\end{array}
$$

$$\forall c, a \bullet ((a, c) \in \mathsf{R} \wedge a \in \mathrm{dom}\, \mathsf{AOp_i}) \Rightarrow c \in \mathrm{dom}\, \mathsf{COp_i})$$
$$\equiv \{ \text{ introduce schema quantification, pre for dom } \}$$
$$\forall \mathit{CState}; \mathit{AState} \bullet R \wedge \mathrm{pre}\, \mathit{AOp_i} \Rightarrow \mathrm{pre}\, \mathit{COp_i}$$

Finally, for correctness we have:

$$(\mathrm{dom}\, \mathsf{AOp_i} \lhd \mathsf{R}) \mathbin{⨾} \mathsf{COp_i} \subseteq \mathsf{AOp_i} \mathbin{⨾} \mathsf{R}$$
$$\equiv \{ \text{ definition of } \subseteq \}$$
$$\forall a, c' \bullet ((a, c') \in (\mathrm{dom}\, \mathsf{AOp_i} \lhd \mathsf{R}) \mathbin{⨾} \mathsf{COp_i} \Rightarrow (a, c') \in \mathsf{AOp_i} \mathbin{⨾} \mathsf{R})$$
$$\equiv \{ \text{ definition of } ⨾ \}$$
$$\forall a, c' \bullet ((\exists c \bullet (a, c) \in \mathrm{dom}\, \mathsf{AOp_i} \lhd \mathsf{R} \wedge (c, c') \in \mathsf{COp_i})$$
$$\Rightarrow (a, c') \in \mathsf{AOp_i} \mathbin{⨾} \mathsf{R})$$
$$\equiv \{ \text{ predicate calculus } \}$$
$$\forall a, c, c' \bullet (((a, c) \in \mathrm{dom}\, \mathsf{AOp_i} \lhd \mathsf{R} \wedge (c, c') \in \mathsf{COp_i})$$
$$\Rightarrow (a, c') \in \mathsf{AOp_i} \mathbin{⨾} \mathsf{R})$$
$$\equiv \{ \text{ definition of } \lhd \}$$
$$\forall a, c, c' \bullet (((a, c) \in \mathsf{R} \wedge a \in \mathrm{dom}\, \mathsf{AOp_i} \wedge (c, c') \in \mathsf{COp_i})$$
$$\Rightarrow (a, c') \in \mathsf{AOp_i} \mathbin{⨾} \mathsf{R})$$
$$\equiv \{ \text{ definition of } ⨾ \text{; introduce schema quantification, pre for dom } \}$$
$$\forall \mathit{AState}; \mathit{CState}; \mathit{CState}' \bullet R \wedge \mathrm{pre}\, \mathit{AOp_i} \wedge \mathit{COp_i}$$
$$\Rightarrow \exists \mathit{AState}' \bullet \mathit{AOp_i} \wedge R'$$

Together, these conditions make up the rules for downward simulation for Z schemas in systems without input or output. It is easy to see that the behavioural variant of correctness is obtained from the corresponding rule in Definition 3.6 completely analogously, resulting in the conjunction with pre $\mathit{AOp_i}$ being removed from the condition above.

Definition 4.3 (Downward simulation) Given Z data types $A = (\mathit{AState}, \mathit{AInit}, \{\mathit{AOp_i}\}_{i \in I})$ and $C = (\mathit{CState}, \mathit{CInit}, \{\mathit{COp_i}\}_{i \in I})$. The relation R on $\mathit{AState} \wedge \mathit{CState}$ is a *downward simulation* from A to C if

$$\forall \mathit{CState}' \bullet \mathit{CInit} \Rightarrow \exists \mathit{AState}' \bullet \mathit{AInit} \wedge R'$$

and for all $i \in I$:

$$\forall \mathit{AState}; \mathit{CState} \bullet \mathrm{pre}\, \mathit{AOp_i} \wedge R \Rightarrow \mathrm{pre}\, \mathit{COp_i}$$
$$\forall \mathit{AState}; \mathit{CState}; \mathit{CState}' \bullet \mathrm{pre}\, \mathit{AOp_i} \wedge R \wedge \mathit{COp_i}$$
$$\Rightarrow \exists \mathit{AState}' \bullet R' \wedge \mathit{AOp_i}$$

If such a simulation exists, we also say that C is a downward simulation of A, also denoted $A \sqsubseteq_{DS} C$, and similarly for corresponding operations of A and C.

The above rules assume the contractual interpretation of ADTs. In the behavioural interpretation, the rule for correctness becomes

$$\forall AState; CState; CState' \bullet R \land COp_i \Rightarrow \exists AState' \bullet R' \land AOp_i$$

When this behavioural correctness condition holds, the applicability condition above is equivalent to

$$\forall AState; CState \bullet R \Rightarrow (\text{pre}\, AOp_i \Leftrightarrow \text{pre}\, COp_i) \qquad \square$$

This formalisation of downward simulation has extended the notion of operation refinement (cf. Definition 2.6) by considering initialisation and also changes to the state space. The intuition behind the applicability and correctness requirements are the same as described after that definition, but now also taking into account the change of state space, and this effect is described by the retrieve relation R. Definition 4.3 therefore tackles all the issues highlighted in Sect. 1.6, i.e., change of state space and reduction of non-determinism. Examples of the use of these rules are found later in this chapter.

Using Definition 4.3 we can now verify Theorem 2.4 (imposing invariants).

Proof of Theorem 2.4 Consider the downward simulation conditions for $A = (S, Init, \{Op_i\}_{i \in I})$ and $C = (S, Init \land Inv', \{Op_i \land Inv \land Inv'\}_{i \in I})$ using $R = Inv$.

Then in order for C to have a satisfiable initialisation,

$$\begin{array}{l} \exists S' \bullet Init \land Inv \\ \quad \equiv \\ \exists Init \bullet Inv' \end{array}$$

which is the first condition for Theorem 2.4.

The initialisation condition evaluates to

$$\forall S' \bullet Init \land Inv' \Rightarrow \exists S' \bullet Init \land Inv'$$

which also requires the concrete initialisation to be satisfiable.

Correctness of the operations is automatically satisfied, cf. also Lemma 2.3. Applicability of the operations requires

$$\begin{array}{l} \text{pre}\, Op_i \land Inv \Rightarrow \text{pre}(Op_i \land Inv \land Inv') \\ \qquad \equiv \{\, Inv \text{ does not constrain after-state} \,\} \\ \text{pre}\, Op_i \land Inv \Rightarrow Inv \land \text{pre}(Op_i \land Inv') \\ \qquad \equiv \{\, \text{calculus} \,\} \\ \text{pre}\, Op_i \land Inv \Rightarrow \text{pre}(Op_i \land Inv') \end{array}$$

which is the second assumption of Theorem 2.4. $\square$

4.3 Deriving Upward Simulation in Z

We now derive similar conditions for upward simulation in an analogous fashion.

For initialisation we have:

$$
\begin{aligned}
&\mathsf{CInit} \mathbin{;} \mathsf{T} \subseteq \mathsf{AInit}\\
&\qquad \equiv \{ \text{ definition of } \subseteq \}\\
&\forall g, a' \bullet (g, a') \in \mathsf{CInit} \mathbin{;} \mathsf{T} \Rightarrow (g, a') \in \mathsf{AInit}\\
&\qquad \equiv \{ \text{ definition of } \mathbin{;} \}\\
&\forall g, a' \bullet (\exists c' \bullet (g, c') \in \mathsf{CInit} \land (c', a') \in \mathsf{T}) \Rightarrow (g, a') \in \mathsf{AInit}\\
&\qquad \equiv \{ \text{ definition of } \mathsf{Init}\text{, using } g = * \}\\
&\forall a' \bullet (\exists c' \bullet c' \in \{CInit \bullet \theta CState'\} \land (c', a') \in \mathsf{T})\\
&\qquad\quad \Rightarrow a' \in \{AInit \bullet \theta AState'\}\\
&\qquad \equiv \{ \text{ predicate calculus, introducing schema quantification } \}\\
&\forall AState'; \ CState' \bullet CInit \land T' \Rightarrow AInit
\end{aligned}
$$

Finalisation produces an important condition:

$$
\begin{aligned}
&\mathsf{CFin} \subseteq \mathsf{T} \mathbin{;} \mathsf{AFin}\\
&\qquad \equiv \{ \text{ definition of } \subseteq \text{ and } \mathbin{;} \}\\
&\forall g, c \bullet (c, g) \in \mathsf{CFin} \Rightarrow \exists a \bullet (c, a) \in \mathsf{T} \land (a, g) \in \mathsf{AFin}\\
&\qquad \equiv \{ \text{ finalisation is total, } g = * \}\\
&\forall c \bullet \exists a \bullet (c, a) \in \mathsf{T}\\
&\qquad \equiv \{ \text{ introducing schema quantification } \}\\
&\forall CState \bullet \exists AState \bullet T
\end{aligned}
$$

i.e., the simulation relation needs to be total on the concrete state, in line with Theorem 3.8. The applicability condition for finalisation reduces to *true*, because $\operatorname{dom} CFin = CState$.

A lemma that will be used both for applicability and correctness is derived first:

$$
\begin{aligned}
&c \notin \operatorname{dom}(\mathsf{T} \mathbin{\rhd\!\!\!\!-} \operatorname{dom} \mathsf{AOp_i})\\
&\qquad \equiv \{ \text{ definition dom } \}\\
&\neg \exists a \bullet (c, a) \in \mathsf{T} \mathbin{\rhd\!\!\!\!-} \operatorname{dom} \mathsf{AOp_i}\\
&\qquad \equiv \{ \text{ definition } \rhd\!\!\!\!- \}\\
&\neg \exists a \bullet (c, a) \in \mathsf{T} \land a \notin \operatorname{dom} \mathsf{AOp_i}\\
&\qquad \equiv \{ \text{ predicate calculus } \}\\
&\forall a \bullet (c, a) \in \mathsf{T} \Rightarrow a \in \operatorname{dom} \mathsf{AOp_i}
\end{aligned}
$$

Applicability gives:

$$
\begin{aligned}
&\overline{\operatorname{dom} \mathsf{COp_i}} \subseteq \operatorname{dom}(\mathsf{T} \mathbin{\rhd\!\!\!\!-} \operatorname{dom} \mathsf{AOp_i})\\
&\qquad \equiv \{ \text{ definition of } \subseteq \text{ and } \bar{\ } \}
\end{aligned}
$$

$$\forall c \bullet c \notin \mathrm{dom}\,\mathsf{COp_i} \Rightarrow c \in \mathrm{dom}(\mathsf{T} \rhd \mathrm{dom}\,\mathsf{AOp_i})$$
$$\equiv \{ \text{ contraposition } \}$$
$$\forall c \bullet c \notin \mathrm{dom}(\mathsf{T} \rhd \mathrm{dom}\,\mathsf{AOp_i})) \Rightarrow c \in \mathrm{dom}\,\mathsf{COp_i}$$
$$\equiv \{ \text{ lemma above } \}$$
$$\forall c \bullet (\forall a \bullet (c, a) \in \mathsf{T} \Rightarrow a \in \mathrm{dom}\,\mathsf{AOp_i}) \Rightarrow c \in \mathrm{dom}\,\mathsf{COp_i}$$
$$\equiv \{ \text{ introducing schema quantification } \}$$
$$\forall \mathit{CState} \bullet (\forall \mathit{AState} \bullet T \Rightarrow \mathrm{pre}\,\mathit{AOp_i}) \Rightarrow \mathrm{pre}\,\mathit{COp_i}$$

For correctness, we have:

$$\mathrm{dom}(\mathsf{T} \rhd \mathrm{dom}\,\mathsf{AOp_i}) \lhd \mathsf{COp_i} \mathbin{;} \mathsf{T} \subseteq \mathsf{T} \mathbin{;} \mathsf{AOp_i}$$
$$\equiv \{ \text{ definition } \subseteq \}$$
$$\forall c, a' \bullet (c, a') \in \mathrm{dom}(\mathsf{T} \rhd \mathrm{dom}\,\mathsf{AOp_i}) \lhd \mathsf{COp_i} \mathbin{;} \mathsf{T} \Rightarrow (c, a') \in \mathsf{T} \mathbin{;} \mathsf{AOp_i}$$
$$\equiv \{ \text{ definition } ; \}$$
$$\forall c, a' \bullet (\exists c' \bullet (c, c') \in \mathrm{dom}(\mathsf{T} \rhd \mathrm{dom}\,\mathsf{AOp_i}) \lhd \mathsf{COp_i} \land (c', a') \in \mathsf{T})$$
$$\Rightarrow (c, a') \in \mathsf{T} \mathbin{;} \mathsf{AOp_i}$$
$$\equiv \{ \text{ predicate calculus; definition of } \lhd \}$$
$$\forall c, a', c' \bullet ((c, c') \in \mathsf{COp_i} \land c \notin \mathrm{dom}(\mathsf{T} \rhd \mathrm{dom}\,\mathsf{AOp_i}) \land (c', a') \in \mathsf{T})$$
$$\Rightarrow (c, a') \in \mathsf{T} \mathbin{;} \mathsf{AOp_i}$$
$$\equiv \{ \text{ lemma above; propositional calculus } \}$$
$$\forall c, a', c' \bullet (\forall a \bullet (c, a) \in \mathsf{T} \Rightarrow a \in \mathrm{dom}\,\mathsf{AOp_i})$$
$$\Rightarrow ((c, c') \in \mathsf{COp_i} \land (c', a') \in \mathsf{T} \Rightarrow (c, a') \in \mathsf{T} \mathbin{;} \mathsf{AOp_i})$$
$$\equiv \{ \text{ introducing schema quantification } \}$$
$$\forall \mathit{AState}'; \mathit{CState}; \mathit{CState}' \bullet (\forall \mathit{AState} \bullet T \Rightarrow \mathrm{pre}\,\mathit{AOp_i})$$
$$\Rightarrow (\mathit{COp_i} \land T' \Rightarrow \exists \mathit{AState} \bullet T \land \mathit{AOp_i})$$

Definition 4.4 (Upward simulation) Given Z data types $A = (\mathit{AState}, \mathit{AInit}, \{\mathit{AOp_i}\}_{i \in I})$ and $C = (\mathit{CState}, \mathit{CInit}, \{\mathit{COp_i}\}_{i \in I})$. Then the relation T on $\mathit{AState} \land \mathit{CState}$ is an *upward simulation* from A to C if

$$\forall \mathit{CState} \bullet \exists \mathit{AState} \bullet T$$
$$\forall \mathit{AState}'; \mathit{CState}' \bullet \mathit{CInit} \land T' \Rightarrow \mathit{AInit}$$

and for all $i \in I$:

$$\forall \mathit{CState} \bullet (\forall \mathit{AState} \bullet T \Rightarrow \mathrm{pre}\,\mathit{AOp_i}) \Rightarrow \mathrm{pre}\,\mathit{COp_i}$$
$$\forall \mathit{AState}'; \mathit{CState}; \mathit{CState}' \bullet (\forall \mathit{AState} \bullet T \Rightarrow \mathrm{pre}\,\mathit{AOp_i}) \Rightarrow$$
$$(\mathit{COp_i} \land T' \Rightarrow \exists \mathit{AState} \bullet T \land \mathit{AOp_i})$$

Under the assumption of totality of T, the applicability condition is equivalently phrased as

$$\forall \mathit{CState} \bullet \exists \mathit{AState} \bullet T \land (\mathrm{pre}\,\mathit{AOp_i} \Rightarrow \mathrm{pre}\,\mathit{COp_i})$$

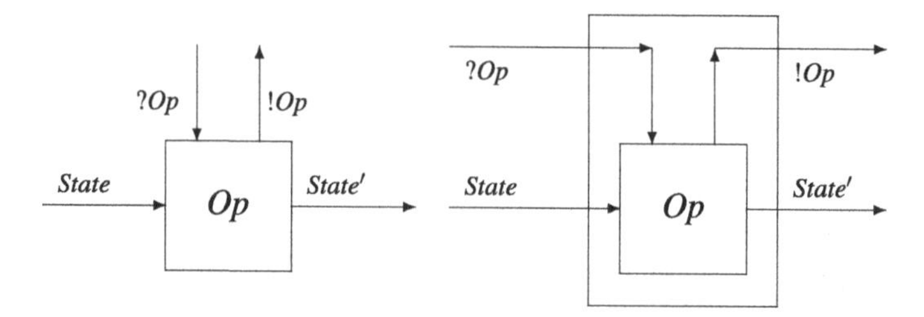

Fig. 4.1 Interpreting operations as heterogeneous binary relations

If such a simulation exists, we also say that C is an upward simulation of A, also denoted $A \sqsubseteq_{US} C$, and similarly for corresponding operations of A and C.

The above rules assume the contractual interpretation of ADTs. In the behavioural interpretation, the rule for correctness becomes

$$\forall AState'; CState; CState' \bullet (COp_i \land T') \Rightarrow \exists AState \bullet T \land AOp_i \qquad \square$$

Although the formalisation of upward simulations is different from downward simulations, upward simulations allow the same sort of changes in a refinement that we discussed earlier. In particular, because the definition is still based upon reduction of non-determinism, preconditions can be weakened and non-determinism in postconditions resolved. There is a difference, however. As we saw in Example 3.4 downward simulations do not allow postponement of non-deterministic choice, and for situations like these upward simulations are needed. In addition, further examples can be constructed that need downward rather than upward simulations.

The essence of the difference lies in the type of simulation that the rules express. In a downward simulation a concrete program is simulated starting in the initial state, and each concrete step is then matched by an abstract one. For this reason downward simulations are sometimes known as *forward* simulations.

On the other hand in an upward simulation an arbitrary point in a concrete program is picked and the simulation works backwards to see if it could be simulated from some abstract initialisation. For this reason upward simulations are sometimes known as *backward* simulations.

This also explains why upward simulations need to be total ($\forall CState \bullet \exists AState \bullet T$) whereas downward simulations do not. Totality is needed because the upward simulation begins at an arbitrary point in the concrete program, and we need to be sure that from any such point we can simulate backwards. Because downward simulations begin at the initialisation totality is not necessary for their retrieve relations.

4.4 Embedding Inputs and Outputs

The previous sections described an embedding of Z ADTs without inputs or outputs in the relational setting. In such a context, Z operations translate naturally to the homogeneous relations of the relational data type. However, when we attempt to consider Z operations *with* input and output as binary relations, we encounter two problems. The definition of pre for such operations tells us that input belongs conceptually with the before-state, and output does not, so it probably belongs with the after-state. Thus, an operation *Op* can be viewed as a relation between $State \times ?Op$ and $State \times !Op$—i.e., in general *not* a homogeneous relation (as required in the setup of Chap. 3), see Fig. 4.1. Moreover, inputs and outputs for a single operation play a *rôle* different from that of the state. The state evolves into a new state through an operation, after which the old state no longer needs to be available. On the other hand, outputs produced should be available for global observation, not disappear once the next output is produced, and something similar holds for inputs.

These problems have been solved long ago in the context of functional programming, where there is no notion of a "global state" which could deliver necessary inputs and absorb produced outputs. The usual encoding of a program which consumes inputs and produces outputs (one by one) is to pass sequences of unused inputs and of produced outputs around as both arguments and results of functions. This is exactly the approach we will be taking for embedding inputs and outputs of Z operations in the framework of data types over homogeneous relations.

Input and output will be modelled by sequences whose elements are produced and consumed one by one by the operations of the ADT. As these inputs and outputs form part of the *observable* behaviour of the ADT, those sequences will need to be part of the *global* state. They also need to be available for the ADT to change, so they will also be part of the *local* state. Initialisation and finalisation will be concerned with copying these sequences between the global and local state.

At this stage we make an assumption that only *appears* restrictive because we are using Z as the meta-language in describing the relational world. We will assume that the *type* of input is the same for *every* operation in a given ADT, and similarly for output. In a more liberally typed meta-notation, we could make this assumption true (without being in any way restrictive) by defining non-existent input to be taken from a unit domain, and the universal input type as the disjoint union of the input types of the operations. In Z we could not do that without introducing the formal noise of injections into (disjoint) unions, which would then dominate the rest of the presentation. (The injections would be introduced as part of the embedding.) Thus, we assume that all operations Op_i have input of type *Input*, and output of type *Output*. Note that these are schema types (whose elements are labelled bindings) whose signatures coincide with those of $?Op_i$ and $!Op_i$, respectively.

First, the global state contains sequences of inputs and outputs (only).

$$\mathsf{G} == \operatorname{seq} Input \times \operatorname{seq} Output$$

The local state contains both of those plus a representation of the Z ADT state, the latter is the same as in the IO-free embedding above.

$$\mathsf{State} == \operatorname{seq} Input \times \operatorname{seq} Output \times State$$

The initialisation transfers the sequence of inputs from the global state to the local state, and picks an initial local ADT state that satisfies the ADT's initialisation.

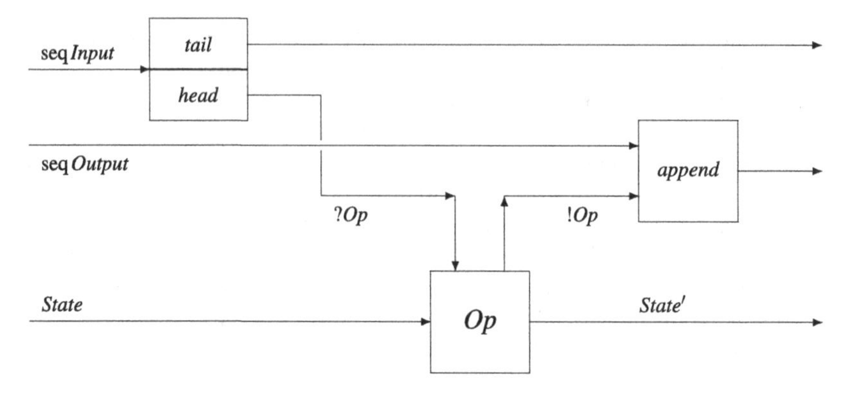

Fig. 4.2 An operation embedded as a homogeneous relation ($append(x, y) = x \frown \langle y \rangle$). The tail of the input sequence is passed on and its head is used as Op's input; the output sequence is passed on with Op's output appended to the end

There is some choice as to how the relational initialisation deals with the output sequence. The output sequence plays a *rôle* analogous to that of an output file—which we could either assume to be opened previously, or to be created in the program. If we had an interest in sequentially composing ADTs, the most obvious choice would be for the output sequence also to be copied across. However, we will assume that the output sequence is only of interest *after* the ADT program, and set it to empty at initialisation. Knowing that all outputs in the local state were actually produced by the ADT may prove an advantage later.

$$\mathsf{Init} == \{Init;\ is : \mathrm{seq}\, Input;\ os : \mathrm{seq}\, Output \bullet (is, os) \mapsto (is, \langle\rangle, \theta State')\}$$

Finalisation is more or less dual to initialisation, but we have to decide what to do with the remaining inputs: copy or discard (or even leave finalisation undefined when any inputs remain). We choose to discard them, but it will make no difference to what follows—we do not consider the sequential composition of ADTs or (complete) programs.

$$\mathsf{Fin} == \{State;\ is : \mathrm{seq}\, Input;\ os : \mathrm{seq}\, Output \bullet (is, os, \theta State) \mapsto (\langle\rangle, os)\}$$

The effect of an operation Op on the modified local state is as follows, see also Fig. 4.2. The first element is taken from the input sequence, and used as the input ($= \theta ?Op$) for the operation. The remainder of the input sequence is left for the following operations. The output produced by the operation ($\theta !Op$) is appended to the output sequence. The state is transformed according to the operation.

$$\begin{aligned}\mathsf{Op}_\mathsf{i} == \{&Op_i;\ is : \mathrm{seq}\, Input;\ os : \mathrm{seq}\, Output \bullet \\ &(\langle \theta ?Op_i \rangle \frown is, os, \theta State) \mapsto (is, os \frown \langle \theta !Op_i \rangle, \theta State')\}\end{aligned}$$

This completes the representation of a standard Z ADT with inputs and outputs by a relational data type, as summarised in the definition below.

Definition 4.5 (Relational interpretation of ADT with IO) The Z ADT (*State*, *Init*, $\{Op_i\}_{i \in I}$) where all Op_i have input of type *Input* and output of type *Output* is interpreted by the relational data type (State, Init, $\{\mathsf{Op_i}\}_{\mathsf{i \in I}}$, Fin), where

$$\begin{aligned}
&\mathsf{State} == \operatorname{seq} Input \times \operatorname{seq} Output \times State \\
&\mathsf{Init} == \{Init;\ is : \operatorname{seq} Input;\ os : \operatorname{seq} Output \bullet (is, os) \mapsto (is, \langle\rangle, \theta State')\} \\
&\mathsf{Op_i} == \{Op_i;\ is : \operatorname{seq} Input;\ os : \operatorname{seq} Output \bullet \\
&\qquad (\langle \theta ?Op_i \rangle \frown is, os, \theta State) \mapsto (is, os \frown \langle \theta !Op_i \rangle, \theta State')\} \\
&\mathsf{Fin} == \{State;\ is : \operatorname{seq} Input;\ os : \operatorname{seq} Output \bullet \\
&\qquad (is, os, \theta State) \mapsto (\langle\rangle, os)\}
\end{aligned}$$

□

Finally, the embedding of retrieve relations in the relational setting is trivial. As they do not refer to inputs and outputs, the only sensible approach is for them to assume identical inputs and outputs. Assuming R is a retrieve relation between *AState* and *CState*:

$$\begin{aligned}
&\mathsf{R} == \{R;\ is : \operatorname{seq} Input;\ os : \operatorname{seq} Output \bullet \\
&\qquad\qquad (is, os, \theta AState) \mapsto (is, os, \theta CState)\}
\end{aligned}$$

Note that this embedding of the retrieve relation confirms the requirement of conformity of inputs and outputs between concrete and abstract ADT.

4.5 Deriving Simulation Rules in Z—Again

For the more complicated relational embedding of Z ADTs defined above, we derive the simulation rules here. These derivations are not terribly illustrative, containing essentially the same steps as those in Sects. 4.2 and 4.3, plus extra steps to account for the copying and discarding of sequences of inputs and outputs. For that reason, we only give the derivations for downward simulation in the contract interpretation. We assume Z data types $(AState, AInit, \{AOp_i\}_{i \in I})$ and $(CState, CInit, \{COp_i\}_{i \in I})$, and retrieve relation R, with their relational interpretations in sans serif. The types of is, os' etc. will be consistently omitted.

For initialisation, first observe that

$$\mathsf{AInit} \fatsemi \mathsf{R} = \{R';\ AInit;\ is;\ os \bullet (is, os) \mapsto (is, \langle\rangle, \theta CState')\}$$

Then we have:

$$\begin{aligned}
&\mathsf{CInit} \subseteq \mathsf{AInit} \fatsemi \mathsf{R} \\
&\qquad \equiv \{ \text{ definition of } \subseteq \} \\
&\forall is, os, c', is', os' \bullet ((is, os), (is', os', c')) \in \mathsf{CInit} \Rightarrow \\
&\qquad\qquad\qquad ((is, os), (is', os', c')) \in \mathsf{AInit} \fatsemi \mathsf{R} \\
&\qquad \equiv \{ \text{ CInit copies input: } is = is', \text{ and deletes output: } os' = \langle\rangle \ \}
\end{aligned}$$

$$\forall is, os, c' \bullet ((is, os), (is, \langle\rangle, c')) \in \mathsf{CInit} \Rightarrow$$
$$((is, os), (is, \langle\rangle, c')) \in \mathsf{AInit} \mathbin{⨾} \mathsf{R}$$
$$\equiv \{ \text{definitions, satisfied for } is, os \}$$
$$\forall c' \bullet c' \in \{CInit \bullet \theta CState'\} \Rightarrow c' \in \{R'; AInit \bullet \theta CState'\}$$
$$\equiv \{ \text{introduce schema quantification} \}$$
$$\forall CState' \bullet CInit \Rightarrow \exists AState' \bullet R' \land AInit$$

Thus, unsurprisingly, the initialisation condition is unchanged in the presence of inputs and outputs.

The given finalisation is total, so the applicability condition for finalisation holds. Observe that

$$\mathsf{R} \mathbin{⨾} \mathsf{CF} = \{R;\ is;\ os \bullet (is, os, \theta AState) \mapsto (\langle\rangle, os)\}$$

Then we have:

$$\mathsf{R} \mathbin{⨾} \mathsf{CF} \subseteq \mathsf{AF}$$
$$\equiv \{ \text{definition of } \subseteq \}$$
$$\forall is, os, a, is', os' \bullet ((is, os, a), (is', os')) \in \mathsf{R} \mathbin{⨾} \mathsf{CF} \Rightarrow$$
$$((is, os, a), (is', os')) \in \mathsf{AF}$$
$$\equiv \{ \mathsf{R} \mathbin{⨾} \mathsf{CF} \text{ deletes input: } is' = \langle\rangle \text{ and copies output: } os' = os \}$$
$$\forall is, os, a \bullet ((is, os, a), (\langle\rangle, os) \in \mathsf{R} \mathbin{⨾} \mathsf{CF} \Rightarrow ((is, os, a), (\langle\rangle, os) \in \mathsf{AF}$$
$$\Leftarrow \{ \text{predicate calculus} \}$$
$$\forall is, os, a \bullet ((is, os, a), (\langle\rangle, os) \in \mathsf{AF}$$
$$\equiv \{ \text{definition of } \mathsf{AF} \}$$
$$true$$

Before we consider the applicability and correctness conditions, let us show that in this embedding pre and dom indeed do not coincide.

$$(is, os, s) \in \mathrm{dom}\, \mathsf{Op_i}$$
$$\equiv \{ \text{definition of dom} \}$$
$$\exists s', is', os' \bullet (is, os, s) \mapsto (is', os', s') \in \mathsf{Op_i}$$
$$\equiv \{ \text{embedding of } Op_i \}$$
$$\exists s', is', os', Op_i \bullet s = \theta State \land s' = \theta State' \land$$
$$is = \langle \theta ?Op_i \rangle \frown is' \land os' = os \frown \langle \theta !Op_i \rangle$$
$$\equiv \{ \text{decomposing sequences } is \text{ and } os' \}$$
$$\exists s', is', o', Op_i \bullet s = \theta State \land s' = \theta State' \land is \neq \langle\rangle \land$$
$$is' = tail\, is \land head\, is = \theta ?Op_i \land o' = \theta !Op_i$$
$$\equiv \{ \text{introducing schema quantification, one point rule for } is' \}$$
$$is \neq \langle\rangle \land \exists \operatorname{pre} Op_i \bullet s = \theta State \land head\, is = \theta ?Op_i$$

The applicability condition uses this analysis, but it is still a lot of work to end up with a condition that adds a single quantification (over the input). Below, we take the first four steps of the corresponding derivation in Sect. 4.2, none of them depending on the particular embedding, in one go:

$$
\begin{aligned}
&\mathrm{ran}(\mathrm{dom}\,\mathsf{AOp_i} \lhd \mathsf{R}) \subseteq \mathrm{dom}\,\mathsf{COp_i}\\
&\quad\equiv \{\text{ definition of } \subseteq, \lhd, \mathrm{dom}; \text{ calculus }\}\\
&\forall c, a, cis, cos, ais, aos \bullet ((ais, aos, a) \mapsto (cis, cos, c) \in \mathsf{R} \land\\
&\qquad (ais, aos, a) \in \mathrm{dom}\,\mathsf{AOp_i})\\
&\qquad \Rightarrow (cis, cos, c) \in \mathrm{dom}\,\mathsf{COp_i}\\
&\quad\equiv \{\ \mathsf{R} \text{ copies input and output, so } cis = ais, cos = aos\ \}\\
&\forall c, a, is, os \bullet ((is, os, a) \mapsto (is, os, c) \in \mathsf{R} \land (is, os, a) \in \mathrm{dom}\,\mathsf{AOp_i}) \Rightarrow\\
&\qquad (is, os, c) \in \mathrm{dom}\,\mathsf{COp_i}\\
&\quad\equiv \{\text{ embedding of } R; \text{ analysis of } \mathrm{dom}\,\mathsf{Op_i} \text{ above }\}\\
&\forall c, a, is, os \bullet ((a, c) \in \{R \bullet \theta AState \mapsto \theta CState\} \land is \neq \langle\rangle \land\\
&\qquad \exists\, \mathrm{pre}\, AOp_i \bullet a = \theta AState \land head\, is = \theta ?AOp_i)\\
&\qquad \Rightarrow (is \neq \langle\rangle \land \exists\, \mathrm{pre}\, COp_i \bullet c = \theta CState\\
&\qquad\qquad \land\ head\, is = \theta ?COp_i)\\
&\quad\equiv \{\text{ calculus: } is \neq \langle\rangle, \text{ remove } os, \text{ replace } is \text{ by } in = head\, is\ \}\\
&\forall c, a, in \bullet ((a, c) \in \{R \bullet \theta AState \mapsto \theta CState\} \land\\
&\qquad \exists\, \mathrm{pre}\, AOp_i \bullet a = \theta AState \land in = \theta ?AOp_i)\\
&\qquad \Rightarrow \exists\, \mathrm{pre}\, COp_i \bullet c = \theta CState\\
&\qquad \land\ in = \theta ?COp_i\\
&\quad\equiv \{\text{ introduce schema quantification }\}\\
&\forall CState; AState; ?AOp_i \bullet R \land \mathrm{pre}\, AOp_i \Rightarrow \mathrm{pre}\, COp_i
\end{aligned}
$$

For correctness, first observe that

$$
\begin{aligned}
\mathsf{AOp_i} \mathbin{⨾} \mathsf{R} = \{R'; AOp_i; is; os \bullet (\langle \theta ?AOp_i \rangle \frown is, os, \theta AState) \mapsto\\
(is, os \frown \langle \theta !AOp_i \rangle, \theta CState')\}
\end{aligned}
$$

The start of the derivation is as for the simpler embedding, taking the first four steps in one go:

$$
\begin{aligned}
&(\mathrm{dom}\,\mathsf{AOp_i} \lhd \mathsf{R}) \mathbin{⨾} \mathsf{COp_i} \subseteq \mathsf{AOp_i} \mathbin{⨾} \mathsf{R}\\
&\quad\equiv \{\text{ definition of } \subseteq, ⨾, \lhd; \text{ predicate calculus }\}\\
&\forall a, ais, aos, c, cis, cos, c', cis', cos' \bullet\\
&\qquad ((ais, aos, a) \mapsto (cis, cos, c) \in \mathsf{R}\ \land\ (ais, aos, a) \in \mathrm{dom}\,\mathsf{AOp_i}\\
&\qquad \land\ (cis, cos, c) \mapsto (cis', cos', c') \in \mathsf{COp_i})\\
&\qquad \Rightarrow (ais, aos, a) \mapsto (cis', cos', c') \in \mathsf{AOp_i} \mathbin{⨾} \mathsf{R}\\
&\quad\equiv \{\text{ embedding of } R\text{: } ais = cis, aos = cos\ \}
\end{aligned}
$$

$$\forall a, is, os, c, c', cis', cos' \bullet ((a, c) \in \{R \bullet (\theta AState, \theta CState)\}$$
$$\wedge (is, os, a) \in \mathrm{dom}\, \mathsf{AOp_i}$$
$$\wedge (is, os, c) \mapsto (cis', cos', c') \in \mathsf{COp_i})$$
$$\Rightarrow (is, os, a) \mapsto (cis', cos', c') \in \mathsf{AOp_i} \operatorname{\text{;}} \mathsf{R}$$

$\equiv$ { embedding of COp_i and $AOp_i \operatorname{\text{;}} R$;
replace cos' by $out = last\, cos'$; remove cis' }

$$\forall a, is, os, c, c', out \bullet$$
$$((a, c) \in \{R \bullet (\theta AState, \theta CState)\}$$
$$\wedge (is, os, a) \in \mathrm{dom}\, \mathsf{AOp_i} \wedge is \neq \langle\rangle$$
$$\wedge \exists COp_i \bullet (head\, is = \theta ?COp_i \wedge out = \theta !COp_i$$
$$\wedge c = \theta CState \wedge c' = \theta CState')$$
$$\Rightarrow is \neq \langle\rangle \wedge \exists R'; AOp_i \bullet (a = \theta AState \wedge c' = \theta CState' \wedge$$
$$head\, is = \theta ?AOp_i \wedge out = \theta !AOp_i)$$

$\equiv$ { analysis of dom above; replace is by $in = head\, is$ }

$$\forall a, in, c, c', out \bullet ((a, c) \in \{R \bullet (\theta AState, \theta CState)\}$$
$$\wedge \exists \mathrm{pre}\, AOp_i \bullet (a = \theta AState \wedge in = \theta ?AOp_i$$
$$\wedge \exists COp_i \bullet (in = \theta ?COp_i \wedge out = \theta !COp_i$$
$$\wedge c = \theta CState \wedge c' = \theta CState')$$
$$\Rightarrow \exists R'; AOp_i \bullet (a = \theta AState \wedge c' = \theta CState') \wedge$$
$$in = \theta ?AOp_i \wedge out = \theta !AOp_i)$$

$\equiv$ { introduce schema quantification }

$$\forall AState; ?AOp_i; CState; CState'; !AOp_i \bullet$$
$$R \wedge \mathrm{pre}\, AOp_i \wedge COp_i \Rightarrow \exists AState' \bullet AOp_i \wedge R'$$

This derivation completes the full downward simulation rule for Z ADTs with inputs and outputs.

Definition 4.6 (Downward simulation) Let $A = (AState, AInit, \{AOp_i\}_{i \in I})$ and $C = (CState, CInit, \{COp_i\}_{i \in I})$ be Z data types, where the operations have conformal inputs and outputs. The relation R on $AState \wedge CState$ is a *downward simulation* from A to C if

$$\forall CState' \bullet CInit \Rightarrow \exists AState' \bullet AInit \wedge R'$$

and for all $i \in I$:

$$\forall AState;\ CState;\ ?AOp_i \bullet \text{pre}\,AOp_i \land R \Rightarrow \text{pre}\,COp_i$$
$$\forall AState;\ CState;\ CState';\ ?AOp_i;\ !AOp_i \bullet$$
$$\text{pre}\,AOp_i \land R \land COp_i \Rightarrow \exists AState' \bullet R' \land AOp_i$$

If such a simulation exists, we also say that C is a downward simulation of A, also denoted $A \sqsubseteq_{DS} C$, and similarly for corresponding operations of A and C.

The above rules assume the contractual interpretation of ADTs. In the behavioural interpretation, the rule for correctness becomes

$$\forall AState;\ CState;\ CState';\ ?AOp_i;\ !AOp_i \bullet$$
$$R \land COp_i \Rightarrow \exists AState' \bullet R' \land AOp_i$$ □

The rules for upward simulation are derived analogously, resulting in the following definition.

Definition 4.7 (Upward simulation) Let $A = (AState, AInit, \{AOp_i\}_{i \in I})$ and $C = (CState, CInit, \{COp_i\}_{i \in I})$ be Z data types, where the operations have conformal inputs and outputs. Then the relation T on $AState \land CState$ is an *upward simulation* from A to C if

$$\forall CState \bullet \exists AState \bullet T$$
$$\forall AState';\ CState' \bullet CInit \land T' \Rightarrow AInit$$

and for all $i \in I$:

$$\forall CState;\ ?AOp_i \bullet (\forall AState \bullet T \Rightarrow \text{pre}\,AOp_i) \Rightarrow \text{pre}\,COp_i$$
$$\forall AState';\ CState;\ CState';\ ?AOp_i;\ !AOp_i \bullet$$
$$(\forall AState \bullet T \Rightarrow \text{pre}\,AOp_i) \Rightarrow (COp_i \land T' \Rightarrow \exists AState \bullet T \land AOp_i)$$

If such a simulation exists, we also say that C is an upward simulation of A, also denoted $A \sqsubseteq_{US} C$, and similarly for corresponding operations of A and C.

The above rules assume the contractual interpretation of ADTs. In the behavioural interpretation, the rule for correctness becomes

$$\forall AState';\ CState;\ CState';\ ?AOp_i;\ !AOp_i \bullet$$
$$(COp_i \land T') \Rightarrow \exists AState \bullet T \land AOp_i$$ □

4.6 Examples

Example 4.3 In this example, the abstract specification evaluates the mean of a set of real numbers in a bag. Operations to count and sum the elements of a bag are given below.

$$
\begin{array}{l}
count : \text{bag}\,\mathbb{R} \to \mathbb{N} \\
sum : \text{bag}\,\mathbb{R} \to \mathbb{R} \\
\hline
count[\![\,]\!] = 0 \land sum[\![\,]\!] = 0 \\
\forall r : \mathbb{R} \bullet count[\![r]\!] = 1 \land sum[\![r]\!] = r \\
\forall b, c : \text{bag}\,\mathbb{R} \bullet count(b \uplus c) = count\,b + count\,c \land \\
\qquad\qquad sum(b \uplus c) = sum\,b + sum\,c
\end{array}
$$

$$
\begin{array}{l}
AState \\
\hline
b : \text{bag}\,\mathbb{R}
\end{array}
\qquad
\begin{array}{l}
AInit \\
\hline
AState' \\
\hline
b' = [\![\,]\!]
\end{array}
$$

$$
\begin{array}{l}
Enter_A \\
\hline
\Delta AState \\
r? : \mathbb{R} \\
\hline
b' = b \uplus [\![r?]\!]
\end{array}
\qquad
\begin{array}{l}
Mean_A \\
\hline
\Xi AState \\
m! : \mathbb{R} \\
\hline
b \neq [\![\,]\!] \land m! = (sum\,b) / count\,b
\end{array}
$$

The concrete specification only maintains a running sum and a count of the items in the bag.

$$
\begin{array}{l}
CState \\
\hline
s : \mathbb{R} \\
n : \mathbb{N}
\end{array}
\qquad
\begin{array}{l}
CInit \\
\hline
CState' \\
\hline
s' = 0 \land n' = 0
\end{array}
$$

$$
\begin{array}{l}
Enter_C \\
\hline
\Delta CState \\
r? : \mathbb{R} \\
\hline
s' = s + r? \\
n' = n + 1
\end{array}
\qquad
\begin{array}{l}
Mean_C \\
\hline
\Xi CState \\
m! : \mathbb{R} \\
\hline
n \neq 0 \land m! = s/n
\end{array}
$$

The required retrieve relation is

$$
\begin{array}{l}
R \\
\hline
AState \\
CState \\
\hline
s = sum\,b \land n = count\,b
\end{array}
$$

Note that it is necessary to use a non-functional retrieve relation here. Intuitively this is because the abstract state maintains information that will never be needed for any observations.

Here we give the verification of downward simulation in full detail as an illustration.

For initialisation, we have:

$$\begin{array}{l}
\forall\, CState' \bullet CInit \Rightarrow \exists\, AState' \bullet AInit \wedge R' \\
\qquad \equiv \{ \text{ definitions } \} \\
\forall\, s' : \mathbb{R};\; n' : \mathbb{N} \bullet s' = 0 \wedge n' = 0 \Rightarrow \\
\quad \exists\, b' : \text{bag}\,\mathbb{R} \bullet b' = [\![\;]\!] \wedge s' = sum\; b' \wedge n' = count\, b' \\
\qquad \equiv \{ \text{ one point rule } \} \\
\forall\, s' : \mathbb{R};\; n' : \mathbb{N} \bullet s' = 0 \wedge n' = 0 \Rightarrow s' = sum[\![\;]\!] \wedge n' = count[\![\;]\!] \\
\qquad \equiv \{ \text{ properties of } [\![\;]\!] \} \\
\forall\, s' : \mathbb{R};\; n' : \mathbb{N} \bullet s' = 0 \wedge n' = 0 \Rightarrow s' = 0 \wedge n' = 0 \\
\qquad \equiv \\
true
\end{array}$$

Before we consider the conditions for the operations, we simplify the expressions for their preconditions. These are actually schemas, although we will only use their predicates later on, as the declarations will turn out to be available in context already.

$$\begin{array}{l}
\text{pre}\, Enter_A \\
\qquad \equiv \{ \text{ definition of pre } \} \\
\exists\, AState';\; !Enter_A \bullet Enter_A \\
\qquad \equiv \{ \; Enter_A \text{ has no outputs; definitions } \} \\
\exists\, b' : \text{bag}\,\mathbb{R} \bullet [\, b, b' : \text{bag}\,\mathbb{R};\; r? : \mathbb{R} \mid b' = b \uplus [\![r]\!] \,] \\
\qquad \equiv \{ \text{ one point rule } \} \\
[\, b : \text{bag}\,\mathbb{R};\; r? : \mathbb{R} \mid b \uplus [\![r?]\!] \in \text{bag}\,\mathbb{R} \,] \\
\qquad \equiv \{ \text{ bag axioms } \} \\
[\, b : \text{bag}\,\mathbb{R};\; r? : \mathbb{R} \,]
\end{array}$$

$$\begin{array}{l}
\text{pre}\, Mean_A \\
\qquad \equiv \{ \text{ definition of pre; definitions } \} \\
\exists\, b' : \text{bag}\,\mathbb{R};\; m! : \mathbb{R} \bullet [\, b, b' : \text{bag}\,\mathbb{R};\; m! : \mathbb{R} \mid b = b' \wedge b \neq [\![\;]\!] \\
\qquad\qquad\qquad\qquad\qquad\qquad\qquad\qquad \wedge\, m! = (sum\; b) / count\, b \,] \\
\qquad \equiv \{ \text{ one point rule for } b' \; \} \\
\exists\, m! : \mathbb{R} \bullet [\, b : \text{bag}\,\mathbb{R};\; m! : \mathbb{R} \mid b \neq [\![\;]\!] \wedge m! = (sum\; b) / count\, b \,] \\
\qquad \equiv \{ \text{ one point rule for } m! \; \} \\
[\, b : \text{bag}\,\mathbb{R} \mid b \neq [\![\;]\!] \wedge (sum\; b) / count\, b \in \mathbb{R} \,] \\
\qquad \equiv \\
[\, b : \text{bag}\,\mathbb{R} \mid b \neq [\![\;]\!] \,]
\end{array}$$

$$
\begin{array}{l}
\text{pre}\, Enter_C \\
\qquad \equiv \{ \text{ definitions } \} \\
\exists s' : \mathbb{R};\ n' : \mathbb{N} \bullet [s, s', r? : \mathbb{R};\ n, n' : \mathbb{N} \mid s' = s + r? \land n' = n + 1] \\
\qquad \equiv \{ \text{ one point rule, twice } \} \\
[s, r? : \mathbb{R};\ n : \mathbb{N} \mid s + r? \in \mathbb{R} \land n + 1 \in \mathbb{N}] \\
\qquad \equiv \{ \text{ properties of } \mathbb{N} \text{ and } \mathbb{R} \} \\
[s, r? : \mathbb{R};\ n : \mathbb{N}]
\end{array}
$$

$$
\begin{array}{l}
\text{pre}\, Mean_C \\
\qquad \equiv \{ \text{ definitions } \} \\
\exists s', m! : \mathbb{R};\ n' : \mathbb{N} \bullet [s, s', m! : \mathbb{R};\ n, n' : \mathbb{N} \mid \\
\qquad\qquad\qquad\qquad s' = s \land n' = n \land n \neq 0 \land m! = s/n] \\
\qquad \equiv \{ \text{ one point rule for } s' \text{ and } n' \} \\
\exists m! : \mathbb{R} \bullet [s, m! : \mathbb{R};\ n : \mathbb{N} \mid n \neq 0 \land m! = s/n] \\
\qquad \equiv \{ \text{ one point rule for } m! \} \\
[s : \mathbb{R};\ n : \mathbb{N} \mid n \neq 0]
\end{array}
$$

Applicability for the *Enter* operation is verified as follows:

$$
\begin{array}{l}
\forall AState;\ CState;\ ?Enter_A \bullet \text{pre}\, Enter_A \land R \Rightarrow \text{pre}\, Enter_C \\
\qquad \equiv \{ \text{ definitions } \} \\
\forall b : \text{bag}\, \mathbb{R};\ s, r? : \mathbb{R};\ n : \mathbb{N} \bullet true \land s = sum\ b \land n = count\, b \Rightarrow true \\
\qquad \equiv \{ \text{ logic } \} \\
true
\end{array}
$$

Applicability for *Mean* is:

$$
\begin{array}{l}
\forall AState;\ CState;\ ?Mean_A \bullet \text{pre}\, Mean_A \land R \Rightarrow \text{pre}\, Mean_C \\
\qquad \equiv \{ \text{ definitions } \} \\
\forall b : \text{bag}\, \mathbb{R};\ s : \mathbb{R};\ n : \mathbb{N} \bullet b \neq [\![\ ]\!] \land s = sum\ b \land n = count\, b \Rightarrow n \neq 0 \\
\qquad \equiv \{ count\, b = 0 \Leftrightarrow b = [\![\ ]\!] \} \\
true
\end{array}
$$

Correctness for *Enter* is:

$$
\begin{array}{l}
\forall AState;\ CState;\ CState';\ ?Enter_A \bullet \\
\quad \text{pre}\, Enter_A \land R \land Enter_C \Rightarrow \exists AState' \bullet R' \land Enter_A \\
\qquad \equiv \{ \text{ definitions } \} \\
\forall b : \text{bag}\, \mathbb{R};\ s, s', r? : \mathbb{R};\ n, n' : \mathbb{N} \bullet \\
\quad s = sum\ b \land n = count\, b \land s' = s + r? \land n' = n + 1 \\
\quad \Rightarrow \exists b' : \text{bag}\, \mathbb{R} \bullet s' = sum\ b' \land n' = count\, b' \land b' = b \uplus [\![r?]\!]
\end{array}
$$

$$
\begin{array}{l}
\qquad \equiv \{ \text{ one point rule } \} \\
\forall b : \text{bag}\,\mathbb{R};\ s, s', r? : \mathbb{R};\ n, n' : \mathbb{N} \bullet \\
\quad s = sum\ b \land n = count\,b \land s' = s + r? \land n' = n + 1 \\
\quad \Rightarrow s' = sum(b \uplus [\![r?]\!]) \land n' = count(b \uplus [\![r?]\!]) \\
\qquad \equiv \{ \text{ properties of } \uplus \ \} \\
true
\end{array}
$$

Finally, correctness for *Mean* is verified as follows:

$$
\begin{array}{l}
\forall AState;\ CState;\ CState';\ !Mean_A \bullet \\
\quad \text{pre}\,Mean_A \land R \land Mean_C \Rightarrow \exists AState' \bullet R' \land Mean_A \\
\qquad \equiv \{ \text{ definitions } \} \\
\forall b : \text{bag}\,\mathbb{R};\ s, s', m! : \mathbb{R};\ n, n' : \mathbb{N} \bullet \\
\quad b \neq [\![\ ]\!] \land s = sum\ b \land n = count\,b \land s' = s \land n' = n \land n \neq 0 \land m! = s/n \Rightarrow \\
\quad \exists b' : \text{bag}\,\mathbb{R} \bullet s' = sum\ b' \land n' = count\,b' \land b \neq [\![\ ]\!] \land m! = (sum\ b)/\,count\,b \\
\qquad \equiv \\
true
\end{array}
$$

This completes the proof that R is a downward simulation between the abstract and concrete data types, and thus that the concrete is a refinement of the abstract. □

Example 4.4 Returning to Example 1.4 (turning a set of numbers into a queue), this is proved to be a refinement as a downward simulation, with retrieve relation

```
┌─R──────────────────────────────
│ A; C
├────────────
│ s = ran q
└────────────────────────────────
```

The initialisation condition comes down to the trivial

$$q = \langle\rangle \land s = \text{ran}\,q \Rightarrow s = \varnothing$$

Applicability is trivial for the addition operations, as their preconditions are *true*. Correctness follows from the property that

$$\text{ran}(s1 \frown s2) = \text{ran}\,s1 \cup \text{ran}\,s2$$

For the removal operations, we have $\text{pre}\,ARemove = [A \mid s \neq \varnothing]$ whereas $preCRemove = [C]$. Applicability is thus again trivial; correctness follows from the same property above. □

Example 4.5 In this simple example we also use a retrieve relation rather than a function. The operation *Pick* outputs a new natural number. In the abstract specification this value is chosen non-deterministically.

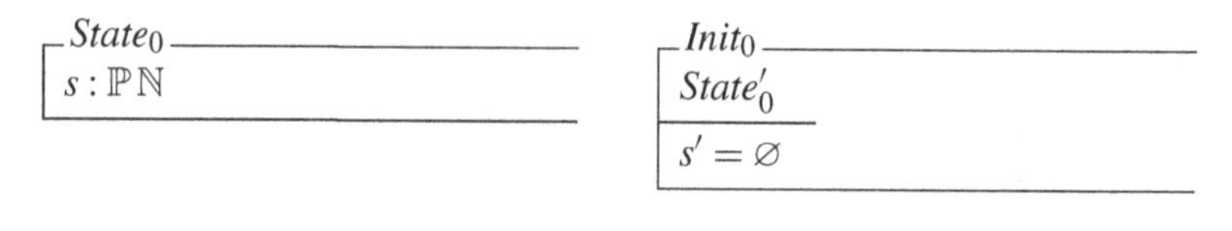

$$\textit{State}_0 \;\hat{=}\; [\, s : \mathbb{P}\,\mathbb{N} \,]$$

$$\textit{Init}_0 \;\hat{=}\; [\, \textit{State}_0' \mid s' = \varnothing \,]$$

$\textit{Pick}_0$

$$\Delta \textit{State}_0$$
$$r! : \mathbb{N}$$

$$s' = s \cup \{r!\}$$
$$r! \notin s$$

In the refinement we use a single natural number n in place of the set s, and the operation *Pick* is no longer non-deterministic.

$$\textit{State}_1 \;\hat{=}\; [\, n : \mathbb{N} \,]$$

$$\textit{Init}_1 \;\hat{=}\; [\, \textit{State}_1' \mid n' = 0 \,]$$

$\textit{Pick}_1$

$$\Delta \textit{State}_1$$
$$r! : \mathbb{N}$$

$$n' = n + 1$$
$$r! = n$$

The retrieve relation between the two state spaces maps n to $max(s) + 1$ if s is not empty, and n to 0 if s is empty. The latter guarantees the initialisation condition. Applicability of *Pick* is trivial, as the concrete operation is total. The crucial property for correctness is

$$max(s \cup \{max(s) + 1\}) = max(s) + 1 \qquad \square$$

Example 4.6 We now re-visit Example 1.5. The retrieve relation linking the two specifications says that every element in the range of *table* is linked to the appropriate element in ran $table_C$, and is given by

T

$$\textit{LeagueTable}$$
$$\textit{LeagueTable}_C$$

$$\forall c : \textit{CLUBS} \bullet \forall u, v, x, y : \mathbb{N} \bullet$$
$$\quad table(c) = (u, v, x, y) \Leftrightarrow (u, v, x, y, c) \in \operatorname{ran} table_C$$

In going from abstract to concrete the operations have changed in a number of ways in addition to the alteration of state:

- $Play_C$ is now total, i.e., we have weakened the precondition of *Play*; and
- $Champion_C$ and $Display_C$ have reduced some non-determinism present in *Champion* and *Display*.

The fact that $Play_C$ is total is easy to see (the *Display* operations are also total and this depends upon elementary properties of ordered sequences).

The reduction of non-determinism is actually represented as a state invariant in $LeagueTable_C$, rather than being encoded directly in the concrete operations. That is, we take the goals scored into account when we store (and update) the league table, rather than just when we display it or pick the champion. However, this fact is only visible via the outputs in $Champion_C$ and $Display_C$.

Finally, because the retrieve relation is in fact a function from concrete to abstract (there is only one *table* for each sequence $table_C$) and also total, the refinement could be verified as either an upward or downward simulation.

For an upward simulation we would have to verify the initialisation:

$$\forall LeagueTable'; LeagueTable_C' \bullet Init_C \wedge T' \Rightarrow Init$$

together with applicability and correctness for each operation. For example, for the *Display* operation applicability is trivial (the operations are total), and correctness would require verification of

$$\begin{array}{l} \forall LeagueTable'; LeagueTable_C; LeagueTable_C'; !Display \bullet \\ \qquad (Display_C \wedge T' \Rightarrow \exists LeagueTable \bullet T \wedge Display) \end{array}$$

□

Example 4.7 In this example we specify a simple addressable memory with operations to *Read* and *Write*. This will then be refined to a description that adds a cache to the state where the *Read* and *Write* operations access and update the cache. The state of the memory *m* is a function from the addresses *A* to data *D*. The first specification is very simple.

$[A, D]$

$$\begin{array}{|l} \hline Mem_0 \\ \hline m_0 : A \rightarrow\!\!\!\!\!\!+ D \\ \hline \end{array}$$

$$\begin{array}{|l} \hline Init_0 \\ \hline Mem_0' \\ \hline m_0' = \varnothing \\ \hline \end{array}$$

$$\begin{array}{|l} \hline Read_0 \\ \hline \Xi Mem_0 \\ a? : A \\ d! : D \\ \hline a? \in \operatorname{dom} m_0 \\ d! = m_0(a?) \\ \hline \end{array}$$

$$\begin{array}{|l} \hline Write_0 \\ \hline \Delta Mem_0 \\ a? : A \\ d? : D \\ \hline m_0' = m_0 \oplus \{a? \mapsto d?\} \\ \hline \end{array}$$

In the refinement we add a (fixed size) cache which is used to satisfy requests if possible; if not the appropriate value from the memory is used. Changes are written back to the memory straight away. The cache holds a subset of the available space, and this is represented by using a partial function from A to D.

$$n : \mathbb{N}_+$$

$$\begin{array}{|l} \hline \textit{Mem}_1 \\ \hline m_1 : A \twoheadrightarrow D \\ c_1 : A \twoheadrightarrow D \\ \hline \end{array}$$

$$\begin{array}{|l} \hline \textit{Init}_1 \\ \hline \textit{Mem}_1' \\ \hline m_1' = \varnothing \wedge c_1' = \varnothing \\ \hline \end{array}$$

Now when a write is performed we create a free space in the cache if necessary, which is when the cache is not full yet and the address is not already in the cache. Creating cache space is done by picking one value to write back to the main memory. Subsequently we update the cache. Thus, the write operation consists of two parts: the first checks the cache, flushing it if necessary, the second writes the value to the cache. When we flush the cache we non-deterministically choose an address to write back to the main memory.

$$\begin{array}{|l} \hline \textit{Flush}_1 \\ \hline \Delta \textit{Mem}_1 \\ \hline \# \operatorname{dom} c_1 < n \Rightarrow \Xi \textit{Mem}_1 \\ \# \operatorname{dom} c_1 = n \Rightarrow \\ \quad (\exists x : A \bullet x \in \operatorname{dom} c_1 \wedge \\ \qquad m_1' = m_1 \oplus \{x \mapsto c_1(x)\} \\ \qquad \{x\} \mathbin{⩤} c_1' = \{x\} \mathbin{⩤} c_1 \\ \qquad \operatorname{dom} c_1' = \operatorname{dom} c_1 \setminus \{x\}) \\ \hline \end{array}$$

$$\begin{array}{|l} \hline \textit{CheckCache}_1 \\ \hline \Delta \textit{Mem}_1 \\ a? : A \\ \hline a? \in \operatorname{dom} c_1 \Rightarrow \Xi \textit{Mem}_1 \\ a? \notin \operatorname{dom} c_1 \Rightarrow \textit{Flush}_1 \\ \hline \end{array}$$

$$\begin{array}{|l} \hline \textit{WrCache}_1 \\ \hline \Delta \textit{Mem}_1 \\ a? : A \\ d? : D \\ \hline m_1' = m_1 \\ c_1' = c_1 \oplus \{a? \mapsto d?\} \\ \hline \end{array}$$

$$\textit{Write}_1 == \textit{CheckCache}_1 \mathbin{⨾} \textit{WrCache}_1$$

Note that the semantics of schema composition ensures that the inputs $a? : A$ in the constituent operations are identified. The read operation is implemented in a similar fashion.

$$\begin{array}{|l} \multicolumn{1}{l}{\underline{\quad RdCache_1 \quad}} \\ \Delta Mem_1 \\ a? : A \\ \hline m_1' = m_1 \\ a? \notin \operatorname{dom} c_1 \Rightarrow c_1' = c_1 \oplus \{a? \mapsto m_1(a?)\} \\ a? \in \operatorname{dom} c_1 \Rightarrow c_1' = c_1 \\ \hline \end{array}$$

$$\begin{aligned} Read_1 == {} & CheckCache_1 \fatsemi RdCache_1 \\ & \fatsemi [\Xi Mem_1;\ a? : A;\ d! : D \mid d! = c_1(a?)] \end{aligned}$$

To verify the refinement we use the following retrieve relation which says that the correct data value for a given address is found by first looking in the cache, and only if that address is not there should we use the main memory.

$$\begin{array}{|l} \multicolumn{1}{l}{\underline{\quad Ret \quad}} \\ Mem_0 \\ Mem_1 \\ \hline m_0 = m_1 \oplus c_1 \\ \hline \end{array}$$

□

Example 4.8 Consider the specification of a system process scheduler (adapted from [7] and rewritten in Z). The system consists of processes either *ready* to be scheduled or *waiting* to become ready and, optionally, a single *active* process. These processes are identified by a unique $PID == \mathbb{N}$. A process cannot be both ready and waiting, and the active process is neither ready nor waiting. In addition, there must be an active process whenever there are processes ready to be scheduled. The scheduling is described abstractly by picking any process from those which are ready. The system can be in two modes: *user* or *super*.

The specification describes *ready* and *waiting* as sets and contains four operations. *New* introduces another process, *Ready* puts a process into the ready state, and *Swap* changes the active process. The operation *Boot* enables a restart if the system is in *super* mode (*nil* stands for an inactive process).

$$ADMIN ::= user \mid super$$

$$\begin{array}{|l} \multicolumn{1}{l}{\underline{\quad State \quad}} \\ active : PID \\ ready : \mathbb{P}\, PID \\ waiting : \mathbb{P}\, PID \\ admin : ADMIN \\ \hline ready \cap waiting = \varnothing \\ active \notin (ready \cup waiting) \\ nil \notin (ready \cup waiting) \\ active = nil \Rightarrow ready = \varnothing \\ \hline \end{array}$$

$$\begin{array}{|l} \multicolumn{1}{l}{\underline{\quad Init \quad}} \\ State' \\ \hline active' = nil \\ ready' \cup waiting' = \varnothing \\ admin' = user \\ \hline \end{array}$$

$$
\begin{array}{|l}
\hline
\multicolumn{1}{l}{\textit{New}} \\
\Delta State \\
p? : PID \\
\hline
p? \neq active \land p? \neq nil \\
p? \notin (ready \cup waiting) \\
waiting' = waiting \cup \{p?\} \\
active' = active \\
ready' = ready \\
admin = user \land admin' = user \\
\hline
\end{array}
$$

$$
\begin{array}{|l}
\hline
\multicolumn{1}{l}{\textit{Ready}} \\
\Delta State \\
q? : PID \\
\hline
q? \in waiting \\
waiting' = waiting \setminus \{q?\} \\
active = nil \Rightarrow (ready' = ready \land active' = q?) \\
active \neq nil \Rightarrow (ready' = ready \cup \{q?\} \land active' = active) \\
admin = user \land admin' = user \\
\hline
\end{array}
$$

$$
\begin{array}{|l}
\hline
\multicolumn{1}{l}{\textit{Swap}} \\
\Delta State \\
\hline
active \neq nil \\
waiting' = waiting \cup \{active\} \\
ready = \varnothing \Rightarrow (active' = nil \land ready' = \varnothing) \\
ready \neq \varnothing \Rightarrow (active' \in ready \land ready' = ready \setminus \{active'\}) \\
admin = user \land admin' = user \\
\hline
\end{array}
$$

$$
\begin{array}{|l}
\hline
\multicolumn{1}{l}{\textit{Boot}} \\
\Delta State \\
\hline
admin = super \\
active = nil \\
waiting = \varnothing \land ready = \varnothing \\
admin' = user \land active' \neq nil \\
(ready' = \varnothing \land waiting' \neq \varnothing) \lor (ready' \neq \varnothing \land waiting' = \varnothing) \\
\hline
\end{array}
$$

An implementation uses sequences instead of sets to record the ready and waiting processes. It has also chosen a particular scheduling strategy, namely to take the process at the head of the ready queue. The implementation has therefore resolved some of the non-determinism in the abstract specification and also changed the state space. The specification is given as follows:

CState

$active : PID$
$cready : \text{iseq}\, PID$
$cwaiting : \text{seq}\, PID$
$admin : ADMIN$

$\text{ran}\, cready \cap \text{ran}\, cwaiting = \varnothing$
$active \notin (\text{ran}\, cready \cup \text{ran}\, cwaiting)$
$nil \notin (\text{ran}\, cready \cup \text{ran}\, cwaiting)$
$active = nil \Rightarrow cready = \langle\,\rangle$

CInit

$CState'$

$active' = nil$
$cready' = \langle\,\rangle \wedge cwaiting' = \langle\,\rangle$
$admin' = user$

CNew

$\Delta CState$
$p? : PID$

$p? \neq active$
$p? \notin (\text{ran}\, cready \cup \text{ran}\, cwaiting)$
$\text{ran}\, cwaiting' = \text{ran}\, cwaiting \cup \{p?\}$
$active' = active$
$cready' = cready$
$admin = user \wedge admin' = user$

CReady

$\Delta CState$
$q? : PID$

$q? \in \text{ran}\, cwaiting$
$\text{ran}\, cwaiting' = \text{ran}\, cwaiting \setminus \{q?\}$
$active = nil \Rightarrow (cready' = cready \wedge active' = q?)$
$active \neq nil \Rightarrow (\text{ran}\, cready' = \text{ran}\, cready \cup \{q?\} \wedge$
$\qquad active' = active)$
$admin = user \wedge admin' = user$

CSwap

$\Delta CState$

$active \neq nil$
$\text{ran}\, cwaiting' = \text{ran}\, cwaiting \cup \{active\}$
$cready = \langle\,\rangle \Rightarrow (active' = nil \wedge cready' = \langle\,\rangle)$
$cready \neq \langle\,\rangle \Rightarrow (active' = head\, cready \wedge cready' = tail\, cready)$
$admin = user \wedge admin' = user$

$$
\begin{array}{|l}
\hline
\textit{CBoot} \\
\Delta \textit{CState} \\
\hline
\textit{admin} = \textit{super} \\
\textit{active} = \textit{nil} \\
\textit{cwaiting} = \langle\rangle \land \textit{cready} = \langle\rangle \\
\textit{admin}' = \textit{user} \land \textit{active}' \neq \textit{nil} \\
\textit{cready}' = \langle\rangle \land \textit{cwaiting}' \neq \langle\rangle \\
\hline
\end{array}
$$

This specification is a downward simulation of the abstract scheduler where the retrieve relation is given by

$$
\begin{array}{|l}
\hline
\textit{Ret} \\
\textit{State} \\
\textit{CState} \\
\hline
\textit{ready} = \operatorname{ran} \textit{cready} \\
\textit{waiting} = \operatorname{ran} \textit{cwaiting} \\
\hline
\end{array}
$$

The initialisation condition boils down to the trivial $\operatorname{ran}\langle\rangle = \varnothing$. Applicability and correctness of most operations is obvious, because in most places the concrete specification is obtained from the abstract by the substitutions in the retrieve relation. Correctness of *CSwap* requires $\textit{head}\,\textit{cready} \in \operatorname{ran} \textit{cready}$, which is obvious. It also requires $\operatorname{ran}(\textit{tail}\,\textit{cready}) = (\operatorname{ran} \textit{cready}) \setminus \{\textit{head}\,\textit{cready}\}$, i.e., there are no further occurrences of the head in the tail of the list. This is why we needed to define *cready* to be an injective sequence (no duplicates) rather than an arbitrary one. □

All of these examples can in fact be verified using a downward simulation; we shall meet an example where an upward simulation is necessary in Chap. 5.

4.7 Reflections on the Embedding

A particular embedding of Z operations was chosen before, here we consider some of the alternatives and consequences.

4.7.1 Alternative Embeddings

In the preceding sections, we have provided separate (analogous) embeddings of Z operation schemas as partial relations, first for operations without IO, and then for operations with IO. Figure 4.3 shows these embeddings, and also hints at alternative paths to follow.

First, we could derive refinement rules for operations with input and output directly at the relational level, as done by Woodcock and Davies [14]. This approach

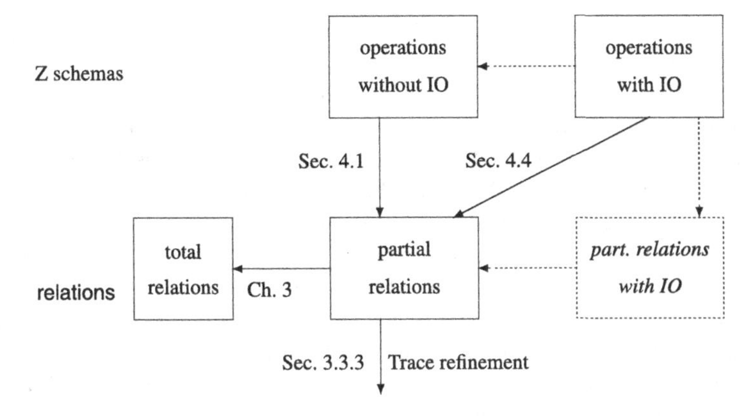

Fig. 4.3 Embeddings of Z schemas into relations: alternatives

turns out to be essentially the same as the one taken in this book, with identical results. There is some advantage in being able to use algebraic laws about products and identities as opposed to predicate calculus with hints stating "the embedding of R copies inputs" etc. However, the disadvantage to the purely relational approach is that a formalism for "relations with IO" would need to be invented in order to make all the details of the derivation visible.

Another alternative that appears attractive is to define input and output as syntactic sugar in the Z notation itself. Using essentially the same construction, the state would contain implicit input and output sequences, with input and output parameters being aliases for the next elements of those sequences. However, the initialisation of the input sequence would need to be dealt with very carefully, as it is not a non-deterministic choice, but (at the level of the embedding) copying part of the global state. One way of addressing this is to move towards Z ADTs which explicitly mention the global state, as done by Stepney, Cooper and Woodcock [11].

Finally, note that the soundness of simulations for *partial* operations means that we can also apply the theory of simulations directly to partial relations resulting from embedding Z, as described in Sect. 3.3.3. This will give downward and upward simulation refinement conditions for trace refinement, which are identical to the conditions in the behavioural approach *except* for omitting applicability conditions.

4.7.2 Programs, Revisited

In order to derive the refinement conditions for operations with IO, we made the simplifying assumption that all operations had input of the same type *Input*. This was mostly necessary to ensure that the sequence of inputs in the global and local state was homogeneous. In the conditions derived, this sequence and the type(s) of its elements no longer played a *rôle*. We also indicated that the assumption of all inputs having the same type could be satisfied in every case by constructing a disjoint sum of the input types of all operations.

Now consider the sequence of inputs, all taken from this disjoint sum. In the standard construction of disjoint sums, each of its elements consists of a tuple: a label plus a value. These labels uniquely identify the operation which is to consume this particular input value, and therefore could be taken from the ADT's operation index set I. Now by projecting the input sequence on the *labels* only, we obtain a familiar sequence: this projection could be said to be *the* program to be executed, i.e., in this embedding, the environment does not only drive the ADT by executing its operations one by one, but it also needs to provide, at initialisation, complete information about which operations it will be requesting and in which order. Thus, the input sequence contains the program. In process algebras, such as LOTOS and CSP, actions with inputs are viewed semantically as *collections* of different actions, one for each possible input value. If we adopted this interpretation in Z as well, we could even say that the input sequence *is* the program. This also provides a different angle on identical inputs being part of the conformity requirement between ADTs: the operation is not characterised by its index only, but the input value is also required. Therefore corresponding programs between ADTs need to have identical inputs, as well as the same indices.

4.8 Proving and Disproving Refinement

The main effort in a refinement proof is often in constructing the right simulation—or in proving that no "right" simulation can exist. The proof rules give no explicit guidance for doing so, but we can provide a few hints and strategies.

First, should we look for an upward simulation or a downward simulation? In general, it is best to look for a downward simulation first—because the downward simulation rules have a slightly simpler shape. This means that if a suitable simulation can be found which is actually both an upwards and a downwards simulation, we have a slightly simpler proof. In practical examples it appears that there is only one circumstance in which upward simulation is always brought in: when the "concrete" data type resolves non-determinism in a later operation than in the "abstract" data type (as in the infamous counter-example in Chap. 3).

How to construct a downward simulation? It is clear that the applicability condition, which only has R on the left of the implication, is easier to satisfy for "small" retrieve relations. In particular, it is satisfied for the empty relation, and so is correctness, where the relation occurs both on the left and on the right of the implication. However, the initialisation condition requires a non-empty relation. So, if no further information is available, a downward simulation can be incrementally constructed, by first linking each of the concrete initial states to some abstract initial state(s), and then adding pairs to the retrieve relation as required by the correctness condition. This should always be allowed by applicability—if not, an alternative linking needs to be sought, or if none exists, no downward simulation exists. Chapter 3 contains a few examples of this kind of reasoning, in particular in Example 3.4.

The existential quantification over the retrieve relation also means that it is impossible to blame any individual condition for upward or downward simulation not holding—only for a particular relation not being a simulation. Between any two data types, the full relation between the states will always satisfy the initialisation condition for downward simulation, and the empty relation will always satisfy correctness, and applicability for downward simulation, for all operations.

Proving that *refinement* does not hold is rather more complicated than proving that no downward simulation exists. An upward simulation may still exist, or even a combination of upward and downward simulations. This seems to imply an unfeasibly large search space (of all intermediate specifications); however, the proof of the Completeness Theorem (Theorem 3.7) indicates that it is sufficient to consider only the case where the intermediate specification is obtained by a "powerset construction" from the original. See also Chap. 8 which uses the same construction to obtain a single complete refinement rule.

4.9 Some Pitfalls

4.9.1 Name Capture

Although the use of names makes specifications clearer most of the time, it can lead to a mix up between (Z) syntax and (refinement) semantics. This may happen when we are comparing two data types whose state schemas have names in common.

First, we may end up with syntax errors.

Example 4.9 Consider the state schemas below.

$BIT == \{0, 1\}$

Both could be part of data types for Boolean logic but we would never be able to define a retrieve relation between them, as it would have to include two clashing definitions of x, i.e., a syntax error. □

Worse, our attempt at defining a retrieve relation may fail due to name capture without any syntax problems. Consider the following ADTs, each outputting the sequence $1, 0, 1, 0, \ldots$.

Example 4.10

$$\begin{array}{l} S1 \\ \hline x : \mathbb{N} \end{array} \qquad \begin{array}{l} S2 \\ \hline x : \mathbb{N} \end{array}$$

$$\begin{array}{l} Init1 \\ \hline S1' \\ \hline x' = 0 \end{array} \qquad \begin{array}{l} Init2 \\ \hline S1' \\ \hline x' = 1 \end{array}$$

$$\begin{array}{l} Op1 \\ \hline \Delta S1 \\ x! : \mathbb{N} \\ \hline x' = 1 - x \\ x! = x' \end{array} \qquad \begin{array}{l} Op2 \\ \hline \Delta S2 \\ x! : \mathbb{N} \\ \hline x' = 1 - x \\ x! = x \end{array}$$

Defining a retrieve relation with signature $S1 \wedge S2$ would not lead to a successful proof of refinement, as the required relation is that x (in $S1$) is 1 whenever x (in $S2$) is 0, and *vice versa*. This is not expressed by $x = 1 - x$, which does not hold for any natural number. □

To avoid these problems, many of our examples use subscripts 0 and 1, or A and C. This glosses over the fact that, due to the possibility of name capture, the Z simulation rules may be said to be incomplete with respect to the simulation relations on the underlying relational interpretations. Clearly the relational interpretations of the above examples could be proved refinement-equivalent, using retrieve relations essentially like the one in Example 4.2 for the Boolean types, and based on mappings $\{\langle\!|\, x == 0 \,|\!\rangle \mapsto \langle\!|\, x == 1 \,|\!\rangle, \langle\!|\, x == 1 \,|\!\rangle \mapsto \langle\!|\, x == 0 \,|\!\rangle\}$ for Example 4.10. However, such mappings cannot occur as the result of the embedding of retrieve schemas given in this chapter.

Incompleteness of the rules in this respect is actually not a problem. Clearly any such "impossible" refinement may be proved in two steps, one of which is purely a renaming of components, if the underlying relational refinement could be proved in a single simulation step. In the worst case, still no more than two steps (an upward simulation and a downward simulation) will be required, as the names of the state components in the intermediate specification can always be chosen to be disjoint from those in the other specifications.

Observe that requiring concrete and abstract states to have disjoint component names (as some textbooks do) avoids these kinds of problems, but also makes refinements where components *do not* change unnecessarily complicated. Proving reflexivity of simulations in Z, and establishing or imposing invariants (as in Sect. 2.4) would take two refinement steps, rather than one, with such a restriction.

4.9.2 Incompleteness of Behavioural Rules

The examples used in this chapter have all employed the traditional Z refinement rules, called the "contract" approach in this book. The incompleteness of the refinement rules for the alternative "behavioural" approach as discussed in Sect. 3.3.2 carries over to Z specifications. The specifications $A = (S, AInit, \{Dec\})$ and $C = (S, CInit, \{Dec\})$ given below are mutual refinements in their behavioural relational embeddings, but the refinement of A by C cannot be proved using the behavioural simulation rules in Z—see [4] for the details and proof.

$$\begin{array}{|l}\hline S \\ x : \mathbb{N} \\ \hline\end{array}$$

$$\begin{array}{|l}\hline Dec \\ \Delta S \\ \hline x' = x - 1 \\ \hline\end{array}$$

$$\begin{array}{|l}\hline AInit \\ S' \\ \hline x' \in \{1, 3\} \\ \hline\end{array}$$

$$\begin{array}{|l}\hline CInit \\ S' \\ \hline x' \in \{1, 2, 3\} \\ \hline\end{array}$$

4.9.3 Interaction with Underspecification

As discussed in Sect. 1.3, Z also allows particular kinds of underspecification, such as through the use of axiomatic definitions, and given sets. Thus, Z specifications in general have multiple models. As a consequence, underspecification by itself *also* characterises a refinement relation, defined as the inclusion of models. For example, the specification containing only

$$\begin{array}{|l} squadsize : \mathbb{N} \\ \hline squadsize = 25 \end{array}$$

is a "model refinement" of the one containing only

$$\begin{array}{|l} squadsize : \mathbb{N} \\ \hline squadsize \geq 11 \end{array}$$

Model inclusion also applies to given sets (and, by extension, to free types). Thus,

$$\begin{array}{l} \mathbb{B} == \{0, 1\} \\ true == 1 \\ false == 0 \end{array}$$

is a "model refinement" of

$$\mathbb{B} ::= true \mid false$$

The interactions between this type of model refinement and the notions of refinement discussed in the rest of this monograph are explored in detail in [3]. We highlight one problematic issue here: the use of negation (and the Σ signature notation based on it) in schema calculus interacts very unhelpfully with model refinement. As an example, consider the schema

$$\begin{array}{|l} \multicolumn{1}{l}{B} \\ \hline b : \mathbb{B} \\ \hline \end{array}$$

and the predicate (schema with empty signature)

$$\exists (B \vee \neg B) \bullet \neg B$$

For the version of $\mathbb{B}$ as a free type, this schema (asking whether b's type has additional inhabitants beyond $\mathbb{B}$) is equivalent to *false*. However, for the implementation of $\mathbb{B}$ as a subset of numbers, i.e. a valid model refinement, we have

$$\begin{array}{|l} \multicolumn{1}{l}{\neg B} \\ \hline b : \mathbb{Z} \\ \hline b \notin \{0, 1\} \\ \hline \end{array}$$

and thus the predicate is *true* in this case.

4.10 Bibliographical Notes

The embedding of Z ADTs in the relational setting and the embedding of inputs and outputs as used here were first described by Woodcock and Davies [14]. The description in this chapter presents much more of the detail, and adds the behavioural interpretation, which has also been described by Bolton, Davies and Woodcock [5].

A more general Z ADT, which mentions the global state explicitly, is described by Stepney, Cooper and Woodcock [11, 13]. We will discuss this approach further in Chap. 10, as it is closely related to the generalisations introduced there.

The main operators in the Z schema calculus are not monotonic with respect to operation or data refinement. This is discussed by Groves in [8, 9] which also identifies conditions under which components of schema expressions can be safely replaced by their refinements.

References

1. Araki, K., Galloway, A., & Taguchi, K. (Eds.) (1999). *International Conference on Integrated Formal Methods 1999 (IFM'99)*. York: Springer.

2. Bert, D., Bowen, J. P., Henson, M. C., & Robinson, K. (Eds.) (2002). *ZB 2002: Formal Specification and Development in Z and B, 2nd International Conference of B and Z Users. Lecture Notes in Computer Science: Vol. 2272*. Berlin: Springer.
3. Boiten, E. A. (2002) Loose specification and refinement in Z. In Bert et al. [2] (pp. 226–241).
4. Boiten, E. A., & Derrick, J. (2010). Incompleteness of relational simulations in the blocking paradigm. *Science of Computer Programming*, *75*(12), 1262–1269.
5. Bolton, C., Davies, J., & Woodcock, J. C. P. (1999) On the refinement and simulation of data types and processes. In Araki et al. [1] (pp. 273–292).
6. Bowen, J. P., Fett, A., & Hinchey, M. G. (Eds.) (1998). *ZUM'98: the Z Formal Specification Notation. Lecture Notes in Computer Science: Vol. 1493*. Berlin: Springer.
7. Dick, J., & Faivre, A. (1993). Automating the generation and sequencing of test cases from model-based specifications. In J. C. P. Woodcock & P. G. Larsen (Eds.), *FME'93: Industrial-Strength Formal Methods. Lecture Notes in Computer Science: Vol. 428* (pp. 268–284). Berlin: Springer.
8. Groves, L. (2002). Refinement and the Z schema calculus. *Electronic Notes in Theoretical Computer Science*, *70*(3), 70–93.
9. Groves, L. (2005) Practical data refinement for the Z schema calculus. In Treharne et al. [12] (pp. 393–413).
10. Grundy, J., Schwenke, M., & Vickers, T. (Eds.) (1998). *International Refinement Workshop & Formal Methods Pacific '98*, Canberra, September 1998. *Discrete Mathematics and Theoretical Computer Science*. Berlin: Springer.
11. Stepney, S., Cooper, D., & Woodcock, J. C. P. (1998) More powerful data refinement in Z. In Bowen et al. [6] (pp. 284–307).
12. Treharne, H., King, S., Henson, M. C., & Schneider, S. A. (Eds.) (2005). *ZB 2005: Formal Specification and Development in Z and B, 4th International Conference of B and Z Users. Lecture Notes in Computer Science: Vol. 3455*. Berlin: Springer.
13. Woodcock, J. C. P. (1998) Industrial-strength refinement. In Grundy et al. [10] (pp. 33–44).
14. Woodcock, J. C. P., & Davies, J. (1996). *Using Z: Specification, Refinement, and Proof.* New York: Prentice Hall.

Chapter 5
Calculating Refinements

In the examples we have looked at so far we have used simulations to verify refinements. This has involved writing down the concrete specification, postulating a retrieve relation, and then verifying the conditions necessary for a refinement to hold.

In this chapter we discuss an alternative approach to refinement where we move the emphasis from verification to calculation. That is, instead of writing down the concrete operations and verifying that they are refinements, we will *calculate* the operations and initialisation. To do this, all that is needed is the abstract specification, a description of the concrete state space and a retrieve relation which links the abstract to concrete. The result of this calculation will be the most general data refinement of the abstract specification with respect to the concrete state space and retrieve relation used.

In fact *any* refinement can be expressed as the most general data refinement with respect to the particular retrieve relation followed by operation refinement (which may weaken preconditions or resolve non-determinism). Hence using a calculational approach we can clearly separate out the refinement due to the retrieve relation (which can be calculated) from the reduction of non-determinism and the weakening of preconditions in an operation refinement.

There are clear advantages in moving effort from verification to calculation in terms of complexity and automation of the process—provided the calculations are simple enough. We shall be interested then in finding the simplest means to calculate both upward and downward simulations of a given specification. Whether we can perform a simple calculation will turn out to depend upon the retrieve relation used. In particular, if the retrieve relation is functional (from concrete to abstract) then the calculations become extremely simple to perform.

The following simple example of a protocol specification illustrates the general approach.

Excerpts from pp. 2–4, 6–8 of [8] are reprinted within this chapter with permission from Elsevier.

J. Derrick, E.A. Boiten, *Refinement in Z and Object-Z*,
DOI 10.1007/978-1-4471-5355-9_5, © Springer-Verlag London 2014

Example 5.1 Consider the following very abstract version of the Multiprocessor Shared-Memory Information Exchange (MSMIE), which is a protocol used in embedded software designed for nuclear installations. In our simple description we assume that information is being transferred from a single slave processor to a collection of master processors.

At the first very abstract level our state consists of a set of processors (ms) which are reading information written by a single slave processor. The protocol uses multiple buffering to ensure that data is not prematurely overwritten.

The buffers are used to ensure that neither reading nor writing processes have to wait for a buffer to become available. At any one time one of the buffers will be available for writing, one for reading and the third will either contain the newest information or be idle. However, in our very abstract first specification we do not model the status of the buffers directly but use a Boolean (b_A) which simply records whether a read has ever occurred on the buffers.

We consider here just three of the possible operations: *Slave*, *Release* and *Acquire*. *Slave* is executed when a write finishes and resets the status of the buffer to *true*. *Acquire* is executed when a read begins and the set of reading processors is then updated. *Release* is invoked when a read ends, and the operation removes the reader from the set of reading processors (ms). Initially no read has occurred and the set ms is empty.

$[MNAME]$

$State_A$

$b_A : \mathbb{B}$
$ms : \mathbb{P}\, MNAME$

$\neg b_A \Rightarrow ms = \varnothing$

$Init_A$

$State'_A$

$\neg b'_A \wedge ms' = \varnothing$

$Slave_A$

$\Delta State_A$

$b'_A \wedge ms' = ms$

$Acquire_A$

$\Delta State_A$
$l? : MNAME$

$b_A \wedge b'_A$
$l? \notin ms \wedge ms' = ms \cup \{l?\}$

$Release_A$

$\Delta State_A$
$l? : MNAME$

$l? \in ms \wedge ms' = ms \setminus \{l?\}$
$(ms' = \varnothing \wedge b_A) \Rightarrow b'_A$
$(ms' \neq \varnothing) \Rightarrow (b_A = b'_A)$

This specification is exceptionally abstract, and we wish to refine it to include a description of the status of each buffer. In this more detailed design the protocol is described more explicitly by the three buffers. Each of the three buffers has a status: s, m, i or n, which represent: assigned to slave, assigned to master, idle and newest respectively. The buffers can be in one of only four configurations: *sii*, *sin*, *sim* and *snm*, and we use these as values in our state space. The more detailed design can be written as

$$CONFIG ::= sii \mid sin \mid sim \mid snm$$

$$\begin{array}{|l}
\textit{State}_C \\
\hline
b_C : CONFIG \\
ms : \mathbb{P}\, MNAME \\
\hline
b_C \in \{sii, sin\} \Leftrightarrow ms = \varnothing \\
\hline
\end{array}$$

$$\begin{array}{|l}
\textit{Init}_C \\
\hline
State'_C \\
\hline
b'_C = sii \wedge ms' = \varnothing \\
\hline
\end{array}$$

$$\begin{array}{|l}
\textit{Slave}_C \\
\hline
\Delta State_C \\
\hline
ms' = ms \\
b_C \in \{sii, sin\} \Rightarrow b'_C = sin \\
b_C \in \{sim, snm\} \Rightarrow b'_C = snm \\
\hline
\end{array}$$

$$\begin{array}{|l}
\textit{Acquire}_C \\
\hline
\Delta State_C \\
l? : MNAME \\
\hline
b_C \neq sii \wedge l? \notin ms \\
ms' = ms \cup \{l?\} \\
b'_C \in \{sin, sim, snm\} \\
\hline
\end{array}$$

$$\begin{array}{|l}
\textit{Release}_C \\
\hline
\Delta State_C \\
l? : MNAME \\
\hline
l? \in ms \wedge ms' = ms \setminus \{l?\} \\
(ms' = \varnothing \wedge b_C \neq sii) \Rightarrow b'_C \neq sii \\
(ms' \neq \varnothing) \Rightarrow b_C = b'_C \\
\hline
\end{array}$$

The retrieve relation between the two specifications is R:

$$\begin{array}{|l}
R \\
\hline
State_A \\
State_C \\
\hline
\neg b_A \Leftrightarrow b_C = sii \\
\hline
\end{array}$$

and it is possible to verify the downward simulation conditions for the initialisation and every operation.

However, instead of this verification we could calculate the concrete operations from the retrieve relation and state spaces. This will save the need for any verification, the concrete operations will automatically be refinements. For example, the concrete *Acquire* operation can be calculated by setting:

$$Acquire == \exists \Delta State_A \bullet R \land Acquire_A \land R'$$

which upon expansion is the same as $Acquire_C$. □

When the retrieve relation is a function from concrete to abstract the calculations will always be this simple. However, if we wish to use a relation instead of a function, the necessary calculations become slightly more complex, and in some cases we are not even guaranteed that any refinement will be possible. The next sections derive these results in more detail.

5.1 Downward Simulations

In this section we develop rules for calculating downward simulations. To do so we work in the relational setting as discussed in Chap. 3, giving the results in Z as corollaries. Suppose we are given a specification of an abstract data type $\mathsf{A} = (\mathsf{AState}, \mathsf{AInit}, \{\mathsf{AOp_i}\}_{\mathsf{i} \in \mathsf{I}}, \mathsf{AFin})$, a concrete state space CState together with a retrieve relation R between AState and CState. We aim to calculate the most general refinement of A, that is calculate the initialisation, finalisation and concrete operations.

The calculations can be found by considering the most general solutions to the simulation requirements in Definition 3.6 given above. Therefore, the most general (i.e., weakest) solution for a downward simulation is given by:

$$\begin{aligned}
&\widehat{\mathsf{CInit}} = \widehat{\mathsf{AInit}} \,_{9}^{\circ}\, \widetilde{\mathsf{R}} \\
&\widetilde{\mathsf{R}} \,_{9}^{\circ}\, \widehat{\mathsf{CFin}} = \widehat{\mathsf{AFin}} \\
&\widetilde{\mathsf{R}} \,_{9}^{\circ}\, \widehat{\mathsf{COp_i}} = \widehat{\mathsf{AOp_i}} \,_{9}^{\circ}\, \widetilde{\mathsf{R}} \quad \text{for each index } \mathsf{i} \in \mathsf{I}
\end{aligned}$$

which have explicit solutions:

$$\begin{aligned}
&\widehat{\mathsf{CInit}} = \widehat{\mathsf{AInit}} \,_{9}^{\circ}\, \widetilde{\mathsf{R}} \\
&\widehat{\mathsf{CFin}} = \widehat{\mathsf{AFin}} / \widetilde{\mathsf{R}} \\
&\widehat{\mathsf{COp_i}} = (\widehat{\mathsf{AOp_i}} \,_{9}^{\circ}\, \widetilde{\mathsf{R}}) / \widetilde{\mathsf{R}} \quad \text{for each index } \mathsf{i} \in \mathsf{I}
\end{aligned}$$

where $\mathsf{X}/\mathsf{R} = \overline{(\mathsf{R}^{-1} \,_{9}^{\circ}\, \overline{\mathsf{X}})}$, see p. 14. In order to apply this to Z we need to simplify these conditions and to extract the calculation on the underlying partial relations.

Extracting the calculations for the initialisation and finalisation is easy since we know that $\widehat{\mathsf{CInit}} = \widehat{\mathsf{AInit}} \,_{9}^{\circ}\, \widetilde{\mathsf{R}}$ if and only if $\mathsf{CInit} = \mathsf{AInit} \,_{9}^{\circ}\, \mathsf{R}$, and $\widehat{\mathsf{CFin}} = \widehat{\mathsf{AFin}} / \widetilde{\mathsf{R}}$ if and only if $\mathsf{CFin} = \mathsf{AFin}/\mathsf{R}$. Therefore the weakest concrete initialisation and finalisation are given by

$$\begin{aligned}
&\mathsf{CInit} = \mathsf{AInit} \,_{9}^{\circ}\, \mathsf{R} \\
&\mathsf{CFin} = \mathsf{AFin}/\mathsf{R}
\end{aligned}$$

These are the correct initialisation and finalisation, if they exist. However, some relations R will be too small or too large to define refinements, and thus for CInit and CFin to be correct refinements we need to check that the concrete initialisation is satisfiable and that finalisation applicability holds:

$$\mathsf{AInit} \mathbin{⨾} \mathsf{R} \neq \varnothing$$
$$\operatorname{ran}((\operatorname{dom} \mathsf{AFin}) \lhd \mathsf{R}) \subseteq \operatorname{dom}(\mathsf{AFin}/\mathsf{R})$$

To calculate the concrete operations, note that $\widehat{\mathsf{COp_i}} = (\widehat{\mathsf{AOp_i}} \mathbin{⨾} \widetilde{\mathsf{R}})/\widetilde{\mathsf{R}}$ can be re-written as two conditions.

$$(\operatorname{dom} \mathsf{AOp_i} \lhd \mathsf{R} \mathbin{⨾} \mathsf{COp_i}) = \mathsf{AOp_i} \mathbin{⨾} \mathsf{R}$$
$$\operatorname{ran}((\operatorname{dom} \mathsf{AOp_i}) \lhd \mathsf{R}) = \operatorname{dom} \mathsf{COp_i}$$

Therefore $\mathsf{COp_i}$ is given by the weakest solution which is:

$$\mathsf{COp_i} = \operatorname{ran}((\operatorname{dom} \mathsf{AOp_i}) \lhd \mathsf{R}) \lhd ((\mathsf{AOp_i} \mathbin{⨾} \mathsf{R})/(\operatorname{dom} \mathsf{AOp_i} \lhd \mathsf{R}))$$

However, for a partial operation we also need to check applicability, and only if this concrete operation satisfies the applicability condition does a downward simulation exist. This can be summarised as the following theorem.

Theorem 5.1 *The weakest data type that is a downward simulation of* A *with respect to* R *is given by*

$$\mathsf{CInit} = \mathsf{AInit} \mathbin{⨾} \mathsf{R}$$
$$\mathsf{CFin} = \mathsf{AFin}/\mathsf{R}$$
$$\mathsf{COp_i} = \operatorname{ran}((\operatorname{dom} \mathsf{AOp_i}) \lhd \mathsf{R}) \lhd ((\mathsf{AOp_i} \mathbin{⨾} \mathsf{R})/(\operatorname{dom} \mathsf{AOp_i} \lhd \mathsf{R}))$$

whenever

$$\mathsf{AInit} \mathbin{⨾} \mathsf{R} \neq \varnothing$$
$$\operatorname{ran}((\operatorname{dom} \mathsf{AOp_i}) \lhd \mathsf{R}) \subseteq \operatorname{dom} \mathsf{COp_i}$$
$$\operatorname{ran}((\operatorname{dom} \mathsf{AFin}) \lhd \mathsf{R}) \subseteq \operatorname{dom}(\mathsf{AFin}/\mathsf{R})$$

If the latter do not hold then no downward simulation is possible for this A *and* R. □

We can describe these results in the schema calculus. Note that because for embeddings of Z ADTs AFin is total the requirement $\operatorname{ran}((\operatorname{dom} \mathsf{AFin}) \lhd \mathsf{R}) \subseteq \operatorname{dom}(\mathsf{AFin}/\mathsf{R})$ is trivially satisfied.

Corollary 5.1 *Let* $(AState, AInit, \{AOp\})$ *be an ADT, CState a state space, and R a retrieve relation between AState and CState. Let*

$$CInit == \exists AState' \bullet AInit \land R'$$
$$COp == (\exists AState \bullet \operatorname{pre} AOp \land R) \land$$
$$\qquad (\forall AState \bullet \operatorname{pre} AOp \land R \Rightarrow \exists AState' \bullet AOp \land R')$$

Then a downward simulation exists for this abstract operation and retrieve relation if, and only if,

$$\exists AInit;\ CState' \bullet R'$$
$$\text{pre}\, AOp \wedge R \Rightarrow \text{pre}\, COp$$

Furthermore, the weakest such downward simulation is given by $(CState, CInit, \{COp\})$. □

In other words, we can always calculate downward simulations when they exist. However, we can do significantly better than this under certain circumstances. In particular, if the retrieve relation is functional (from concrete to abstract) then we can use simpler calculations as in Example 5.1. In fact, the retrieve relation does not have to be completely functional, it is sufficient that it is functional on a restricted domain. The following result proves that the simplification to the calculation can be made under these conditions.

Theorem 5.2 *For* A, CState *and* R *as above, let*

$$\mathsf{COp} = \text{ran}(\text{dom}\,\mathsf{AOp} \lhd \mathsf{R}) \lhd ((\mathsf{AOp} \mathbin{;} \mathsf{R})/(\text{dom}\,\mathsf{AOp} \lhd \mathsf{R}))$$

Then $\mathsf{COp} \subseteq \mathsf{R}^{-1} \mathbin{;} \mathsf{AOp} \mathbin{;} \mathsf{R}$. *Furthermore if* $\text{ran}((\text{dom}\,\mathsf{AOp}) \lhd \mathsf{R}) \subseteq \text{dom}\,\mathsf{COp}$ *and* $\text{dom}\,\mathsf{AOp} \lhd \mathsf{R} \mathbin{;} \mathsf{R}^{-1} \subseteq \mathsf{id}$ *then* $\mathsf{COp} = \mathsf{R}^{-1} \mathbin{;} \mathsf{AOp} \mathbin{;} \mathsf{R}$.

Proof We first of all show that $\mathsf{COp} \subseteq \mathsf{R}^{-1} \mathbin{;} \mathsf{AOp} \mathbin{;} \mathsf{R}$. Let $(a, b) \in \mathsf{COp}$. Then

$$\begin{array}{l}(\exists s \bullet (s, a) \in (\text{dom}\,\mathsf{AOp} \lhd \mathsf{R})) \\ \qquad \wedge\ (\forall s \bullet (s, a) \notin (\text{dom}\,\mathsf{AOp} \lhd \mathsf{R}) \vee (s, b) \in (\mathsf{AOp} \mathbin{;} \mathsf{R}))\end{array}$$

Hence there exists an s such that $(s, a) \in (\text{dom}\,\mathsf{AOp} \lhd \mathsf{R})$ and $(s, b) \in (\mathsf{AOp} \mathbin{;} \mathsf{R})$, and therefore $(a, b) \in \mathsf{R}^{-1} \mathbin{;} \mathsf{AOp} \mathbin{;} \mathsf{R}$.

Next we show that whenever $\text{ran}((\text{dom}\,\mathsf{AOp}) \lhd \mathsf{R}) \subseteq \text{dom}\,\mathsf{COp}$ then $\mathsf{R}^{-1} \mathbin{;} \mathsf{AOp} \mathbin{;} \mathsf{R} \subseteq \mathsf{COp}$. To do so suppose that $(a, b) \in \mathsf{R}^{-1} \mathbin{;} \mathsf{AOp} \mathbin{;} \mathsf{R}$, then there exists s such that $(s, a) \in (\text{dom}\,\mathsf{AOp} \lhd \mathsf{R})$ and $(s, b) \in (\mathsf{AOp} \mathbin{;} \mathsf{R})$. We have to show that $\forall u \bullet (u, a) \notin (\text{dom}\,\mathsf{AOp} \lhd \mathsf{R}) \vee (u, b) \in (\mathsf{AOp} \mathbin{;} \mathsf{R})$. Consider any u with $(u, a) \in (\text{dom}\,\mathsf{AOp} \lhd \mathsf{R})$, it suffices to show that $(u, b) \in (\mathsf{AOp} \mathbin{;} \mathsf{R})$.

Since $(u, a) \in (\text{dom}\,\mathsf{AOp} \lhd \mathsf{R})$ and $(a, b) \in \mathsf{R}^{-1} \mathbin{;} \mathsf{AOp} \mathbin{;} \mathsf{R}$, we find that $(u, b) \in \text{dom}\,\mathsf{AOp} \lhd \mathsf{R} \mathbin{;} \mathsf{R}^{-1} \mathbin{;} \mathsf{AOp} \mathbin{;} \mathsf{R} \subseteq \mathsf{AOp} \mathbin{;} \mathsf{R}$ since $\text{dom}\,\mathsf{AOp} \lhd \mathsf{R} \mathbin{;} \mathsf{R}^{-1} \subseteq \mathsf{id}$.

Therefore $\forall u \bullet (u, a) \notin (\text{dom}\,\mathsf{AOp} \lhd \mathsf{R}) \vee (u, b) \in (\mathsf{AOp} \mathbin{;} \mathsf{R})$. We also know that $\exists s \bullet (s, a) \in (\text{dom}\,\mathsf{AOp} \lhd \mathsf{R})$ and $(s, b) \in (\mathsf{AOp} \mathbin{;} \mathsf{R})$. Thus by the definition of COp, $(a, b) \in \mathsf{COp}$. □

In fact if we inspect this proof we can see there are a number of sufficient conditions for the simplification to hold. The key part is to be able to deduce that

$$\text{dom}\,\mathsf{AOp} \lhd \mathsf{R} \mathbin{;} \mathsf{R}^{-1} \mathbin{;} \mathsf{AOp} \mathbin{;} \mathsf{R} \subseteq \mathsf{AOp} \mathbin{;} \mathsf{R}$$

Clearly $\text{dom}\,\mathsf{AOp} \lhd \mathsf{R} \fatsemi \mathsf{R}^{-1} \subseteq \mathsf{id}$ is one sufficient condition, and another is if $\mathsf{R}^{-1} \fatsemi \mathsf{AOp} \fatsemi \mathsf{R}$ is deterministic. The latter condition holds, for example, if AOp is deterministic and R is functional from abstract to concrete. Note also that there is no requirement of surjectivity on the retrieve relation, i.e., not every abstract state needs to be linked to a concrete state.

The consequences of this theorem are the following. For the simpler calculation $\mathsf{COp} = \mathsf{R}^{-1} \fatsemi \mathsf{AOp} \fatsemi \mathsf{R}$ to be used

- it is not necessary that R^{-1} is a function, being functional on the smaller set $\text{ran}(\text{dom}\,\mathsf{AOp} \lhd \mathsf{R})$ is sufficient;
- even if this is not the case, $\mathsf{R}^{-1} \fatsemi \mathsf{AOp} \fatsemi \mathsf{R}$ being deterministic is a sufficient condition.

As we shall see in a moment these are sufficient, but not necessary, conditions, i.e., there are occasions where the simplified calculation can be used even when the above conditions on R fail. For application to Z specifications the simplification takes the following form.

Corollary 5.2 *Given an ADT, a state space and a retrieve relation as in Corollary* 5.1. *Suppose that R is a function from CState to AState for all concrete states such that* $\exists AState \bullet \text{pre}\,AOp \land R$. *Then the most general downward simulation can be calculated as*:

$$CInit == \exists AState' \bullet AInit \land R'$$
$$COp == \exists AState; AState' \bullet (R \land AOp \land R')$$

whenever a downward simulation exists (which is guaranteed when R is functional from concrete to abstract and $\exists AInit; CState' \bullet R'$). □

Example 5.2 The retrieve relation in Example 5.1 is functional from concrete to abstract. Therefore we can calculate the weakest concrete initialisation and operations using Corollary 5.2. We calculated *Acquire* above, the schemas *Init* and *Slave* are

$$Init == \exists\, State'_A \bullet Init_A \land R'$$
$$Slave == \exists\, \Delta State_A \bullet (R \land Slave_A \land R')$$

which simplify to

Init
$State'_C$
$b'_C = sii \land ms' = \varnothing$

Slave
$\Delta State_C$
$ms' = ms$
$b'_C \in \{sin, sim, snm\}$

Slave is the weakest refinement of $Slave_A$, and indeed we can easily see that the operation $Slave_C$ is a refinement of *Slave*. □

5.1.1 Non-functional Retrieve Relations

Not every retrieve relation is functional, and we have to work slightly harder to derive the concrete operations in these cases.

Example 5.3 Consider an abstract specification consisting of two sequences s and t (both initially empty). There are two operations: $push_A$ and pop_A. The $push_A$ operation has two inputs and pushes $m?$ into one of the sequences according to whether $i?$ is 1 or 0. The pop_A operation non-deterministically pops one of the sequences when at least one is non-empty, and outputs an error message if they are both empty. This operation is specified as the disjunction of two operations pop_{OkA} and pop_{ErrorA}. The specification is as follows.

$$REPORT ::= error \mid no\_error$$

$$\begin{array}{|l}\hline AState \\ s, t : \mathrm{seq}\,\mathbb{N} \\ \hline\end{array}$$

$$\begin{array}{|l}\hline AInit \\ AState' \\ \hline s' = \langle\rangle \wedge t' = \langle\rangle \\ \hline\end{array}$$

$$\begin{array}{|l}\hline push_A \\ \Delta AState \\ m? : \mathbb{N} \\ i? : \{0, 1\} \\ \hline (i? = 0 \wedge t' = t \frown \langle m? \rangle \wedge s' = s) \vee \\ (i? = 1 \wedge s' = s \frown \langle m? \rangle \wedge t' = t) \\ \hline\end{array}$$

$$\begin{array}{|l}\hline pop_{OkA} \\ \Delta AState \\ n! : \mathbb{N} \\ report! : REPORT \\ \hline (t = t' \frown \langle n! \rangle \wedge s' = s) \vee (s = s' \frown \langle n! \rangle \wedge t' = t) \\ report! = no\_error \\ \hline\end{array}$$

$$\begin{array}{|l}\hline pop_{ErrorA} \\ \Xi AState \\ report! : REPORT \\ \hline s = \langle\rangle \wedge t = \langle\rangle \wedge report! = error \\ \hline\end{array}$$

$$pop_A == pop_{OkA} \vee pop_{ErrorA}$$

We can refine this to a concrete specification whose state space consists of a single sequence u. Thus the two separate sequences were actually unnecessary in terms of the observable behaviour which consists of output values that are just some valid merge of the input streams. So the non-determinism in the pop_{OkA} operation about which sequence is popped can be replaced by the non-determinism of taking any valid merge of s and t in a single sequence. The concrete specification will have the following state space.

```
┌─ CState ────────────────────────────
│ u : seq ℕ
└─────────────────────────────────────
```

The retrieve relation is

```
┌─ R ─────────────────────────────────
│ AState
│ CState
├──────────
│ u_merge(s, t)
└─────────────────────────────────────
```

where the predicate in the retrieve relation defines a merge of the two sequences s and t, and, for example, has the properties: $u_{merge}(s, \langle\rangle)$ iff $u = s$ and $u_{merge}(s, t)$ iff $u_{merge}(t, s)$. This relation is not functional: for every u there are many choices of s and t with $u_{merge}(s, t)$.

Calculating the concrete initialisation gives $CInit == [CState' \mid u' = \langle\rangle]$. In order to calculate the concrete push operation we note that since $push_A$ is total, the retrieve relation R is not functional (from concrete to abstract) on $\exists AState \bullet (\text{pre}\, push_A \wedge R)$ which is the whole of $CState$. Furthermore, although $push_A$ is deterministic, R is not functional from abstract to concrete. Therefore we cannot automatically use the simple calculation $push_C == \exists t, s, t', s' : \text{seq}\,\mathbb{N} \bullet (R \wedge push_A \wedge R')$. Thus we have to calculate $push_C$ by

$$\begin{aligned} push_C == &\ (\exists s, t : \text{seq}\,\mathbb{N} \bullet \text{pre}\, push_A \wedge R) \\ &\wedge (\forall s, t \bullet \text{pre}\, push_A \wedge R \Rightarrow \exists s', t' \bullet push_A \wedge R') \end{aligned}$$

which evaluates to

```
┌─ push_C ────────────────────────────
│ ΔCState
│ m? : ℕ
├──────────
│ ∃ t, s : seq ℕ • u_merge(t, s) ∧ u'_merge(t ⁀ ⟨m?⟩, s)
└─────────────────────────────────────
```

For the pop operation we note that the retrieve relation is functional on $\exists AState \bullet (\text{pre}\, pop_{ErrorA} \wedge R)$, i.e., R links $u = \langle\rangle$ to only one abstract state (namely when both

s and t are also empty). Therefore the simple calculation

$$pop_{ErrorC} == \exists t, s, t', s' : \operatorname{seq} \mathbb{N} \bullet (R \wedge pop_{ErrorA} \wedge R')$$

can be used. This evaluates to

$$\begin{array}{|l}
\hline
pop_{ErrorC} \\
\Xi CState \\
report! : REPORT \\
\hline
u = \langle\rangle \wedge report! = error \\
\hline
\end{array}$$

Incidentally, this example shows that the condition of functionality is not necessary since, for example, the calculation

$$pop_{OkC} == \exists t, s, t', s' : \operatorname{seq} \mathbb{N} \bullet (R \wedge pop_{OkA} \wedge R')$$

evaluates to the same schema as the more complex calculation, namely

$$\begin{array}{|l}
\hline
pop_{OkC} \\
\Delta CState \\
n! : \mathbb{N} \\
report! : REPORT \\
\hline
\exists t, s : \operatorname{seq} \mathbb{N} \bullet u_{merge}(t \frown \langle n! \rangle, s) \wedge u'_{merge}(t, s) \\
report! = no_error \\
\hline
\end{array}$$

This is the weakest refinement, one obvious further refinement of this would be the operation which pops the output off the end of the sequence.

$$\begin{array}{|l}
\hline
pop \\
\Delta CState \\
n! : \mathbb{N} \\
report! : REPORT \\
\hline
u = u' \frown \langle n! \rangle \\
report! = no_error \\
\hline
\end{array}$$

Aside The astute reader will note that the calculation of $push_C$ should in fact be

$$\begin{array}{|l}
\hline
push_C \\
\Delta CState \\
m? : \mathbb{N} \\
i? : \{0, 1\} \\
\hline
\exists t, s : \operatorname{seq} \mathbb{N} \bullet u_{merge}(t, s) \wedge u'_{merge}(t \frown \langle m? \rangle, s) \\
\hline
\end{array}$$

However, because the input $i?$ is redundant (it has no visible effect whatsoever) we have removed it. Strictly speaking this is not a valid refinement in the theory presented so far, but Chap. 10 will show why this is a correct refinement step to perform. □

In this example we found that although we could not use the simpler calculation straight away, the more complex one produced the same evaluation. The following example shows that this is not always the case.

Example 5.4 Continuing with the same example as in Example 5.3 we add a further abstract operation Op_A. For the sake of simplicity the domain of the operation is restricted to a couple of values.

Op_A
$\Delta AState$

$(s' = s \vee s = \langle 1 \rangle \wedge s' = \langle 0 \rangle) \wedge t' = t$
$(s, t) \in \{(\langle\,\rangle, \langle 1 \rangle), (\langle 1 \rangle, \langle\,\rangle)\}$

With the given retrieve relation a refinement is still possible for this operation, and the complex and simple calculations are

Op_C
$\Delta CState$

$u = u' \wedge u = \langle 1 \rangle$

$\exists\, \Delta State_A \bullet (R \wedge Op_A \wedge R')$
$\Delta CState$

$u' = \langle 1 \rangle \wedge u \in \{\langle 0 \rangle, \langle 1 \rangle\}$

The complex calculation is the correct weakest refinement, but $\exists\, \Delta State_A \bullet (R \wedge Op_A \wedge R')$ fails to be a refinement because it has included the mapping $\langle 0 \rangle \mapsto \langle 1 \rangle$. This has arisen because some of the non-determinism has been resolved by the retrieve relation. □

Figure 5.1 illustrates typical scenarios when the simpler calculation fails to be a refinement. In each case the correct weakest refinement COp is shown. The calculation $\mathsf{COp} = \mathsf{R}^{-1} \operatorname{;} \mathsf{AOp} \operatorname{;} \mathsf{R}$ would have included a mapping between a and b which invalidates the refinement.

Our final example in this section illustrates that for some operations no refinements exist at all for the chosen retrieve relation.

Example 5.5 This example extends Example 4.3 (p. 110). The abstract specification evaluates the mean of a set of real numbers in a bag.

$AState$
$b : \text{bag}\, \mathbb{R}$

$AInit$
$AState'$

$b' = [\![\,]\!]$

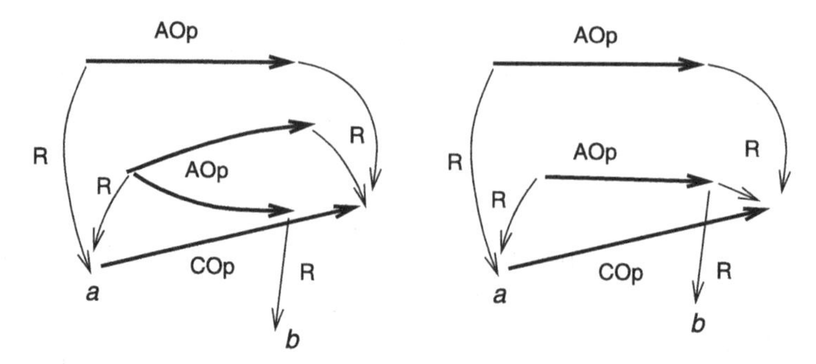

Fig. 5.1 Calculating the weakest refinement

$$
\begin{array}{|l}
\textit{Enter}_A \\
\hline
\Delta \textit{AState} \\
r? : \mathbb{R} \\
\hline
b' = b \uplus [\![r?]\!] \\
\hline
\end{array}
\qquad
\begin{array}{|l}
\textit{Mean}_A \\
\hline
\Xi \textit{AState} \\
m! : \mathbb{R} \\
\hline
b \neq [\![\,]\!] \land m! = (\textit{sum}\ b) / \textit{count}\ b \\
\hline
\end{array}
$$

We can refine this to a concrete specification which consists of a running total s and number n of items in the bag. The concrete state space and (non-functional) retrieve relation are:

$$
\begin{array}{|l}
\textit{CState} \\
\hline
s : \mathbb{R} \\
n : \mathbb{N} \\
\hline
\end{array}
$$

$$
\begin{array}{|l}
R \\
\hline
\textit{AState} \\
\textit{CState} \\
\hline
s = \textit{sum}\ b \land n = \textit{count}\ b \\
\hline
\end{array}
$$

The calculations of concrete operations proceed as normal. For example, the concrete *Mean* operation is

$$
\begin{array}{|l}
\textit{Mean}_C \\
\hline
\Xi \textit{CState} \\
m! : \mathbb{R} \\
\hline
n \neq 0 \land m! = s/n \\
\hline
\end{array}
$$

Consider, however, the abstract operation to subtract a fixed member from the bag ($b \uminus [\![3]\!]$ will take off one item from the bag b if possible, and leaves the bag unchanged otherwise).

$Minus_A$
$\Delta AState$

$b' = b \uplus [\![3]\!]$

There is no valid downward simulation of $Minus_A$ with respect to the chosen retrieve relation. To see this we have to calculate the candidate possibility $Minus_C$, and show that $\mathrm{pre}\,Minus_A \wedge R$ does not imply $\mathrm{pre}\,Minus_C$.

The domain of $Minus_C$ excludes the element $(s, n) = (5, 3)$, since some three element bags which sum to 5 contain the element 3 and some do not. However, $(5, 3)$ is in $\mathrm{pre}\,Minus_A \wedge R$ since $[\![1, 1, 3]\!]$ is in the domain of $Minus_A$ and $[\![1, 1, 3]\!]$ is linked to $(5, 3)$ via the retrieve relation. Thus we can conclude that no valid refinement of $Minus_A$ exists for this R. □

5.2 Upward Simulations

Turning to the case of upward simulations, we can produce analogous results to those in the previous section. The most general solution for an upward simulation (cf. Definition 3.7) is given by

$$\widehat{\mathsf{CInit}} = \widetilde{\mathsf{T}} \setminus \widehat{\mathsf{AInit}}$$
$$\widehat{\mathsf{CFin}} = \widetilde{\mathsf{T}} \fatsemi \widehat{\mathsf{AFin}}$$
$$\widehat{\mathsf{COp_i}} = \widetilde{\mathsf{T}} \setminus (\widetilde{\mathsf{T}} \fatsemi \widehat{\mathsf{AOp_i}}) \quad \text{for each index } \mathsf{i} \in \mathsf{I}$$

where $\mathsf{T} \setminus \mathsf{X} = \overline{(\overline{\mathsf{X}} \fatsemi \mathsf{T}^{-1})}$, see p. 14. As before we have to extract the calculations for the underlying partial relations. For example, the calculations for the initialisation and finalisation are given by

$$\mathsf{CInit} = \mathsf{T} \setminus \mathsf{AInit}$$
$$\mathsf{CFin} = \mathsf{T} \fatsemi \mathsf{AFin}$$

As in the case of downward simulation we need to check such an T defines a refinement, and this requires that

$$\mathsf{T} \setminus \mathsf{AInit} \neq \varnothing$$
$$\mathsf{T} \text{ is total on } \mathsf{CState}$$

In order to calculate the concrete operations, we replace the equation $\widehat{\mathsf{COp_i}} = \widetilde{\mathsf{T}} \setminus (\widetilde{\mathsf{T}} \fatsemi \widehat{\mathsf{AOp_i}})$ by two equivalent conditions.

$$\mathrm{dom}(\mathsf{T} \rhd (\mathrm{dom}\,\mathsf{AOp_i})) \lhd \mathsf{COp_i} \fatsemi \mathsf{T} = \mathsf{T} \fatsemi \mathsf{AOp_i}$$
$$\overline{\mathrm{dom}\,\mathsf{COp_i}} \subseteq \mathrm{dom}(\mathsf{T} \rhd (\mathrm{dom}\,\mathsf{AOp_i}))$$

Therefore the weakest solution, if one exists, is given by

$$\mathrm{dom}(\mathsf{T} \rhd (\mathrm{dom}\,\mathsf{AOp_i})) \lhd \mathsf{COp_i} = \mathsf{T} \setminus (\mathsf{T} \,{}_{9}^{o}\, \mathsf{AOp_i})$$

and hence

$$\mathsf{COp_i} = \mathrm{dom}(\mathsf{T} \rhd (\mathrm{dom}\,\mathsf{AOp_i})) \lhd (\mathsf{T} \setminus (\mathsf{T} \,{}_{9}^{o}\, \mathsf{AOp_i}))$$

This will be a refinement whenever this $\mathsf{COp_i}$ satisfies the applicability condition $\overline{\mathrm{dom}\,\mathsf{COp_i}} \subseteq \overline{\mathrm{dom}(\mathsf{T} \rhd (\mathrm{dom}\,\mathsf{AOp_i}))}$. If $\mathsf{COp_i}$ is total this condition is true since in this case $\overline{\mathrm{dom}\,\mathsf{COp}} = \varnothing$. If the applicability condition fails then no upward simulation is possible for this A and T. We can summarise this as follows.

Theorem 5.3 *The weakest data type that is an upward simulation of* A *with respect to* T *is given by*

$$\begin{aligned}
&\mathsf{CInit} = \mathsf{T} \setminus \mathsf{AInit} \\
&\mathsf{CFin} = \mathsf{T} \,{}_{9}^{o}\, \mathsf{AFin} \\
&\mathsf{COp_i} = \mathrm{dom}(\mathsf{T} \rhd (\mathrm{dom}\,\mathsf{AOp_i})) \lhd (\mathsf{T} \setminus (\mathsf{T} \,{}_{9}^{o}\, \mathsf{AOp_i}))
\end{aligned}$$

whenever

$$\begin{aligned}
&\mathsf{T} \setminus \mathsf{AInit} \neq \varnothing \\
&\overline{\mathrm{dom}\,\mathsf{COp_i}} \subseteq \overline{\mathrm{dom}(\mathsf{T} \rhd (\mathrm{dom}\,\mathsf{AOp_i}))} \\
&\mathsf{T} \textit{ is total on } \mathsf{CState}
\end{aligned}$$

If the latter do not hold then no upward simulation is possible for this A *and* T. □

This can be rephrased in terms of the Z schema calculus

Corollary 5.3 *Let (AState, AInit, {AOp}) be an abstract specification, CState a concrete state space and T a retrieve relation between CState and AState. Let*

$$\begin{aligned}
&CInit == \forall AState' \bullet (T' \Rightarrow AInit) \\
&COp == \forall AState \bullet (T \Rightarrow \mathrm{pre}\, AOp) \\
&\qquad\qquad\qquad \wedge (\forall AState' \bullet (T' \Rightarrow \exists AState \bullet T \wedge AOp))
\end{aligned}$$

Then an upward simulation exists for this abstract operation and retrieve relation if and only if

$$\begin{aligned}
&\exists CState' \bullet \forall AState' \bullet (T' \Rightarrow AInit) \\
&(T \Rightarrow \mathrm{pre}\, AOp) \Rightarrow \mathrm{pre}\, COp \\
&T \textit{ is total on } CState
\end{aligned}$$

Furthermore, the weakest such upward simulation is given by (CState, CInit, {COp}).

□

Therefore, as with downward simulations, we can always calculate the most general upward simulation whenever it exists. There is also a similar simplification in the calculation under certain circumstances. As we saw in Chap. 3, upward simulations are total (this follows from the totality of the implicit finalisation and the condition that $\mathsf{CFin} \subseteq \mathsf{T} \mathbin{⨾} \mathsf{AFin}$), and if an upward simulation is functional then it also defines a downward simulation. Therefore if the upward simulation is functional the same calculation as in Corollary 5.2 should be used. However, as with downward simulations the simplification can be made if T is functional on a restricted domain.

Theorem 5.4 *With* A, CState *and* T *as above, let*

$$\mathsf{COp} = \mathrm{dom}(\mathsf{T} \rhd (\mathrm{dom}\,\mathsf{AOp})) ⩤ (\mathsf{T} \setminus (\mathsf{T} \mathbin{⨾} \mathsf{AOp}))$$

Suppose further that T *satisfies* $\mathrm{dom}\,\mathsf{AOp} \lhd \mathsf{T}^{-1} \mathbin{⨾} \mathsf{T} \subseteq \mathsf{id}$ *and* $\mathrm{ran}\,\mathsf{AOp} \lhd \mathsf{T}^{-1} \mathbin{⨾} \mathsf{T} \subseteq \mathsf{id}$. *Then* $\mathsf{COp} = \mathsf{T} \mathbin{⨾} \mathsf{AOp} \mathbin{⨾} \mathsf{T}^{-1}$.

Proof Let $(a, b) \in \mathsf{COp}$. Then we have that $a \notin \mathrm{dom}(\mathsf{T} \rhd (\mathrm{dom}\,\mathsf{AOp}))$ and $(a, b) \in (\mathsf{T} \setminus (\mathsf{T} \mathbin{⨾} \mathsf{AOp}))$. Hence,

$$\forall \beta \bullet (a, \beta) \in \mathsf{T} \Rightarrow \beta \in \mathrm{dom}\,\mathsf{AOp} \quad \text{and}$$
$$\forall c \bullet (b, c) \in \mathsf{T} \Rightarrow (a, c) \in (\mathsf{T} \mathbin{⨾} \mathsf{AOp})$$

By the assumption of totality there exists at least one c with $(b, c) \in \mathsf{T}$, and hence $(a, c) \in (\mathsf{T} \mathbin{⨾} \mathsf{AOp})$. Thus $(a, b) \in \mathsf{T} \mathbin{⨾} \mathsf{AOp} \mathbin{⨾} \mathsf{T}^{-1}$.

For the converse we need the assumption of functionality. Let $(a, b) \in \mathsf{T} \mathbin{⨾} \mathsf{AOp} \mathbin{⨾} \mathsf{T}^{-1}$. To show that $(a, b) \in \mathsf{COp}$ we need to show that:

$$a \notin \mathrm{dom}(\mathsf{T} \rhd (\mathrm{dom}\,\mathsf{AOp})) \quad \text{and}$$
$$(a, b) \in (\mathsf{T} \setminus (\mathsf{T} \mathbin{⨾} \mathsf{AOp}))$$

For the former this amounts to showing that $\forall y \bullet (a, y) \notin \mathsf{T} \vee y \in \mathrm{dom}\,\mathsf{AOp}$. However, since T is a function and $(a, b) \in \mathsf{T} \mathbin{⨾} \mathsf{AOp} \mathbin{⨾} \mathsf{T}^{-1}$ there is precisely one y with $(a, y) \in \mathsf{T}$ and for this y we know that $y \in \mathrm{dom}\,\mathsf{AOp}$.

Showing that $(a, b) \in (\mathsf{T} \setminus (\mathsf{T} \mathbin{⨾} \mathsf{AOp}))$ amounts to showing that $\forall c \bullet (b, c) \in \mathsf{T} \Rightarrow (a, c) \in (\mathsf{T} \mathbin{⨾} \mathsf{AOp})$, and again by the assumptions of functionality of T on appropriate domains this is easily seen to be true. Hence $(a, b) \in \mathsf{COp}$. □

These sufficient conditions can be encapsulated as the following corollary.

Corollary 5.4 *Given an ADT A, a state space CState and a retrieve relation T as above. Suppose that T is functional on* $\exists AState \bullet (T \wedge \mathrm{pre}\,AOp)$ *and* $\exists AState; AState' \bullet (T' \wedge AOp)$. *Then the most general upward simulation can be calculated as*:

$$CInit == \forall AState' \bullet (T' \Rightarrow AInit)$$
$$COp == \exists AState; AState' \bullet (T \wedge AOp \wedge T')$$

whenever an upward simulation exists for this retrieve relation. □

Fig. 5.2 A split node

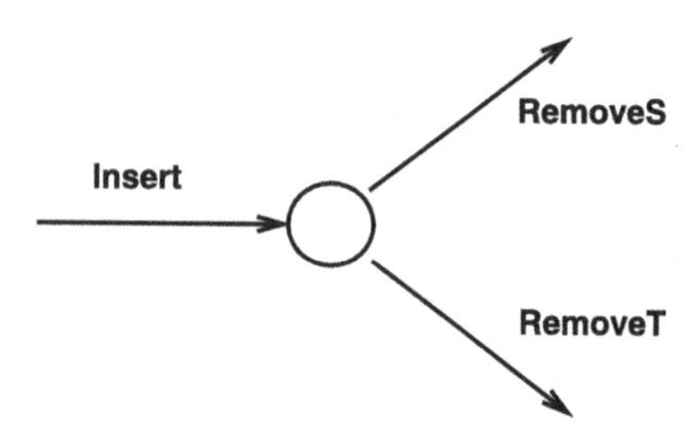

Example 5.6 In this example we use the state spaces from Example 5.3 to describe two specifications of a split node. Such a node receives messages $m?$ via an *Insert* operation and also outputs messages $n!$. Each message is non-deterministically transmitted down one of two paths by *RemoveS* or *RemoveT*, see Fig. 5.2.

Our abstract specification makes the choice of where to output a message when the message is received by non-deterministically appending it to one of two sequences.

$$\begin{array}{|l} \multicolumn{1}{l}{AState} \\ \hline s, t : \operatorname{seq} \mathbb{N} \\ \hline \end{array} \qquad \begin{array}{|l} \multicolumn{1}{l}{AInit} \\ \hline AState' \\ \hline s' = \langle\rangle \land t' = \langle\rangle \\ \hline \end{array}$$

$$\begin{array}{|l} \multicolumn{1}{l}{Insert_A} \\ \hline \Delta AState \\ m? : \mathbb{N} \\ \hline (t' = t \frown \langle m? \rangle \land s' = s) \lor (s' = s \frown \langle m? \rangle \land t' = t) \\ \hline \end{array}$$

$$\begin{array}{|l} \multicolumn{1}{l}{RemoveT_A} \\ \hline \Delta AState \\ n! : \mathbb{N} \\ \hline (t = t' \frown \langle n! \rangle \land s' = s) \\ \hline \end{array} \qquad \begin{array}{|l} \multicolumn{1}{l}{RemoveS_A} \\ \hline \Delta AState \\ n! : \mathbb{N} \\ \hline (s = s' \frown \langle n! \rangle \land t' = t) \\ \hline \end{array}$$

Our concrete specification will have the state space *CState* with the same retrieve relation as in Example 5.3 (which is total on *CState*).

$$\begin{array}{|l} \multicolumn{1}{l}{CState} \\ \hline u : \operatorname{seq} \mathbb{N} \\ \hline \end{array}$$

The retrieve relation defines an upward simulation between the two state spaces and we can use this to calculate operations which are the weakest upward simulation of the abstract specification. For example,

$$\begin{aligned} CInit == & [CState' \mid \forall s, t : \operatorname{seq} \mathbb{N} \bullet u'_{merge}(s, t) \Rightarrow s = \langle\rangle \land t = \langle\rangle] \\ = & [CState' \mid u' = \langle\rangle] \end{aligned}$$

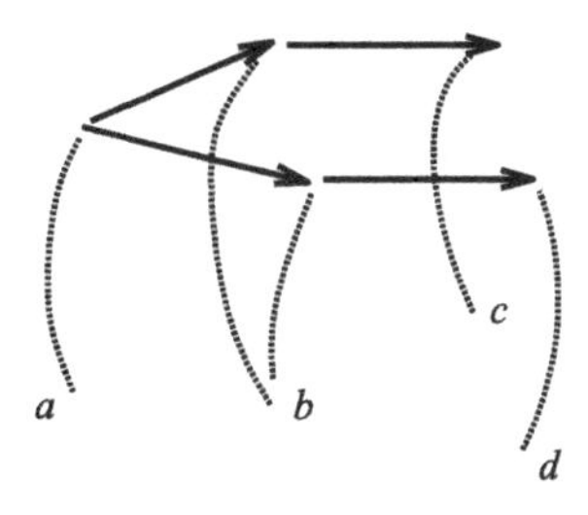

Fig. 5.3 An abstract specification and retrieve relation

For the concrete insert operation we calculate

$$\begin{aligned} Insert_C == \forall AState \bullet (R \Rightarrow \text{pre}\, Insert_A) \\ \wedge (\forall AState' \bullet (R' \Rightarrow \exists AState \bullet R \wedge Insert_A)) \end{aligned}$$

which evaluates to

$$\begin{array}{|l} \textit{Insert}_C \\ \hline \Delta CState \\ m? : \mathbb{N} \\ \hline \exists t, s : \text{seq}\, \mathbb{N} \bullet u_{merge}(t, s) \wedge u'_{merge}(t \frown \langle m? \rangle, s) \\ \hline \end{array}$$

Similarly $RemoveS_C$ and $RemoveT_C$ both evaluate to

$$\begin{array}{|l} \hline \Delta CState \\ n! : \mathbb{N} \\ \hline u \neq \langle\rangle \wedge \exists t, s : \text{seq}\, \mathbb{N} \bullet u_{merge}(t \frown \langle n! \rangle, s) \wedge u'_{merge}(t, s) \\ \hline \end{array}$$

Again note that these have further refinements to

$$Insert == [\Delta CState;\ m? : \mathbb{N} \mid u' = u \frown \langle m? \rangle]$$
$$RemoveS = RemoveT == [\Delta CState;\ n! : \mathbb{N} \mid u = u' \frown \langle n! \rangle]$$

As is characteristic with upward simulations the difference between the two specifications is in the point of non-determinism: the non-determinism in the abstract specification has been resolved later in the concrete specification. □

To understand the difference between the constructions given by the downward and upward simulations consider the simplified representation of Example 5.6 given in Fig. 5.3.

The calculations of the concrete operations given by the rules for downward and upward simulations are represented in Fig. 5.4.

The downward simulation calculation would fail to be a refinement in the case of Example 5.6 because there are no transitions for $RemoveS_C$ and hence $\text{pre}\, RemoveS_A \wedge R \not\Rightarrow \text{pre}\, RemoveS_C$, with a similar failing for $RemoveT_C$.

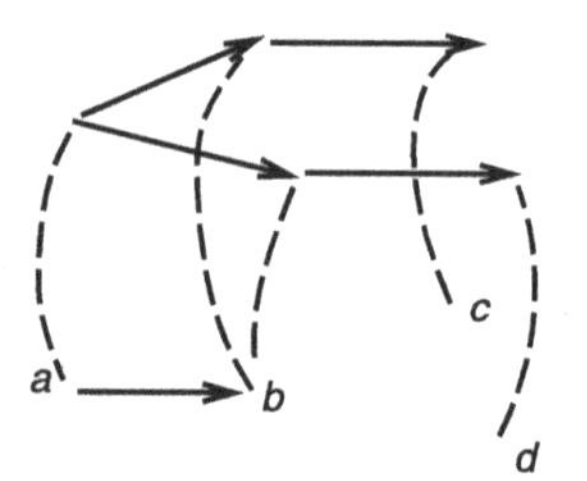

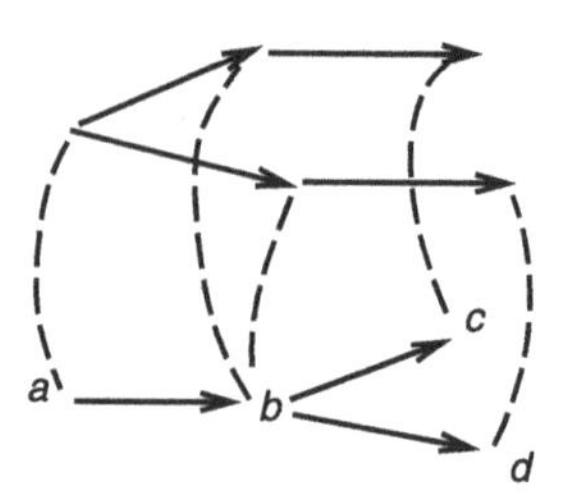

Fig. 5.4 The difference between downward and upward simulation calculations

5.3 Calculating Common Refinements

Using calculational methods, we can also solve another problem: given two specifications, does a refinement of *both* of them exist? This problem is especially relevant in the context of viewpoint specification [2, 10], where it is called *viewpoint consistency* [4, 5]. It turns out that this question can be answered in a constructive way for Z: if such a refinement[1] exists, we can construct it, given a relation between the two specifications' state spaces. This relation, given as a Z schema, is called a *correspondence* relation.

The construction is a two-step process. First, the state space of the common refinement is constructed. The construction is a variant of the product of the state spaces, most closely resembling that of the outer join in (database) relational algebra [7]: apart from pairs of values related by the correspondence, values not related by the correspondence are also included, joined with a "null" value. This construction guarantees that all operations of either of the viewpoint ADTs can be refined to operations on the combined state space.

In the second step, corresponding operations from the two ADTs, now operating on the combined state space, are combined using a (partial) least operation refinement combinator. Whenever this operator is not applicable, the original operations were inconsistent.

For the full details of the construction, and a proof that the construction results in the most general common refinement, we refer to [4]. In this book we will only present the theorem for most general common operation refinement, and an example of the construction of the most general data refinement in a relatively simple case.

Theorem 5.5 (Most general common operation refinement) *The most general common operation refinement of operations OpA and OpB, operating on the same state State exists whenever*

$$\text{pre}\, OpA \wedge \text{pre}\, OpB \Rightarrow \text{pre}(OpA \wedge OpB)$$

and is then given by

[1] Approaches to this have so far concentrated on downward simulation only, but the retrieve relations constructed are functional.

$$
\begin{array}{|l}
\cup(OpA, OpB) \\
\hline
\Delta State \\
\hline
\text{pre}\, OpA \Rightarrow OpA \\
\text{pre}\, OpB \Rightarrow OpB \\
\hline
\end{array}
$$

This assumes the contractual approach; in the behavioural approach, the most general common operation refinement is $OpA \land OpB$, *provided that* $\text{pre}\, OpA = \text{pre}\, OpB$ *holds in addition to the above condition.* □

Example 5.7 Two cohabiting people, Pat and Jim, go shopping together, but they have (at first sight) rather different approaches to shopping. Pat brings a wallet with money in it, and knows what is on the shopping list. Jim brings the shopping list, and can decide to stop shopping at any point.

We have *ITEM*s that can be bought in the shop, at a *price*.

$$[ITEM]$$

$$| \; price : ITEM \rightarrow \mathbb{N}_1$$

The ADTs that represent Pat and Jim's views of shopping are given below.

$$
\begin{array}{|l}
PatShop \\
\hline
wallet : \mathbb{N} \\
to\_get, basket : \mathbb{F}\, ITEM \\
\hline
\end{array}
$$

$$
\begin{array}{|l}
PatInit \\
\hline
PatShop' \\
\hline
basket' = \varnothing \\
\hline
\end{array}
$$

$$
\begin{array}{|l}
PatBuy \\
\hline
\Delta PatShop \\
\hline
\exists\, i : ITEM \bullet \\
\quad i \in to\_get \\
\quad wallet' = wallet - price(i) \\
\quad basket' = basket \cup \{i\} \\
\quad to\_get' = to\_get \setminus \{i\} \\
\hline
\end{array}
$$

$$
\begin{array}{|l}
JimShop \\
\hline
list : \text{iseq}\, ITEM \\
basket : \mathbb{F}\, ITEM \\
done : \mathbb{B} \\
\hline
\end{array}
$$

$$
\begin{array}{|l}
JimInit \\
\hline
JimShop' \\
\hline
basket' = \varnothing \\
\hline
\end{array}
$$

$$
\begin{array}{|l}
JimBuy \\
\hline
\Delta JimShop \\
\hline
\neg done \\
list \neq \langle\rangle \\
basket' = basket \cup \{head\, list\} \\
list' = tail\, list \\
\hline
\end{array}
$$

The relation between the state spaces could be that Jim and Pat agree what is on the shopping list (and, implicitly, what is in the basket), and that Jim stops buying things exactly when they cannot afford the item at the head of the list.

$$\begin{array}{|l}\hline \textit{PatJimShop} \\ \textit{PatShop} \\ \textit{JimShop} \\ \hline \textit{to\_get} = \operatorname{ran} \textit{list} \\ \textit{list} \neq \langle\rangle \Rightarrow \textit{done} = (\textit{price}(\textit{head list}) > \textit{wallet}) \\ \hline \end{array}$$

As the correspondence relation *PatJimShop* is total on both *PatShop* and *JimShop*, it is also the state space of the common refinement.

The retrieve relation between either of the ADTs and the common refinement is also represented by *PatJimShop*. Thus, in both cases it is a function from concrete to abstract. We can therefore use Corollary 5.1 to determine the most general downward simulations of the operations and initialisations.[2] In this case, these amount to $[PatJimShop' \mid JimInit]$ and $[PatJimShop' \mid PatInit]$ for the initialisations, and $AJB == \Delta PatJimShop \wedge JimBuy$ and $APB == \Delta PatJimShop \wedge PatBuy$ for the operations. The initialisation of the weakest common refinement is the conjunction of the adapted initialisations, which is clearly satisfiable.[3]

The operations, with the "interesting" consequences of the invariants of $\Delta PatJimShop$ listed explicitly, are:

$$\begin{array}{|l}\hline APB \\ \Delta PatJimShop \\ \hline list \neq \langle\rangle \\ \exists\, i : ITEM \bullet \\ \quad price(i) \leq wallet \\ \quad i \in \operatorname{ran} list \\ \quad basket' = basket \cup \{i\} \\ \quad \operatorname{ran} list' = to\_get' \\ \quad to\_get' = to\_get \setminus \{i\} \\ \quad wallet' = wallet - price(i) \\ \hline \end{array}$$

$$\begin{array}{|l}\hline AJB \\ \Delta PatJimShop \\ \hline list \neq \langle\rangle \\ \neg done \\ price(head\, list) \leq wallet \\ \exists\, i : ITEM \bullet \\ \quad i = head\, list \\ \quad basket' = basket \cup \{i\} \\ \quad list' = tail\, list \\ \quad to\_get' = to\_get \setminus \{i\} \\ \hline \end{array}$$

The operation *AJB* expresses "buy the first thing on our list as long as we can afford that", without actually modelling the financial consequences of the purchase. The operation *APB* expresses "buy something off the list as long as we can afford it". Formally, we have

$$\begin{aligned} \operatorname{pre} AJB &= list \neq \langle\rangle \wedge \neg done \\ &= list \neq \langle\rangle \wedge price(head\ list) \leq wallet \\ \operatorname{pre} APB &= list \neq \langle\rangle \wedge \exists\, i : \operatorname{ran} list \bullet price(i) \leq wallet \end{aligned}$$

[2] In fact Corollary 5.1 does not account for the possibility that *AState* is textually included in *CState*, as is the case here—in that case, the existential quantifications over *AState* and *AState'* disappear.

[3] Observe that *JimInit* containing $\neg done'$ would disallow initial states with an empty wallet. This would not be an inconsistency, strictly speaking. However, if we went on to model *wallet* and *list* as inputs provided by the environment at initialisation (cf. Sect. 4.7.1), it would cause an inconsistency.

Thus, $\text{pre}\,AJB \Rightarrow \text{pre}\,APB$. It is *not* the case that $AJB \Rightarrow APB$ because AJB does not constrain $wallet'$. When both preconditions hold, the conjunction of the operations is clearly satisfiable, by buying the first item on the list, with the normal effect on the wallet. Outside the joint precondition, i.e., when only APB is applicable, we get APB's effect of picking any affordable item on the list.

$$
\begin{array}{l}
\cup(APB, AJB) \\
\hline
\Delta PatJimShop \\
\hline
list \neq \langle\rangle \\
\exists i : \text{ran}\, list \bullet \\
\quad price(head\, list) \leq wallet \Rightarrow i = head\, list \wedge list' = tail\, list \\
\quad basket' = basket \cup \{i\} \\
\quad to\_get' = to\_get \setminus \{i\} \\
\quad wallet' = wallet - price(i) \\
\hline
\end{array}
$$

Informally, Pat agrees to start by always buying the first item on Jim's list, as long as that is affordable. Jim agrees not to stop shopping at an arbitrary time, but rather when the first item on the list is one that Pat's wallet cannot afford. After that, Pat tries to buy some more items off the list that they *can* still afford.

An inconsistency might have arisen if Jim's shopping list had not been injective, because Pat, whose view of the required shopping is a set rather than a bag, is unwilling to buy anything twice. □

5.4 Bibliographical Notes

There has been extensive work on program development by transformation and calculation. The principal work on calculating simulations in a relational setting is due to He Jifeng and Hoare [13]. In [13] refinement calculations in the context of total relations were considered. Prior to this, rules for calculating downward simulations of Z specifications were given by Josephs in [15]. It was commented that the simplification of the calculation to $\exists AState; AState' \bullet (R \wedge AOp \wedge R')$ could be derived when R defines a (partial) *surjective function* from *CState* to *AState*.

In [8] Derrick and Boiten considered the calculation of operations and showed that this hypothesis could be relaxed. In particular, surjectivity was not necessary and it was sufficient to consider the retrieve relation to be a function on a restricted domain. This paper also derived the calculations for upward simulations of Z specifications. The specification of the MSMIE protocol from Example 5.1 was originally specified in CCS in [6] by Bruns and Anderson.

The most general operation refinement in Z was first defined by Ainsworth, Cruickshank, Wallis and Groves [1], and called *amalgamation* there. In a setting of abstract binary relations, it was also used by Frappier et al. [12] (under the name of *demonic join*) in the context of feature interaction. D'Souza and Wills [9] developed the same construction for OCL.

We extended this to the most general *data* refinement (also called "unification") in [4]. Boiten and Bujorianu [3] explored how intuitive understandings of this operation in UML could be used to justify properties of UML refinement.

Viewpoint specification in Z has also been extensively studied by Zave and M. Jackson [17, 18] and D. Jackson [14]. In [4] we explored how D. Jackson's viewpoint composition operator relates to unification.

References

1. Ainsworth, M., Cruickshank, A. H., Wallis, P. J. L., & Groves, L. J. (1994). Viewpoint specification and Z. *Information and Software Technology*, *36*(1), 43–51.
2. Boiten, E. A., Bowman, H., Derrick, J., & Steen, M. W. A. (1996) Issues in multiparadigm viewpoint specification. In Finkelstein and Spanoudakis [11] (pp. 162–166).
3. Boiten, E. A., & Bujorianu, M. C. (2003). Exploring UML refinement through unification. In J. Jürjens, B. Rumpe, R. France, & E. B. Fernandez (Eds.), *Critical Systems Development with UML—Proceedings of the UML'03 Workshop, Number TUM-I0323*, September 2003 (pp. 47–62). München: Technische Universität München.
4. Boiten, E. A., Derrick, J., Bowman, H., & Steen, M. W. A. (1999). Constructive consistency checking for partial specification in Z. *Science of Computer Programming*, *35*(1), 29–75.
5. Bowman, H., Boiten, E. A., Derrick, J., & Steen, M. W. A. (1996) Viewpoint consistency in ODP, a general interpretation. In Najm and Stefani [16] (pp. 189–204).
6. Bruns, G., & Anderson, S. (1992). *The formalization of a communications protocol* (Technical report). LFCS/Adelard SCCS, April 1992.
7. Codd, E. F. (1979). Extending the database relational model to capture more meaning. *ACM Transactions on Database Systems*, *4*(4).
8. Derrick, J., & Boiten, E. A. (1999). Calculating upward and downward simulations of state-based specifications. *Information and Software Technology*, *41*, 917–923.
9. D'Souza, D. F., & Wills, A. C. (1998). *Objects, Components, and Frameworks with UML: the Catalysis Approach. Object Technology Series*. Reading: Addison-Wesley.
10. Finkelstein, A., Kramer, J., Nuseibeh, B., Finkelstein, L., & Goedicke, M. (1992). Viewpoints: a framework for integrating multiple perspectives in system development. *International Journal on Software Engineering and Knowledge Engineering*, *2*(1), 31–58.
11. Finkelstein, A. & Spanoudakis, G. (Eds.) (1996). *SIGSOFT '96 International Workshop on Multiple Perspectives in Software Development (Viewpoints '96)*. New York: ACM.
12. Frappier, M., Mili, A., & Desharnais, J. (1997). Defining and detecting feature interactions. In R. S. Bird & L. Meertens (Eds.), *IFIP TC2 WG 2.1 International Workshop on Algorithmic Languages and Calculi* (pp. 212–239). London: Chapman & Hall.
13. He, J., & Hoare, C. A. R. (1990). Prespecification and data refinement. In *Data Refinement in a Categorical Setting. Technical Monograph PRG-90*. Oxford: Oxford University Computing Laboratory.
14. Jackson, D. (1995). Structuring Z specifications with views. *ACM Transactions on Software Engineering and Methodology*, *4*(4), 365–389.
15. Josephs, M. B. (1988). The data refinement calculator for Z specifications. *Information Processing Letters*, *27*, 29–33.
16. Najm, E. & Stefani, J. B. (Eds.) (1996). *First IFIP International Workshop on Formal Methods for Open Object-Based Distributed Systems*, Paris, March 1996. London: Chapman & Hall.
17. Zave, P., & Jackson, M. (1993). Conjunction as composition. *ACM Transactions on Software Engineering and Methodology*, *2*(4), 379–411.
18. Zave, P., & Jackson, M. (1996). Where do operations come from? A multiparadigm specification technique. *IEEE Transactions on Software Engineering*, *22*(7), 508–528.

Chapter 6
Promotion

In this chapter we discuss a technique for structuring specifications known as *promotion*. The purpose of promotion is to provide an elegant way of composing specifications in order to build multiple indexed instances of a single component. To do so the component is described as a local state together with operations acting on that state, a global state is then defined which consists of multiple instances of this local state together with global operations defined in terms of the local operations and a special promotion schema. (Note that these notions of "local" and "global" state are unrelated to those used in e.g., Chaps. 3 and 10.)

The use of promotion together with the schema calculus allows a nice separation of concerns between the local and global systems by separating out the common part of the global operations into the promotion schema. For this reason promotion is useful in many situations.

There is also an elegant relationship between promotion and refinement: under certain circumstances the promotion of a refinement is a refinement of a promotion. That is, if we refine the local state, the promotion of it is a refinement of the original promotion. This is a useful compositionality property, which allows for piecewise development.

In the second half of the chapter we discuss the conditions needed for this result to hold, and look at some commonly arising cases. We begin with a simple example of the use of the technique.

6.1 Example: Multiple Processors

Consider the specification of a process scheduler given in Example 4.8 (p. 117) which included a state space, initialisation and operation *New*:

$[PID]$

J. Derrick, E.A. Boiten, *Refinement in Z and Object-Z*,
DOI 10.1007/978-1-4471-5355-9_6, © Springer-Verlag London 2014

$$\begin{array}{|l}
\hline \textit{State} \\
active : PID \\
ready : \mathbb{P}\, PID \\
waiting : \mathbb{P}\, PID \\
admin : ADMIN \\
\hline
ready \cap waiting = \varnothing \\
active \notin (ready \cup waiting) \\
nil \notin (ready \cup waiting) \\
active = nil \Rightarrow ready = \varnothing \\
\hline
\end{array}$$

$$\begin{array}{|l}
\hline \textit{Init} \\
State' \\
\hline
active' = nil \\
ready' \cup waiting' = \varnothing \\
admin' = user \\
\hline
\end{array}$$

$$\begin{array}{|l}
\hline \textit{New} \\
\Delta State \\
p? : PID \\
\hline
p? \neq active \\
p? \notin (ready \cup waiting) \\
waiting' = waiting \cup \{p?\} \\
active' = active \\
ready' = ready \\
admin = user \wedge admin' = user \\
\hline
\end{array}$$

This describes a single component—a scheduler. We now wish to specify a collection of schedulers, and we do so by defining the following global state

$[I]$

$$\begin{array}{|l}
\hline \textit{GState} \\
f : I \rightarrow State \\
\hline
\end{array}$$

This defines a number of schedulers named via the function f and index set I: for every i, $f(i)$ is a different scheduler. We also need to define operations in the global specification. Some of our global operations will describe what happens to individual schedulers, and to specify this we use a *promotion schema* to encapsulate the common part of all the operations, i.e., the process of updating a single scheduler. The other part of the operation, what updating is actually carried out, will be given by a local operation. The promotion schema we use is given by

$$
\begin{array}{|l}
\hline
\textit{Update} \\
\Delta GState \\
\Delta State \\
i? : I \\
\hline
\theta State = f(i?) \\
f' = f \oplus \{i? \mapsto \theta State'\} \\
\hline
\end{array}
$$

This specifies that the named scheduler $i?$ is updated, since the expression $\theta State = f(i?)$ says that the values of the variables from *State* are the same as the values in $f(i?)$, and the second predicate ensures that no other scheduler changes. However, it does not say exactly *how* that scheduler is altered.

To complete the global operation we need to describe how the local state changes, i.e., how $\theta State'$ is related to $\theta State$, and this is given by the local operations. Putting the pieces together we can define an operation to create a new process in a particular scheduler by writing

$$GlobalNew == \exists \Delta State \bullet Update \wedge New$$

The promoted version of the other local scheduling operations are defined in a similar manner. The promotion schemas are sometimes called *update* or *framing* schemas for obvious reasons.

Of course not all global operations will be defined in terms of changes to just one component, but for a certain class of operations, promotion allows the global operation to be factored into two parts in a succinct fashion.

The example above uses a very simple global state consisting of a total function, however, in general the global state might be more complex, for example, it may consist of a partial function, sequence or relation between indexes and local state together with optional extra structure.

6.2 Example: A Football League

As an example of a slightly more complex promotion we consider the description of a collection of associated football clubs. We look at a club from the perspective of its membership list, which we represent as a function from a set of identifiers to names of potential fans. Membership allows reduced prices for home matches. However, in addition to membership, in order to buy a ticket for an away match an away club card is needed, so the set of fans who have bought away cards is also recorded. We also record $points_A$ which represents the points scored so far this season.

The specification for a single club includes an operation $Join_A$ for becoming a member, and an operation Buy_A to purchase an away club card.

$$[PEOPLE]$$
$$ID == \mathbb{N}$$

$$
\begin{array}{|l}
\hline L_A \\
members_A : ID \rightarrowtail\!\!\!\!\!\!+\; PEOPLE \\
away_A : \mathbb{P}\, ID \\
points_A : \mathbb{N} \\
\hline
away_A \subseteq \operatorname{dom} members_A \\
\hline
\end{array}
\qquad
\begin{array}{|l}
\hline InitL_A \\
L'_A \\
\hline
members'_A = \varnothing \\
points'_A = 0 \\
\hline
\end{array}
$$

$$
\begin{array}{|l}
\hline Join_A \\
\Delta L_A \\
app? : PEOPLE \\
id! : ID \\
\hline
app? \notin \operatorname{ran} members_A \\
id! \notin \operatorname{dom} members_A \\
members'_A = members_A \cup \{id! \mapsto app?\} \\
points'_A = points_A \\
away'_A = away_A \\
\hline
\end{array}
$$

$$
\begin{array}{|l}
\hline Buy_A \\
\Delta L_A \\
app? : PEOPLE \\
\hline
points'_A = points_A \\
members'_A = members_A \\
\exists x : ID \bullet (x \mapsto app? \in members_A \land away'_A = away_A \cup \{x\}) \\
\hline
\end{array}
$$

A league of clubs can then be described by a global state indexed over some set I.

$$
\begin{array}{|l}
\hline G_A \\
clubs_A : I \twoheadrightarrow\!\!\!\!\!\!+\; L_A \\
europe_A : \mathbb{P}\, I \\
\hline
europe_A \subseteq \operatorname{dom} clubs_A \\
\hline
\end{array}
$$

where $europe_A$ is the collection of clubs playing in European competitions this season. The promotion schema for the football league is similar to the previous example:

$$
\begin{array}{|l}
\hline
\textit{Update}_A \\
\Delta G_A \\
\Delta L_A \\
i? : I \\
\hline
i? \in \operatorname{dom} clubs_A \\
\theta L_A = clubs_A(i?) \\
clubs'_A = clubs_A \oplus \{i? \mapsto \theta L'_A\} \\
europe'_A = europe_A \\
\hline
\end{array}
$$

The promoted operations are defined as before. For example, the operation to buy an away card at a particular club $i?$ is given by

$$GlobalBuy_A == \exists\, \Delta L_A \bullet Update_A \wedge Buy_A$$

Occasionally clubs are added to or deleted from the league, and we model this by using the following two promotion schemas

$$
\begin{array}{|l}
\hline
\textit{Create}_A \\
\Delta G_A \\
L'_A \\
i? : I \\
\hline
i? \notin \operatorname{dom} clubs_A \\
clubs'_A = clubs_A \cup \{i? \mapsto \theta L'_A\} \\
europe'_A = europe_A \\
\hline
\end{array}
$$

$$
\begin{array}{|l}
\hline
\textit{Del}_A \\
\Delta G_A \\
L_A \\
i? : I \\
\hline
i? \in \operatorname{dom} clubs_A \\
\theta L_A = clubs_A(i?) \\
clubs'_A = \{i?\} \lhd clubs_A \\
europe'_A = europe_A \\
\hline
\end{array}
$$

in order to promote the operations as follows

$$GNew_A == \exists\, L'_A \bullet Create_A \wedge InitL_A$$
$$GDel_A == \exists\, L_A \bullet Del_A \wedge Term_A$$

Here $Term_A$ records the state from which clubs can be removed from the league (having neither members nor points).

$$
\begin{array}{|l}
\hline
Term_A \\
L_A \\
\hline
members_A = \varnothing \\
points_A = 0 \\
\hline
\end{array}
$$

Not every global operation can be defined in terms of local operations and a promotion schema. For example, at the end of the season clubs are promoted (nothing to do with Z or the schema calculus!) and relegated to and from the league. To model relegation we need to compare points with the other clubs—only the three lowest are relegated—and thus the operation to relegate one of the three bottom clubs is best described at a global level as

$$
\begin{array}{|l}
\hline
Relegate_A \\
\Delta G_A \\
i? : I \\
\hline
i? \in \mathrm{dom}\, clubs_A \\
clubs'_A = \{i?\} \lhd clubs_A \\
\#\{j : \mathrm{dom}\, I \mid clubs_A(j).points \leq clubs_A(i?).points\} \leq 3 \\
\hline
\end{array}
$$

6.3 Free Promotions and Preconditions

We need to be able to calculate the precondition of a promoted operation, preconditions being particularly relevant in verifying refinements. Clearly we could expand out a global operation and then calculate its precondition directly. However, can we simplify this process by calculating the precondition of a promoted operation in terms of the precondition of the local operation? That is, if Op is an operation acting on local state L, and G is the global state with promotion schema U, when is it the case that

$$\mathrm{pre}(\exists\, \Delta L \bullet U \wedge Op) \Leftrightarrow \exists\, L \bullet \mathrm{pre}\, U \wedge \mathrm{pre}\, Op$$

The answer to this question will depend on whether the promotion is *free* or whether it is *constrained*.

Definition 6.1 With the notation introduced above, a promotion is free precisely when

$$(\exists\, L' \bullet \exists\, G' \bullet U) \Rightarrow (\forall\, L' \bullet \exists\, G' \bullet U) \qquad \square$$

That is, a promotion is free if the promotion schema U does not further constrain the ability to place a local after-state L' if it is possible to place one at all. Free promotions occur when no constraints are placed on the local state by the promotion schema or by the global invariant in G. Although this is a rather strong condition, it will allow us to distribute the preconditions through a promotion.

Example 6.1 The collection of schedulers defined in Sect. 6.1 is a free promotion, as is the football league described in Sect. 6.2. These facts are easily verified. □

The key point about free promotions is the following result due to Lupton [3].

Theorem 6.1 *With the notation above, if the promotion schema U is free then*

$$\text{pre}(\exists \Delta L \bullet U \wedge Op) \Leftrightarrow \exists L \bullet \text{pre}\, U \wedge \text{pre}\, Op$$

Proof One direction of the equivalence is obvious. For the other, we note

$$\exists L \bullet \text{pre}\, U \wedge \text{pre}\, Op$$

$$\equiv \{ \text{definition of pre} \}$$

$$\exists L \bullet (\exists G'; L' \bullet U) \wedge (\exists L' \bullet Op)$$

$$\Rightarrow \{ \text{free} \}$$

$$\exists L \bullet (\forall L' \bullet \exists G' \bullet U) \wedge (\exists L' \bullet Op)$$

$$\Rightarrow \{ \text{logic} \}$$

$$\exists L; G'; L' \bullet U \wedge Op$$

□

Using this result we can easily calculate the preconditions of the global operations defined in the examples above. For example, consider the operation $GlobalBuy_A$ from the example in Sect. 6.2. Since the promotion is free we know that

$$\text{pre}\, GlobalBuy_A \Leftrightarrow \exists L_A \bullet \text{pre}\, Update_A \wedge \text{pre}\, Buy_A$$

Calculating the components in the right-hand side we find:

```
—pre Update_A ——————————————————————
| G_A
| L_A
| i? : I
|——————————
| i? ∈ dom clubs_A
| θL_A = clubs_A(i?)
————————————————————————————————————
```

$$\begin{array}{|l}
\hline
\text{pre } BuyA \\
L_A \\
app? : PEOPLE \\
\hline
app? \in \text{ran } members_A \\
\hline
\end{array}$$

Hence pre $GlobalBuy_A$ is given by

$$\begin{array}{|l}
\hline
\text{pre } GlobalBuy_A \\
G_A \\
i? : I \\
app? : PEOPLE \\
\hline
i? \in \text{dom } clubs_A \\
app? \in \text{ran}(clubs_A(i?).members_A) \\
\hline
\end{array}$$

6.4 The Refinement of a Promotion

As we have seen, promotion provides a nice separation of concerns at the specification level. What we would like to do is to provide a similar separation of concerns for refinements of promotions in the following sense. Suppose that we are given a local state L_A with operation Op_A promoted to a global state G_A by promotion schema U_A. Suppose further that we have a concrete local state L_C with operation Op_C promoted to a global state G_C by promotion schema U_C. If the concrete local specification is a refinement of the abstract local specification, then will the concrete promotion be a refinement of the abstract promotion? That is, is refinement monotonic (or compositional) for this specification constructor, and can we thus apply piecewise refinement?

It turns out that the answer to this question is sometimes positive, but not always so. The purpose of this section is to outline sufficient conditions for this to hold in the case of a downward simulation. In effect these results are about piecewise refinement, where a context allows the refinement of a component to produce a refinement of the whole specification. We return to some of these issues again in the context of Object-Z in Chap. 17.

For the remainder of this chapter we use the following notation. Subscripts A and C refer to abstract and concrete, L denotes the local space and G the global. U is the promotion schema. Suppose the downward simulation, between (L_C, Op_C) and (L_A, Op_A), is verified by a retrieve relation R, then the task is to find a suitable retrieve relation P_R which defines a downward simulation between the global specifications:

$$\begin{array}{ccc} L_A & \xrightarrow{Op_A} & L'_A \\ R\downarrow & & \downarrow R' \\ L_C & \xrightarrow[Op_C]{} & L'_C \end{array} \qquad \begin{array}{ccc} G_A & \xrightarrow{\exists\,\Delta L_A \bullet (U_A \land Op_A)} & G'_A \\ P_R\downarrow & & \downarrow P'_R \\ G_C & \xrightarrow[\exists\,\Delta L_C \bullet (U_C \land Op_C)]{} & G'_C \end{array}$$

6.4.1 Example: Refining Multiple Processors

Example 6.2 Consider the specification of multiple processors given in Sect. 6.1. As we saw in Chap. 4 a single processor is refined by the following more concrete version

$$\begin{array}{|l} \hline State_C \\ \hline active : PID \\ cready : \mathrm{iseq}\, PID \\ cwaiting : \mathrm{seq}\, PID \\ admin : ADMIN \\ \hline \mathrm{ran}\, cready \cap \mathrm{ran}\, cwaiting = \varnothing \\ active \notin (\mathrm{ran}\, cready \cup \mathrm{ran}\, cwaiting) \\ nil \notin (\mathrm{ran}\, cready \cup \mathrm{ran}\, cwaiting) \\ active = nil \Rightarrow cready = \langle\rangle \\ \hline \end{array}$$

$$\begin{array}{|l} \hline Init_C \\ \hline State'_C \\ \hline active' = nil \\ cready' = \langle\rangle \land cwaiting' = \langle\rangle \\ admin' = user \\ \hline \end{array}$$

$$\begin{array}{|l} \hline New_C \\ \hline \Delta State_C \\ p? : PID \\ \hline p? \neq active \\ p? \notin (\mathrm{ran}\, cready \cup \mathrm{ran}\, cwaiting) \\ \mathrm{ran}\, cwaiting' = \mathrm{ran}\, cwaiting \cup \{p?\} \\ active' = active \\ cready' = cready \\ admin = user \land admin' = user \\ \hline \end{array}$$

using the retrieve relation

$$
\begin{array}{|l}
\hline
R \\
State \\
State_C \\
\hline
ready = \operatorname{ran} cready \\
waiting = \operatorname{ran} cwaiting \\
\hline
\end{array}
$$

A concrete global state and promotion schema for multiple concrete processors can be defined

$$
\begin{array}{|l}
\hline
GState_C \\
f_C : I \rightarrow State_C \\
\hline
\end{array}
$$

$$
\begin{array}{|l}
\hline
Update_C \\
\Delta GState_C \\
\Delta State_C \\
i? : I \\
\hline
\theta State_C = f_C(i?) \\
f_C' = f_C \oplus \{i? \mapsto \theta State_C'\} \\
\hline
\end{array}
$$

and concrete global operations promoted in the usual fashion. Then the concrete global specification is a downward simulation of the abstract global specification, and this can be verified by the retrieve relation P_R:

$$
\begin{array}{|l}
\hline
P_R \\
GState \\
GState_C \\
\hline
\forall n : I \bullet \exists R \bullet \theta State = f(n) \wedge \theta State_C = f_C(n) \\
\hline
\end{array}
$$

This relation states that two global states are related whenever the local states are related by the local retrieve relation R. Using this retrieve relation the verification of the global refinement is not difficult. □

This example illustrates a particular case where refinement distributes through promotion. We now define sufficient conditions on the retrieve relations R and P_R and promotions for this to be the case in general.

6.4.2 Refinement Conditions for Promotion

The following definition contains the conditions that will turn out to be sufficient for refining promotions.

Definition 6.2 Promotion schemas U_A and U_C freely promote downward simulation with respect to R if there is a relation P_R such that

A. $P_R \wedge \mathrm{pre}\, U_A \wedge \mathrm{pre}\, U_C \Rightarrow R$
B. $P_R \wedge \mathrm{pre}\, U_A \wedge U_C \Rightarrow (\forall L'_A \bullet R' \Rightarrow (\exists G'_A \bullet U_A \wedge P'_R))$
C. $P_R \wedge \mathrm{pre}\, U_A \Rightarrow (\exists L_C \bullet \mathrm{pre}\, U_C)$
D. $\mathrm{pre}\, U_C \Rightarrow (\forall L'_C \bullet \exists G'_C \bullet U_C)$ □

In this definition, condition A defines the necessary relationship between the two retrieve relations R and P_R, and in effect places minimum conditions on a correct definition of P_R given the underlying local retrieve relation R. Conditions B and C are refinement conditions on promotions, B is a correctness condition, and C is an applicability condition. Finally, D requires that the concrete promotion is free, and this ensures that there are no global invariants that constrain the way local states are joined up to form a global state.

With this definition in place we can now prove (assuming the notation introduced above):

Theorem 6.2 *If U_A and U_C freely promote downward simulation with respect to the local retrieve relation R, then there is a downward simulation between the promoted global operations $\exists \Delta L_A \bullet (U_A \wedge Op_A)$ and $\exists \Delta L_C \bullet (U_C \wedge Op_C)$ verified by retrieve relation P_R.*

Proof We have to show both applicability and correctness for the global refinement. We begin with applicability, i.e.,

$$\mathrm{pre}(\exists \Delta L_A \bullet (U_A \wedge Op_A)) \wedge P_R \Rightarrow \mathrm{pre}(\exists \Delta L_C \bullet (U_C \wedge Op_C))$$

$$\mathrm{pre}(\exists \Delta L_A \bullet (U_A \wedge Op_A)) \wedge P_R$$
$$\Rightarrow \{\ \text{logic}\ \}$$
$$\exists L_A \bullet (\mathrm{pre}\, U_A \wedge \mathrm{pre}\, Op_A) \wedge P_R$$
$$\Rightarrow \{\ \text{logic}\ \}$$
$$\exists L_A \bullet (\mathrm{pre}\, U_A \wedge P_R \wedge \mathrm{pre}\, Op_A) \wedge P_R$$
$$\Rightarrow \{\ \text{by C}\ \}$$
$$\exists L_C; L_A \bullet \mathrm{pre}\, U_C \wedge \mathrm{pre}\, Op_A \wedge \mathrm{pre}\, U_A \wedge P_R$$
$$\Rightarrow \{\ \text{by A}\ \}$$

$$\exists L_C \bullet \text{pre}\, Op_A \land R \land \text{pre}\, U_C$$
$$\Rightarrow \{ \text{ by local refinement } \}$$
$$\exists L_C \bullet \text{pre}\, Op_C \land \text{pre}\, U_C$$
$$\Rightarrow \{ \text{ by D } \}$$
$$\text{pre}(\exists \Delta L_C \bullet (U_C \land Op_C))$$

For correctness we have to show the following

$$\text{pre}(\exists \Delta L_A \bullet (U_A \land Op_A)) \land P_R \land \exists \Delta L_C \bullet (U_C \land Op_C)$$
$$\Rightarrow \exists G'_A \bullet P'_R \land (\exists \Delta L_A \bullet (U_A \land Op_A))$$

$$\text{pre}(\exists \Delta L_A \bullet (U_A \land Op_A)) \land P_R \land \exists \Delta L_C \bullet (U_C \land Op_C)$$
$$\Rightarrow \{ \text{ logic } \}$$
$$\exists L_A \bullet (\text{pre}\, U_A \land \text{pre}\, Op_A) \land P_R \land \exists \Delta L_C \bullet (U_C \land Op_C)$$
$$\equiv \{ \text{ logic } \}$$
$$\exists L_A;\ \Delta L_C \bullet (\text{pre}\, U_A \land \text{pre}\, Op_A \land P_R \land U_C \land Op_C)$$
$$\Rightarrow \{ \text{ by A } \}$$
$$\exists L_A;\ \Delta L_C \bullet (\text{pre}\, U_A \land \text{pre}\, Op_A \land P_R \land U_C \land Op_C \land R)$$
$$\Rightarrow \{ \text{ by B } \}$$
$$\exists L_A;\ \Delta L_C \bullet \forall L'_A \bullet (R' \Rightarrow (\exists G'_A \bullet U_A \land P'_R)) \land Op_C \land R \land \text{pre}\, Op_A$$
$$\Rightarrow \{ \text{ by refinement } \}$$
$$\exists L_A;\ \Delta L_C \bullet \forall L'_A \bullet (R' \Rightarrow (\exists G'_A \bullet U_A \land P'_R)) \land (\exists L'_A \bullet Op_A \land R'))$$
$$\Rightarrow \{ \text{ logic } \}$$
$$\exists \Delta L_A;\ \Delta L_C \bullet (R' \Rightarrow (\exists G'_A \bullet U_A \land P'_R)) \land Op_A \land R'))$$
$$\Rightarrow \{ \text{ logic } \}$$
$$\exists \Delta L_A \bullet ((\exists G'_A \bullet U_A \land P'_R) \land Op_A)$$
$$\Rightarrow \{ \text{ logic } \}$$
$$\exists G'_A \bullet \exists \Delta L_A \bullet U_A \land P'_R \land Op_A$$

□

To complete the proof we need to refine the global initial states, and the corresponding definition and theorem are in fact just special cases of the definition and theorem above. In particular, the conditions on the promotion schema $Create_A$ and $Create_C$ used to promote the initialisations are as follows:

A. $Create_A$ (resp. $Create_C$) does not contain L_A (resp. L_C) in its signature.
B. $P_R \wedge \text{pre}\, Create_A \wedge Create_C \Rightarrow (\forall L'_A \bullet R' \Rightarrow (\exists G'_A \bullet Create_A \wedge P'_R))$
C. $P_R \wedge \text{pre}\, Create_A \Rightarrow \text{pre}\, Create_C$
D. $Create_C$ is a free promotion

With these results in place we could verify that the promoted concrete specification in Sect. 6.1 is a refinement of the promoted abstract specification by verifying conditions A to D above. Such a verification is easy, if tedious. However, many promotion schemas and global states take a common form (just look at the first two examples in this chapter), and we can take advantage of this to reduce the proof obligations that need to be verified when our promotions conform to this format. The next section illustrates how we can do this.

6.5 Commonly Occurring Promotions

Suppose that the global abstract and concrete states are given by

G_A

$f_A : I \rightarrow\!\!\!\!\!\!+ \; L_A$

$pred_A$

G_C

$f_C : I \rightarrow\!\!\!\!\!\!+ \; L_C$

$pred_C$

where $pred_A$ and $pred_C$ are predicates, and that the promotion schemas U_A and U_C have the following format

U_A

ΔG_A
ΔL_A
$i? : I$

$i? \in \text{dom}\, f_A$
$\theta L_A = f_A(i?)$
$f'_A = f_A \oplus \{i? \mapsto \theta L'_A\}$

$$
\begin{array}{|l}
\hline
U_C \\
\Delta G_C \\
\Delta L_C \\
i? : I \\
\hline
i? \in \operatorname{dom} f_C \\
\theta L_C = f_C(i?) \\
f'_C = f_C \oplus \{i? \mapsto \theta L'_C\} \\
\hline
\end{array}
$$

Given a refinement between the local specifications verified by retrieve relation R, let P_R be the relation

$$
\begin{array}{|l}
\hline
P_R \\
G_A \\
G_C \\
\hline
\operatorname{dom} f_A = \operatorname{dom} f_C \\
\forall n : \operatorname{dom} f_A \bullet \exists R \bullet \theta L_A = f_A(n) \wedge \theta L_C = f_C(n) \\
\hline
\end{array}
$$

It is then very easy to show that the promotion schemas U_A and U_C satisfy conditions A and C of Definition 6.2, that is,

A. $P_R \wedge \operatorname{pre} U_A \wedge \operatorname{pre} U_C \Rightarrow R$
C. $P_R \wedge \operatorname{pre} U_A \Rightarrow (\exists L_C \bullet \operatorname{pre} U_C)$

Whether or not B and D hold depends on the constraints $pred_A$ and $pred_C$ in the global states. In particular, we shall see that if both predicates are true, then B and D will always hold for promotions in this format.

In general, to show D the crucial property that needs to hold is that

$$\forall L'_C \bullet \exists G'_C \bullet f'_C = f_C \oplus \{i? \mapsto \theta L'_C\}$$

That is, does the definition f'_C violate the predicate in G'_C ? If not, then D holds, so that the concrete promotion is free as required.

To show that B holds we need to prove

$$P_R \wedge \operatorname{pre} U_A \wedge U_C \Rightarrow (\forall L'_A \bullet R' \Rightarrow (\exists G'_A \bullet U_A \wedge P'_R))$$

Calculating the components, $P_R \wedge \operatorname{pre} U_A \wedge U_C$ is

$$
\begin{array}{|l}
G_A;\ G_C;\ L_A \\
\Delta G_C \\
\Delta L_C \\
i? : I \\
\hline
i? \in \operatorname{dom} f_A \\
i? \in \operatorname{dom} f_C \\
\operatorname{dom} f_A = \operatorname{dom} f_C \\
\theta L_A = f_A(i?) \\
\theta L_C = f_C(i?) \\
f'_C = f_C \oplus \{i? \mapsto \theta L'_C\} \\
\forall i : \operatorname{dom} f_A \bullet \exists R \bullet \theta L_C = f_C(i) \wedge \theta L_A = f_A(i)
\end{array}
$$

and $\exists G'_A \bullet U_A \wedge P'_R$ is

$$
\begin{array}{|l}
G_A \\
G'_C \\
\Delta L_A \\
i? : I \\
\hline
i? \in \operatorname{dom} f_A \\
\theta L_A = f_A(i?) \\
\exists f'_A \bullet (f'_A = f_A \oplus \{i? \mapsto \theta L'_A\} \\
\qquad \operatorname{dom} f'_A = \operatorname{dom} f'_C \\
\qquad \forall i : \operatorname{dom} f'_A \bullet \exists R' \bullet \theta L'_C = f'_C(i) \wedge \theta L'_A = f'_A(i))
\end{array}
$$

Let us assume $\theta L'_A \mapsto \theta L'_C \in R'$ for an arbitrary L'_A. The key part of the proof is to show we can find an appropriate f'_A. In fact we have to set $f'_A = f_A \oplus \{i? \mapsto \theta L'_A\}$ and thus for B to hold it is necessary that this satisfies the predicate $pred'_A$. If it does the remaining condition is that

$$\forall i : \operatorname{dom} f'_A \bullet \exists R' \bullet \theta L'_C = f'_C(i) \wedge \theta L'_A = f'_A(i)$$

or in other words that $f'_A(i) \mapsto f'_C(i) \in R'$ for all $i \in \operatorname{dom} f'_A$. Now if $i = i?$, then $f'_A(i) = \theta L'_A$ and $f'_C(i) = \theta L'_C$. Thus $f'_A(i) \mapsto f'_C(i) \in R'$ by hypothesis. On the other hand if $i \neq i?$, then $f'_A(i) = f_A(i)$ and $f'_C(i) = f_C(i)$, and again $f'_A(i) \mapsto f'_C(i) \in R'$.

The obvious corresponding result also holds for the initialisation and termination promotion schemas. With these results in place we can use them to simplify the proofs needed for promotion refinements in a number of situations.

Example 6.3 The global state spaces and promotion schema of the example in Sects. 6.1 and 6.4.1 are in the format given above. Therefore with no further verification we know that the concrete global specification is a refinement of the abstract global specification. □

Example 6.4 Consider the specification of a football league given in Sect. 6.2. We can refine the local specification of a single club by replacing the members function by a sequence to give a concrete version of a club.

L_C

$members_C : \mathrm{iseq}\, PEOPLE$
$away_C : \mathrm{iseq}\, ID$
$points_C : \mathbb{N}$

$\mathrm{ran}\, away_C \subseteq \mathrm{dom}\, members_C$

$InitL_C$

L'_C

$members'_C = \langle\,\rangle$
$points'_C = 0$

$Term_C$

L_C

$members_C = \langle\,\rangle$
$points_C = 0$

$Join_C$

ΔL_C
$app? : PEOPLE$
$id! : ID$

$app? \notin \mathrm{ran}\, members_A$
$id! = \#members_A + 1$
$members'_A = members_A \frown \langle app? \rangle$
$points'_A = points_A$
$away'_A = away_A$

Buy_C

ΔL_C
$app? : PEOPLE$

$points'_C = points_C$
$members'_C = members_C$
$\exists x : \mathbb{N} \bullet (members_C(x) = app? \wedge away'_C = away_C \frown \langle x \rangle)$

The concrete football league will promote this specification using the global and promotion schemas G_C and U_C which allow the description of global operations such as $GlobalBuy_C$, $GNew_C$ etc.

$$
\begin{array}{|l}
\hline
G_C \\
clubs_C : I \twoheadrightarrow L_C \\
europe_C : \mathbb{P}\, I \\
\hline
europe_C \subseteq \mathrm{dom}\, clubs_C \\
\hline
\end{array}
$$

$$
\begin{array}{|l}
\hline
Update_C \\
\Delta G_C \\
\Delta L_C \\
i? : I \\
\hline
i? \in \mathrm{dom}\, clubs_C \\
\theta L_C = clubs_C(i?) \\
clubs'_C = clubs_C \oplus \{i? \mapsto \theta L'_C\} \\
europe'_C = europe_C \\
\hline
\end{array}
$$

Then to show that the concrete football league is a refinement of the abstract league we can apply the above result. The global state spaces are almost in the format given above, except we have the addition of a single state variable *europe*.

It is easy to show that the proofs go through where we augment the predicate in P_R with $europe_A = europe_C$. For example, it is clear that $clubs'_A = clubs_A \oplus \{i? \mapsto \theta L'_A\}$ and $europe'_A$ satisfy the invariant in G'_A and that conditions A to D hold. Therefore the concrete league is a refinement of the abstract league. □

Example 6.5 We again consider the specification of a football league. The description of an abstract and concrete single club are the same as in Sect. 6.2 and Example 6.4. However, we will use different global state spaces. This time the collection of clubs will be represented as a sequence instead of a function, ordered by a club's position in the league (i.e., highest points first).

$$
\begin{array}{|l}
\hline
League_A \\
clubs_A : \mathrm{seq}\, L_A \\
\hline
\forall i, j : \mathrm{dom}\, clubs_A \mid i < j \bullet (clubs_A(i).points_A \geq clubs_A(j).points_A) \\
\hline
\end{array}
$$

$$
\begin{array}{|l}
\hline
League_C \\
clubs_C : \mathrm{seq}\, L_C \\
\hline
\forall i, j : \mathrm{dom}\, clubs_C \mid i < j \bullet (clubs_C(i).points_C \geq clubs_C(j).points_C) \\
\hline
\end{array}
$$

Promotion schemas $Update_A$ and $Update_C$ will be syntactically the same as in Example 6.2 and 6.4 apart from the omission of the predicates relating to $europe_A$ and $europe_C$.

Now when we attempt to show that refinement distributes through promotion by verifying conditions A–D we find that D fails, i.e., the concrete promotion is not free (in fact neither is the abstract promotion). To see why the promotions are constrained, consider the expression $clubs'_C = clubs_C \oplus \{i? \mapsto \theta L'_C\}$. For the promotion to be free this has to satisfy the invariant in $League'_C$ for every possible local after-state $\theta L'_C$. The problem is that the invariant requires the point order to be preserved, however, nothing about $\theta L'_C$ guarantees this, a local operation might increase the points invalidating the $clubs'_C$ from satisfying $League'_C$.

In fact, the promotion can be adjusted so that it becomes free by altering the promotion schema U_C. Indeed the requirement $clubs'_C = clubs_C \oplus \{i? \mapsto \theta L'_C\}$ is too strong since it preserves the ordering of the clubs in the league. Instead we could relax this to simply require that

$$items(clubs'_C) = items(clubs_C \oplus \{i? \mapsto \theta L'_C\})$$

This allows any reordering of $clubs_C$ necessary to preserve the invariant, and the promotion would then be free. □

The conditions in Definition 6.2 are sufficient to guarantee refinements of promotions, however, they are not necessary conditions. For example, even if the concrete promotion is constrained, a global refinement might still hold, but to find out whether this is the case the operations have to be checked on a case by case basis.

Example 6.6 Consider the promotions of the local away club card *Buy* operation to the global state of Example 6.5 given by $LeagueBuy_A == \exists \Delta L_A \bullet Update_A \land Buy_A$ and $LeagueBuy_C == \exists \Delta L_C \bullet Update_C \land Buy_C$. Although the concrete promotion is constrained, $LeagueBuy_C$ is a correct refinement of $LeagueBuy_A$, and this can be verified directly by expanding the definitions of these operations and checking the downward simulation conditions.

However, not all refinements are promoted. For example, consider the *Match* operation, which describes the change to a club's points as a result of playing a match. In the abstract local specification this is represented by $Match_A$ which non-deterministically increments a club's points (non-deterministically because the form of a club on paper appears to have no relation to the outcome of a match!):

$Match_A$

ΔL_A

$\exists x : \{0, 1, 3\} \bullet points'_A = points_A + x$
$members'_A = members_A$
$away'_A = away_A$

A rather cunning implementation refines this to the following, which always increments the points by 3:

$$\begin{array}{|l} \textit{Match}_C \\ \hline \Delta L_C \\ \hline \textit{points}'_C = \textit{points}_C + 3 \\ \textit{members}'_C = \textit{members}_C \\ \textit{away}'_C = \textit{away}_C \\ \hline \end{array}$$

$Match_C$ is a downward simulation of $Match_A$. However, $LeagueMatch_A == \exists \Delta L_A \bullet Update_A \wedge Match_A$ is not refined by $LeagueMatch_C == \exists \Delta L_C \bullet Update_C \wedge Match_C$. This refinement fails since in the abstract we can keep the points unchanged, so preserving the sortedness invariant in $League_A$, but in the concrete we lose that option making $LeagueMatch_C$ undefined when $LeagueMatch_A$ is not. □

The refinements we have described so far have kept the same structure between the abstract and concrete global state spaces. However, Theorem 6.2 can also be used to verify refinements where the data structures in the abstract and concrete global state spaces differ.

Example 6.7 Sticking with the football league theme, consider the abstract description of a single club given by L_A, $InitL_A$, etc. We will now consider two promotions using different global states. Suppose our set of football clubs is given by the enumerated type I consisting of 20 clubs.

$$I ::= leeds \mid manutd \mid sunderland \mid \cdots \mid sheffwed$$

Consider the global state spaces

$$\begin{array}{|l} \textit{PremLeague}_A \\ \hline \textit{clubs}_A : I \rightarrow L_A \\ \hline \end{array}$$

$$\begin{array}{|l} \textit{PremLeague}_C \\ \hline \textit{clubs}_C : \operatorname{seq} L_A \\ \hline \# \operatorname{dom} \textit{clubs}_C = 20 \\ \hline \end{array}$$

together with promotion schemas as before.

The local specifications used in the promotions are identical, and we use Definition 6.2 to verify a refinement between the promotions. To do so we define a mapping M between the index set I and $\mathbb{N}$ (i.e., the domain of $clubs_C$) with $M : I \rightarrowtail \mathbb{N}$ given by $M(leeds) = 1, M(manutd) = 2$ etc. The correct retrieve relation P_R will be given by

$$
\begin{array}{|l}
\hline
P_R \\
PremLeague_A \\
PremLeague_C \\
\hline
\mathrm{dom}\, clubs_C = M(\!|\, \mathrm{dom}\, clubs_A \,|\!) \\
\forall i : \mathrm{dom}\, clubs_A \bullet \theta clubs_A(i) = \theta clubs_C(M(i)) \\
\hline
\end{array}
$$

Conditions A–D in Definition 6.2 are easily verified and therefore the global concrete specification is a refinement of the global abstract specification. □

6.6 Calculating Refinements

The techniques for calculating refinements discussed in Chap. 5 can be combined with the use of promotion in an obvious fashion. For example, given an abstract local specification, concrete local state space and retrieve relation R, the concrete initialisation and operations can be calculated as discussed in Chap. 5. If R and the promotion schemas satisfy Definition 6.2, then promotions of all the calculated operations will be refinements of promotions of the abstract. In particular, if our promotions conform to the format described in Sect. 6.5 then we know that they freely promote any downward simulation so that promotion will distribute over any calculated refinement.

6.7 Upward Simulations of Promotions

When a refinement is verified by an upward simulation we can also distribute the refinement through a promotion under circumstances similar to those discussed in Sects. 6.4 and 6.5.

For example, suppose the global abstract and concrete states are given by

$$
\begin{array}{|l}
\hline
G_A \\
f_A : I \rightarrow\!\!\!\!\!+ L_A \\
\hline
\end{array}
\qquad
\begin{array}{|l}
\hline
G_C \\
f_C : I \rightarrow\!\!\!\!\!+ L_C \\
\hline
\end{array}
$$

and that the promotion schemas U_A and U_C have the following format

$$
\begin{array}{|l}
\multicolumn{1}{l}{U_A} \\
\hline
\Delta G_A \\
\Delta L_A \\
i? : I \\
\hline
i? \in \operatorname{dom} f_A \\
\theta L_A = f_A(i?) \\
f_A' = f_A \oplus \{i? \mapsto \theta L_A'\} \\
\hline
\end{array}
$$

$$
\begin{array}{|l}
\multicolumn{1}{l}{U_C} \\
\hline
\Delta G_C \\
\Delta L_C \\
i? : I \\
\hline
i? \in \operatorname{dom} f_C \\
\theta L_C = f_C(i?) \\
f_C' = f_C \oplus \{i? \mapsto \theta L_C'\} \\
\hline
\end{array}
$$

Given an upward simulation between the local specifications verified by retrieve relation T, define P_T by

$$
\begin{array}{|l}
\multicolumn{1}{l}{P_T} \\
\hline
G_A \\
G_C \\
\hline
\operatorname{dom} f_A = \operatorname{dom} f_C \\
\forall n : \operatorname{dom} f_A \bullet \exists T \bullet \theta L_A = f_A(n) \land \theta L_C = f_C(n) \\
\hline
\end{array}
$$

Then there is an upward simulation between the promoted global operations $\exists \Delta L_A \bullet (U_A \land Op_A)$ and $\exists \Delta L_C \bullet (U_C \land Op_C)$ verified by retrieve relation P_T.

6.8 Bibliographical Notes

Promotion as a technique for structuring Z specifications appeared as early as the first edition of [2], and there are extensive examples documented (see, for example, [1, 3, 7]). The definition of freeness is due to Woodcock [6] and work on refinement of promotions goes back to [3]. In [3] Lupton defines the conditions which are sufficient for the distribution of refinement through promotion and proves the result for downward simulations. Upward simulations were not considered in [3]. Mahony [4] has considered how the structuring technique of promotion can be embedded within the refinement calculus notation.

References

1. Barden, R., Stepney, S., & Cooper, D. (1994). *Z in Practice. BCS Practitioner Series*. New York: Prentice Hall.
2. Hayes, I. J. (Ed.) (1987). *Specification Case Studies. International Series in Computer Science*. New York: Prentice Hall. 2nd ed., 1993.
3. Lupton, P. J. (1990) Promoting forward simulation. In Nicholls [5] (pp. 27–49).
4. Mahony, B. P. (1999). The least conjunctive refinement and promotion in the refinement calculus. *Formal Aspects of Computing*, *11*(1), 75–105.
5. Nicholls, J. E. (Ed.) (1990). *Workshops in Computing. Z User Workshop*, Oxford. Berlin: Springer.
6. Woodcock, J. C. P. (1989). Mathematics as a management tool: proof rules for promotion. In *CSR Sixth Annual Conference on Large Software Systems*. Amsterdam: Elsevier.
7. Woodcock, J. C. P., & Davies, J. (1996). *Using Z: Specification, Refinement, and Proof*. New York: Prentice Hall.

Chapter 7
Testing and Refinement

In this chapter we look at the relationship between testing and refinement, and in particular look at how to test refinements and implementations based upon tests generated from abstract specifications.

Testing and formal specifications might appear at first sight to be strange bedfellows, but a specification, whether formal or informal, acts as the benchmark against which any implementation is tested. Specifications can also be used to generate tests for an implementation, and formal methods are important because they offer a possibility of automating this test generation process. For this reason there has been considerable work on model-based testing and developing test generation techniques from formal specifications, including those written in Z and Object-Z.

The essential problem in deriving tests from a specification is how to be able to produce a finite set of relevant tests from a specification which potentially describes an infinite state space. To tackle this many approaches aim to derive a finite state machine (FSM) from the specification. The states in the FSM are equivalence classes of states in the specification, computed on the basis of properties the developer considers important to test, and the transitions between states will be instances of the operations which make that state change. The FSM can then be used to construct test suites, which are structured sequences of test cases covering all the test cases and states.

This process produces tests based on the abstract specification, but clearly a specification is likely to be refined in a non-trivial manner before its implementation, and therefore it is necessary to relate the abstract tests to the actual implementation. The problem is that the abstract tests may be insufficient or even incomparable to the concrete implementation. For example, the abstract tests may be defined in terms of sets whereas the concrete implementation uses lists, and in order to use the abstract tests to test the implementation it is necessary to relate the values in the state spaces (e.g., using the retrieve relation). Similarly non-determinism in the abstract specifi-

Excerpts from pp. 27–35, 38, 40–44, 46 of [13] are reprinted within this chapter with permission from John Wiley & Sons Limited.

J. Derrick, E.A. Boiten, *Refinement in Z and Object-Z*,
DOI 10.1007/978-1-4471-5355-9_7, © Springer-Verlag London 2014

cation might be resolved in an implementation and new or different tests might be necessary because of this.

All of this means that tests generated from a specification will only be usable if the specification is the implementation specification (i.e., the one from which the coding is done directly), or if the refinements are extremely simple (e.g., one using a functional retrieve relation with little data refinement).

To deal effectively with these situations a method is needed to test an implementation based upon the tests generated from the abstract specification. This is achieved by calculating concrete tests for a refinement from the abstract tests using the calculational methods from Chap. 5, and by doing so a new concrete finite state machine can be calculated in a relatively simple manner from the abstract FSM.

In this chapter we will discuss both how to derive tests from a formal specification, and also how tests can be refined for use with an implementation. Section 7.1 describes briefly how to derive tests from a Z specification. Sections 7.2 and 7.3 then consider how concrete tests can be calculated from abstract ones for refinements which are downward simulations, and finally Sect. 7.4 looks at upward simulations.

7.1 Deriving Tests from Specifications

There are many approaches to deriving tests from formal specifications (see Sect. 7.5), and in this chapter we shall concentrate on one which produces a FSM by analysing the states and operations together.

The basic test case generation technique we use consists of a partition analysis, which reduces the specification of each operation into a Disjunctive Normal Form (DNF). Each element in the DNF represents an individual test case for the operation. From these test cases a partition of the system state is performed resulting in a set of disjoint states, each of which is either the before-state or after-state of at least one test to be performed. This partition then serves as a basis for the construction of a finite state machine which is then used to derive the test suites.

This process can be described as the following procedure:

1. Perform a partition analysis on all operations to generate the test cases based upon what the tester considers important. These are the transitions in the FSM.
2. From each test case obtain its before-state and after-state (by existentially quantifying variables not being considered); each of these will be an equivalence class of ADT states.
3. Perform a DNF partition analysis on the states from Step 2. This gives the states in the FSM.
4. Construct the FSM by resolving transitions against states.

The DNF partition analysis of an operation defined in Step 1 produces a set of tests by splitting an operation schema into a disjunction of schemas given by the DNF. Each schema in this DNF represents a single test case. Each test case will be disjoint, allowing them all to be treated separately. This is a semi-automatic process rather than an automatic one because for any non-trivial specification there are many possible choices for DNF; the choice taken depends upon the aspects that are considered important by the developers.

Example 7.1 Consider the specification of the MSMIE protocol given in Example 5.1 (p. 130). Suppose that we decide the important states to be tested are when b_A is true and the set ms is empty. To generate the test cases the predicate of each operation is written as a DNF with respect to these properties in the before-state. This results in the following test cases:

$$\begin{array}{l} \textit{Init}_A \\ \hline \textit{State}'_A \\ \hline \neg b'_A \wedge ms' = \varnothing \end{array}$$

$$\begin{array}{l} \textit{Slave}_A 1 \\ \hline \Delta \textit{State}_A \\ \hline \neg b_A \wedge b'_A \wedge ms = \varnothing \wedge ms' = \varnothing \end{array}$$

$$\begin{array}{l} \textit{Slave}_A 2 \\ \hline \Delta \textit{State}_A \\ \hline b_A \wedge b'_A \wedge ms = \varnothing \wedge ms' = \varnothing \end{array}$$

$$\begin{array}{l} \textit{Slave}_A 3 \\ \hline \Delta \textit{State}_A \\ \hline b_A \wedge b'_A \wedge ms \neq \varnothing \wedge ms' = ms \end{array}$$

$$\begin{array}{l} \textit{Acquire}_A 1 \\ \hline \Delta \textit{State}_A \\ l? : \textit{MNAME} \\ \hline b_A \wedge b'_A \\ ms = \varnothing \wedge ms' = \{l?\} \end{array}$$

$$\begin{array}{l} \textit{Acquire}_A 2 \\ \hline \Delta \textit{State}_A \\ l? : \textit{MNAME} \\ \hline b_A \wedge b'_A \\ ms \neq \varnothing \wedge l? \notin ms \\ ms' = ms \cup \{l?\} \end{array}$$

$$\begin{array}{l} \textit{Release}_A 1 \\ \hline \Delta \textit{State}_A \\ l? : \textit{MNAME} \\ \hline l? \in ms \wedge ms' = ms \setminus \{l?\} \\ (ms' = \varnothing \wedge b_A) \Rightarrow b'_A \\ (ms' \neq \varnothing) \Rightarrow (b_A = b'_A) \end{array}$$

Note that the invariant on *State* already excludes states with $\neg b_A \wedge ms \neq 0$. □

The construction performed by Step 1 has two useful properties: coverage and disjointedness. That is, each operation equals the disjunction of its test cases (coverage) and these tests are disjoint. In general a collection of tests $\{AOp_i\}_{i \in T}$ is said to cover an operation *AOp* acting on state space *AState* if

$$AOp = \exists\, i : T \bullet AOp_i$$

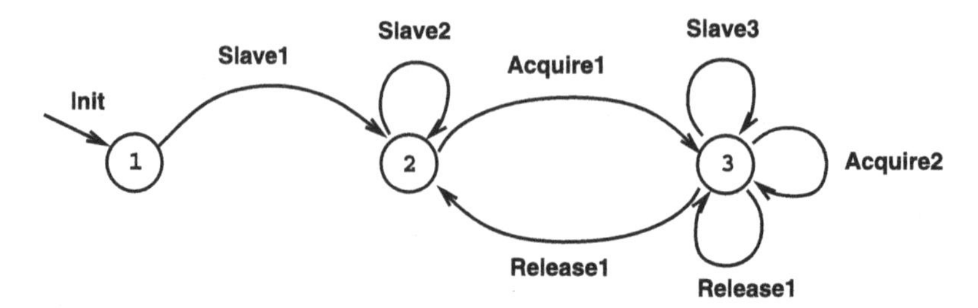

Fig. 7.1 The FSM for the protocol

and the tests are disjoint, if, for all $i \neq j \in T$

$$\neg \exists AState;\ AState' \bullet AOp_i \wedge AOp_j$$

In the protocol example above it is easy to see that $\{Slave_A1, Slave_A2, Slave_A3\}$ form a disjoint covering for $Slave_A$.

Steps 2 and 3 in the procedure then calculate the states that result from these test cases. The before-state for a test case AOp_i is given by $\exists AState';\ ?AOp_i;\ !AOp_i \bullet AOp_i$, and the after-state will be given by $\exists AState;\ ?AOp_i;\ !AOp_i \bullet AOp_i$. After performing a DNF partition analysis on the states that result from this consideration of the test cases we are left with the states of the FSM machine that will be used to test implementations of the specification. The FSM can now be constructed by calculating the individual transitions between the available states.

Example 7.2 We can calculate the states that result from the protocol test cases as follows. The before-state for the $Slave_A1$ test case is given by $\exists State'_A \bullet Slave_A1$, and the after-state will be given by $\exists State_A \bullet Slave_A1$. After calculating the potential before- and after-states for all the protocol operations the following states remain:

$State_A1$	$State_A2$	$State_A3$
$State_A$	$State_A$	$State_A$
$\neg b_A$ $ms = \varnothing$	b_A $ms = \varnothing$	b_A $ms \neq \varnothing$

Step 4 now builds the FSM by resolving transitions against states. For example, a transition of $Slave_A1$ exists from state $State_A1$ to $State_A2$ precisely when the following evaluates to true: $\exists State_A;\ State'_A;\ ?Slave_A1;\ !Slave_A1 \bullet State_A1 \wedge Slave_A1 \wedge State_A2'$. After completing this process we obtain the FSM shown in Fig. 7.1. □

Example 7.3 An automatic teller machine (ATM) enables users to withdraw cash on insertion of a card and valid PIN. The ATM we specify has operations to insert a card, check the PIN, withdraw money and cancel the process. To describe these the customers' bank accounts are represented by a partial function *accts* from account numbers to amounts, and the PIN for an account n is given by $pins(n)$. The card currently inside the machine is represented by *card* (0 represents no card present), and a Boolean *trans* is used to determine whether a transaction is allowed to proceed.

$$CARD == \mathbb{N}$$
$$OPTIONS ::= passwd \mid insert\_card \mid withdraw\_cash \mid sorry$$

ATM

$$accts : \mathbb{N} \pfun \mathbb{N}$$
$$pins : \mathbb{N} \pfun \mathbb{N}$$
$$card : \mathbb{N}$$
$$trans : \mathbb{B}$$

$$\mathrm{dom}\, accts \subseteq \mathrm{dom}\, pins$$

Init

$$ATM'$$

$$card' = 0$$
$$\neg trans'$$

None of the operations we specify here changes *pins*, so we define *ATMOp* as

ATMOp

$$\Delta ATM$$

$$pins' = pins$$

When a card is inserted it is checked to see whether it is a known account and, if it is, the user must input his/her PIN. If, and only if, this PIN is valid can money be withdrawn.

Insert

$$ATMOp$$
$$card? : CARD$$
$$screen! : OPTIONS$$

$$card = 0 \wedge trans' = trans \wedge accts' = accts$$
$$((card? \in \mathrm{dom}\, accts \wedge card' = card? \wedge screen! = passwd) \vee$$
$$(card? \notin \mathrm{dom}\, accts \wedge card' = card \wedge screen! = sorry))$$

Passwd

$$ATMOp$$
$$pin? : \mathbb{N}$$
$$screen! : OPTIONS$$

$$card \neq 0 \wedge \neg trans \wedge card' = card \wedge accts' = accts$$
$$((pins(card) = pin? \wedge trans' \wedge screen! = withdraw\_cash) \vee$$
$$(pins(card) \neq pin? \wedge \neg trans' \wedge screen! = passwd))$$

At any stage we can cancel the transaction and the *Cancel* operation takes us back into one of the previous states.

$$
\begin{array}{l}
\textit{Cancel} \\
\hline
\textit{ATMOp} \\
\textit{screen}! : \textit{OPTIONS} \\
\hline
\textit{card} \neq 0 \land \textit{accts}' = \textit{accts} \\
((\neg \textit{trans} \land \textit{card}' = 0 \land \neg \textit{trans}' \land \textit{screen}! = \textit{insert\_card}) \lor \\
(\textit{trans} \land \textit{trans}' \land \textit{card}' = \textit{card} \land \textit{screen}! = \textit{withdraw\_cash}) \lor \\
(\textit{trans} \land \neg \textit{trans}' \land \textit{card}' = 0 \land \textit{screen}! = \textit{insert\_card}))
\end{array}
$$

The withdraw operation at this level of abstraction is non-deterministic. Sometimes the bank gives out money, but at other times it just keeps the card, and to the user it is unclear why this happens.

$$
\begin{array}{l}
\textit{Withdraw} \\
\hline
\textit{ATMOp} \\
\textit{amount}?, \textit{money}! : \mathbb{N} \\
\hline
\textit{trans} \land \textit{accts}' = \textit{accts} \oplus \{\textit{card} \mapsto \textit{accts}(\textit{card}) - \textit{money}!\} \\
((\textit{card}' = 0 \land \neg \textit{trans}' \land \textit{money}! = 0) \lor \\
(\textit{card}' = \textit{card} \land \textit{trans}' \land 0 \leq \textit{money}! \leq \textit{amount}?))
\end{array}
$$

We wish to test our ATM on the basis of whether the card is known and whether the PIN is correct, i.e., according to whether $\textit{card} = 0$ and $\textit{trans} = \textit{true}$. Performing a DNF partition on the operations produces the following tests:

$$
\begin{array}{l}
\textit{Insert1} \\
\hline
\textit{ATMOp} \\
\textit{card}? : \textit{CARD} \\
\textit{screen}! : \textit{OPTIONS} \\
\hline
\neg \textit{trans} \land \textit{card} = 0 \land \textit{trans}' = \textit{trans} \land \textit{accts}' = \textit{accts} \\
((\textit{card}? \in \operatorname{dom} \textit{accts} \land \textit{card}' = \textit{card}? \land \textit{screen}! = \textit{passwd}) \lor \\
(\textit{card}? \notin \operatorname{dom} \textit{accts} \land \textit{card}' = \textit{card} \land \textit{screen}! = \textit{sorry}))
\end{array}
$$

$$
\begin{array}{l}
\textit{Insert2} \\
\hline
\textit{ATMOp} \\
\textit{card}? : \textit{CARD} \\
\textit{screen}! : \textit{OPTIONS} \\
\hline
\textit{trans} \land \textit{card} = 0 \land \textit{trans}' = \textit{trans} \land \textit{accts}' = \textit{accts} \\
((\textit{card}? \in \operatorname{dom} \textit{accts} \land \textit{card}' = \textit{card}? \land \textit{screen}! = \textit{passwd}) \lor \\
(\textit{card}? \notin \operatorname{dom} \textit{accts} \land \textit{card}' = \textit{card} \land \textit{screen}! = \textit{sorry}))
\end{array}
$$

Passwd1

$$
\begin{array}{|l}
\hline
ATMOp \\
pin? : \mathbb{N} \\
screen! : OPTIONS \\
\hline
card \neq 0 \land \neg trans \land card' = card \land accts' = accts \\
((pins(card) = pin? \land trans' \land screen! = withdraw\_cash) \lor \\
(pins(card) \neq pin? \land \neg trans' \land screen! = passwd)) \\
\hline
\end{array}
$$

Cancel1

$$
\begin{array}{|l}
\hline
ATMOp \\
screen! : OPTIONS \\
\hline
card \neq 0 \land accts' = accts \\
\neg trans \land card' = 0 \land \neg trans' \land screen! = insert\_card \\
\hline
\end{array}
$$

Cancel2

$$
\begin{array}{|l}
\hline
ATMOp \\
screen! : OPTIONS \\
\hline
card \neq 0 \land accts' = accts \\
((trans \land trans' \land card' = card \land screen! = withdraw\_cash) \lor \\
(trans \land \neg trans' \land card' = 0 \land screen! = insert\_card)) \\
\hline
\end{array}
$$

Withdraw1

$$
\begin{array}{|l}
\hline
ATMOp \\
amount?, money! : \mathbb{N} \\
\hline
trans \land accts' = accts \oplus \{card \mapsto accts(card) - money!\} \\
((card' = 0 \land \neg trans' \land money! = 0) \lor \\
(card' = card \land trans' \land 0 \leq money! \leq amount?)) \\
\hline
\end{array}
$$

Steps 2 and 3 produce four before- and after-states for these transitions:

State1

$$
\begin{array}{|l}
\hline
ATM \\
\hline
card = 0 \land \neg trans \\
\hline
\end{array}
$$

State2

$$
\begin{array}{|l}
\hline
ATM \\
\hline
card \neq 0 \land \neg trans \\
\hline
\end{array}
$$

State3

$$
\begin{array}{|l}
\hline
ATM \\
\hline
card \neq 0 \land trans \\
\hline
\end{array}
$$

State4

$$
\begin{array}{|l}
\hline
ATM \\
\hline
card = 0 \land trans \\
\hline
\end{array}
$$

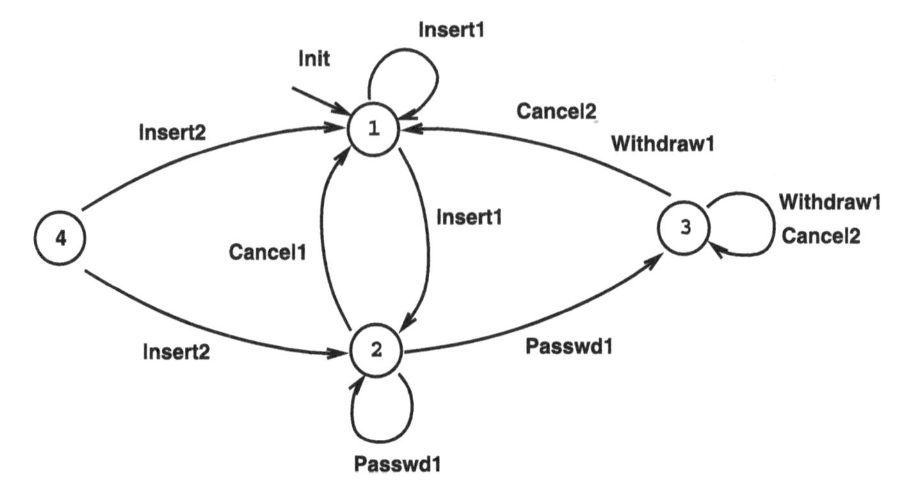

Fig. 7.2 The FSM for the ATM

Finally in Step 4 we resolve the test cases against the states to derive the FSM shown in Fig. 7.2.

Notice that in fact *State*4 is unreachable, i.e., we can never reach a state with $card = 0$ and $trans = true$ through a sequence of operations as defined. Although refinement allows arbitrary behaviour to be introduced in unreachable states (establish reachability invariant, then weaken precondition), we assume that even those states carry some meaning in the specifier's mind; thus we do not prune unreachable states. □

7.2 Testing Refinements and Implementations

In addition to deriving tests from a formal specification, any non-trivial specification will be refined further before its implementation.

We would like to be able to test a concrete implementation against the abstract specification, by using the test cases and the derived FSM. However, the transitions and states in the FSM will be described in terms of the abstract state space and not the concrete realisation. In addition, the process of refinement potentially resolves or moves the non-determinism in the operations, for example, the concrete operations might have weaker preconditions and stronger postconditions. Hence the FSM generated from the abstract specification cannot be used to test a concrete implementation except under the simplest of refinements.

In the remainder of this chapter we discuss how we can test refinements where, instead of deriving the FSM from the concrete specification, we calculate a concrete FSM from the abstract FSM using methods taken from Chap. 5.

The technique employed to generate tests for a refinement is very simple. Given an abstract specification with operation *AOp* and a covering disjoint set of tests

$\{AOp_i\}_{i \in T}$; a concrete specification with operation COp which refines AOp, and a retrieve relation R, a set of tests $\{COp_i\}_{i \in T}$ is generated where each test COp_i is the weakest refinement calculated from R and AOp_i. These new concrete tests will become the transitions in the new FSM. The states are generated using the retrieve relation in a similar fashion.

The two cases of downward and upward simulations are considered separately. In each case the following questions are explored:

- How are the new transitions COp_i generated?
  - Do these tests $\{COp_i\}_{i \in T}$ cover COp?
  - Are these tests $\{COp_i\}_{i \in T}$ disjoint?
- How are the new states and the FSM generated?

In the remainder of this section we will discuss these questions when the refinement is a downward simulation, in Sect. 7.4 we will look at upward simulations.

7.2.1 Calculating Concrete Tests

Let AOp be an operation with $\{AOp_i\}_{i \in T}$ being its disjoint set of tests. Let R be a retrieve relation which defines a downward simulation between states $AState$ and $CState$. Then a set of concrete tests can be specified by (cf. Chap. 5) either

$$COp_i == \exists AState; AState' \bullet R \wedge AOp_i \wedge R'$$

or

$$\begin{aligned} COp_i == {} & (\exists AState \bullet \text{pre}\, AOp_i \wedge R) \wedge \\ & (\forall AState \bullet \text{pre}\, AOp_i \wedge R \Rightarrow \exists AState' \bullet AOp_i \wedge R') \end{aligned}$$

where we can use the first, simpler, form of calculation depending on whether R is a function from $CState$ to $AState$ on $\exists AState \bullet \text{pre}\, AOp \wedge R$, see Corollary 5.2. Of course when a simplification is possible the two definitions of COp_i coincide.

These will be our test cases for the original concrete operation COp, and we have the following result.

Theorem 7.1 *Let AOp be an abstract operation with $\{AOp_i\}_{i \in T}$ being its disjoint set of tests. Let COp be a downward simulation of AOp. Let R be the retrieve relation. Let COp_i be the concrete tests given above. Then*

$$(\exists i : T \bullet COp_i) \sqsubseteq_{DS} COp$$

and if COp is the weakest downward simulation of AOp then $COp = \exists i : T \bullet COp_i$. □

The purpose of this is to enable one to use abstract tests together with the retrieve relation to calculate a new concrete partition analysis for the refinement. If the concrete is the weakest refinement of the abstract then these concrete tests exactly cover the concrete operations. In addition, for a functional retrieve relation disjoint abstract tests will generate disjoint concrete tests.

Theorem 7.2 *Let $\{AOp_i\}_{i \in T}$ be disjoint test cases, R a functional (from concrete to abstract) retrieve relation and COp_i calculated from AOp_i. Then $\{COp_i\}_{i \in T}$ are disjoint.* □

Note that disjointedness does not imply inequality, since two tests with false predicates are considered disjoint.

Example 7.4 Consider the refinement of the MSMIE protocol given in Example 5.1. We will calculate test cases for the refinement directly from the abstract FSM derived in Example 7.2. The retrieve relation is functional, therefore the concrete tests are given by the simplified calculations, for example,

$$Slave_C1 == \exists\, State_C;\ State'_C \bullet R \wedge Slave_A1 \wedge R'$$

and the complete set of tests for $Slave_C$ is:

$Slave_C1$

$\Delta State_C$

$b_C = sii \wedge b'_C \neq sii$
$ms = \varnothing \wedge ms' = ms$

$Slave_C2$

$\Delta State_C$

$b_C \neq sii \wedge b'_C \neq sii$
$ms = \varnothing \wedge ms' = ms$

$Slave_C3$

$\Delta State_C$

$b_C \neq sii \wedge b'_C \neq sii$
$ms \neq \varnothing \wedge ms' = ms$

As expected the tests produced by this means form an exact covering of the concrete operations since the concrete specification was the weakest refinement of the abstract. Therefore $(\exists\, i : \{1, 2, 3\} \bullet Slave_Ci) = Slave_C$. In addition, since the retrieve relation is functional, all the calculated tests are disjoint and can be calculated easily. □

Example 7.5 Our ATM from Example 7.3 is being implemented as a concrete teller machine (CTM). The CTM is a simple refinement of the ATM where the state space remains unchanged, however some of the non-determinism in the operations is refined out. For example, the *Cancel* operation above was non-deterministic if the user entered a valid card and PIN, and we refine this non-determinism to offer a simple cancellation. Similarly the *Withdraw* operation is refined here to one that eats the card only if there is insufficient money remaining in the account. We also weaken the precondition of *Cancel* to ensure that nothing happens if it is applied in the initial state. The concrete versions of these two operations are given by

$$\begin{array}{|l}
\multicolumn{1}{l}{\underline{\quad Cancel_{Cnew} \quad}} \\
\Delta ATM \\
screen! : OPTIONS \\
\hline
card \neq 0 \wedge accts' = accts \\
((\neg trans \wedge card' = 0 \wedge \neg trans' \wedge screen! = insert_card) \vee \\
(trans \wedge \neg trans' \wedge card' = 0 \wedge screen! = insert_card)) \\
\hline
\end{array}$$

$$Cancel_{Cerr} == [\Xi ATM \mid card = 0]$$
$$Cancel_C == Cancel_{Cerr} \vee Cancel_{Cnew}$$

$$\begin{array}{|l}
\multicolumn{1}{l}{\underline{\quad Withdraw_C \quad}} \\
\Delta ATM \\
amount?, money! : \mathbb{N} \\
\hline
trans \wedge accts' = accts \oplus \{card \mapsto accts(card) - money!\} \\
((card' = 0 \wedge \neg trans' \wedge money! = 0 \wedge amount? \geq accts(card)) \vee \\
(card' = card \wedge trans' \wedge money! = amount? \wedge \\
\ amount? < accts(card))) \\
\hline
\end{array}$$

We also weaken the precondition of the *Passwd* operation; for security reasons it is important that it only allows a transaction with a valid PIN and account. We therefore refine it to

$$Passwd_C == Passwd \vee [\Xi ATM \mid card = 0 \wedge \neg trans]$$

The *Insert* operation is left unchanged. The concrete specification is then a simple operation refinement of the ATM. We can therefore calculate a set of concrete tests as in the previous example, but with an identity retrieve relation this set of tests is the same as the abstract set. Theorem 7.2 guarantees disjointedness, and Theorem 7.1 guarantees that $(\exists i \bullet COp_i) \sqsubseteq_{DS} COp$ for all concrete operations. However, the concrete specification is not the weakest refinement of the abstract and, except for the *Insert* operation, we do not have $COp = \exists i \bullet COp_i$. This means that some additional tests may be necessary whilst some concrete calculations will be redundant. Section 7.3 below explains the necessary calculations. □

7.2.2 Calculating Concrete States

We have now seen how to take transitions in an abstract FSM and calculate corresponding transitions to test an implementation with a concrete FSM. To construct the concrete FSM it is also necessary to define the states of the FSM and then describe the transitions between the states.

Suppose that an abstract FSM has n disjoint states $AState_1, \ldots, AState_n$, each one being a before- or after-state of an abstract transition AOp_i $(1 \leq i \leq m)$. The concrete states $CState_1, \ldots, CState_n$ are calculated by taking

$$CState_i == \exists AState \bullet AState_i \land R$$

Each concrete state $CState_i$ will be a potential before- or after-state of a concrete transition. However, some abstract states might collapse (i.e., become identified or reduce to *false*) when their concrete counterparts are calculated. In fact it is not hard to see that the concrete states will be disjoint whenever the retrieve relation is a function, but that a general relation will not necessarily produce disjoint concrete states.

Example 7.6 The concrete states can now be calculated for the MSMIE protocol using the abstract states from Example 7.2. Since R is functional the concrete states will be disjoint, and they are:

$State_C1$	$State_C2$	$State_C3$
$State_C$	$State_C$	$State_C$
$b_C = sii$	$b_C \neq sii$	$b_C \neq sii$
$ms = \varnothing$	$ms = \varnothing$	$ms \neq \varnothing$

□

7.3 Building the Concrete Finite State Machine

A set of potential states $CState_j$ has been identified, as has a set of potential tests COp_i. It is now necessary to resolve the test cases against the states, and see if any new transitions are enabled due to the refinement process.

Because the concrete tests are given by calculations it is clear that in the weakest refinement there exists a transition COp_i between $CState_j$ and $CState_k$ precisely when there exists a transition AOp_i between $AState_j$ and $AState_k$.

Whether these are the only transitions depends on whether or not the retrieve relation is functional and whether or not the concrete specification is the weakest refinement of the abstract with respect to the retrieve relation. We consider these possibilities in turn.

7.3.1 Using a Functional Retrieve Relation

If the retrieve relation is functional, then disjoint abstract states and transitions produce disjoint concrete states and transitions. In addition if the concrete operations are the weakest refinement of the abstract ones, the concrete tests cover the concrete operations, no further non-determinism has been resolved and there exists a transition COp_i between $CState_j$ and $CState_k$ precisely when there exists a transition AOp_i between $AState_j$ and $AState_k$.

Therefore the concrete FSM will be isomorphic to the abstract FSM (isomorphic but not identical as the states and transitions are defined in terms of the concrete state space).

Example 7.7 The concrete MSMIE protocol was the weakest refinement of the abstract, and the retrieve relation is functional, therefore the concrete states and transitions for the operations are identified precisely by direct reference to those in the abstract FSM. Hence the concrete FSM is isomorphic to that in Fig. 7.1. □

When the concrete specification is not the weakest refinement, the process of refinement can potentially add new transitions or disable existing ones because refinement can both weaken an operation's precondition but also reduce any non-determinism in an operation by strengthening its postcondition.

When an operation's postcondition is strengthened, the set of potential outcomes is reduced. If one of these outcomes was a transition on its own, the concrete FSM can dispense with this transition. The actual transitions can be calculated by taking $COp_i \wedge COp$ where COp_i is the calculated test and COp the more deterministic concrete operation. In general a transition is not always lost, whether it is or not depends on the partition analysis chosen.

When an operation's precondition is weakened under refinement, states and transitions have to be added, in general, to the concrete FSM, possibly leading to calculation of new portions of the graph. By weakening an operation's precondition under refinement, portions of the graph which were unreachable may now become directly accessible. For this reason we do not prune unreachable states in the initial abstract finite state machine.

Example 7.8 In the operation refinement of the ATM the concrete states will be identical to the abstract states. However, the concrete ATM is not the weakest refinement of the abstract and, in particular, the precondition of *Passwd* was weakened and the postconditions of *Cancel* and *Withdraw* were strengthened.

The weakened precondition of *Passwd* means that additional tests are needed to cover the concrete operation fully, and the extra tests must cover the weakened part of the precondition whilst being disjoint from the remaining tests. Here we simply use

$$
\begin{array}{|l}
\hline
\textit{Passwd2} \\
\Xi ATM \\
\hline
card = 0 \wedge \neg trans \\
\hline
\end{array}
$$

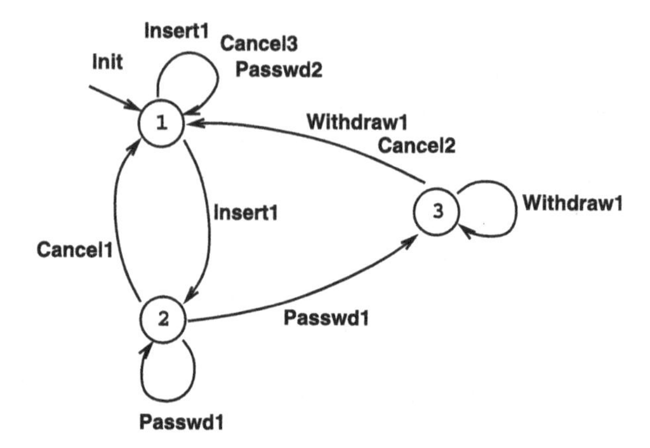

Fig. 7.3 The concrete FSM for the ATM

For the concrete tests for $Cancel_C$ and $Withdraw_C$ we calculate $Cancel1 \wedge Cancel_C$, $Cancel2 \wedge Cancel_C$, $Withdraw1 \wedge Withdraw_C$. Evaluating these we find

$$Cancel1 \wedge Cancel_C = Cancel1$$

$Cancel2 \wedge Cancel_C$

$$\begin{array}{l} \Delta ATM \\ screen! : OPTIONS \\ \hline card \neq 0 \wedge accts' = accts \\ trans \wedge \neg trans' \wedge card' = 0 \wedge screen! = insert\_card \end{array}$$

$Withdraw1 \wedge Withdraw_C$

$$\begin{array}{l} \Delta ATM \\ amount?, money! : \mathbb{N} \\ \hline trans \wedge accts' = accts \oplus \{card \mapsto accts(card) - money!\} \\ ((card' = 0 \wedge \neg trans' \wedge money! = 0 \wedge amount? \geq accts(card)) \vee \\ (card' = card \wedge trans' \wedge money! = amount? \wedge \\ \ amount? < accts(card))) \end{array}$$

We also need an additional test for the weakened precondition of *Cancel*, and we take

$$Cancel3 == Cancel_{Cerr}$$

Thus the FSM for the concrete ATM is as shown in Fig. 7.3. Since all operations are now total we can safely remove the unreachable state 4, since it now can never be reached in any further operation refinement. □

Example 7.9 Consider the specification of the process scheduler and its refinement given in Example 4.8 (p. 117). We choose to test the abstract specification by considering first whether $active = nil$ and then if we require more tests whether $ready, waiting = \varnothing$. The resulting test cases are given by *Init* plus the following tests cases from the operations.

$$
\begin{array}{l}
\textit{New1} \\
\hline
\Delta State \\
p? : PID \\
\hline
active' = nil \wedge active = nil \\
ready' = \varnothing \wedge ready = \varnothing \\
p? \notin waiting \\
waiting' = waiting \cup \{p?\} \\
admin = user \wedge admin' = user \\
\hline
\end{array}
$$

$$
\begin{array}{l}
\textit{New2} \\
\hline
\Delta State \\
p? : PID \\
\hline
p? \neq active \\
p? \notin (ready \cup waiting) \\
waiting' = waiting \cup \{p?\} \\
active' = active \wedge active \neq nil \\
ready' = ready \\
admin = user \wedge admin' = user \\
\hline
\end{array}
$$

$$
\begin{array}{l}
\textit{Ready1} \\
\hline
\Delta State \\
q? : PID \\
\hline
q? \in waiting \\
waiting' = waiting \setminus \{q?\} \\
active = nil \\
ready' = ready \\
active' = q? \\
admin = user \wedge admin' = user \\
\hline
\end{array}
$$

$$
\begin{array}{l}
\textit{Ready2} \\
\hline
\Delta State \\
q? : PID \\
\hline
q? \in waiting \\
waiting' = waiting \setminus \{q?\} \\
active \neq nil \\
ready' = ready \cup \{q?\} \\
ready = \varnothing \\
active' = active \\
admin = user \wedge admin' = user \\
\hline
\end{array}
$$

$$
\begin{array}{l}
\textit{Ready3} \\
\hline
\Delta State \\
q? : PID \\
\hline
q? \in waiting \\
waiting' = waiting \setminus \{q?\} \\
active \neq nil \\
ready' = ready \cup \{q?\} \\
ready \neq \varnothing \\
active' = active \\
admin = user \wedge admin' = user \\
\hline
\end{array}
$$

$$
\begin{array}{l}
\textit{Swap1} \\
\hline
\Delta State \\
\hline
active \neq nil \\
waiting = \varnothing \\
ready = \varnothing \wedge ready' = \varnothing \\
active' = nil \\
waiting' = waiting \cup \{active\} \\
admin = user \wedge admin' = user \\
\hline
\end{array}
$$

$$\begin{array}{|l}\hline \textit{Swap2} \\ \Delta State \\ \hline active \neq nil \\ waiting = \varnothing \\ ready \neq \varnothing \\ active' \in ready \\ ready' = ready \setminus \{active'\} \\ waiting' = waiting \cup \{active\} \\ admin = user \wedge admin' = user \\ \hline \end{array}$$

$$\begin{array}{|l}\hline \textit{Swap3} \\ \Delta State \\ \hline active \neq nil \\ waiting \neq \varnothing \\ ready = \varnothing \wedge ready' = \varnothing \\ active' = nil \\ waiting' = waiting \cup \{active\} \\ admin = user \wedge admin' = user \\ \hline \end{array}$$

$$\begin{array}{|l}\hline \textit{Swap4} \\ \Delta State \\ \hline active \neq nil \\ waiting \neq \varnothing \\ ready \neq \varnothing \\ active' \in ready \\ ready' = ready \setminus \{active'\} \\ waiting' = waiting \cup \{active\} \\ admin = user \wedge admin' = user \\ \hline \end{array}$$

$$\begin{array}{|l}\hline \textit{Boot} \\ \Delta State \\ \hline admin = super \\ active = nil \\ waiting = \varnothing \wedge ready = \varnothing \\ admin' = user \wedge active' \neq nil \\ ((ready' = \varnothing \wedge waiting' \neq \varnothing) \vee \\ (ready' \neq \varnothing \wedge waiting' = \varnothing)) \\ \hline \end{array}$$

Next the states that result from these test cases are calculated:

$$\begin{array}{|l}\hline \textit{State1} \\ State \\ \hline active = nil \\ ready = \varnothing \\ waiting = \varnothing \\ admin = user \\ \hline \end{array}$$

$$\begin{array}{|l}\hline \textit{State2} \\ State \\ \hline active = nil \\ ready = \varnothing \\ waiting \neq \varnothing \\ admin = user \\ \hline \end{array}$$

$$\begin{array}{|l}\hline \textit{State3} \\ State \\ \hline active \neq nil \\ ready = \varnothing \\ waiting = \varnothing \\ admin = user \\ \hline \end{array}$$

$$\begin{array}{|l}\hline \textit{State4} \\ State \\ \hline active \neq nil \\ ready = \varnothing \\ waiting \neq \varnothing \\ admin = user \\ \hline \end{array}$$

$$\begin{array}{|l}\hline \textit{State5} \\ State \\ \hline active \neq nil \\ ready \neq \varnothing \\ waiting = \varnothing \\ admin = user \\ \hline \end{array}$$

$$\begin{array}{|l}\hline \textit{State6} \\ State \\ \hline active \neq nil \\ ready \neq \varnothing \\ waiting \neq \varnothing \\ admin = user \\ \hline \end{array}$$

$$\begin{array}{|l}\hline \textit{State7} \\ State \\ \hline active = nil \\ ready = \varnothing \\ waiting = \varnothing \\ admin = super \\ \hline \end{array}$$

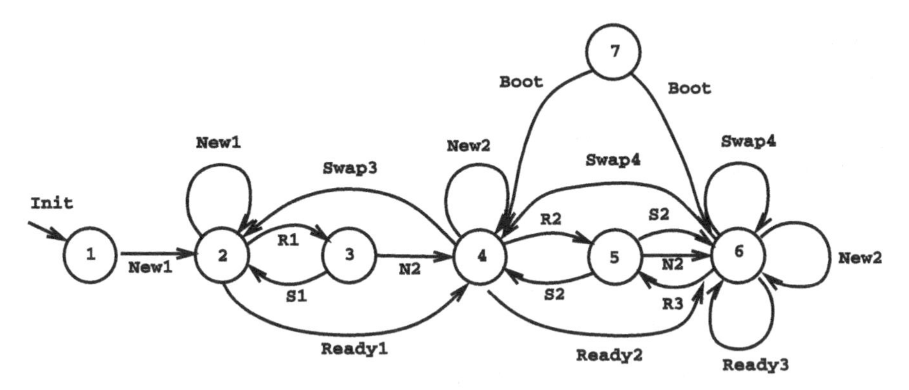

Fig. 7.4 The FSM for the scheduler

Resolving the transitions produces the FSM for the abstract scheduler specification as shown in Fig. 7.4. To calculate tests for the refinement we note that the retrieve relation is functional and we calculate concrete tests as before. For example, the concrete tests for the *New* operation are:

$$
\begin{array}{l}
\textit{CNew1} \\
\hline
\Delta \textit{CState} \\
p? : \textit{PID} \\
\hline
\textit{active}' = \textit{nil} \wedge \textit{active} = \textit{nil} \\
\textit{cready}' = \langle\rangle \wedge \textit{cready} = \langle\rangle \\
p? \notin \operatorname{ran} \textit{cwaiting} \\
\operatorname{ran} \textit{cwaiting}' = \operatorname{ran} \textit{cwaiting} \cup \{p?\} \\
\textit{admin} = \textit{user} \wedge \textit{admin}' = \textit{user} \\
\hline
\end{array}
$$

$$
\begin{array}{l}
\textit{CNew2} \\
\hline
\Delta \textit{CState} \\
p? : \textit{PID} \\
\hline
p? \neq \textit{active} \\
p? \notin (\operatorname{ran} \textit{cready} \cup \operatorname{ran} \textit{cwaiting}) \\
\operatorname{ran} \textit{cwaiting}' = \operatorname{ran} \textit{cwaiting} \cup \{p?\} \\
\textit{active}' = \textit{active} \wedge \textit{active} \neq \textit{nil} \\
\operatorname{ran} \textit{cready}' = \operatorname{ran} \textit{cready} \\
\textit{admin} = \textit{user} \wedge \textit{admin}' = \textit{user} \\
\hline
\end{array}
$$

It can be seen that $\exists\, i \bullet \textit{CNewi} \sqsubseteq_{DS} \textit{CNew}$, however, *CNew* is not the weakest refinement because the weakest refinement only requires that $\operatorname{ran} \textit{cready}' = \operatorname{ran} \textit{cready}$ whereas *CNew* resolved this non-determinism to require that $\textit{cready}' = \textit{cready}$ (i.e., the order in the ready sequence must be preserved). Again an exact covering can be constructed by replacing the test *CNew*2 by the test $\textit{CNew2} \wedge \textit{CNew}$.

The concrete states can now be calculated for the scheduler, for example, *CState*1 will be

$$\begin{array}{|l}
\hline
CState1 \\
\hline
CState \\
\hline
active = nil \\
cready = \langle\,\rangle \\
cwaiting = \langle\,\rangle \\
admin = user \\
\hline
\end{array}$$

Finally we build the new FSM. In the concrete scheduler the operation *CReady* was the weakest refinement of *Ready*. Therefore the concrete tests and transitions for the *CReady* operation are identified precisely by direct reference to those due to *Ready* in the abstract FSM. However, the postcondition in *CBoot* strengthens that of *Boot*, resulting in only one test for *CBoot*.

If we further refine the scheduler by replacing the *CSwap* operation by $CCS == CSwap \lor CSI$ where *CSI* is given by

$$\begin{array}{|l}
\hline
CSI \\
\hline
\Delta CState \\
\hline
active = nil \land active' = nil \\
cready = \langle\,\rangle \land cready' = \langle\,\rangle \land cwaiting = \langle\,\rangle \land cwaiting' = \langle\,\rangle \\
admin = user \\
admin' = super \\
\hline
\end{array}$$

This new operation has weakened the precondition by being enabled initially. Now the concrete tests for the swap operation do not cover *CCS* exactly. Under these circumstances it is necessary to include additional tests (in fact just one test here, *CSI* itself) and recalculate new states and transitions. New states might arise if the after-state of the additional test is not included in the existing concrete partition, which can happen if $\exists\, i \bullet CStatei \neq CState$.

In fact, in this example this new test *CSI* adds only one new transition from *CState*1 to *CState*7, and the FSM can be adjusted accordingly. The final FSM is shown in Fig. 7.5 where we see the previously unreachable state 7 can now be reached via *CSI*. □

7.3.2 Using a Non-functional Retrieve Relation

If the retrieve relation is not functional and the concrete specification is the weakest refinement of the abstract then no new transitions can be enabled, but the concrete FSM potentially has fewer states and transitions than the abstract because disjoint

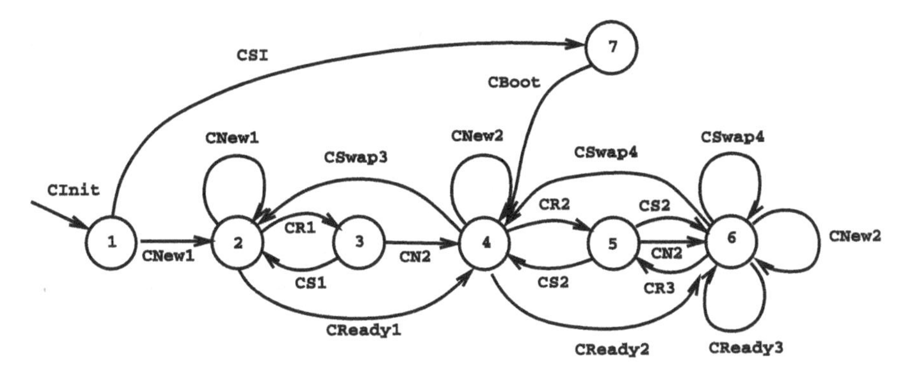

Fig. 7.5 The FSM for the concrete scheduler

states and transitions can be identified under refinement. However, this is the result of the calculation, and no further additions to the process are needed.

Example 7.10 Consider the following specification of a *BadQueue*, in which the *Get* operation is only defined if the last operation was a *Put*.

$$[X]$$

$$\begin{array}{l} \underline{\mathit{BadQueue}} \\ \mathit{items} : \mathrm{seq}\, X \\ \mathit{counter} : \mathbb{N} \\ \mathit{lastput} : \mathbb{B} \end{array}$$

$$\begin{array}{l} \underline{\mathit{Init}} \\ \mathit{BadQueue}' \\ \hline \mathit{items}' = \langle\,\rangle \\ \mathit{counter}' = 0 \\ \neg \mathit{lastput}' \end{array}$$

$$\begin{array}{l} \underline{\mathit{Put}} \\ \Delta \mathit{BadQueue} \\ \mathit{elem?} : X \\ \hline \mathit{items}' = \mathit{items} \frown \langle \mathit{elem?} \rangle \\ \mathit{counter}' = \mathit{counter} + 1 \\ \mathit{lastput}' \end{array}$$

$$\begin{array}{l} \underline{\mathit{Get}} \\ \Delta \mathit{BadQueue} \\ \mathit{elem!} : X \\ \hline \mathit{items} = \mathit{items}' \frown \langle \mathit{elem!} \rangle \\ \mathit{counter}' = \mathit{counter} - 1 \\ \mathit{counter} > 0 \\ \mathit{lastput} \wedge \neg \mathit{lastput}' \end{array}$$

This can be refined by the *Queue* ADT which is described as

$$\begin{array}{l} \underline{\mathit{Queue}} \\ \mathit{items} : \mathrm{seq}\, X \end{array}$$

$$\begin{array}{l} \underline{\mathit{Init}} \\ \mathit{Queue}' \\ \hline \mathit{items}' = \langle\,\rangle \end{array}$$

$$\begin{array}{|l}\hline \textit{Put} \\ \Delta Queue \\ elem? : X \\ \hline items' = items \frown \langle elem? \rangle \\ \hline \end{array}$$

$$\begin{array}{|l}\hline \textit{Get} \\ \Delta Queue \\ elem! : X \\ \hline items = items' \frown \langle elem! \rangle \\ \hline \end{array}$$

with (non-functional) retrieve relation

$$\begin{array}{|l}\hline R \\ Queue \\ BadQueue \\ \hline counter = \#items \\ \hline \end{array}$$

We decide to test *BadQueue* on the basis of whether $counter = 0$ and $lastput = true$. This results in the FSM shown in Fig. 7.6 where the abstract test cases for *Put* are:

$$\begin{array}{|l}\hline Put1 \\ \Delta BadQueue \\ elem? : X \\ \hline counter = 0 \\ counter' = 1 \\ items' = \langle elem? \rangle \\ \neg lastput \land lastput' \\ \hline \end{array}$$

$$\begin{array}{|l}\hline Put2 \\ \Delta BadQueue \\ elem? : X \\ \hline counter > 0 \\ counter' = counter + 1 \\ items' = items \frown \langle elem? \rangle \\ lastput \land lastput' \\ \hline \end{array}$$

$$\begin{array}{|l}\hline Put3 \\ \Delta BadQueue \\ elem? : X \\ \hline counter > 0 \\ counter' = counter + 1 \\ items' = items \frown \langle elem? \rangle \\ \neg lastput \land lastput' \\ \hline \end{array}$$

We calculate the concrete test cases in the standard fashion, so, for example, the test cases for the *Put* operation become

$$\begin{array}{|l}\hline CPut1 \\ \Delta Queue \\ elem? : X \\ \hline items = \langle \rangle \\ items' = \langle elem? \rangle \\ \hline \end{array}$$

$$\begin{array}{|l}\hline CPut2 \\ \Delta Queue \\ elem? : X \\ \hline items \neq \langle \rangle \\ items' = items \frown \langle elem? \rangle \\ \hline \end{array}$$

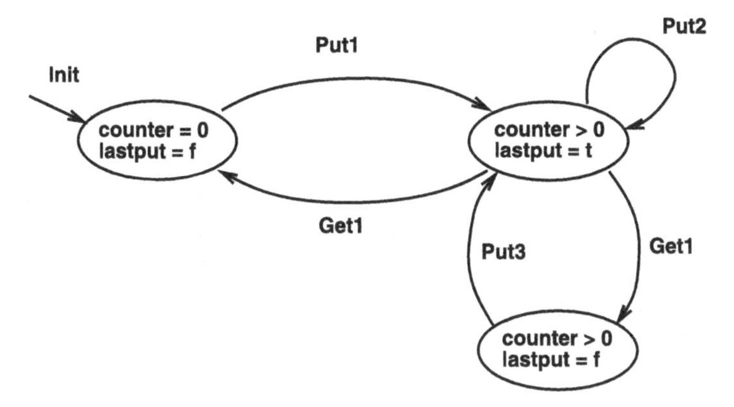

Fig. 7.6 The FSM for *BadQueue*

Fig. 7.7 The FSM for *Queue*

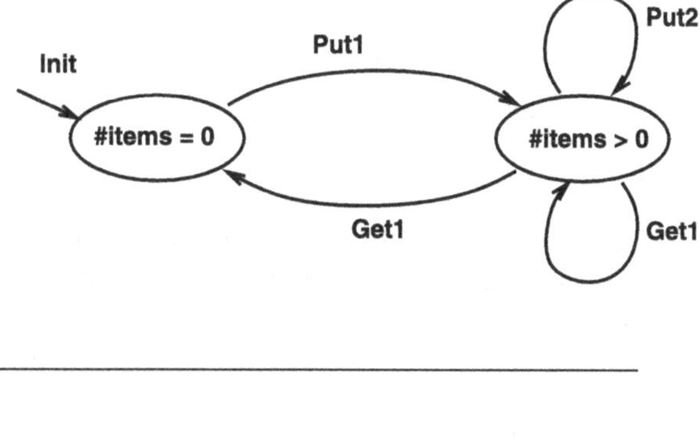

$CPut3$

$\Delta Queue$
$elem? : X$

$items \neq \langle\rangle$
$items' = items \frown \langle elem? \rangle$

What has happened is that the abstract *Put*2 and *Put*3 have collapsed to a single non-disjoint test under the non-functional retrieve relation. Similarly the second and third abstract states both collapse to a single concrete state resulting in the concrete FSM depicted in Fig. 7.7. □

7.4 Refinements Due to Upward Simulations

When a refinement is due to an upward simulation we can obtain analogous results to those in Sects. 7.2 and 7.3. In order to calculate a concrete FSM from an abstract FSM for an upward simulation the same methodology is adopted as before, that is, to use the weakest refinement calculation to generate the transitions and the states in the concrete FSM.

To generate concrete tests from the abstract test cases $\{AOp_i\}_{in \in T}$ the following formula is used (see Corollary 5.3).

$$
\begin{aligned}
COp_i == {} & (\forall AState \bullet S \implies \text{pre}\, AOp_i) \land \\
& \forall AState' \bullet (S' \implies \exists AState \bullet S \land AOp_i)
\end{aligned}
$$

Since it is known that S defines a refinement (no need to check applicability), each COp_i is a refinement of AOp_i. Coverage and disjointedness properties are similar to those for downward simulations.

Theorem 7.3 *Let AOp be an abstract operation with* $\{AOp_i\}_{in \in T}$ *its disjoint set of tests. Let COp be an upward simulation of AOp. Let S be the retrieve relation. Let* COp_i *be the concrete tests given above. Then*

$$
(\exists i : TCOp_i) \sqsubseteq_{US} COp
$$

and if COp is the weakest upward simulation of AOp then $COp = \exists i : T \bullet COp_i$. □

The disjointedness properties are also symmetric. When the retrieve relation is a function, the formulae for calculating tests are the same as for downward simulations. Therefore, as was the case then, disjoint abstract tests will produce disjoint concrete tests. However, in general refined tests are not disjoint.

Having calculated the test cases it is necessary to calculate the states for a concrete FSM. To do so calculate $CStatei == \exists AState \bullet AState_i \land S$ for each state $AState_i$ in the abstract FSM. This gives, for a weakest refinement, all the possible concrete states in the new FSM. However, in general, some of these states might be redundant because a transition has been disabled under the upward simulation. Of course the non-functionality of the retrieve relation will, in general, collapse some of the states to produce non-disjoint concrete states.

Example 7.11 Consider the specification and refinement of a split node given in Example 5.6 (p. 144). Let us test the abstract specification on the basis of whether s and t are empty. This will produce an FSM with four states (see Fig. 7.8) with tests such as

$$
\begin{array}{|l}
\hline
\textit{Insert}_A 1 \\
\Delta AState \\
m? : \mathbb{N} \\
\hline
s = \varnothing \land t = \varnothing \\
(t' = \langle m? \rangle \land s' = s) \lor (s' = \langle m? \rangle \land t' = t) \\
\hline
\end{array}
$$

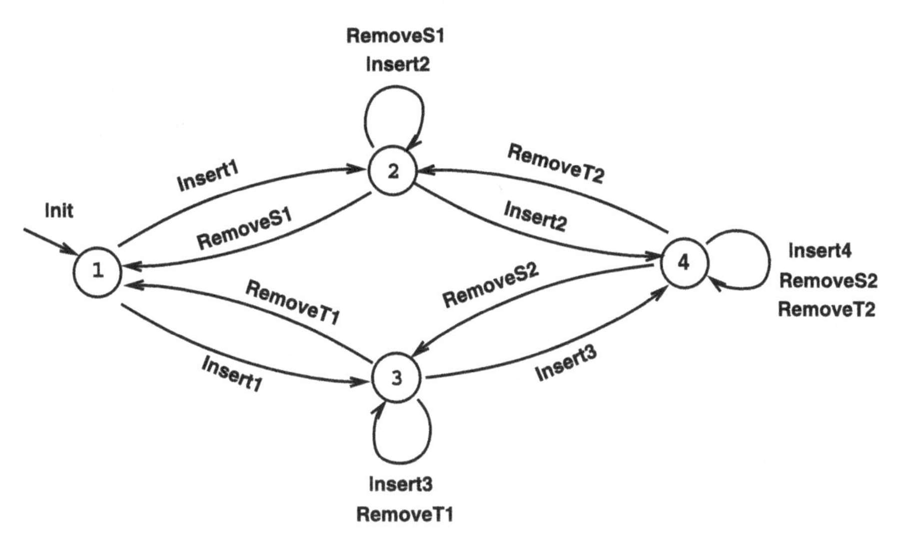

Fig. 7.8 The FSM for the split node

$Insert_A 2$

$$\Delta AState$$
$$m? : \mathbb{N}$$

$$s = \varnothing \wedge t \neq \varnothing$$
$$(t' = t \frown \langle m? \rangle \wedge s' = s) \vee (s' = \langle m? \rangle \wedge t' = t)$$

$RemoveT_A 1$

$$\Delta AState$$
$$n! : \mathbb{N}$$

$$(t = t' \frown \langle n! \rangle \wedge s' = \varnothing \wedge s = \varnothing)$$

$RemoveS_A 1$

$$\Delta AState$$
$$n! : \mathbb{N}$$

$$(s = s' \frown \langle n! \rangle \wedge t' = \varnothing \wedge t = \varnothing)$$

Based upon the retrieve relation we can calculate a concrete FSM for the purposes of testing the refinement, which is described in terms of a single sequence u. The retrieve relation is non-functional and this causes both states and transitions to collapse in the calculation. So, for example, states 2, 3 and 4 reduce to the single concrete state where $u \neq \langle\rangle$. Similarly we are left with just three concrete tests after their calculation and simplification, namely,

$Insert_C E$

$$\Delta CState$$
$$m? : \mathbb{N}$$

$$u' = \langle m? \rangle$$

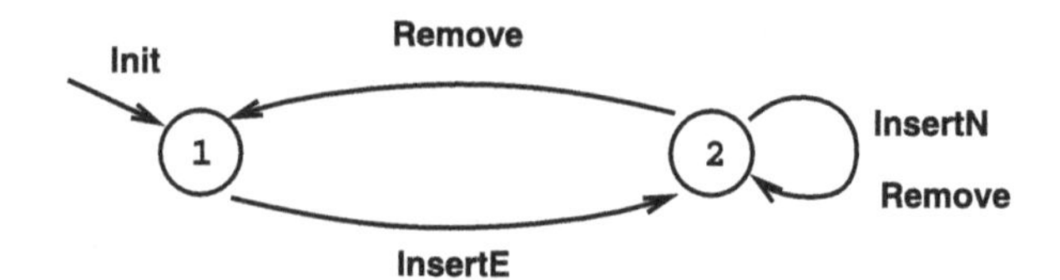

Fig. 7.9 The concrete FSM for the split node

$$
\begin{array}{|l}
\hline
\textit{Insert}_C N \\
\Delta CState \\
m? : \mathbb{N} \\
\hline
\exists t, s : \operatorname{seq} \mathbb{N} \bullet u_{merge}(t, s) \land u'_{merge}(t \frown \langle m? \rangle, s) \land u \neq \langle\rangle \\
\hline
\end{array}
$$

$$
\begin{array}{|l}
\hline
\textit{Remove}_C \\
\Delta CState \\
n! : \mathbb{N} \\
\hline
u \neq \langle\rangle \land \exists t, s : \operatorname{seq} \mathbb{N} \bullet u_{merge}(t \frown \langle n! \rangle, s) \land u'_{merge}(t, s) \\
\hline
\end{array}
$$

where $Remove_C$ tests both $RemoveS_C$ and $RemoveT_C$. Figure 7.9 shows the final concrete FSM. □

Example 7.12 A rather erratic coffee machine is described by specifying four operations: *Tea*, *Coffee*, *Coin* and *Bang*. When a coin is inserted the machine nondeterministically chooses your selection for you, except that the machine is faulty. Fortunately, by hitting the machine (*Bang*) the fault is cured (not that we condone violence against vending machines). The variable *cups* represents the number of cups currently inside the machine.

$$SELECT ::= nul \mid tea \mid coffee$$

$$
\begin{array}{|l}
\hline
State_A \\
select : SELECT \\
fault : \mathbb{B} \\
cups : \mathbb{N} \\
\hline
\end{array}
$$

$$
\begin{array}{|l}
\hline
Init_A \\
State'_A \\
\hline
select' = nul \land \neg fault' \\
cups' \neq 0 \\
\hline
\end{array}
$$

$$
\begin{array}{|l}
\hline
Coin \\
\Delta State_A \\
coin? : \mathbb{N} \\
\hline
select = nul \\
select' \in \{tea, coffee\} \\
fault' \land cups' = cups \\
\hline
\end{array}
$$

$$
\begin{array}{|l}
\hline
Bang \\
\Delta State_A \\
\hline
select' = select \\
fault \land \neg fault' \\
cups' \neq 0 \land cups' = cups \\
\hline
\end{array}
$$

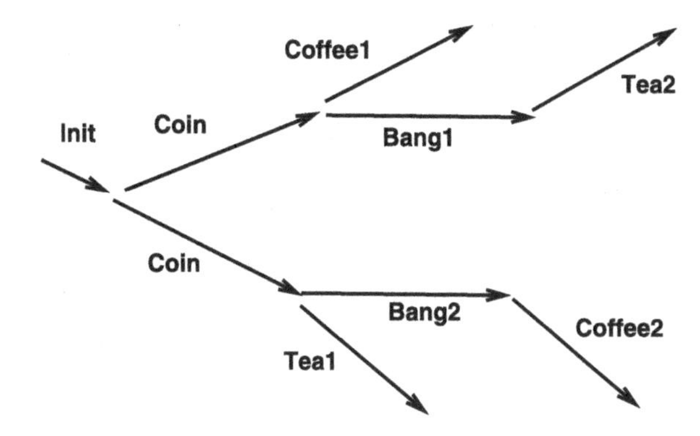

Fig. 7.10 The FSM for the abstract coffee machine

Tea

$\Delta State_A$
$cup! : SELECT$

$cup! = tea$
$((select = coffee \wedge fault \wedge cups \neq 0) \vee$
$(select = tea \wedge \neg fault \wedge cups \neq 0))$
$cups' = 0 \wedge fault' = fault$

Coffee

$\Delta State_A$
$cup! : SELECT$

$cup! = coffee$
$((select = tea \wedge fault \wedge cups \neq 0) \vee$
$(select = coffee \wedge \neg fault \wedge cups \neq 0))$
$cups' = 0 \wedge fault' = fault$

It has a seemingly slightly better behaved refinement. The refinement has the same state space and initialisation. *Coin* and *Bang* are also unchanged, the only differences are to *Tea* and *Coffee*.

Tea

$\Delta State_A$
$cup! : SELECT$

$cup! = tea$
$select = tea \wedge fault' = fault$
$cups \neq 0 \wedge cups' = 0$

Coffee

$\Delta State_A$
$cup! : SELECT$

$cup! = coffee$
$select = coffee \wedge fault' = fault$
$cups \neq 0 \wedge cups' = 0$

This refinement is an upward simulation of the abstract specification. We can test the abstract coffee machine according to what state the machine is in, and derive the FSM in Fig. 7.10, where the abstract test cases have the obvious description.

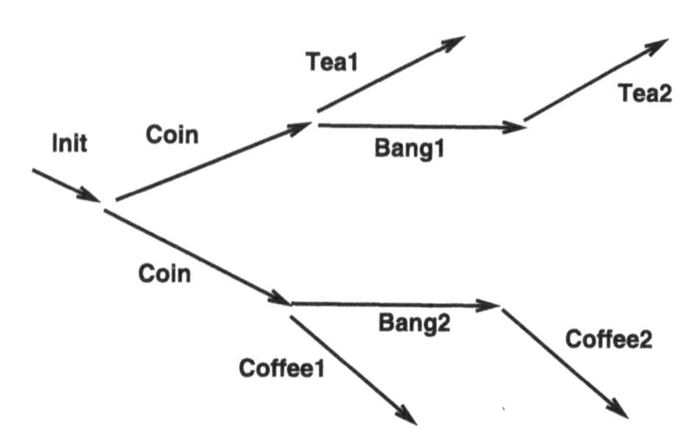

Fig. 7.11 The FSM for the concrete coffee machine

The concrete specification is not the weakest refinement of the abstract, and in order to produce a set of concrete tests that exactly cover the implementation we restrict the calculated tests by taking, for example, $CBang1 \wedge Bang_C$. The FSM that would result from testing the concrete specification directly is given in Fig. 7.11. □

7.5 Bibliographical Notes

There has been considerable theoretical and practical work on how to automatically and semi-automatically generate test cases from formal specifications, and how scheduling of these tests can be achieved. Different types of formalisms have developed different ways to do this, for example, there has been considerable research on testing specifications in the context of process algebras [6, 7, 11, 17]. A number of techniques have also been developed for state-based languages, see for example [9, 10, 14, 22, 27, 29].

The approach we focused on in this chapter has its roots in the work of Dick and Faivre [14], who describe a means to automate test generation and sequencing from VDM specifications. This work has also been applied to Z specifications in [22, 28], and to Object-Z specifications in [24]. In [22] Hörcher describes an application of this methodology to a portion of the Cabin Intercommunication Data System for the Airbus A330/340 aircraft. In [28] this methodology is combined with the classification tree method [15] and is used to test an adaptive control system. Tool support for the technique is described in [14] and [31], and has been applied to B by Van Aertryck [31] and Behnia and Waeselynck [1].

The idea of testing an implementation using a retrieve relation was first raised in [14]. The problem was also discussed by Stocks [30] and Stepney [29]. Hörcher discusses the problem in more depth in [22], and comments that retrieve relations may be used in the production of test oracles.

Early work in this area includes that by Hayes [16], which amongst other things discusses using an implementation of a retrieve function to test an implementation against a more abstract implementation. Additional work on testing from Object-Z specifications includes [8], some discussion on tool support is given in [25]. General work on generating test data from specifications includes [26].

The work in this chapter was first described by us in [13] and [12]: [12] considered the partition analysis of individual operations and includes proofs of Theorems 7.1 and 7.2; [13] described how to calculate the concrete FSM from the abstract. Although the results here are described with reference to the construction of finite state machines similar to that described in [14], they are also applicable to other approaches, such as those described in [18, 24].

More recent work includes a survey article on *Using formal specifications to support testing* in ACM Computing Surveys [19], which provides a good introduction to many of the aspects of testing and how formal methods can support these activities, it includes references to tool support and a number of larger case studies. The book *Formal Methods and Testing* [20] contains papers from the Formal Methods and Testing (FORTEST) network—a network funded by the UK EPSRC that investigated the relationships between formal (and semi-formal) methods and software testing.

References

1. Behnia, S., & Waeselynck, H. (1999) Test criteria definition for B models. In Wing et al. [32] (pp. 509–529).
2. Bloomfield, R., Marshall, L., & Jones, R. (Eds.) (1988). *VDM'88. Lecture Notes in Computer Science: Vol. 328*. Berlin: Springer.
3. Bowen, J. P., Fett, A., & Hinchey, M. G. (Eds.) (1998). *ZUM'98: the Z Formal Specification Notation. Lecture Notes in Computer Science: Vol. 1493*. Berlin: Springer.
4. Bowen, J. P. & Hinchey, M. G. (Eds.) (1995). *ZUM'95: the Z Formal Specification Notation. Lecture Notes in Computer Science: Vol. 967*. Limerick: Springer.
5. Bowen, J. P. & Nicholls, J. E. (Eds.) (1992). *Seventh Annual Z User Workshop*. London: Springer.
6. Brinksma, E. (1988). A theory for the derivation of tests. In S. Aggarwal & K. Sabnani (Eds.), *Protocol Specification, Testing and Verification, VIII*, Atlantic City, USA, June 1988 (pp. 63–74). Amsterdam: North-Holland.
7. Brinksma, E., Scollo, G., & Steenbergen, C. (1986). Process specification, their implementation and their tests. In B. Sarikaya & G. v. Bochmann (Eds.), *Protocol Specification, Testing and Verification, VI*, Montreal, Canada, June 1986 (pp. 349–360). Amsterdam: North-Holland.
8. Carrington, D., Maccoll, I., Mcdonald, J., Murray, L., & Strooper, P. (2000). From Object-Z specifications to ClassBench test suites. *Journal on Software Testing, Verification and Reliability*, *10*(2), 111–137.
9. Carrington, D., & Stocks, P. (1998) A tale of two paradigms: Formal methods and software testing. In Bowen et al. [3] (pp. 51–68).
10. Cusack, E., & Wezeman, C. (1992) Deriving tests for objects specified in Z. In Bowen and Nicholls [5] (pp. 180–195).
11. de Nicola, R., & Hennessy, M. (1984). Testing equivalences for processes. *Theoretical Computer Science*, *34*(3), 83–133.
12. Derrick, J., & Boiten, E. A. (1998) Testing refinements by refining tests. In Bowen et al. [3] (pp. 265–283).
13. Derrick, J., & Boiten, E. A. (1999). Testing refinements of state-based formal specifications. *Software Testing, Verification & Reliability*, *9*, 27–50.

14. Dick, J., & Faivre, A. (1993). Automating the generation and sequencing of test cases from model-based specifications. In J. C. P. Woodcock & P. G. Larsen (Eds.), *FME'93: Industrial-Strength Formal Methods. Lecture Notes in Computer Science: Vol. 428* (pp. 268–284). Berlin: Springer.
15. Grochtmann, M., & Grimm, K. (1993). Classification trees for partition testing. *Software Testing, Verification & Reliability*, *3*, 63–82.
16. Hayes, I. J. (1986). Specification directed module testing. *IEEE Transactions on Software Engineering*, *SE-12*(1), 124–133.
17. Heerink, L., & Tretmans, J. (1997) Refusal testing for classes of transition systems with inputs and outputs. In Mizuno et al. [23] (pp. 23–38).
18. Hierons, R. M. (1997). Testing from a Z specification. *Software Testing, Verification & Reliability*, *7*(1), 19–33.
19. Hierons, R. M., Bogdanov, K., Bowen, J. P., Cleaveland, R., Derrick, J., Dick, J., Gheorghe, M., Harman, M., Kapoor, K., Krause, P., Lüttgen, G., Simons, A. J. H., Vilkomir, S., Woodward, M. R., & Zedan, H. (2009). Using formal specifications to support testing. *ACM Computing Surveys*, *41*(2), 9:1–9:76.
20. Hierons, R. M., Bowen, J. P., & Harman, M. (Eds.) (2008). *Formal Methods and Testing, an Outcome of the FORTEST Network, Revised Selected Papers*. *Lecture Notes in Computer Science: Vol. 4949*. Berlin: Springer.
21. Hinchey, M. G. & Liu, S. (Eds.) (1997). *First International Conference on Formal Engineering Methods (ICFEM'97)*. Hiroshima, Japan, November 1997. Los Alamitos: IEEE Comput. Soc.
22. Hörcher, H.-M. (1995) Improving software tests using Z specifications. In Bowen and Hinchey [4] (pp. 152–166).
23. Mizuno, T., Shiratori, N., Higashino, T., & Togashi, A. (Eds.) (1997) *FORTE/PSTV'97*, Osaka, Japan, November 1997. London: Chapman & Hall.
24. Murray, L., Carrington, D., MacColl, I., McDonald, J., & Strooper, P. (1998) Formal derivation of finite state machines for class testing. In Bowen et al. [3] (pp. 42–59).
25. Murray, L., Carrington, D., Maccoll, I., & Strooper, P. (1999). Tinman—a test derivation and management tool for specification-based class testing. In *International Conference on Technology of Object-Oriented Languages (TOOLS'99)* (pp. 222–233). Los Alamitos: IEEE Comput. Soc.
26. Offutt, J., Liu, S., Abdurazik, A., & Ammann, P. (2003). Generating test data from state-based specifications. *The Journal of Software Testing, Verification and Reliability*, *13*, 25–53.
27. Scullard, G. T. (1988) Test case selection using VDM. In Bloomfield et al. [2] (pp. 178–186).
28. Singh, H., Conrad, M., & Sadeghipour, S. (1997) Test case design based on Z and the classification-tree method. In Hinchey and Liu [21] (pp. 81–90).
29. Stepney, S. (1995) Testing as abstraction. In Bowen and Hinchey [4] (pp. 137–151).
30. Stocks, P. (1993). *Applying formal methods to software testing*. PhD thesis, Department of Computer Science, University of Queensland, St. Lucia, Australia.
31. van Aertryck, L., Benveniste, M., & Le Metayer, D. (1997) Casting: A formally based software test generation method. In Hinchey and Liu [21] (pp. 101–110).
32. Wing, J. M., Woodcock, J. C. P., & Davies, J. (Eds.) (1999). *FM'99 World Congress on Formal Methods in the Development of Computing Systems*. *Lecture Notes in Computer Science: Vol. 1708*. Berlin: Springer.

Chapter 8
A Single Simulation Rule

In Chap. 3 we discussed the use of simulations to verify data refinements, and we saw that although downward and upward simulations are jointly complete, each simulation method is incomplete on its own: there are valid refinements that cannot be verified by using either downward or upward simulations on their own. However, the two simulation methods taken together are jointly complete, and, in particular, any valid refinement can be verified by using one downward simulation followed by one upward simulation.

In this chapter we discuss how we can make use of this joint completeness to derive a single complete simulation method, called a powersimulation. However, using a retrieve relation that simply links concrete to abstract states, it is not possible to derive a single complete simulation rule. Therefore we have to move from a retrieve relation which maps concrete states to abstract states to a possibility mapping which maps concrete states to *sets* of abstract states. We will look at the relational theory of this approach in the next section, before showing how this can all be expressed in the Z schema calculus in Sect. 8.2. Of course given any simulation method there is also the possibility of adopting the calculational approach from Chap. 5, and in Sect. 8.3 we discuss how to calculate the weakest powersimulation with respect to a given possibility mapping.

We begin as usual with an example, illustrating the need for both upward and downward simulations in an arbitrary refinement.

Example 8.1 Continuing with the theme of erratic coffee machines from Example 7.12, we specify a faulty abstract vending machine (*AVM*) with three operations. *Coin* allows money to be inserted into the machine, and *Dispense* serves us a drink. Finally, *Bang*-ing the machines cures the fault, or introduces one if the machine was not faulty to start with.

$SELECT ::= null \mid tea \mid coffee$

Excerpts from pp. 2, 5–12 of [4] are reprinted within this chapter by permission from Oxford University Press.

J. Derrick, E.A. Boiten, *Refinement in Z and Object-Z*,
DOI 10.1007/978-1-4471-5355-9_8, © Springer-Verlag London 2014

AVM

$fault, coin : \mathbb{B}$
$status : SELECT$

$Init_A$

AVM'

$\neg fault' \wedge \neg coin'$
$status' = null$

$Coin_A$

ΔAVM

$status = null \wedge status' \in \{tea, coffee\}$
$\neg fault \wedge fault' = fault$
$\neg coin \wedge coin'$

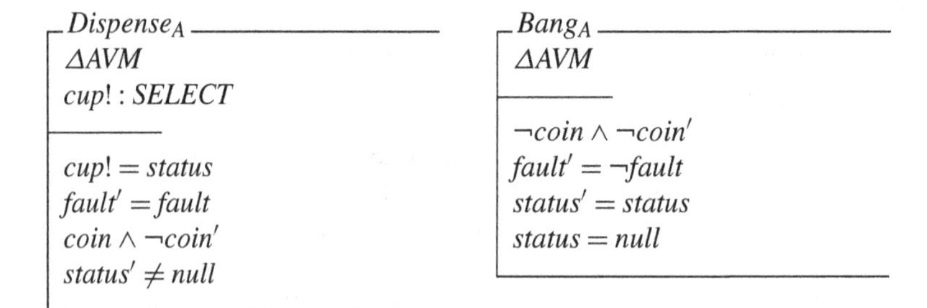

The following concrete vending machine is a refinement of *AVM* according to Definition 3.5.

CVM

$coin : \mathbb{B}$
$status : SELECT$

$Init_C$

CVM'

$\neg coin' \wedge status' = null$

$Coin_C$

ΔCVM

$status = null \wedge status' = status$
$\neg coin \wedge coin'$

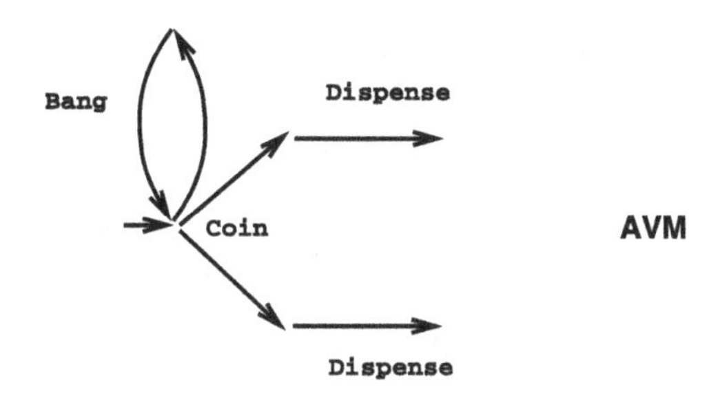

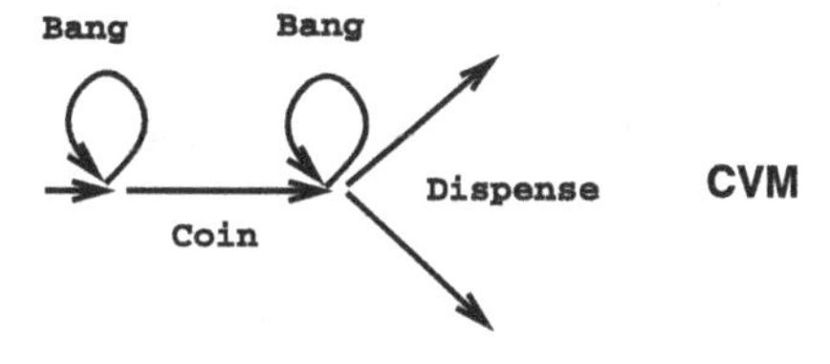

Fig. 8.1 The behaviour of the two vending machines

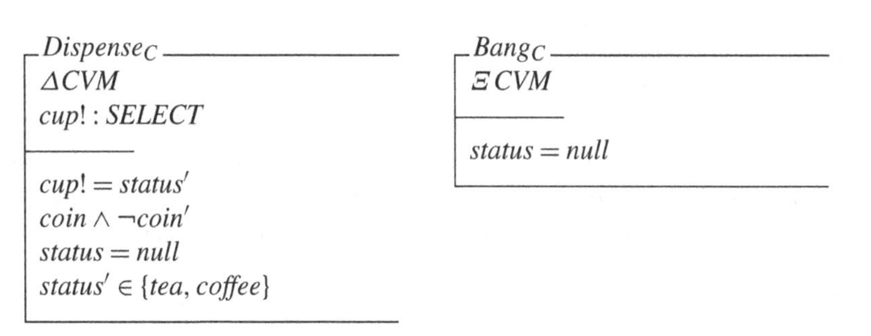

$Dispense_C$

ΔCVM
$cup! : SELECT$

$cup! = status'$
$coin \wedge \neg coin'$
$status = null$
$status' \in \{tea, coffee\}$

$Bang_C$

ΞCVM

$status = null$

Although *CVM* is a refinement of *AVM*, it cannot be verified by using a downward or an upward simulation on its own, and to verify the refinement we have to proceed in two steps. The first step refines *AVM* to an intermediate specification. Having done this the second step is to observe that *CVM* is an upward simulation of this intermediate specification. The picture in Fig. 8.1 illustrates the behaviour of the abstract and concrete systems.

8.1 Powersimulations

In this section we will derive a single complete method in the relational setting used in Chap. 3. The formalisation in Z will be given subsequently in Sect. 8.2.

Throughout the remainder of this section let $\mathsf{A} = (\mathsf{AState}, \mathsf{AInit}, \{\mathsf{AOp_i}\}_{\mathsf{i \in I}}, \mathsf{AFin})$ and $\mathsf{C} = (\mathsf{CState}, \mathsf{CInit}, \{\mathsf{COp_i}\}_{\mathsf{i \in I}}, \mathsf{CFin})$ be (partial) data types such that C refines A.

We give the definition of the single simulation rule first of all and then show how it is derived.

Definition 8.1 (Powersimulation) A powersimulation between abstract data types A and C is a relation $\mathsf{R} : \mathbb{P}\,\mathsf{AState} \leftrightarrow \mathsf{CState}$ such that for all $\mathsf{i} \in \mathsf{I}$:

$$\mathsf{CInit} \subseteq (\mathsf{L} \setminus \mathsf{AInit}) \mathbin{;} \mathsf{R}$$
$$\mathrm{dom}\,\mathsf{AOp_i} \lhd_P (\mathsf{R} \mathbin{;} \mathsf{COp_i}) \subseteq (\mathsf{L} \setminus (\mathsf{L} \mathbin{;} \mathsf{AOp_i})) \mathbin{;} \mathsf{R}$$
$$\mathrm{ran}(\mathrm{dom}\,\mathsf{AOp_i} \lhd_P \mathsf{R}) \subseteq \mathrm{dom}\,\mathsf{COp_i}$$
$$\mathsf{CFin} \subseteq (\mathsf{L} \mathbin{;} \mathsf{AFin})/\mathsf{R}$$

where L is defined by

$$\alpha\,\mathsf{L}\,\beta \quad \text{iff} \quad \beta \in \alpha$$
□

We will see that every valid refinement can be verified by one powersimulation, so powersimulations will provide a complete method. In addition, powersimulations are sound since a powersimulation will, in fact, be just a combination of an upward and downward simulation. It will therefore give us a single complete method for refinement. However, the definition does not provide much intuition as to where it comes from, and in the rest of this section we show how it is derived.

To derive the definition we will use the construction used in the joint completeness proof. Remember that the completeness result is the following: if data type A is refined by C, then there is an intermediate data type CA with an upward simulation from CA to A and a downward simulation from CA to C. The data type CA is given constructively and is equivalent to A (there are simulations both ways). It is also canonical, that is, all operations and the initialisation are functions, so that all non-determinism present in A has been factored out. Note that there is often more than one way to verify a refinement. For example, with the vending machines above we used a downward simulation first of all, followed by an upward simulation.

To derive the powersimulation conditions we first totalise the partial relations in A and C and then construct the intermediate data type CA.

From the data type A the construction of CA begins by defining the concrete state space as $\mathsf{CAState} = \mathbb{P}(\mathsf{AState}_\perp)$. It then defines an upward simulation L between A and CA. This is given by

$$\alpha\,(\mathsf{L})\,\beta \quad \text{iff} \quad \beta \in \alpha$$

This relation is then used to define the relations CAInit, $\{\mathsf{CAOp_i}\}_{\mathsf{i} \in \mathsf{I}}$, CAFin in CA as follows. For ease of presentation we will fix an $\mathsf{i} \in \mathsf{I}$, and consider just one operation. Let AOp, COp and CAOp be corresponding operations in the data types. So these operations are relations between the following domains: $\widehat{\mathsf{AOp}} : \mathsf{AState}_\perp \leftrightarrow \mathsf{AState}_\perp$ and $\widehat{\mathsf{COp}} : \mathsf{CState}_\perp \leftrightarrow \mathsf{CState}_\perp$. Because A was a partial data type we totalise its relations, however, the relations in CA are total by construction, thus $\mathsf{CAOp} : \mathsf{CAState} \leftrightarrow \mathsf{CAState}$.

The relations in CA: CAInit, CAOp, CAFin can now be calculated as the weakest solutions to the following equations.

$$\mathsf{CAInit} \fatsemi \mathsf{L} = \widehat{\mathsf{AInit}}$$
$$\mathsf{CAFin} = \mathsf{L} \fatsemi \widehat{\mathsf{AFin}}$$
$$\mathsf{CAOp} \fatsemi \mathsf{L} = \mathsf{L} \fatsemi \widehat{\mathsf{AOp}}$$

Having done this the construction also uses a downward simulation ρ between CA and C. That is, there will be a downward simulation between CA and C verified by a retrieve relation ρ. Therefore, the data types satisfy the following set of equations.

$$\widehat{\mathsf{CInit}} \subseteq \mathsf{CAInit} \fatsemi \rho$$
$$\rho \fatsemi \widehat{\mathsf{CFin}} \subseteq \mathsf{CAFin}$$
$$\rho \fatsemi \widehat{\mathsf{COp}} \subseteq \mathsf{CAOp} \fatsemi \rho$$

These two sets of equations are given in terms of the totalised relations. To make use of them in Z we are going to derive equivalent conditions on the underlying partial relations. In order to do so we define additional domain restriction and subtraction operators $\lhd_P$ and $⩤_P$ which are analogous to the standard domain restriction and subtraction operators $\lhd$ and $⩤$. However, as our simulation construction involves powersets in the state space of CA we are interested in restricting to *subsets* as opposed to *elements* (e.g., as in $\lhd$) of $\mathrm{dom}\,\mathsf{AOp}$.

Definition 8.2

$$\mathrm{dom}\,\mathsf{AOp} \lhd_P \mathsf{CAOp} = \{(\alpha, \beta) \mid (\alpha, \beta) \in \mathsf{CAOp} \land \alpha \subseteq \mathrm{dom}\,\mathsf{AOp}\}$$
$$\mathrm{dom}\,\mathsf{AOp} ⩤_P \mathsf{CAOp} = \{(\alpha, \beta) \mid (\alpha, \beta) \in \mathsf{CAOp} \land \alpha \not\subseteq \mathrm{dom}\,\mathsf{AOp}\} \qquad \Box$$

With this in place we can prove the following lemma which derives conditions on the underlying partial relations that are equivalent to the condition $\mathsf{CAOp} \fatsemi \mathsf{L} = \mathsf{L} \fatsemi \widehat{\mathsf{AOp}}$.

Lemma 8.1 $\mathsf{CAOp} \fatsemi \mathsf{L} \subseteq \mathsf{L} \fatsemi \widehat{\mathsf{AOp}}$ *iff* $\mathrm{dom}\,\mathsf{AOp} \lhd_P \mathsf{CAOp} \subseteq \mathsf{L} \setminus (\mathsf{L} \fatsemi \mathsf{AOp})$

Proof To begin we note that

$$\begin{aligned}\mathsf{L} \fatsemi \widehat{\mathsf{AOp}} &= \mathsf{L} \fatsemi (\mathsf{AOp} \cup \overline{\mathrm{dom}\,\mathsf{AOp}_\perp} \times \mathsf{AState}_\perp)\\ &= \mathsf{L} \fatsemi \mathsf{AOp} \cup \mathsf{L} \fatsemi (\overline{\mathrm{dom}\,\mathsf{AOp}_\perp} \times \mathsf{AState}_\perp)\end{aligned}$$

Therefore

$$\mathsf{CAOp} \fatsemi \mathsf{L} \subseteq \mathsf{L} \fatsemi \widehat{\mathsf{AOp}}$$
$$\equiv$$
$$\mathsf{CAOp} \fatsemi \mathsf{L} \subseteq \mathsf{L} \fatsemi \mathsf{AOp} \cup \mathsf{L} \fatsemi (\overline{\mathrm{dom}\,\mathsf{AOp}_\perp} \times \mathsf{AState}_\perp)$$

$$\equiv$$

$$(\text{dom}\,\mathsf{AOp} \lhd_P \mathsf{CAOp}) \fatsemi \mathsf{L} \subseteq \mathsf{L} \fatsemi \mathsf{AOp}$$

$$\equiv$$

$$\text{dom}\,\mathsf{AOp} \lhd_P \mathsf{CAOp} \subseteq \mathsf{L} \setminus (\mathsf{L} \fatsemi \mathsf{AOp}) \qquad \square$$

In a similar fashion the next lemma considers the downward simulation equation $\rho \fatsemi \widehat{\mathsf{COp}} \subseteq \mathsf{CAOp} \fatsemi \rho$. In doing so we need to define a restriction $\underline{\rho} : \mathbb{P}\,\mathsf{AState} \leftrightarrow \mathsf{CState}$ of the retrieve relation ρ purely to defined values.

Definition 8.3 Given a relation ρ we define $\underline{\rho} : \mathbb{P}\,\mathsf{AState} \leftrightarrow \mathsf{CState}$ by

$$\underline{\rho} = (\mathbb{P}\,\mathsf{AState}) \lhd \rho \rhd \mathsf{CState} \qquad \square$$

Lemma 8.2 $\rho \fatsemi \widehat{\mathsf{COp}} \subseteq \mathsf{CAOp} \fatsemi \rho$ *is equivalent to the following conditions*:

$$\text{dom}\,\mathsf{AOp} \lhd_P (\underline{\rho} \fatsemi \mathsf{COp}) \subseteq (\text{dom}\,\mathsf{AOp} \lhd_P \mathsf{CAOp}) \fatsemi \underline{\rho}$$
$$\text{ran}(\text{dom}\,\mathsf{AOp} \lhd_P \underline{\rho}) \subseteq \text{dom}\,\mathsf{COp}$$

Proof Suppose for the moment that ρ satisfies a condition, namely that

$$\begin{aligned} \alpha(\rho)\bot &\Rightarrow \bot \in \alpha \\ \bot \in \alpha &\Rightarrow \alpha(\rho)\beta \quad \text{for all } \beta \in \mathsf{CState}_\bot \end{aligned} \tag{8.1}$$

Then $\rho \fatsemi \widehat{\mathsf{COp}} \subseteq \mathsf{CAOp} \fatsemi \rho$ if, and only if,

$$\rho \fatsemi \widehat{\mathsf{COp}}$$

$$\subseteq$$

$$(\text{dom}\,\mathsf{AOp} \lhd_P \mathsf{CAOp} \fatsemi \rho) \cup (\text{dom}\,\mathsf{AOp} \ntriangleleft_P \mathsf{CAOp} \fatsemi \rho)$$

$$= \{\text{by (8.1)}\}$$

$$(\text{dom}\,\mathsf{AOp} \lhd_P \mathsf{CAOp} \fatsemi \rho) \cup \text{dom}(\text{dom}\,\mathsf{AOp} \ntriangleleft_P \mathsf{CAOp}) \times \mathsf{CState}_\bot$$

Therefore

$$\rho \fatsemi \widehat{\mathsf{COp}} \subseteq \mathsf{CAOp} \fatsemi \rho$$

$$\equiv$$

$$\text{dom}\,\mathsf{AOp} \lhd_P (\rho \fatsemi \widehat{\mathsf{COp}}) \subseteq (\text{dom}\,\mathsf{AOp} \lhd_P \mathsf{CAOp}) \fatsemi \underline{\rho}$$

$$\equiv$$

$$\text{dom}\,\mathsf{AOp} \lhd_P (\underline{\rho} \fatsemi \widehat{\mathsf{COp}}) \subseteq (\text{dom}\,\mathsf{AOp} \lhd_P \mathsf{CAOp}) \fatsemi \underline{\rho}$$

$$\equiv$$

$$\text{dom}\,\mathsf{AOp} \lhd_P (\underline{\rho} \fatsemi \mathsf{COp}) \subseteq (\text{dom}\,\mathsf{AOp} \lhd_P \mathsf{CAOp}) \fatsemi \underline{\rho}$$

and

$$\mathrm{dom}\,\mathsf{AOp} \lhd_P (\underline{\rho} \fatsemi \overline{(\mathrm{dom}\,\mathsf{COp}_\perp} \times \mathsf{CState}_\perp)) \subseteq (\mathrm{dom}\,\mathsf{AOp} \lhd_P \mathsf{CAOp}) \fatsemi \underline{\rho}$$

Note that the latter condition holds if, and only if, $\mathrm{ran}(\mathrm{dom}\,\mathsf{AOp} \lhd_P \underline{\rho}) \subseteq \mathrm{dom}\,\mathsf{COp}$. □

We have assumed that ρ satisfies (8.1), we have to justify this or show that any simulation relation ρ can in fact be replaced by one which does. It is easy to see that the latter option is always possible. For example, suppose that $\alpha(\rho)\perp$. Then we have $(\alpha, \beta) \in \mathsf{CAOp} \fatsemi \rho$ for all $\beta \in \mathsf{CState}_\perp$, since $\rho \fatsemi \widehat{\mathsf{COp}} \subseteq \mathsf{CAOp} \fatsemi \rho$. Therefore we can assume without harm that $\perp \in \alpha$. The other condition can also be shown to be safely assumed in a similar way.

Corollary 8.1 *The conditions in the upward and downward simulations* $\rho \fatsemi \widehat{\mathsf{COp}} \subseteq \mathsf{CAOp} \fatsemi \rho$ *and* $\mathsf{CAOp} \fatsemi \mathsf{L} = \mathsf{L} \fatsemi \widehat{\mathsf{AOp}}$ *are equivalent to*

$$\mathrm{dom}\,\mathsf{AOp} \lhd_P (\underline{\rho} \fatsemi \mathsf{COp}) \subseteq (\mathsf{L} \setminus (\mathsf{L} \fatsemi \mathsf{AOp})) \fatsemi \underline{\rho}$$
$$\mathrm{ran}(\mathrm{dom}\,\mathsf{AOp} \lhd_P \underline{\rho}) \subseteq \mathrm{dom}\,\mathsf{COp}$$ □

In a similar way we can consider the initialisation and finalisation conditions which, because the abstract and concrete initialisations and finalisations are total, have a simple form.

Lemma 8.3 *The initialisation conditions* $\mathsf{CAInit} \fatsemi \mathsf{L} = \widehat{\mathsf{AInit}}$ *and* $\widehat{\mathsf{CInit}} \subseteq \mathsf{CAInit} \fatsemi \rho$ *are jointly equivalent to*

$$\mathsf{CInit} \subseteq (\mathsf{L} \setminus \mathsf{AInit}) \fatsemi \rho$$

The finalisation conditions $\mathsf{CAFin} = \mathsf{L} \fatsemi \widehat{\mathsf{AFin}}$ *and* $\rho \fatsemi \widehat{\mathsf{CFin}} \subseteq \mathsf{CAFin}$ *are equivalent to*

$$\mathsf{CFin} \subseteq (\mathsf{L} \fatsemi \mathsf{AFin})/\rho$$ □

Finally, if we define R by $\mathsf{R} = \underline{\rho}$ we see that we have proved the following result.

Theorem 8.1 *A powersimulation is a sound and complete method for verifying refinements between (finitary) data types.* □

8.2 Application to Z

In order to derive a single complete simulation in Z we will apply the relational conditions we derived in the previous section to the Z schema calculus.

To do this we need to use the θ notation. Remember that if S is the name of a schema, then θS denotes the characteristic binding of components from S, where components are bound to the values of variables. This allows a schema to be considered as a set of bindings, and because of this we can take their powersets. This is crucial to us here since a powersimulation involves sets of abstract values and we need to represent that in the schema calculus. So for a schema *AState*, $\mathbb{P}\,AState$ is the powerset of the set of its bindings. Likewise $\mathrm{pre}\,AOp$ describes a schema (the precondition of operation *AOp*), and since this can be considered as a set of bindings we can take its powerset.

Suppose the Z data types A and C are defined as usual. To derive the powersimulation conditions our retrieve relation will be given as a relation $r : \mathbb{P}\,AState \leftrightarrow CState$. Let us consider two operations *AOp* and *COp*. As we saw in Chap. 4 we can think of the schemas as defining relations which are given by

$$\begin{aligned}
&\mathsf{AOp} == \{AOp \bullet \theta AState \mapsto \theta AState'\}\\
&\mathsf{COp} == \{COp \bullet \theta CState \mapsto \theta CState'\}\\
&\mathsf{AInit} == \{AInit \bullet * \mapsto \theta AState'\}\\
&\mathsf{CInit} == \{CInit \bullet * \mapsto \theta CState'\}
\end{aligned}$$

We are now able to express each condition in the powersimulation in terms of schemas. The derivation for the initialisation condition is:

$$\mathsf{CInit} \subseteq (\mathsf{L} \setminus \mathsf{AInit}) \mathbin{⨾} \mathsf{R}$$

$$\equiv$$

$$\forall c : CState \bullet c \in \mathsf{CInit} \Rightarrow c \in (\mathsf{L} \setminus \mathsf{AInit}) \mathbin{⨾} \mathsf{R}$$

$$\equiv$$

$$\forall CState \bullet \theta CState \in \mathsf{CInit} \Rightarrow \theta CState \in (\mathsf{L} \setminus \mathsf{AInit}) \mathbin{⨾} \mathsf{R}$$

$$\equiv$$

$$\begin{aligned}
&\forall CState \bullet \theta CState \in \{CInit \bullet \theta CState'\}\\
&\qquad\qquad \Rightarrow \exists \gamma : \mathbb{P}\,AState \bullet r(\gamma) = \theta CState' \land \gamma \in (\mathsf{L} \setminus \mathsf{AInit})
\end{aligned}$$

$$\equiv$$

$$\forall CState' \bullet CInit \Rightarrow \exists \gamma : \mathbb{P}\,AInit \bullet r(\gamma) = \theta CState'$$

In a similar way the applicability and correctness conditions can be given in terms of schemas.

$$\begin{aligned}
&\mathrm{dom}\,\mathsf{AOp} \lhd_P (\underline{\mathsf{R}} \mathbin{⨾} \mathsf{COp}) \subseteq (\mathsf{L} \setminus (\mathsf{L} \mathbin{⨾} \mathsf{AOp})) \mathbin{⨾} \underline{\mathsf{R}} \quad \text{iff}\\
&\forall \gamma_1 : \mathbb{P}(\mathrm{pre}\,AOp);\ CState;\ CState' \bullet r(\gamma_1) = \theta CState \land COp \Rightarrow\\
&\exists \gamma_2 : \mathbb{P}\,AState \bullet r(\gamma_2) = \theta CState' \land\\
&\qquad\qquad\qquad \forall AState' : \gamma_2 \bullet \exists AState : \gamma_1 \bullet AOp
\end{aligned}$$

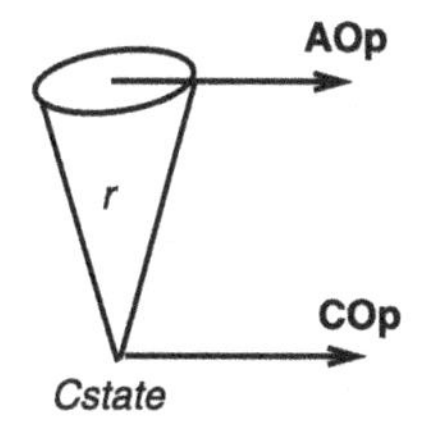

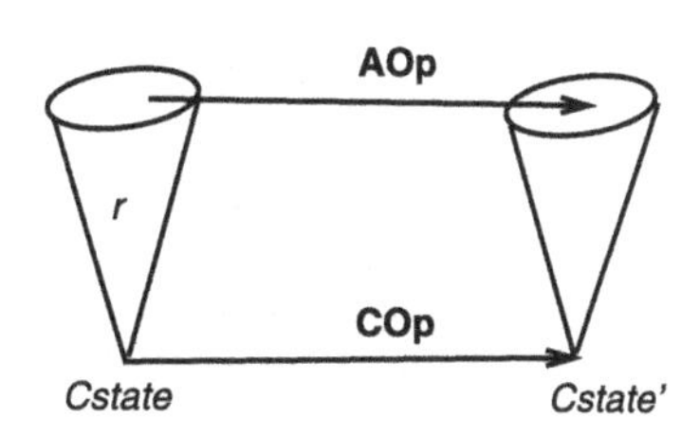

Fig. 8.2 Applicability and correctness requirements for a powersimulation

and

$$\begin{array}{l}\text{ran}(\text{dom}\,\mathsf{AOp} \lhd_P \underline{\mathsf{R}}) \subseteq \text{dom}\,\mathsf{COp} \quad \text{iff} \\ \qquad \forall\, CState \bullet \forall\, \gamma : \mathbb{P}(\text{pre}\, AOp) \bullet r(\gamma) = \theta CState \Rightarrow \text{pre}\, COp\end{array}$$

As before we do not need an explicit finalisation requirement for powersimulations. The nice aspect of this characterisation is that since expressions such as $AState$, $\mathbb{P}AState$, pre AOp and $\mathbb{P}(\text{pre}\, AOp)$ denote sets or powersets of bindings, the explicit conditions for powersimulations we have derived are formulated within the schema calculus itself.

Strictly speaking we should consider inputs and outputs now. However, remember that the standard approach is to embed them as components in the state space, and we could do that here. When we do we find that the form of the rules remains the same apart from the need to quantify over all possible inputs and outputs, but we choose to elide them in the following definition which characterises powersimulations for use in Z.

Definition 8.4 (Powersimulation) A powersimulation between Z data types $A = (AState, AInit, \{AOp_i\}_{i\in I})$ and $C = (CState, CInit, \{COp_i\}_{i\in I})$ is a relation $r : \mathbb{P}AState \leftrightarrow CState$ such that:

$$\forall\, CState' \bullet CInit \Rightarrow \exists\, \gamma : \mathbb{P}AInit \bullet r(\gamma) = \theta CState'$$

and for all $i \in I$:

$$\begin{array}{l}\forall\, CState \bullet \forall\, \gamma : \mathbb{P}(\text{pre}\, AOp_i) \bullet r(\gamma) = \theta CState \Rightarrow \text{pre}\, COp_i \\ \forall\, \gamma_1 : \mathbb{P}(\text{pre}\, AOp_i);\ CState;\ CState' \bullet r(\gamma_1) = \theta CState \wedge COp_i \Rightarrow \\ \qquad \exists\, \gamma_2 : \mathbb{P}AState \bullet r(\gamma_2) = \theta CState' \wedge \\ \qquad\qquad\qquad \forall\, AState' : \gamma_2 \bullet \exists\, AState : \gamma_1 \bullet AOp_i\end{array}$$

□

Figure 8.2 illustrates the applicability and correctness requirements for a powersimulation. In the correctness condition, if γ_1 and γ_2 are linked to concrete states related by COp, then every value in γ_2 must be related to some value in γ_1 by the abstract operation AOp.

Example 8.2 We can use a powersimulation to verify the refinement of vending machines given in Example 8.1. To do so we define a relation $R : \mathbb{P}AVM \leftrightarrow CVM$

by listing its elements (abbreviating the bindings to just values):

$$(\{\langle\!|\, fault, \neg coin, null \,|\!\rangle, \langle\!|\, \neg fault, \neg coin, null \,|\!\rangle\}, \langle\!|\, \neg coin, null \,|\!\rangle)$$
$$(\{\langle\!|\, \neg fault, coin, tea \,|\!\rangle, \langle\!|\, \neg fault, coin, coffee \,|\!\rangle\}, \langle\!|\, coin, null \,|\!\rangle)$$
$$(\{\langle\!|\, \neg fault, \neg coin, tea \,|\!\rangle\}, \langle\!|\, \neg coin, tea \,|\!\rangle)$$
$$(\{\langle\!|\, \neg fault, \neg coin, coffee \,|\!\rangle\}, \langle\!|\, \neg coin, coffee \,|\!\rangle)$$

We now have to prove that the powersimulation conditions are satisfied. That is,

$$\forall CState' \bullet CInit \Rightarrow \exists \gamma : \mathbb{P} AInit \bullet R(\gamma) = \theta CState'$$
$$\forall CState \bullet \forall \gamma : \mathbb{P}(\text{pre}\, AOp) \bullet R(\gamma) = \theta CState \Rightarrow \text{pre}\, COp$$
$$\forall \gamma_1 : \mathbb{P}(\text{pre}\, AOp);\ CState;\ CState' \bullet R(\gamma_1) = \theta CState \wedge COp \Rightarrow$$
$$\exists \gamma_2 : \mathbb{P} AState \bullet R(\gamma_2) = \theta CState' \wedge$$
$$\forall AState' : \gamma_2 \bullet \exists AState : \gamma_1 \bullet AOp$$

for the initialisation and every operation. Proving the initialisation condition amounts to showing that

$$\langle\!|\, coin == false, select == null \,|\!\rangle$$
$$\Rightarrow \exists \gamma : \mathbb{P}\, Init_A \bullet R(\gamma) = \langle\!|\, coin == false, select == null \,|\!\rangle$$

This, and the remaining conditions, are easy to prove. □

Example 8.3 Given the lack of resources for higher education, we decide to build a gambling machine to raise some cash. Our initial specification of a one-armed bandit contains a number of operations. We can *Switch* on the machine, insert a *Coin*, *Play* the machine and *Collect* our winnings. The display consists of three wheels each with a choice of picture, and you hit the jackpot if you get three ♡. In this description the choice of display is in fact made when a coin is put in the machine, as $Coin_A$ is the only operation where d'_A can change.

$$WHEEL ::= \spadesuit \mid \heartsuit \mid \diamondsuit \mid \clubsuit$$

$$|\ bonus : \mathbb{N}$$

AGM

$d_A : \text{seq}\, WHEEL$
$on_A, collect : \mathbb{B}$
$cash : \mathbb{N}$

$Init_A$

AGM'

$d'_A = \langle \heartsuit, \heartsuit, \heartsuit \rangle$
$\neg collect'$
$\neg on'_A$
$cash' = 0$

$$\begin{array}{|l}
\multicolumn{1}{l}{\textit{Switch}_A} \\
\Delta AGM \\
\hline
on'_A \\
cash' = cash \\
collect' = collect \\
d'_A = d_A \\
\hline
\end{array}$$

$$\begin{array}{|l}
\multicolumn{1}{l}{\textit{Coin}_A} \\
\Delta AGM \\
\hline
cash' = 50 \\
\neg collect \wedge \neg collect' \\
on_A \wedge on'_A \\
cash = 0 \\
\hline
\end{array}$$

$$\begin{array}{|l}
\multicolumn{1}{l}{\textit{Play}_A} \\
\Delta AGM \\
disp! : \mathrm{seq}\, WHEEL \\
\hline
on_A \wedge on'_A \\
cash \geq 50 \\
collect' \Rightarrow cash' = bonus \\
\neg collect' \Rightarrow cash' = cash - 50 \\
d'_A = d_A \\
\neg collect \\
d_A = \langle \heartsuit, \heartsuit, \heartsuit \rangle \Leftrightarrow collect' \\
disp! = d_A \\
\hline
\end{array}$$

$$\begin{array}{|l}
\multicolumn{1}{l}{\textit{Collect}_A} \\
\Delta AGM \\
money! : \mathbb{N} \\
\hline
on_A \wedge on'_A \\
collect \wedge \neg collect' \\
d'_A = d_A \\
money! = cash \\
cash' = 0 \\
\hline
\end{array}$$

Upon review of the specification we are worried about where the choice of display is made. We decide that this information should not be stored, and we move the point where we choose the display to when the machine is played. We also decide to keep the machine permanently on!

$$\begin{array}{|l}
\multicolumn{1}{l}{CGM} \\
d_C : \mathrm{seq}\, WHEEL \\
on_C, collect : \mathbb{B} \\
cash : \mathbb{N} \\
\hline
on_C \\
\hline
\end{array}$$

$$\begin{array}{|l}
\multicolumn{1}{l}{\textit{Init}_C} \\
CGM' \\
\hline
d'_C = \langle \heartsuit, \heartsuit, \heartsuit \rangle \\
\neg collect' \\
cash' = 0 \\
\hline
\end{array}$$

$$\begin{array}{|l}
\multicolumn{1}{l}{\textit{Switch}_C} \\
\Xi CGM \\
\hline
\end{array}$$

$$\begin{array}{|l}
\multicolumn{1}{l}{\textit{Coin}_C} \\
\Delta CGM \\
\hline
cash' = 50 \\
\neg collect \wedge \neg collect' \\
cash = 0 \\
d'_C = d_C \\
\hline
\end{array}$$

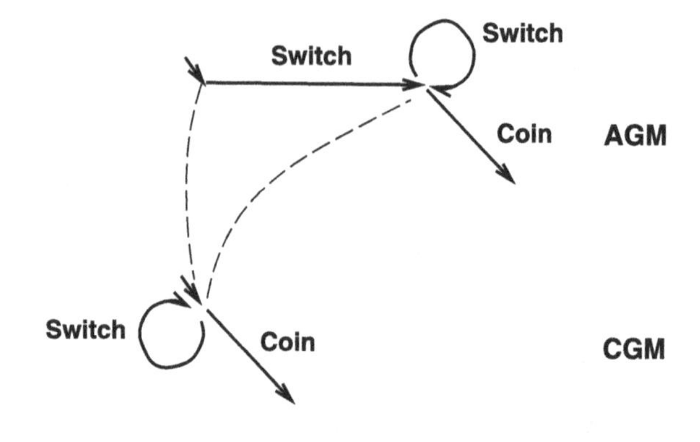

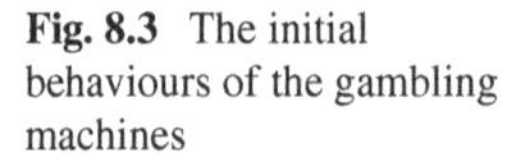
Fig. 8.3 The initial behaviours of the gambling machines

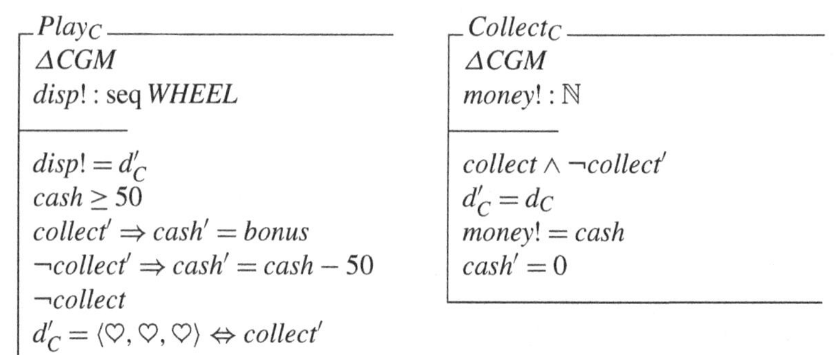

$Play_C$

ΔCGM
$disp! : \text{seq}\ WHEEL$

$disp! = d'_C$
$cash \geq 50$
$collect' \Rightarrow cash' = bonus$
$\neg collect' \Rightarrow cash' = cash - 50$
$\neg collect$
$d'_C = \langle \heartsuit, \heartsuit, \heartsuit \rangle \Leftrightarrow collect'$

$Collect_C$

ΔCGM
$money! : \mathbb{N}$

$collect \wedge \neg collect'$
$d'_C = d_C$
$money! = cash$
$cash' = 0$

CGM is a valid refinement of *AGM*, but needs both a downward and an upward simulation to verify it. The postponement of non-determinism from *Coin* to *Play* is typical of when we need an upward simulation, and a downward simulation cannot be used for this part of the specification. However, an upward simulation is not sufficient to verify the behaviour involving the initialisation and *Switch*. The diagram in Fig. 8.3 shows this portion of behaviour.

Any retrieve relation has to link $on_C = true$ to both $on_A = true$ and $on_A = false$, but with such a setup the initialisation condition in an upward simulation fails.

A powersimulation can be used to verify the refinement in one step. To construct the appropriate relation R between CGM and $\mathbb{P} AGM$, note that consideration of the initial states shows us that $on_C = true$ will be related to the set containing both $on_A = true$ and $on_A = false$, and also to the set consisting of $on_A = false$. The rest of the relation deals with verifying the upward simulation portion. The crucial bit of this is to link up the state at the end of $Coin_C$ to the set of all states at the end of $Coin_A$. Appropriate states of *Play* and *Collect* are also related. With these in place the conditions for a powersimulation follow fairly easily. □

8.3 Calculating Powersimulations

In this section we consider the calculational aspects of powersimulations. Suppose we are given a specification of an abstract data type $\mathsf{A} = (\mathsf{AState}, \mathsf{AInit}, \{\mathsf{AOp_i}\}_{\mathsf{i} \in \mathsf{I}},$

AFin), a concrete state space CState together with a relation R between $\mathbb{P}$AState and CState. In the same way that we calculated weakest simulations in Chap. 5 we can calculate the most general powersimulation of A.

We begin by working in the relational setting, and the following theorem summarises the results. As before we have to check that R is not too small or too large and, in addition, if our data types involve partial relations then we also need to check applicability: only if the calculated ADT satisfies these conditions does a powersimulation exist. The relation L is as defined above.

Theorem 8.2 *The weakest data type that is a powersimulation of* A *with respect to* R *is given by*

$$\begin{aligned}
&\mathsf{CInit} = (\mathsf{L} \setminus \mathsf{AInit}) \fatsemi \mathsf{R}\\
&\mathsf{CFin} = (\mathsf{L} \fatsemi \mathsf{AFin})/\mathsf{R}\\
&\mathsf{COp_i} = \mathrm{ran}(\mathrm{dom}\,\mathsf{AOp_i} \lhd_P \mathsf{R}) \lhd (((\mathsf{L} \setminus (\mathsf{L} \fatsemi \mathsf{AOp_i})) \fatsemi \mathsf{R})/(\mathrm{dom}\,\mathsf{AOp_i} \lhd_P \mathsf{R}))
\end{aligned}$$

whenever

$$\begin{aligned}
&(\mathsf{L} \setminus \mathsf{AInit}) \fatsemi \mathsf{R} \neq \varnothing\\
&\mathrm{ran}((\mathrm{dom}\,\mathsf{AOp_i}) \lhd_P \mathsf{R}) \subseteq \mathrm{dom}\,\mathsf{COp_i}\\
&\mathrm{ran}((\mathrm{dom}(\mathsf{L} \fatsemi \mathsf{AFin})) \lhd \mathsf{R}) \subseteq \mathrm{dom}((\mathsf{L} \fatsemi \mathsf{AFin})/\mathsf{R})
\end{aligned}$$

If the stated conditions do not hold then no powersimulation is possible for this A *and* R. □

Under certain circumstances a simplification is possible. In particular, if the domain of R consists only of singleton sets then the calculation simplifies to $\mathsf{COp_i} = \mathsf{R}^{-1} \fatsemi (\mathsf{L} \setminus (\mathsf{L} \fatsemi \mathsf{AOp_i})) \fatsemi \mathsf{R}$. Furthermore, in this case it is not necessary to check that $\mathrm{ran}((\mathrm{dom}\,\mathsf{AOp_i}) \lhd_P \mathsf{R}) \subseteq \mathrm{dom}\,\mathsf{COp_i}$.

Proposition 8.1 *Let*

$$\mathsf{COp} = \mathrm{ran}(\mathrm{dom}\,\mathsf{AOp} \lhd_P \mathsf{R}) \lhd (((\mathsf{L} \setminus (\mathsf{L} \fatsemi \mathsf{AOp})) \fatsemi \mathsf{R})/(\mathrm{dom}\,\mathsf{AOp} \lhd_P \mathsf{R}))$$

Then $\mathsf{COp} \subseteq \mathsf{R}^{-1} \fatsemi (\mathsf{L} \setminus (\mathsf{L} \fatsemi \mathsf{AOp})) \fatsemi \mathsf{R}$. *Furthermore if* $\mathrm{ran}((\mathrm{dom}\,\mathsf{AOp}) \lhd_P \mathsf{R}) \subseteq \mathrm{dom}\,\mathsf{COp}$ *and the domain of* R *consists of singleton sets then* $\mathsf{COp} = \mathsf{R}^{-1} \fatsemi (\mathsf{L} \setminus (\mathsf{L} \fatsemi \mathsf{AOp})) \fatsemi \mathsf{R}$.

Proof Let $(a, b) \in \mathsf{COp}$. Then

$$\begin{aligned}
&\exists s \bullet (s, a) \in (\mathrm{dom}\,\mathsf{AOp} \lhd_P \mathsf{R}) \land\\
&\forall s \bullet (s, a) \notin (\mathrm{dom}\,\mathsf{AOp} \lhd_P \mathsf{R}) \lor (s, b) \in ((\mathsf{L} \setminus (\mathsf{L} \fatsemi \mathsf{AOp})) \fatsemi \mathsf{R})
\end{aligned}$$

Hence there exists an s such that $(s, a) \in (\mathrm{dom}\,\mathsf{AOp} \lhd_P \mathsf{R})$ and $(s, b) \in ((\mathsf{L} \setminus (\mathsf{L} \fatsemi \mathsf{AOp})) \fatsemi \mathsf{R})$. Therefore $(a, b) \in \mathsf{R}^{-1} \fatsemi (\mathsf{L} \setminus (\mathsf{L} \fatsemi \mathsf{AOp})) \fatsemi \mathsf{R}$.

Conversely, assume that the members of the domain of R are singletons and that $\mathrm{ran}((\mathrm{dom}\,\mathsf{AOp}) \lhd_P \mathsf{R}) \subseteq \mathrm{dom}\,\mathsf{COp}$. Let $(a, b) \in \mathsf{R}^{-1} \fatsemi (\mathsf{L} \setminus (\mathsf{L} \fatsemi \mathsf{AOp})) \fatsemi \mathsf{R}$. Then there exists s_1 such that $(s_1, a) \in \mathsf{R}$ and $(s_1, b) \in ((\mathsf{L} \setminus (\mathsf{L} \mathbin{⨾} \mathsf{AOp})) \mathbin{⨾} \mathsf{R})$. The latter is equivalent to

$$\exists s_2 \bullet (s_2, b) \in \mathsf{R} \land \forall t \bullet t \notin s_2 \lor \exists u \bullet u \in s_1 \land (u, t) \in \mathsf{AOp}$$

The condition on the domain of R implies that $s_1 \subseteq \operatorname{dom} \mathsf{AOp}$, and therefore that $(a, b) \in \mathsf{COp}$. □

If we look at the proof we see that in fact the simplification of COp to $\mathsf{COp} = \mathsf{R}^{-1} \mathbin{⨾} (\mathsf{L} \setminus (\mathsf{L} \mathbin{⨾} \mathsf{AOp})) \mathbin{⨾} \mathsf{R}$ only requires that $\forall s \in \operatorname{dom} \mathsf{R} \bullet (s \cap \operatorname{dom} \mathsf{AOp} = \varnothing) \lor (s \subseteq \operatorname{dom} \mathsf{AOp})$ for every abstract operation. However, the simpler condition on the domain of R is easier to verify. When we rewrite this result in the Z schema calculus it looks like this.

Corollary 8.2 *Given an abstract specification, a concrete state space and a relation* $r : \mathbb{P} AState \leftrightarrow CState$. *The most general powersimulation with respect to r can be calculated as*:

$$\begin{array}{l} CInit == \exists \gamma : \mathbb{P} AInit \bullet r(\gamma) = \theta CState' \\ COp_i == \exists \gamma_1, \gamma_2 : \mathbb{P} AState \bullet \\ \quad r(\gamma_1) = \theta CState \land r(\gamma_2) = \theta CState' \land \gamma_1 \subseteq \operatorname{pre} AOp_i \land \\ \quad \forall AState' : \gamma_2 \bullet \exists AState : \gamma_1 \bullet AOp_i \end{array}$$

whenever a powersimulation exists, i.e., when

$$\begin{array}{l} \exists CState' \bullet CInit \\ \forall CState \bullet \forall \gamma : \mathbb{P}(\operatorname{pre} AOp_i) \bullet r(\gamma) = \theta CState \Rightarrow \operatorname{pre} COp_i \end{array}$$

Furthermore, if dom *r consists solely of singleton sets, then the calculation of* COp_i *can be simplified to*

$$\begin{array}{l} COp_i == \exists \gamma_1, \gamma_2 : \mathbb{P} AState \bullet r(\gamma_1) = \theta CState \land r(\gamma_2) = \theta CState' \land \\ \qquad\qquad\qquad \forall AState' : \gamma_2 \bullet \exists Astate : \gamma_1 \bullet AOp_i \end{array}$$ □

Example 8.4 From the specification of *AGM* and the relation *R* in Example 8.3 we can calculate the most general concrete operations. Consider the *Switch* operation. The domain of *R* contains sets γ_1 and γ_2 with

$$R(\gamma_i) = \langle\!| \, on_C == true, \ldots \, |\!\rangle$$

where γ_1 contains assignments of on_A to both *true* and *false*, and γ_2 contains only those with $on_A = true$. Using these sets in the calculation of the weakest refinement of *Switch* we find that this calculation evaluates to $Switch_C$. □

Example 8.5 We can turn the upward simulation between the specifications in Example 5.6 (p. 144) to a powersimulation defined by a relation $R : \mathbb{P} AState \leftrightarrow CState$

by taking

$$(X, u) \in R \text{ iff } \forall (s, t) \in X \bullet u_{merge}(s, t)$$

The weakest powersimulation can be calculated, and when evaluated they are, unsurprisingly, the same evaluations as in Example 5.6. □

Example 8.6 In a similar fashion we can turn the downward simulation R from Example 7.10 (p. 191) into a powersimulation r by defining

$$r = \{(\{x\}, y) \mid (x, y) \in R\}$$

Using this relation the weakest powersimulation calculations reduce to the weakest downward simulation calculations given in Example 7.10. □

8.4 Bibliographical Notes

The use of possibility mappings was originally proposed as a single complete method of refinement for I/O automata by Lynch in [7], and soundness and completeness for automata are discussed by Merrit in [9]. The use of possibility mappings in the context of transition systems is given by Gerth in [6] where the resultant rule is called failure simulation and is in essence the same as the relational characterisation we derived above. Other complete refinement methods include Abadi and Lamport's history and prophesy variable approach [1], the relationship between this and possibility mappings is discussed in [9]. Lynch and Vaandrager [8] provide an overview of simulation methods for untimed and timed automata which surveys the relationship between many of these approaches.

A single complete method can also be derived by using predicate transformers [5] in its formulation. In [5] Gardiner and Morgan show how upward and downward simulations are special cases of their method, which is therefore complete.

The use of possibility mappings in a relational context was discussed briefly in [3] and first appeared in the form given in this chapter in [4].

References

1. Abadi, M., & Lamport, L. (1991). The existence of refinement mappings. *Theoretical Computer Science*, *2*(82), 253–284.
2. de Bakker, J. W., de Roever, W.-P., & Rozenberg, G. (Eds.) (1990). *REX Workshop on Stepwise Refinement of Distributed Systems*, Nijmegen, 1989. *Lecture Notes in Computer Science: Vol. 430*. Berlin: Springer.
3. de Roever, W.-P., & Engelhardt, K. (1998). *Data Refinement: Model-Oriented Proof Methods and Their Comparison*. Cambridge: Cambridge University Press.
4. Derrick, J. (2000). A single complete refinement rule for Z. *Journal of Logic and Computation*, *10*(5), 663–675.

5. Gardiner, P. H. B., & Morgan, C. C. (1993). A single complete rule for data refinement. *Formal Aspects of Computing*, *5*, 367–382.
6. Gerth, R. (1990) Foundations of compositional program refinement. In de Bakker et al. [2] (pp. 777–807).
7. Lynch, N. A. (1990) Multivalued possibility mappings. In de Bakker et al. [2] (pp. 519–543).
8. Lynch, N. A., & Vaandrager, F. (1992). Forward and backward simulations for timing-based systems. In J. W. de Bakker, W.-P. de Roever, C. Huizing, & G. Rozenberg (Eds.), *Real-Time: Theory in Practice, REX Workshop*, Mook, The Netherlands, June 1991. *Lecture Notes in Computer Science: Vol. 600* (pp. 397–446). Berlin: Springer.
9. Merritt, M. (1990) Completeness theorems for automata. In de Bakker et al. [2] (pp. 544–560).

Part II
Interfaces and Operations: ADTs Viewed in an Environment

The derivation of refinement conditions as presented in Chaps. 3 and 4 relies on particular assumptions about the relation between the abstract data type and its environment: the state is not observable; the ADT's interface to the environment consists of exactly the given operations (no fewer, no more), with input and output parameters exactly as specified.

In particular applications and specification styles, these assumptions are too restrictive. This part of the book considers refinement rules based on various relaxations of the assumptions, and their application contexts.

In Chap. 9, we consider some different specification styles, expressed as variants on the standard ADT where some of the state variables are not viewed as being hidden. By interpreting these variants in terms of standard ADTs we obtain refinement rules which provide an extra level of flexibility without requiring further generalisations of the basic setup as described in Chap. 3.

Chapter 10 gives a more liberal interpretation of the basic setup, which allows inputs and outputs to be changed, added or deleted in data refinement steps.

Chapter 11 describes a notion of refinement for systems in which certain operations are viewed as "internal" operations which are executed by the system rather than by the observer. From a generalisation of the definition of data refinement (Definition 3.5) we obtain versions of the simulation conditions which allow the refinement, introduction and removal of such operations, leading up to a discussion of divergence.

Chapter 12 considers the consequences of allowing operations to be decomposed in refinement, which involves another generalisation of Definition 3.5. This is often known as non-atomic or action refinement in the literature. The final sections discuss applications to verifying concurrent data structures.

In Chap. 13, we apply these techniques in an extended case study, describing a variety of implementations of a watch.

The final chapter in this part, Chap. 14, briefly considers two further ways in which the operation index set may evolve during formal development: alphabet extension and alphabet translation, together with a generalisation which looks at approximate refinement between abstract and concrete systems, once again these involve generalisations of Definition 3.5.

Chapter 9
Refinement, Observation and Modification

In Chap. 1 we defined the notion of a "standard Z ADT", and so far we have tacitly assumed that anyone developing systems using Z would use this specification style. Of course, this assumption is by no means realistic. Much of the power and attraction of Z derives from Z being little more than logic and set theory with convenient options for naming and structuring—which can be (and is) used for anything for which logic and set theory can be used. As a consequence, although the semantics of a Z specification is fixed, its *interpretation* can be rather flexible.

In this chapter we will look at some mild variations on the "standard Z ADT", which all stay relatively close to the "states-and-operations" style of specification. The assumption that we will challenge in these specification styles is the representation hiding implied by the "abstract", "black box" nature of the state. Put differently, we are looking at data types which are not entirely abstract, but which may be partially concrete. In this sense, we nearly return to the concrete data types a.k.a. "repertoires" considered in a relational context in Chap. 2.

Example 9.1 Consider the following specification, of a UK traffic light (i.e., in between the "red" and "green" states, it displays amber and red lights together).

$$\begin{array}{|l}
\hline
\textit{TrafficLight} \\
\hline
red, green, amber : \mathbb{B} \\
\hline
\end{array}$$

The traffic light consists of three lights of different colours, which may be on or off.

$$\begin{array}{|l}
\hline
\textit{Init} \\
\hline
TrafficLight' \\
\hline
red' \land \neg green' \land \neg amber' \\
\hline
\end{array}$$

Excerpts from pp. 48–54 of [2] are reprinted within this chapter with kind permission of Springer Science and Business Media.

J. Derrick, E.A. Boiten, *Refinement in Z and Object-Z*,
DOI 10.1007/978-1-4471-5355-9_9, © Springer-Verlag London 2014

Initially only the red light is on. (This seems a safe way of starting off a traffic light, considering they normally occur in groups.)

$$
\begin{array}{|l}
\textit{Trans} \\
\hline
\Delta \textit{TrafficLight} \\
\hline
\textit{amber}' = \neg \textit{amber} \\
\textit{red}' = (\textit{amber} \neq \textit{red}) \\
\textit{green}' = \textit{amber} \wedge \textit{red} \\
\hline
\end{array}
$$

The amber light always changes at a transition; the red one will be on if red or amber was before, but not both; it is only green after amber plus red.

The intention of the specification is immediately clear, however, the semantics of the ADT (*TrafficLight*, *Init*, {*Trans*}) is *actually* completely trivial. The specification has no outputs, and allows any sequence of (identical) operations with no observable difference in outcome. □

We will admit that we have written many such specifications, even for papers on refinement. There are also many examples of excellent Z specifications with no outputs in the literature, e.g., the example of a visual editor in [12]. We do not even think anyone should be ashamed of that! If our examples are intuitively clear but formally nearly meaningless, we should consider adapting the formal theory before throwing away the nice examples.

A "fix" in the example above is of course to add an operation with outputs to the ADT, e.g.,

$$
\begin{array}{|l}
\textit{SeeTheLight} \\
\hline
\Xi \textit{TrafficLight} \\
\textit{amber}!, \textit{red}!, \textit{green}! : \mathbb{B} \\
\hline
\textit{amber}! = \textit{amber} \wedge \textit{red}! = \textit{red} \wedge \textit{green}! = \textit{green} \\
\hline
\end{array}
$$

or to add the same outputs and predicate to the *Trans* operation. Our reluctance to do the latter might be related to the fact that in real life traffic lights are observed more often than they change (certainly when one is in a hurry!). The problem with an operation like *SeeTheLight* is that its content is already fully implied by (our understanding of) the rest of the data type, so its explicit definition constitutes "formal noise".

This issue has been resolved below for Z by the definition of "grey box data types", and in Object-Z by allowing state components in the visibility list.

Apart from the fact that it sometimes "feels more natural" to have a visible (part of the) state, another reason for having "less abstract" data types can be found in the link to programming languages. For the final stages of formal development, it is convenient if the features of the specification language approximate those of the programming language. Representation hiding is *recommended* to programmers in most modern languages ("accessors and mutators"), but enforced by only few practical languages. A formal specification which contains a visible state component is closer to a Java class with a public attribute than one which contains explicit operations to that effect.

Note however that "visibility" in programming languages tends to imply modifiability, unlike in specification languages. Languages like C++ and Java have a variety of visibility attributes with corresponding restrictions and rules for inheritance—however, neither system is felt to be perfect by programmers. This we take as additional evidence that visibility is a subtle issue. Also, some of the visibility restrictions will need to have their counterparts on the formal specification side. For example, visible state components may not be removed in data refinement, much like public attributes need to remain public in inherited subclasses. (The relation between refinement and inheritance is further explored in Chap. 16.)

The next section will define the notion of a grey box data type, and give an interpretation in terms of ("black box") ADTs as used previously. From that interpretation, we then derive data refinement rules for grey box data types. Similarly, we define the notion of a display box data type, which generalises the grey box data type by allowing arbitrary observation operations. ("accessors"), not just of state components.

9.1 Grey Box Data Types

A grey box data type is a variant on the "black box" ADT used in previous chapters, for which some of the components of the state schema are declared to be observable or modifiable. We have been unable to come up with a use for state components which are modifiable, but not observable. As a consequence, the state of a grey box data type should, in general, be partitioned into three parts, all of them obtained from the complete state through schema projection or hiding. The private components are those which are not observable. They are specified only implicitly, *viz.* as those components not in the other two subsets. Both the readable components and the modifiable components can be observed, the modifiable components can be changed as well.

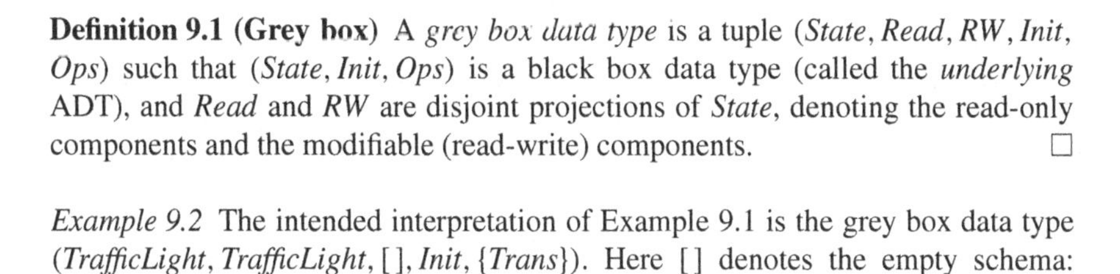

Definition 9.1 (Grey box) A *grey box data type* is a tuple $(State, Read, RW, Init, Ops)$ such that $(State, Init, Ops)$ is a black box data type (called the *underlying* ADT), and $Read$ and RW are disjoint projections of $State$, denoting the read-only components and the modifiable (read-write) components. □

Example 9.2 The intended interpretation of Example 9.1 is the grey box data type $(TrafficLight, TrafficLight, [\,], Init, \{Trans\})$. Here $[\,]$ denotes the empty schema: there are no modifiable components present. □

Note that we have not yet specified what it *means* for a component to be observable or modifiable—this will follow from the interpretation of a grey box as a black box below.

Making components modifiable is non-trivial, and involves a number of design choices. One of the arguments for not using public attributes in object-oriented programming languages (but mutators instead) is to prevent invariants becoming invalid. In a specification notation with partial operations, it is easy to prevent modifications that invalidate an invariant concerned with a *single* component (for example, that a sequence is sorted). The real problem is invariants involving multiple components, not all of which may be modifiable/public. For example, it is unwise, in a queue class, to make the queue *contents* directly modifiable when there is also a private variable that maintains the *size* of the queue. In programming languages, such undesirable modifications may lead to inconsistent states. They may be advised against, but their absence cannot be enforced. In a specification language, we may not need to forbid such changes, but instead allow for other components to change whenever such a modification takes place. However, it is not clear in general *which* other components should be allowed to change in order to re-establish the invariant. This is an instance of the well-known *framing* problem, discussed and partially solved in [3], and in the specific context of Z specification in [12]. In terms of the refinement calculus [14], the question we need to ask is: which variables apart from x are in the frame F in our desired specification[1] F:$[p, (x = x? \wedge p)]$, where p is the state predicate?

It would be nice if we could require a minimal change of the other variables: as few as possible variables to change, and those only if necessary. The following example shows that this does not lead to a unique solution.

Example 9.3 Consider the state schema

Seventeen
$x, y, z : \mathbb{N}$
$x + y + z = 17$

Now if we wanted to make only x modifiable, at least one of y and z would have to change if x did. As the predicate is symmetric in y and z, there is no way to determine which of the two should be allowed to change with x. □

For that reason, we take a pragmatic and not overly ambitious approach. Variables that are not declared to be modifiable will not be allowed to be changed indirectly. Modifiable components will only be allowed to be changed explicitly. However, we do allow simultaneous change of collections of modifiable components. This solution satisfies at least two desirable properties: it induces no restrictions if the state predicate is *true* (i.e., when there is no invariant to be violated), and it results in moderately simple specifications and refinement rules further on.

A convenient notation for expressing the black box interpretation is the following generalisation of Z schema decoration.

[1] In words, this specification statement expresses "change variables in F in such a way that, if p held initially, $x = x? \wedge p$ holds afterwards".

Definition 9.2 For schemas $S1$ and $S2$, the schema $S1?_{S2}$ denotes the schema obtained from $S1$ by decorating every component from $S2$ with a "?"; similarly for $S1!_{S2}$. Using the binding formation operator θ this could be expressed as:

$$S1?_{S2} == [\, S1;\ S2? \mid \theta S2 = \theta S2? \,] \setminus S2$$

We also use this notation with $S2$ a bracketed list of components rather than a schema. □

Observe that $S?$ is the same schema as $S?_S$.

After these considerations, we can define the meaning of a grey box, by interpreting it in terms of a black box data type. For simplicity, we have only included a *single* modification operator here that allows simultaneous modification of *all* modifiable components. In principle, we could provide $2^n - 1$ modification operations if there are n modifiable components.

Definition 9.3 (Interpretation of a grey box) The black box interpretation of a grey box $(State, Read, RW, Init, Ops)$ is the ADT $(State, Init, Ops \cup \{Mod\} \cup Obs)$ where Obs contains for every component x of $Read \land RW$ the observation operation

$$Obsx == \Xi State \land (\Xi State)!_{(x)}$$

and the modification operation is defined by

$$Mod == [\Delta State;\ RW? \mid (\Xi State)?_{RW}]$$ □

Example 9.4 Continuing from Example 9.3, consider the grey box data type $(Seventeen, [\,], RW, Init, \{\})$ where $RW == (\exists y, z : \mathbb{N} \bullet Seventeen) = [x : \mathbb{N} \mid x \leq 17]$ and $Init == [Seventeen']$.

For the observation operation (only one, since $RW \land [\,]$ has a single component x), we have

$$\begin{array}{|l}
\hline (\Xi Seventeen)!_{(x)} \\
x!, y, z : \mathbb{N} \\
x', y', z' : \mathbb{N} \\
\hline
x! + y + z = 17 \\
x' = x! \land y' = y \land z' = z \\
\hline
\end{array}$$

so the complete observation operation is the conjunction of that with $\Xi Seventeen$, which is

$$\begin{array}{|l}
\hline Obsx \\
x!, x, y, z : \mathbb{N} \\
x', y', z' : \mathbb{N} \\
\hline
x + y + z = 17 \\
x' = x! \land x' = x \land y' = y \land z' = z \\
\hline
\end{array}$$

(note that the equations $y' = y$ and $z' = z$ are contributed twice).

Observe that $RW? \equiv [x? : \mathbb{N} \mid x? \leq 17]$, and we have

$$\begin{array}{|l}
\hline
(\Xi\, Seventeen)?_{RW} \\
x?, y, z : \mathbb{N} \\
x', y', z' : \mathbb{N} \\
\hline
x? + y + z = 17 \\
x' = x? \wedge y' = y \wedge z' = z \\
\hline
\end{array}$$

so the complete modification operation is the extension of that with a declaration of x and the predicate of *Seventeen*, which is

$$\begin{array}{|l}
\hline
Mod \\
\Delta Seventeen \\
x? : \mathbb{N} \\
\hline
x? + y + z = 17 \\
x' = x? \wedge y' = y \wedge z' = z \\
\hline
\end{array}$$

A closer look reveals that $\Delta Seventeen \wedge x? + y + z = 17$ implies that $x = x?$, i.e., x can only be modified to the value it already had, as y and z are not allowed to change with x. This is the formal counterpart to the intuitive idea expressed in Example 9.3: x cannot be changed by itself in any meaningful way. □

Example 9.5 Continuing with the traffic light example, the interpretation of the grey box given in Example 9.2 would not contain the operation *SeeTheLight* suggested in Example 9.1, but rather three individual operations, each observing one of the lamps. The conjunction of these three would be *SeeTheLight* indeed. □

Example 9.6 An (imaginary) ancient machine for displaying four bit numbers has four switches, a handle, and a display. When you turn the handle, the display changes to the number represented by the current setting of the switches. This is specified by the grey box $(Anc, Anc \upharpoonright (disp), Anc \setminus (disp), Init, \{Handle\})$ where

$$BIT == \{0, 1\}$$

$$\begin{array}{|l}
\hline
Anc \\
sw1, sw2, sw4, sw8 : BIT \\
disp : \mathbb{N} \\
\hline
\end{array}
\qquad
\begin{array}{|l}
\hline
Init \\
Anc' \\
\hline
disp' = 0 \\
\hline
\end{array}$$

$$\begin{array}{|l}
\hline
Handle \\
\Delta Anc \\
\hline
\Xi(Anc \setminus (disp)) \\
disp' = sw1 + 2{*}sw2 + 4{*}sw4 + 8{*}sw8 \\
\hline
\end{array}$$

The interpretation as a grey box contains observation operations

$$\begin{array}{|l}
\multicolumn{1}{l}{\textit{Obsdisp}} \\
\hline
\Xi Anc \\
disp! : \mathbb{N} \\
\hline
disp! = disp \\
\hline
\end{array}
\qquad
\begin{array}{|l}
\multicolumn{1}{l}{\textit{Obssw1}} \\
\hline
\Xi Anc \\
sw1! : BIT \\
\hline
sw1! = sw1 \\
\hline
\end{array}$$

(and similarly for the other switches), plus a modification operation

$$\begin{array}{|l}
\multicolumn{1}{l}{\textit{Mod}} \\
\hline
\Delta Anc \\
sw1?, sw2?, sw4?, sw8? : BIT \\
\hline
disp' = disp \\
sw1' = sw1? \land sw2' = sw2? \land sw4' = sw4? \land sw8' = sw8? \\
\hline
\end{array}$$

□

A completely different approach to observability would be to encode grey box data types directly into the relational data types of Chap. 3. Observability of a component could then also be ensured by including it in the finalisation. This would be equivalent to having a single observation operation at the end of a program. As every operation could potentially be the last operation of a program, this would require availability for observation of the component after every operation. Semantically this would *not* be equivalent to the interpretation given above, as that allows a *sequence* of observations to be output (and thus transferred to the global state) rather than just the observation at the end of the program. The consequences for refinement would be very similar, though—both approaches need to ensure that it is possible to observe the relevant state component(s) at any point.

9.2 Refinement of Grey Box Data Types

Now that we have given an interpretation of grey box data types in terms of black box data types, we could actually conclude the discussion of grey box data types at this point. Black box data types come with a refinement relation. We can use this relation to verify whether refinement holds between grey box data types by applying it to their interpretations.

Refinement is often presented as an issue of verification only, where we get two ADTs and a retrieve relation, and we need to establish *post-hoc* whether these satisfy the relevant conditions. On the other hand, as we have argued previously in Chap. 5, sometimes we also want to approach refinement in a more constructive or calculational sense. For grey boxes, the verificational approach based on black box interpretations is clearly extremely non-constructive. As a consequence, it is hard to use for development, and it may be difficult to reconstruct the problem when a refinement fails to hold.

As an alternative, we will present a refinement relation defined directly between grey box data types. Clearly this refinement relation will be sound with respect to the black box refinement of interpretations. However, it will not and cannot be complete, as we shall illustrate later. Also, we will only concentrate on downward simulation in the remainder of this chapter.

Operation refinement and other simple refinements of grey boxes are not very interesting. The rules for the operations are just the same as for black box operation refinement, which follows from the fact that every operation in the grey box becomes an operation in the black box and the fact that the rules for operation refinement are really independent between the various operations. No invariants can be established or imposed over modifiable components other than those already apparent from their declarations. Observation operations will preserve any invariant, by not changing the state.

In the interpretation as black boxes, non-trivial operation refinement of operations *Obsx* is not possible because these operations are already total and deterministic. Operation refinements of the modification operation *Mod* are possible, however they will not normally result in black boxes that represent grey boxes.

Components being directly observable has radical consequences for data refinement: in order for the concrete observation operations to be projections as above, the observable components (i.e., including the modifiable ones) should be *preserved* unchanged in data refinement. This can be explained as follows. First, the type of $x!$ in *Obsx* cannot change in data refinement, as inputs and outputs are not changed in data refinement. Second, the predicate of *Obsx can* change in data refinement, but when it no longer has the shape $x = x!$ it cannot be an observation operation in the black box interpretation of a grey box. Thus, *Obsx* will have to retain variable x in data refinement and, as a consequence, so will the state.

Section 9.3 will remove this restriction, by defining observing operations which do not necessarily project out state components.

Data refinement between two grey boxes will thus, in the most general case, be between $(AState, Read, RW, AInit, AOps)$ and $(CState, Read, RW, CInit, COps)$ (i.e., identical *Read* and *RW* components), using a *retrieve relation R* whose signature is $AState \wedge CState$ (*AState* and *CState* must share all their observable components in *Read* and *RW*—recall that modifiable components are expected to be observable too). The rules for initialisation, and between *AOps* and *COps*, will be the same as those for the underlying black box. The rules as they derive from the implicit operations are investigated below (considering only downward simulation).

Observation Operations Observations can always be made, so the precondition of an observation operation is *true* for all states. As a consequence, the applicability condition $\text{pre}\, Obsx_A \wedge R \Rightarrow \text{pre}\, Obsx_C$ for observation operations *Obsx* reduces to *true*.

The correctness condition for observation operations also reduces to *true* because the state is unchanged, and the output of both the concrete and the abstract observation operation equals a component that was unchanged in data refinement.

Thus, the presence of observable components does not lead to any refinement conditions (apart from such components having to be present in both abstract and concrete state).

Modification Operation The analysis for the modification operation is slightly more involved. We can make two crucial observations. First, modification is a *deterministic* operation. Second, its precondition is independent from the modified components in the before-state, only from the remaining components plus the inputs. It only requires the state invariant to be satisfied by the remaining components and the inputs with ? decoration removed. Using the notation introduced in Definition 9.2, the predicate of pre *Mod* is $State?_{RW}$.

Applicability then becomes (eliding the quantification)

$$AState?_{RW} \land R \Rightarrow CState?_{RW}$$

and correctness between Mod_A and Mod_C (using determinacy of Mod_A)

$$R \land Mod_A \land Mod_C \Rightarrow R'$$

—informally, changing the same modifiable variables to the same values in two linked states should result in linked states afterwards. We thus have the following.

Definition 9.4 (Grey box data refinement) The grey box data type $(CState, Read, RW, CInit, COps)$ is a (downward simulation) data refinement of $(AState, Read, RW, AInit, AOps)$ if a retrieve relation R exists whose signature is $AState \land CState$ such that

underlying black boxes
$(CState, CInit, COps)$ is a black box data refinement of $(AState, AInit, AOps)$ using retrieve relation R.

modifiability
Any modification in the concrete type is possible when it is possible in the abstract type:

$$\forall AState; CState; RW? \bullet AState?_{RW} \land R \Rightarrow CState?_{RW}$$

correct modification
For the modification operations of the two types, Mod_A and Mod_C, we have:

$$\forall AState; CState; AState'; CState'; RW? \bullet$$
$$R \land Mod_A \land Mod_C \Rightarrow R'$$
□

First we will present (contrived) examples of data refinements that fail to hold due to either of the grey box specific conditions; these examples demonstrate that the two new conditions are independent. Then we will give an example of a correct data refinement.

Example 9.7 The grey box data type $(CState, [\,], [x : \mathbb{N}], CInit, COps)$ is *not* a data refinement of $(AState, [\,], [x : \mathbb{N}], AInit, AOps)$ for the given retrieve relation R, for any $CInit$ and $COps$:

$$\begin{array}{|l}\hline AState \\ x, y : \mathbb{N} \\ \hline x = y \lor x = y + 1 \\ \hline\end{array} \qquad \begin{array}{|l}\hline CState \\ x, z : \mathbb{N} \\ \hline x = z \\ \hline\end{array} \qquad \begin{array}{|l}\hline R \\ AState \\ CState \\ \hline y = z \\ \hline\end{array}$$

It fails on the modifiability condition, because in the abstract state where $x = y = 2$, x may be modified to 3 (i.e., $x? = 3$ is allowed by the precondition of the interpreting operation), whereas in the corresponding concrete state, i.e., where $x = z = 2$, it may not. □

Example 9.8 Consider the grey boxes $(S, S \upharpoonright (x), [\,], SInit, Ops)$ and $(T, T \upharpoonright (x), [\,], TInit, Ops)$.

$$\begin{array}{|l}\hline S \\ x, y : \mathbb{N} \\ \hline x = y \lor x = y + 1 \\ \hline\end{array} \qquad \begin{array}{|l}\hline T \\ x, z : \mathbb{N} \\ \hline x = z \lor x + 1 = z \\ \hline\end{array}$$

$$\begin{array}{|l}\hline SInit \\ S' \\ \hline x' = 1 \\ \hline\end{array} \qquad \begin{array}{|l}\hline TInit \\ T' \\ \hline x' = 1 \\ \hline\end{array}$$

with retrieve relation $R == [S;\ T \mid x = y]$. When $x, y = 5$ in S, x may be modified to 6 (leaving y unchanged). A related state in T is $x = 5, z = 6$ and also here x may be modified to 6 leaving z unchanged. However, the resulting states are unrelated. Thus, in this case refinement fails on the condition of modification correctness. □

Example 9.9 The users of the ancient machine in Example 9.6 decide to use their machine for banking purposes, and in order to model (modest) overdrafts they also require negative numbers. For that purpose, they replace their bit switches by "trit" switches, whose values can be 0, 1, or -1. As long as nobody puts any switch in the -1 position, the new machine should behave exactly like the old one, i.e., it should be a data refinement.

The new machine is given by the grey box $(Anc2, Anc2 \upharpoonright (disp), Anc2 \setminus (disp), Init2, \{Handle2\})$ where

$$TRIT == \{0, 1, -1\}$$

$$\begin{array}{|l}\hline Anc2 \\ sw1, sw2, sw4, sw8 : TRIT \\ disp : \mathbb{Z} \\ \hline\end{array} \qquad \begin{array}{|l}\hline Init2 \\ Anc2' \\ \hline disp' = 0 \\ \hline\end{array}$$

$$
\begin{array}{|l}
\hline
\textit{Handle2} \\
\Delta Anc2 \\
\hline
\Xi(Anc2 \setminus (disp)) \\
disp' = sw1 + 2*sw2 + 4*sw4 + 8*sw8 \\
\hline
\end{array}
$$

Using *Anc* itself as the retrieve relation, the underlying black boxes are clearly related by data refinement. Both modification operations are total, so the modifiability condition is satisfied. Correct modification follows from the fact that the concrete modification operation coincides with the abstract one on their common domain, and that the retrieve relation is the identity on that domain. (We do not need to consider the observation operations in the interpretations at all because they will be refinements thanks to the grey box formalisation.) □

The data refinement rule for grey boxes given above is clearly not complete—even if we added a similar rule for upward simulation, they would not be jointly complete. This is due to the fact that the function interpreting grey boxes as black boxes is not injective. Observation operations can be given implicitly, by declaring components to be visible in the grey box, or explicitly, by including explicit operations of the correct form. In particular, any black box can be taken as the underlying black box of a grey box with no modifiable or readable components.

A more serious issue with grey box refinement is that it requires, throughout stepwise development, that every observed variable must remain present as a state variable, even if the information that is to be "observed" could also be constructed from the state in another way. This problem is overcome by using so-called display boxes instead.

9.3 Display Boxes

Grey boxes go some way towards separating out "mutators" (which change the state, and possibly return a value) and "accessors" (which return a value but do not change the state) of an ADT. Modifiability can be used for simple mutators ("set" methods without data hiding); observability is used for simple accessors ("get" methods without data hiding); everything else has to be an operation on the underlying black box.

This section introduces an alternative type of ADT which allows all accessors to be separated out from the other operations. Rather than calling them accessors, we call them *displays*, suggesting an interpretation in behavioural terms: rather than being a true part of the possible sequences of actions of the ADT, they indicate observations that can be made throughout.

Allowing data hiding in displays has a number of consequences, which lead to major differences with grey box data types. These are due to the fact that data hiding requires an *explicit* description of the link between state and observation. Making it explicit allows this link to be modified when the state components change in a data refinement step. Unfortunately, the possibility for intuitively obvious implicit modification operations gets lost. Because the link between state and observation need not be injective nor surjective, providing an alternative observation as an input will not in general lead to a uniquely defined new state. Examples of this will be given later.

Displays will be written as schemas which do not refer to the after-state $State'$, and which have no inputs. Their interpretation is to add $\Xi State$ to displays. A cynic might wonder whether the invention of a new notion of data type is a worthwhile effort if it saves writing down a single Ξ. However, there is some advantage in being able to separate out the operations which have $\Xi State$ in refinement steps.

Definition 9.5 (Display box) A *display box data type* is a tuple $(State, Ds, Init, Ops)$ such that $(State, Init, Ops)$ is a black box data type (the *underlying black box*), and every element D of the set Ds is a schema on $State$ and some other ("output") type, such that D is total, i.e., $D \upharpoonright State = State$. □

The interpretation as a black box is very close to the display box: it just involves making $\Xi State$ explicit in every display.

Definition 9.6 (Interpretation of a display box) The display box $(State, Ds, Init, Ops)$ is interpreted as the black box $(State, Init, Ops \cup Disps)$ where $Disps$ contains for every element D of Ds the operation $D \wedge \Xi State$. □

Example 9.10 Consider the following model of a WWW browser.

$[PAGE]$
$URL == STRING$
$STATE ::= ok \mid failc \mid failg \mid connecting$

$$\begin{array}{|l}
www : URL \pfun PAGE \\
none : URL \\
null : PAGE \\
\hline
none \notin \operatorname{dom} www
\end{array}$$

$$\begin{array}{|l}
\multicolumn{1}{l}{\textit{Browser}} \\
\hline
page : PAGE \\
showing, req : URL \\
s : STATE \\
\hline
\end{array}$$

$$\begin{array}{|l}
\textit{Init} \\
\hline
Browser' \\
\hline
page' = null \\
showing' = none \land req' = none \\
s' = ok \\
\hline
\end{array}$$

$$\begin{array}{|l}
\textit{Request} \\
\hline
\Delta Browser \\
r? : URL \\
\hline
s' = connecting \\
page' = page \\
showing' = showing \\
req' = r? \\
\hline
\end{array}$$

$$\begin{array}{|l}
\textit{GetPage} \\
\hline
\Delta Browser \\
s = connecting \land req' = req \\
(s' = failc \land page' = page \land showing' = showing) \\
\quad \lor \\
(s' = failg \land page' = page \land showing' = showing \land req \notin \operatorname{dom} www) \\
\quad \lor \\
(s' = ok \land showing' = req \land req \in \operatorname{dom} www \land page' = www(req)) \\
\hline
\end{array}$$

As a black box, this would be a semantically fairly trivial example, allowing any number of requests with any URL parameter, with at most one *GetPage* after every one of them.

We can make it more meaningful by including two displays, one for the screen, using

$$[SCREEN]$$

$$\mid\ render : PAGE \rightarrow SCREEN$$

and one for the status line. This assumes *STRING* to be sequences of characters, and *STRING* literals denotable with double quotes.

$$\begin{array}{|l}
\textit{Screen} \\
\hline
Browser \\
s! : SCREEN \\
\hline
s! = render(page) \\
\hline
\end{array}$$

$$
\begin{array}{|l}
\hline
\textit{StatusLine} \\
\hline
\textit{Browser} \\
l! : \textit{STRING} \\
\hline
s = ok \Rightarrow l! = \text{“}\texttt{Displaying }\text{”} \frown \textit{showing} \\
s = \textit{failc} \Rightarrow l! = \text{“}\texttt{Connection to}\text{”} \frown \textit{req} \frown \\
\qquad\qquad \text{“}\texttt{ failed, displaying}\text{”} \frown \textit{showing} \\
s = \textit{failg} \Rightarrow l! = \text{“}\texttt{Page }\text{”} \frown \textit{req} \frown \\
\qquad\qquad \text{“}\texttt{ does not exist, displaying}\text{”} \frown \textit{showing} \\
s = \textit{connecting} \Rightarrow l! = \text{“}\texttt{Connecting to}\text{”} \frown \textit{req} \frown \\
\qquad\qquad \text{“}\texttt{, displaying}\text{”} \frown \textit{showing} \\
\hline
\end{array}
$$

Now $(\textit{Browser}, \{\textit{Screen}, \textit{StatusLine}\}, \textit{Init}, \{\textit{Request}, \textit{GetPage}\})$ is a display box data type with the required complex observable behaviour.

Observe that intuitively there is a significant difference between the operations *Request* and *GetPage*: one is initiated by the user, and the other (then) executed by the system. This sort of interpretation is formalised in Chap. 11. □

Example 9.11 The original ancient machine from Example 9.6 is given by the display box $(\textit{Anc}, \{\textit{Sw}1, \textit{Sw}2, \textit{Sw}3, \textit{Sw}4, \textit{Disp}\}, \textit{Init}, \{\textit{Handle}\})$ where $\textit{Sw}1 == [\textit{Anc};\ \textit{sw}1! : \textit{BIT} \mid \textit{sw}1! = \textit{sw}1]$, etc., and $\textit{Disp} == [\textit{Anc};\ \textit{disp}! : \mathbb{N} \mid \textit{disp}! = \textit{disp}]$. □

Example 9.12 The traffic light from Example 9.1 could also be described by the display box $(\textit{TrafficLight}, \{\textit{See}\}, \textit{Init}, \{\textit{Trans}\})$ where

$$
\begin{array}{|l}
\hline
\textit{See} \\
\hline
\textit{TrafficLight} \\
\textit{amber}!, \textit{red}!, \textit{green}! : \mathbb{B} \\
\hline
\textit{amber}! = \textit{amber} \land \textit{red}! = \textit{red} \land \textit{green}! = \textit{green} \\
\hline
\end{array}
$$

□

As in the case of grey boxes, we need to define refinement for display boxes, by translating back to refinements of interpreting black boxes. For this purpose, we will employ the technique of *calculating* most general data refinements as described in Chap. 5. The correctness and applicability conditions for the most general data refinement of a display operation reduce to *true*.

Definition 9.7 (Display box refinement) The display box $(\textit{AState}, \textit{ADs}, \textit{AInit}, \textit{AOps})$ is (downward simulation) data refined by display box $(\textit{CState}, \textit{CDs}, \textit{CInit}, \textit{COps})$ using retrieve relation R with signature $\textit{AState} \land \textit{CState}$ if

underlying black box
$(\textit{AState}, \textit{AInit}, \textit{AOps})$ is data refined by $(\textit{CState}, \textit{CInit}, \textit{COps})$ using retrieve relation R.

displays
The displays in *ADs* and *CDs* can be matched in pairs (AD, CD) such that when *WCD* is the weakest downward simulation of *AD* for *R*, then *CD* is an operation refinement of $WCD \lor \neg \operatorname{pre} WCD$. □

The disjunction with $\neg$ pre *WCD* constructs the weakest *total* operation which is a downward simulation.

Example 9.13 Consider the display box data type $(State, \{Lights\}, Init, \{Step\})$ where

$$STATE ::= stop \mid go \mid careful \mid rev$$

$$\begin{array}{|l}
\multicolumn{1}{l}{State} \\
\hline
s : STATE \\
\hline
\end{array}$$

$$\begin{array}{|l}
\multicolumn{1}{l}{Init} \\
\hline
State' \\
\hline
s' = stop \\
\hline
\end{array}$$

$$\begin{array}{|l}
\multicolumn{1}{l}{Step} \\
\hline
\Delta Trans \\
\hline
s = stop \Rightarrow s' = rev \\
s = rev \Rightarrow s' = go \\
s = go \Rightarrow s' = careful \\
s = careful \Rightarrow s' = stop \\
\hline
\end{array}$$

$$\begin{array}{|l}
\multicolumn{1}{l}{Lights} \\
\hline
State \\
red!, green!, amber! : \mathbb{B} \\
\hline
red! = ((s = stop) \lor (s = rev)) \\
green! = (s = go) \\
amber! = ((s = careful) \lor (s = rev)) \\
\hline
\end{array}$$

This display box is refinement equivalent to the one given in Example 9.12, with the retrieve relation *R* very similar to *Lights*:

$$\begin{array}{|l}
\multicolumn{1}{l}{R} \\
\hline
State; TrafficLight \\
\hline
red = ((s = stop) \lor (s = rev)) \\
green = (s = go) \\
amber = ((s = careful) \lor (s = rev)) \\
\hline
\end{array}$$

Indeed, *Lights* is the weakest total downward simulation of *See*. Conversely, *See* is not the weakest total downward simulation of *Lights*, as it makes specific choices of output for unreachable states (like all lights on, all lights off, or green + amber), which are unconstrained in the weakest simulation. □

The last example illustrates a clear advantage of display boxes over grey boxes: they allow either direct *or* indirect observations of variables, and thus a much broader range of data refinements.

However, there is also a downside to using display boxes: defining modifiable displays is problematic. This is an instance of the well-known and extensively studied view update problem in databases [1], and related to the problem of linking displays and updates in visualisation systems [13, 15]. Displays are defined in terms of state variables, but it is usually not clear how an explicit change in a display should be translated back to changes in those variables.

Example 9.14 A relevant part of the internal state of a drinks machine is the money contained in it. As in Example 2.2, p. 60, the coins that the machine can handle form a fixed set *coins*, represented by the set of their values. The coins actually held in the machine could then be recorded as:

CoinsHeld

$held : \mathrm{seq}\, coins$

Part of the drinks company's interface to the drinks machine might be a display as follows:

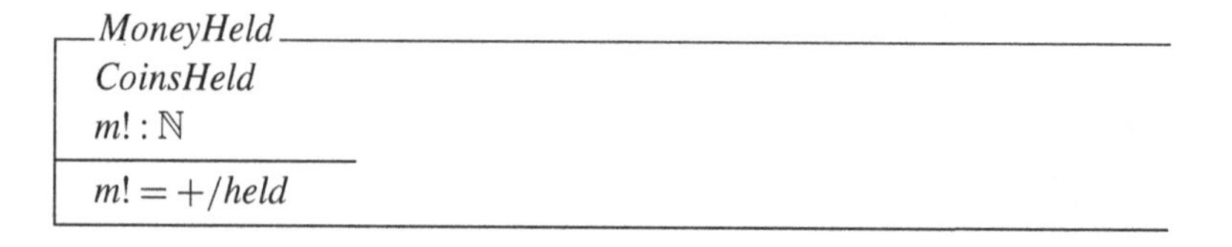

One might think that the drinks company could also do with a modifiable version of this display (modelling the opening of the money drawers, and leaving some arbitrary amount), but this would not be well-defined, as it would not specify which *coins* would be present afterwards. In addition, it may not be a total operation, depending on which sort of amounts could be expressed in the given selection of coins (cf. Example 2.2). □

We could introduce a data type with updateable displays by introducing the restriction that updateable displays are *injective*. However, this would result in seriously constrained data refinement rules, and thus we have omitted this alternative. Another option would be to accept that such display modifications are not always total or deterministic, but that would imply that display modifications have no special properties for refinement at all.

As a final remark we note that display box refinement is not "complete" with respect to the refinement on interpreting black boxes, for the same reasons as for grey boxes.

9.4 Bibliographical Notes

We first presented these ideas for allowing implicit observations and modifications at the 1998 International Refinement Workshop [2], as the very last talk of the workshop. Before that, many of the other presentations had touched on issues of framing and displays. Groves [10] dealt with the framing problem for modification by providing explicit frames, based on an approach by Mahoney. Woodcock illustrated his paper [17] with a web browser example whose outputs were changing constantly. De Groot and Robinson needed snapshots ("probe points") of the ADT state for the refinement analogue of error tracing [5].

An extensive discussion of the framing problem in a Z context may be found in a paper by Daniel Jackson [12].

Approaches to modelling human-computer interaction using formal methods are closely related to our treatment of displays and observations. In particular, work by Duke and Harrison [7, 8] and Sufrin and He [16] discusses notions of observation and refinement.

"Normal variables" in the treatment by de Roever and Engelhardt [6] play a similar *rôle* to modifiable components in grey boxes. By constraining simulation relations to those which are unaffected by normal variables, they minimise the adverse effects that could be caused by operations on normal variables. From the observation that Z inputs and outputs would *also* be modelled by normal variables in their approach, it seems we could make the link between IO and observations/modifications in Z more direct as well, i.e. by having a relational model which contains normal variables.

There is an another concept of "grey box" specification and refinement, required for the formal modelling of components with call-backs, which is unrelated to the concepts in this chapter [4].

References

1. Bertino, E., & Guerrini, G. (1996) Viewpoints in object database systems. In Finkelstein and Spanoudakis [9] (pp. 289–293).
2. Boiten, E. A., & Derrick, J. (1998) Grey box data refinement. In Grundy et al. [11] (pp. 45–59).
3. Borgida, A., Mylopoulos, J., & Reiter, R. (1993). And nothing else changes: the frame problem in procedure specifications. In *Proc. 15th International Conference on Software Engineering*, Baltimore, Maryland, May 1993 (pp. 303–314). Los Alamitos: IEEE Comput. Soc.
4. Büchi, M., & Weck, W. (1999). *The greybox approach: When blackbox specifications hide too much* (Technical Report TUCS-TR-297). Turku Centre for Computer Science.
5. de Groot, M., & Robinson, K. (1998) Correctness in refinement developments. In Grundy et al. [11] (pp. 117–132).
6. de Roever, W.-P., & Engelhardt, K. (1998). *Data Refinement: Model-Oriented Proof Methods and Their Comparison*. Cambridge: Cambridge University Press.
7. Duke, D. J., & Harrison, M. D. (1995). Event model of human-system interaction. *Software Engineering Journal*, *10*(1), 3–12.
8. Duke, D. J., & Harrison, M. D. (1995). Mapping user requirements to implementations. *Software Engineering Journal*, *10*(1), 13–20.

9. Finkelstein, A. & Spanoudakis, G. (Eds.) (1996). *SIGSOFT '96 International Workshop on Multiple Perspectives in Software Development (Viewpoints '96)*. New York: ACM.
10. Groves, L. (1998) Adapting program derivations using program conjunction. In Grundy et al. [11] (pp. 145–164).
11. Grundy, J., Schwenke, M., & Vickers, T. (Eds.) (1998). *International Refinement Workshop & Formal Methods Pacific '98*, Canberra, September 1998. *Discrete Mathematics and Theoretical Computer Science*. Berlin: Springer.
12. Jackson, D. (1995). Structuring Z specifications with views. *ACM Transactions on Software Engineering and Methodology*, *4*(4), 365–389.
13. Klinker, G. J. (1993). An environment for telecollaborative data exploration. In *Proceedings Visualization '93—Sponsored by the IEEE Computer Society* (pp. 110–117).
14. Morgan, C. C. (1994). *Programming from Specifications* (2nd ed.). *International Series in Computer Science*. New York: Prentice Hall.
15. Roberts, J. C. (1998). On encouraging multiple views for visualization. In *Information Visualization IV'98*, London, July 1998. Los Alamitos: IEEE Comput. Soc.
16. Sufrin, B., & He, J. (1990). Specification, refinement, and analysis of interactive processes. In M. D. Harrison & H. W. Thimbleby (Eds.), *Formal Methods in Human Computer Interaction* (pp. 153–200). Cambridge: Cambridge University Press.
17. Woodcock, J. C. P. (1998) Industrial-strength refinement. In Grundy et al. [11] (pp. 33–44).

Chapter 10
IO Refinement

In previous chapters we have not put much emphasis on inputs and outputs of operations. Indeed, often the refinement rules were presented for the simplified case of operations having no inputs or outputs. The justification for this was found in Sect. 4.4, where a fairly tedious derivation led to the conclusion that their presence does not fundamentally complicate matters concerning refinement. Summarising the development there: input and output can be incorporated in the input-output relation represented by the ADT by including sequences of input (to be consumed) and output (produced) in the state. Inputs and outputs then only play a *rôle* in the refinement rules at the operation schema level by the fact that they need to be universally (or sometimes existentially) quantified over. The observability of inputs and outputs is encoded in initialisation and finalisation, but induces few restrictions apart from inputs and outputs being unchanged in any refinement step.

However, conceptually inputs and outputs have just as much potential for description at various levels of abstraction and, consequently, for formal development, as the state. One often wishes to overlook that a natural number being input actually consists of a number of keystrokes—or that the list of tuples resulting from a search operation will actually appear as a beautifully formatted table which, in turn, is represented by millions of pixels on a computer screen. In real life we accept additions of inputs and outputs as "refinements" as well: a dialogue which asks an extra question; a copier with more buttons; a till receipt which, besides mentioning the price paid, also tells us how many bonus points we have gained. These are all acceptable, provided there is a way (using the new "inputs") to produce the results we expected to get, and a way to extract the information we received previously from the extended "output".

It has been known in the Z community for a long time that a sensible interpretation exists for refinements where inputs and outputs do change. This can be seen from, for example, the "birthday book" example in Spivey's reference manual [17]. This first adds an output to the *AddBirthday* example; this is called a "strengthening". Then, in an "implementation" step, an output set is replaced by a sequence. Woodcock and Davies [21] also include a few examples of refinements where inputs have changed.

J. Derrick, E.A. Boiten, *Refinement in Z and Object-Z*,
DOI 10.1007/978-1-4471-5355-9_10, © Springer-Verlag London 2014

In this chapter, refinement where inputs and outputs can change will be presented as a generalisation of refinement as described in Chap. 4, called *IO refinement*. In particular, we will derive a rule for IO refinement from a modified embedding of Z ADTs into relational data types. However, the resulting rule contains so many variables that it is hard to use. For that reason we will concentrate on derived rules which introduce small scale changes.

10.1 Examples of IO Refinement: "Safe" and "Unsafe"

For the reasons outlined above, we would like to be able to change inputs and outputs of any operation in refinement, and also to add new inputs and outputs that did not exist previously. However, not all such changes can be semantically meaningful, and we would like our refinement rules to reflect this. This section presents a number of "safe" and a number of "unsafe" examples of IO refinement. The informal explanations given of these should provide further illustration for the kind of situations where IO refinement is required.

Example 10.1 Consider the specification of an automated teller machine as in Example 7.3 (p. 176), in particular its *Withdraw* operation, which was given as

$$
\begin{array}{|l}
\textit{Withdraw} \\
\hline
\textit{ATMOp} \\
\textit{amount?}, \textit{money!} : \mathbb{N} \\
\hline
\textit{trans} \wedge \textit{accts}' = \textit{accts} \oplus \{\textit{card} \mapsto \textit{accts}(\textit{card}) - \textit{money!}\} \\
((\textit{card}' = 0 \wedge \neg \textit{trans}' \wedge \textit{money!} = 0) \vee \\
\qquad (\textit{card}' = \textit{card} \wedge \textit{trans}' \wedge 0 \leq \textit{money!} \leq \textit{amount?})) \\
\hline
\end{array}
$$

Instead of the output *money* being a number, it would be nice if we could refine it to being a sequence of coins (cf. Example 2.2), whose value adds up to the amount intended. So given a specific set of coin values *COINS*, e.g.,

$$\textit{COINS} == \{1, 2, 5, 10, 20, 50, 100, 200\}$$

we could refine *Withdraw* to

$$
\begin{array}{|l}
\textit{WithdrawC} \\
\hline
\textit{ATMOp} \\
\textit{amount?} : \mathbb{N} \\
\textit{money!} : \operatorname{seq} \textit{COINS} \\
\hline
\mathbf{let}\, \textit{money} == +/\textit{money!} \bullet \\
\qquad \textit{trans} \wedge \textit{accts}' = \textit{accts} \oplus \{\textit{card} \mapsto \textit{accts}(\textit{card}) - \textit{money}\} \\
\qquad ((\textit{card}' = 0 \wedge \neg \textit{trans}' \wedge \textit{money} = 0) \vee \\
\qquad (\textit{card}' = \textit{card} \wedge \textit{trans}' \wedge 0 \leq \textit{money} \leq \textit{amount?})) \\
\hline
\end{array}
$$

This should be an acceptable refinement, because essentially it only changes the representation of the output. This change of representation is non-deterministic (many sequences of coins add up to the same amount), but injective (any such sequence represents a fixed amount). Moreover, the different output does not break the abstraction by using information about non-observable state components. □

Example 10.2 However, one could imagine output refinements that do break abstraction. It does not seem "safe" to add an output to *Withdraw* as follows:

$$
\begin{array}{|l}
\textit{WithdrawX} \\
\hline
ATMOp \\
amount?, money! : \mathbb{N} \\
pins! : \mathbb{N} \twoheadrightarrow \mathbb{N} \\
\hline
trans \wedge accts' = accts \oplus \{card \mapsto accts(card) - money!\} \\
pins! = pins \\
((card' = 0 \wedge \neg trans' \wedge money! = 0) \vee \\
(card' = card \wedge trans' \wedge 0 \leq money! \leq amount?)) \\
\hline
\end{array}
$$

That is, the customer receives free with any withdrawal a list of everyone's PIN numbers. Although the original output can be reconstructed from this extended output (by just forgetting about the PIN codes), one should worry about this sort of refinement step, which makes a previously unobservable[1] state component observable.

However, this is only *apparently* a problem that is due to output refinement. Using ordinary refinement, we could have refined *Withdraw* to include a predicate such as

$$amount? \geq accts(card) \Rightarrow money! = pins(scrooge)$$

which gives Scrooge's PIN code away—so any perceived security problem is related to non-determinism [14], not to output refinement. Indeed, adding an output *pins*! which gave out an unconstrained collection of numbers would not appear to be a problematic output refinement—but then ordinary refinement could deliver the version above. □

Example 10.3 A final modification of *Withdraw* could be to replace the actual amount of money by an executive summary of its value, e.g.,

$$MONEY ::= none \mid some \mid lots$$

[1] Given PIN codes and account numbers are modelled as *unbounded* natural numbers, the value of *pins* cannot even be established by trial and error in finite time.

$$
\begin{array}{|l}
\hline
\textit{WithdrawX} \\
\hline
ATMOp \\
amount? : \mathbb{N} \\
money! : MONEY \\
\hline
\exists m : \mathbb{N} \bullet \\
\quad trans \wedge accts' = accts \oplus \{card \mapsto accts(card) - m\} \\
\quad ((card' = 0 \wedge \neg trans' \wedge m = 0) \vee \\
\quad (card' = card \wedge trans' \wedge 0 \leq m \leq amount?)) \\
\quad m = 0 \Rightarrow money! = none \\
\quad 0 < m \leq 100 \Rightarrow money! = some \\
\quad m > 100 \Rightarrow money! = lots \\
\hline
\end{array}
$$

This should not be acceptable as an output refinement, because it deletes observable information, *viz.* the amount actually handed out by the ATM. □

Example 10.4 In the London Underground example (Sect. 2.5, p. 64) we modelled journeys non-deterministically:

$$
\begin{array}{|l}
\hline
\textit{Travel} \\
\hline
\Delta LU \\
h! : STATION \\
\hline
lu' = lu \\
here \mapsto here' \in connected\ lu \\
h! = here' \\
\hline
\end{array}
$$

This is an acceptable view of an underground traveller for anyone who is not telepathic and who does not have access to the data of the entry gates. To the traveller, though, the trip is not made non-deterministically, but is (mostly) determined by a sequence of lines to be taken. We could model this as follows:

$$
\begin{array}{|l}
\hline
\textit{Itinerary} \\
\hline
\Delta LU \\
h! : STATION \\
l? : \operatorname{seq} LINE \\
\hline
lu' = lu \\
\exists s : \operatorname{seq}_1 STATION \bullet \\
\quad \#s = \#l? + 1 \\
\quad here = head\, s \wedge here' = last\, s \\
\quad \forall i : 2..\#s \bullet \{s(i-1), s(i)\} \subseteq \operatorname{ran}(lu(l?(i))) \\
\hline
\end{array}
$$

The last predicate states that every two stations $s(i-1)$ and $s(i)$ should be connected directly along line $l?(i)$.

We would like *Itinerary* to be an IO refinement of *Travel* which partially resolves the non-determinism present in *Travel* by providing a "control" input parameter. □

10.2 An Embedding for IO Refinement

In this section we will embark on a derivation of the conditions for IO refinement, gradually building up the ingredients and requirements.

The obvious starting point is the type of input and output in the concrete and abstract data types. Let us make similar assumptions to before, for the sake of the derivations only: that all abstract operations have input of type *AInput*, and output of type *AOutput*; all concrete operations have input of type *CInput* and output of type *COutput*. Clearly *CInput* and *AInput* should, in general, be *different* to obtain input refinement, and similarly for *COutput* and *AOutput*.

The assumption of identical input and output per ADT is, as before, not really necessary, and we will drop it in the final versions of the rules. This can be justified by taking, in the following derivations, I-indexed variants of all inputs, outputs and sequences and relations over them, for example a relation $\mathsf{IT} : AInput \leftrightarrow CInput$ becomes an indexed family $(\mathsf{IT}_i : AInput_i \leftrightarrow CInput_i)_{i \in I}$.

Next is the embedded retrieve relation. It should relate abstract state to concrete state as before, and now rather than copying input and output sequences it should provide a more general relation between them. However, not every relation would be suitable here. First, we rule out any relation that would use the (concrete or abstract) state to influence the transformation of input and output. We feel this would introduce an undesirable cross-link between the hidden local state and the observable global state (to which input and output belong in some sense). Also, out of a desire to keep the theory simple, we ensure that transformations of IO cause no extra partialities of operations. This has two implications: the IO transformations need to be total, and they may not reduce the length of, in particular, the input sequence, as this might leave the concrete program in a state where there is no input left to consume. Finally, we only allow the transformations to be defined in terms of element-wise transformations, i.e., they will be defined as $T*$ for some relation T. (This is in line with the close correspondence between IO sequences and programs as identified in Sect. 4.7.2.) The input/output for one concrete operation is then fully determined by the input/output for its abstract counterpart, not by any other inputs or outputs (which would amount to looking in the "future" or in the "past").

Thus, the relational embedding of a retrieve relation R (assumed to be between *AState* and *CState*) together with input and output transformations IT and OT (we will discuss their shape in Z and their interpretations later) is as follows (see Fig. 10.1):

$$\begin{aligned}\mathsf{R} == \{&R; ais; aos; cis; cos \mid (ais, cis) \in \mathsf{IT}* \wedge (aos, cos) \in \mathsf{OT}* \bullet \\ &(ais, aos, \theta AState) \mapsto (cis, cos, \theta CState)\}\end{aligned}$$

The concrete and abstract local states are determined already by this definition of R. However, it is not immediately clear what the global state G should be. Although the input sequence at initialisation should be related to the global state, it cannot be directly contained in it. This is most obvious for the "concrete" ADT, which will, in general, have inputs of a type different from that given originally.

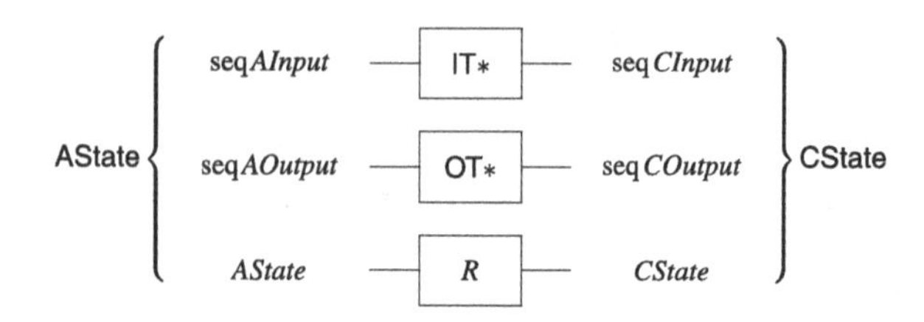

Fig. 10.1 R, the embedding of *R* using *OT* and *IT*

The same holds, however, for the "abstract" ADT, as the terms "abstract" and "concrete" are (here) only relative to the current development step! For the same reason, the global state will not contain the abstract or concrete output sequence directly.

We now have a choice: we can either document how the "current" input and output sequences relate to the global state, or we can assume an "original" type for inputs and outputs, which are included in the global state as before, and relate the current inputs and outputs to that. Although the former approach is more abstract, it has distinct disadvantages on the level of Z ADTs: it requires explicit mention of the global state on the Z level, and it offers no possibility of reconstructing the "original" input and output. For that reason, we will use the "original" IO approach.

Thus, in addition to the ADT we will need to provide two extra relations: an *original input transformer* relation which produces the input sequence, and an *original output transformer* relation between the output sequence and the original output. The types of these relations include the input and output types of the current ADT, so they are in some sense attached to the ADT. This is formalised in the following generalised notion of ADT.

Definition 10.1 (IO-mutable ADT) An *IO-mutable ADT* is a tuple $(State, Init, \{Op_i\}_{i \in I}, OIT, OOT)$, such that $(State, Init, \{Op_i\})$ (called the underlying ADT) is a Z ADT. When Op_i all have input *Input* and output *Output*, types *OInput* and *OOutput* exist such that *OIT* is a schema representing a relation between *OInput* and *Input*, and *OOT* represents a relation between *Output* and *OOutput*. *OIT* is required to be total, and *OOT* must be a function. □

Totality of the original input transformer is to ensure totality of initialisation as required by the relational ADT. Functionality of the output consumption may be seen as the "Every sperm is sacred" [13] view of outputs: whenever they are different, the ADT should preserve their unicity.

As will be clear from the following derivations, *OIT* and *OOT* play very much a subordinate *rôle* to the rest of the extended ADT, and when we are not fundamentally interested in keeping the boundaries of the system identical, we tend to consider the underlying ADT only. In some sense, the main contribution of *OIT* and *OOT* is to ensure that the embeddings of initialisation and finalisation are type-correct. If we ever wanted to reconstruct the inputs of a developed ADT from the originally specified ones, we could do this using (the inverse of) *OIT*. Similarly, we could reconstruct the original outputs using *OOT*. Given that *OIT* and *OOT* should not be prominent, we will in the derivations that follow sometimes *strengthen* the conditions in order to remove dependence on them. This sets the derivations below apart from those in Chap. 4, where the derived conditions were always *equivalent* to the conditions we started with. As a consequence, we make no claims of completeness for our IO refinement conditions—the restricted forms of IO transformation are another cause of incompleteness.

IO-mutable ADTs can be related by refinement if they share their operation index set, and their original input and output types are identical.

Definition 10.2 (Conformity of IO-mutable ADTs) The IO-mutable ADTs $(S_1, Init_1, \{Op1_i\}_{i \in I_1}, OIT_1, OOT_1)$ and $(S_2, Init_2, \{Op2_i\}_{i \in I_2}, OIT_2, OOT_2)$ are *conformal* if $I_1 = I_2$, OIT_1 and OIT_2 have the same source type, and OOT_1 and OOT_2 have the same destination type. □

For two such conformal types, the global state is as in Sect. 4.4 but using the original input and output types.

$$\mathsf{G} == \operatorname{seq} OInput \times \operatorname{seq} OOutput$$

For an IO-mutable ADT with state *State*, and operations with input type *Input* and output type *Output*, the embedding of its local state is as before:

$$\mathsf{State} == \operatorname{seq} Input \times \operatorname{seq} Output \times \{State \bullet \theta State\},$$

however due to transformation of inputs and outputs, the global state is no longer necessarily a part of the local state. Rather than directly copying sequences from one to the other, the initialisation and finalisation now need to take the transformation from/to original IO into account.

$$\begin{aligned}\mathsf{Init} == \{&Init;\ ois : \operatorname{seq} OInput;\ oos : \operatorname{seq} OOutput;\ is : \operatorname{seq} Input \mid \\ &(ois, is) \in \mathsf{OIT}* \bullet (ois, oos) \mapsto (is, \langle\rangle, \theta State')\}\end{aligned}$$

$$\begin{aligned}\mathsf{Fin} == \{&State;\ is : \operatorname{seq} Input;\ os : \operatorname{seq} Output;\ oos : \operatorname{seq} OOutput \mid \\ &(os, oos) \in \mathsf{OOT}* \bullet (is, os, \theta State) \mapsto (\langle\rangle, oos)\}\end{aligned}$$

Clearly finalisation is total whenever OOT is.

The embedding of the operations is unchanged: any transformation of inputs and outputs is dealt with elsewhere: in initialisation, finalisation and the retrieve relation.

$$\begin{aligned}\mathsf{Op_i} == \{&Op_i;\ is : \operatorname{seq} Input;\ os : \operatorname{seq} Output \bullet \\ &(\langle \theta ?Op_i \rangle \frown is, os, \theta State) \mapsto (is, os \frown \langle \theta !Op_i \rangle, \theta State')\}\end{aligned}$$

10.3 Intermezzo: IO Transformers

Before we can complete the embedding of an IO-mutable ADT, we need to determine the syntactic form of the input and output transformations in Z. They will all be represented by a particular style of schemas which we will call *IO transformers*. Informally, an IO transformer is a schema whose components are all either inputs or outputs. We have used a schema like that already (*GiveAmt* in Example 2.2) as an operation on a degenerate state. Here we will use them much like mains plug adapters in real life: providing inputs of the *required* kind on one end, and outputs to match the *available* input type on the other end, or *vice versa*, with a little wiring inside.

The definition of an IO transformer uses the definitions of signatures as given in Chap. 1.

Definition 10.3 (IO transformer) A Z schema S is an IO transformer iff

$$\Sigma S = ?S \wedge !S$$

i.e., the signature of S contains only its input and output components. □

The schema calculus does not provide us with a predefined way of removing decorations from component names when we have no schema declaring only the undecorated names. In fact, all we require here is for ! to be changed into ? and *vice versa*, which is defined below.

Definition 10.4 (IO decorations) For all component names x, let $\overline{x?}$ be the name $x!$, and let $\overline{x!}$ be the name $x?$.

This definition is extended to IO transformers, analogous to the normal Z schema decoration conventions. It also extends to binding formation θ, changing the labels of the generated binding accordingly. □

The overline operator is chosen in analogy with CCS [12]—in Z piping communication (see below) when they are both present with the correct decoration, x and $\overline{x}$ become identified and hidden.

An IO transformer is an input transformer for an operation if its outputs exactly match the operation's inputs, and analogously for output transformers.

Definition 10.5 (Input and output transformer) An IO transformer T is an *input transformer* for an operation Op iff

$$?Op = \overline{!T}$$

and it is an *output transformer* for Op iff

$$!Op = \overline{?T}$$ □

For an IO transformer T, the IO transformer $\overline{T}$ plays a particular *rôle*: not only are T and $\overline{T}$ input and output transformer for each other, but they act as converses in the relational interpretation.

So, how are input and output transformers used with operations? The answer is the use of the Z schema piping operator $\gg$. An input transformer IT is applied to operation Op in $IT \gg Op$. The meaning of this is, informally, the conjunction of Op and IT, equating and hiding the matching inputs of Op and outputs of IT. However, the equating and hiding should in some sense happen "before" the conjunction in order not to capture names of inputs of IT or outputs of Op.

Example 10.5 An input transformer for a function taking a Boolean input, according to the convention that 1 represents *true* and all other numbers represent *false*, is given by

IntBool

$x? : \mathbb{Z}$
$x! : \mathbb{B}$

$x! = (x? = 1)$

which we can apply to the following specification:

Purse

$money : \mathbb{N}$

DoubleOrNothing

$\Delta Purse$
$x? : \mathbb{B}$

$x? \Rightarrow money' = money * 2$
$\neg x? \Rightarrow money' = 0$

IntBool is an input transformer for *DoubleOrNothing*, as $!IntBool = [\, x! : \mathbb{B} \,]$, and thus $\overline{!IntBool} = [\, x? : \mathbb{B} \,] = ?DoubleOrNothing$. The application of the input transformer is

IntBool $\gg$ *DoubleOrNothing*

$\Delta Purse$
$x? : \mathbb{Z}$

$\exists x : \mathbb{B} \bullet$
$\quad x! = (x? = 1)$
$\quad x \Rightarrow money' = money * 2$
$\quad \neg x \Rightarrow money' = 0$

which can obviously be simplified to

IntBool $\gg$ *DoubleOrNothing*

$\Delta Purse$
$x? : \mathbb{Z}$

$(x? = 1) \Rightarrow money' = money * 2$
$(x? \neq 1) \Rightarrow money' = 0$

Note that this schema could not be constructed directly by hiding components of *IntBool* $\wedge$ *DoubleOrNothing*, as that contains a syntax error ($x?$ being both Boolean and integer). □

For every schema, input and output transformers can be defined that act as identities with piping:

Definition 10.6 (Input and output identity) For a schema S its *input identity* is defined by

$$\text{IId}\ S == [\ ?S;\ \overline{?S} \mid \theta ?S = \theta \overline{?S}\ ]$$

and its *output identity* by

$$\text{OId}\ S == [\ !S;\ \overline{!S} \mid \theta !S = \theta \overline{!S}\ ]$$ □

Clearly IId $S \gg S = S$ and $S \gg$ OId $S = S$ for any schema S. Also, the identity is equal to its converse, so $\overline{\text{IId}\ S} = \text{IId}\ S$, and similarly for OId.

Other special kinds of IO transformers are terminators and generators, which only consume output or only produce input.

Definition 10.7 (Terminators and generators) An IO transformer S is

- a *terminator* iff $\Sigma S = ?S$;
- a *generator* iff $\Sigma S = !S$. □

Obviously generators can function as input transformers, and terminators as output transformers, but the reverse is possible as well for operations that have no inputs or outputs.

Example 10.6 The birthday book example from Spivey's Z reference manual [17] is as follows.

$[NAME, DATE]$

```
┌─ BirthdayBook ─────────────────────
│ known : ℙ NAME
│ birthday : NAME ⇸ DATE
├──────────────
│ known = dom birthday
└────────────────────────────────────
```

AddBirthday

$\Delta BirthdayBook$
$name? : NAME$
$date? : DATE$

$name? \notin known$
$birthday' = birthday \cup \{name? \mapsto date?\}$

An indication of success or failure can be added to this as follows:

$$REPORT ::= ok \mid already\_known \mid not\_known$$

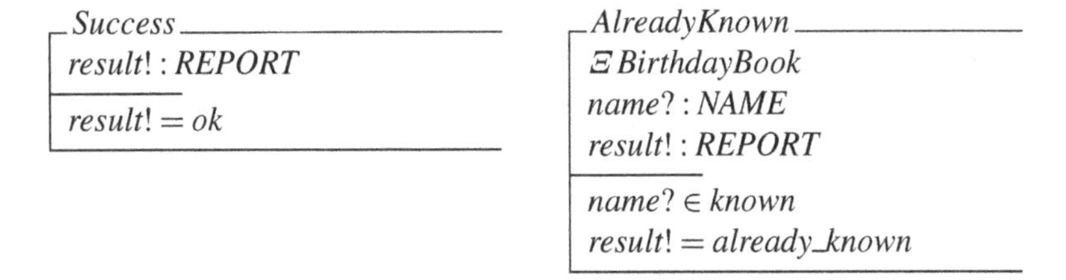

Now a robust (total) version of *AddBirthday* is given by

$$RAddBirthday == (AddBirthDay \wedge Success) \vee AlreadyKnown$$

The schema *Success* is an IO transformer (it has only outputs), and in particular it is a generator. However, it is not being used as an input transformer, but as an output transformer for the operation *AddBirthday* which previously did not have any outputs. (Observe that $AddBirthday \gg Success = AddBirthday \wedge Success$, as there is no matching IO to be hidden.) □

Sometimes it will be necessary to use IO transformers in schema conjunctions, quantifying over both original and transformed output at the same time, rather than using piping. This requires the base names of the IO transformer's inputs and outputs to be disjoint, which means fewer transformations can be expressed this way.

Definition 10.8 (Input and output views) The *input view* of an IO transformer (where all components are changed into inputs) is defined by

$$??IT == \exists\, !IT \bullet IT \wedge \mathrm{OId}\; IT$$

The *output view* of an IO transformer is defined by

$$!!OT == \exists\, ?OT \bullet OT \wedge \mathrm{IId}\; OT$$

□

Example 10.7 It would not be possible to give the input view of *IntBool* above, as its input and output base names are not disjoint. Consider the following variant:

$$\begin{array}{|l}
\multicolumn{1}{l}{\underline{\quad IB \quad}} \\
x? : \mathbb{Z} \\
y! : \mathbb{B} \\
\hline
y! = (x? = 1) \\
\hline
\end{array}$$

The input view of this IO transformer is:

$$\exists\, !IB \bullet IB \wedge \mathrm{OId}\; IB$$
$$\equiv$$
$$\exists\, [\, y! : \mathbb{B}\,] \bullet [\, x? : \mathbb{Z};\ y! : \mathbb{B} \mid y! = (x? = 1)\,] \wedge [\, y!, y? : \mathbb{B} \mid y! = y?\,]$$
$$\equiv$$
$$\exists\, [\, y! : \mathbb{B}\,] \bullet [\, x? : \mathbb{Z};\ y!, y? : \mathbb{B} \mid y! = (x? = 1) \wedge y! = y?\,]$$
$$\equiv$$
$$[\, x? : \mathbb{Z};\ y? : \mathbb{B} \mid y? = (x? = 1)\,]$$

as expected. □

To conclude the intermezzo on IO transformers, we give their embedding in the relational setting. For this embedding, it is important to know how the IO transformer is to be used: input transformers are embedded as relations between input bindings, and output transformers as relations between output bindings.

Definition 10.9 (Embedding of input and output transformers) The relational embedding of an input transformer IT is given by

$$\mathsf{IT} == \{IT \bullet \theta ?IT \mapsto \overline{\theta !IT}\}$$

The relational embedding of an output transformer OT is given by

$$\mathsf{OT} == \{OT \bullet \overline{\theta ?OT} \mapsto \theta !OT\}$$ □

These embeddings are also used for original input and output transformers. With these embeddings we have the useful properties that, for IT an input transformer and OT an output transformer for a given IO transformer TT,

$$\mathsf{IT} \gg \mathsf{TT} = \mathsf{IT} \fatsemi \mathsf{TT}$$
$$\mathsf{TT} \gg \mathsf{OT} = \mathsf{TT} \fatsemi \mathsf{OT}$$

Example 10.8 When used as an input transformer, *IntBool* defined by

$$\begin{array}{|l}
\multicolumn{1}{l}{\underline{\quad IntBool \quad}} \\
x? : \mathbb{Z} \\
x! : \mathbb{B} \\
\hline
x! = (x? = 1) \\
\hline
\end{array}$$

will be embedded as

$$\{\langle\!| x? == 1 |\!\rangle \mapsto \langle\!| x? == true |\!\rangle\} \cup$$
$$\{z : \mathbb{Z} \mid z \neq 1 \bullet \langle\!| x? == z |\!\rangle \mapsto \langle\!| x? == false |\!\rangle\}$$

Its embedding as an output transformer would be obtained by reversing all pairs in the relation, and changing all decorations ? to !. □

10.4 Deriving IO Refinement

Now the embedding is fully defined, we can derive the (downward simulation) IO refinement conditions from it, in a generalisation of the derivation in Sect. 4.4.

10.4.1 Initialisation

The initialisation condition affects both the ADT state and the inputs:

$$
\begin{array}{l}
\mathsf{CInit} \subseteq \mathsf{AInit} \mathbin{;} \mathsf{R} \\
\qquad \equiv \{ \text{ embeddings, definitions of } \subseteq \text{ and } \mathbin{;} \} \\
\forall\, ois, oos, cis, cos, c' \bullet \\
\quad (ois, cis) \in \mathsf{COIT}{*} \land c' \in \{CInit \bullet \theta CState'\} \land cos = \langle\rangle \\
\quad \Rightarrow \\
\quad \exists\, ais, aos, a' \bullet (ois, ais) \in \mathsf{AOIT}{*} \land a' \in \{AInit \bullet \theta AState'\} \\
\qquad\qquad\qquad \land\, aos = \langle\rangle \land (ais, aos, a') \mapsto (cis, cos, c') \in \mathsf{R} \\
\qquad \equiv \{ \text{ embedding } \mathsf{R}\text{; } \mathsf{OT}{*} \text{ links } \langle\rangle \text{ to } \langle\rangle \} \\
\forall\, ois, oos, cis, c' \bullet \\
\quad (ois, cis) \in \mathsf{COIT}{*} \land c' \in \{CInit \bullet \theta CState'\} \\
\quad \Rightarrow \\
\quad \exists\, ais, a' \bullet (ois, ais) \in \mathsf{AOIT}{*} \land a' \in \{AInit \bullet \theta AState'\} \\
\qquad\qquad \land\, (ais, cis) \in \mathsf{IT}{*} \land (a', c') \in \{R' \bullet \theta AState' \mapsto \theta CState'\}
\end{array}
$$

From this, it can be seen that the conditions for the states a' and c' are independent of those for the input sequences. The conditions for the states are identical to those in Sect. 4.4, so they come down to

$$\forall\, CState' \bullet CInit \Rightarrow \exists\, AState' \bullet R' \land AInit$$

The condition for the inputs is

$$\forall\, ois, cis \bullet (ois, cis) \in \mathsf{COIT}{*}$$

$$
\begin{array}{l}
\qquad\qquad \Rightarrow \exists\, ais \bullet (ois, ais) \in \mathsf{AOIT}* \wedge (ais, cis) \in \mathsf{IT}* \\
\qquad \equiv \{ \text{ definition of } \subseteq \text{ and } \fatsemi \} \\
\mathsf{COIT}* \subseteq \mathsf{AOIT}* \fatsemi \mathsf{IT}* \\
\qquad \equiv \{ * \text{ is monotonic and distributes through } \fatsemi \} \\
\mathsf{COIT} \subseteq \mathsf{AOIT} \fatsemi \mathsf{IT} \\
\qquad \Leftarrow \\
COIT = AOIT \gg IT
\end{array}
$$

The final step requires some explanation, particularly as it is formally a strengthening of the requirement. We stated earlier that original input transformations would take a subordinate *rôle* in the IO-mutable ADT, and should preferably not occur in the refinement conditions for the other ADT components. This is what happens here: we *define* the concrete original input transformation to be the abstract one, followed by the current input transformation, and this satisfies the refinement condition. It is enough for us that we can "undo" the input transformation that we require for the operations in some way—here, by just adding it to the original input transformer. In any case, due to *COIT* and *AOIT* both needing to be *total*, *COIT* could only be strictly included in $AOIT \gg IT$ if there was some non-determinism present in the latter—its domain could not be reduced.

Totality of *COIT* is guaranteed by totality of *AOIT* and *IT*.

10.4.2 Finalisation

Also for finalisation, we aim to select the concrete original transformation (for outputs, in this case) in such a way that it can always be constructed from the abstract transformation. However, a minor surprise pops up in this derivation.

$$
\begin{array}{l}
\mathsf{R} \fatsemi \mathsf{CFin} \subseteq \mathsf{AFin} \\
\qquad \equiv \{ \text{ definitions and embeddings } \} \\
\forall\, ais, aos, a, oos, ois \bullet \\
\qquad\quad (\exists\, cis, cos, c \bullet (ais, cis) \in \mathsf{IT}* \wedge (a, c) \in \{R \bullet \theta AState \mapsto \theta CState\} \\
\qquad\qquad\qquad\qquad\qquad \wedge (aos, cos) \in \mathsf{OT}* \wedge (cos, oos) \in \mathsf{COOT}* \wedge ois = \langle\rangle) \\
\qquad\quad \Rightarrow (aos, oos) \in \mathsf{AOOT}* \wedge ois = \langle\rangle \\
\qquad \equiv \{ \text{ } IT \text{ is total; one point rule for } ois\text{; } a, c \text{ unconstrained } \} \\
\forall\, aos, oos \bullet (\exists\, cos \bullet (aos, cos) \in \mathsf{OT}* \wedge (cos, oos) \in \mathsf{COOT}*) \\
\qquad\qquad \Rightarrow (aos, oos) \in \mathsf{AOOT}* \\
\qquad \equiv \{ \text{ definition of } \subseteq \text{ and } \fatsemi \} \\
\mathsf{OT}* \fatsemi \mathsf{COOT}* \subseteq \mathsf{AOOT}* \\
\qquad \equiv \{ * \text{ is monotonic and distributes through } \fatsemi \} \\
\mathsf{OT} \fatsemi \mathsf{COOT} \subseteq \mathsf{AOOT}
\end{array}
$$

The inclusion above is only concerned with non-determinism, not with domains, as *OT* is required to be total. The most general solution to this inequality may not be functional, or it may not satisfy the applicability condition for finalisations. However, defining

$$\mathsf{COOT} == \mathsf{OT}^{-1} \mathbin{;} \mathsf{AOOT}$$

the condition reduces to

$$\mathsf{OT} \mathbin{;} \mathsf{OT}^{-1} \subseteq \mathsf{id}_{\mathsf{AInput}}$$

which is reasonable: the abstract outputs are injectively transformed. Due to totality of *OT*, the inclusion above is strengthened to an equivalence. In terms of schemas, this would be

$$OT >> \overline{OT} = \text{IId}\ OT$$

This choice of *COOT* is indeed functional:

$$\begin{array}{l} \mathsf{COOT}^{-1} \mathbin{;} \mathsf{COOT} \\ \quad \equiv \{ \text{ definition of } COOT \ \} \\ (\mathsf{OT}^{-1} \mathbin{;} \mathsf{AOOT})^{-1} \mathbin{;} \mathsf{OT}^{-1} \mathbin{;} \mathsf{AOOT} \\ \quad \equiv \{ \ (A \mathbin{;} B)^{-1} = B^{-1} \mathbin{;} A^{-1} \ \} \\ \mathsf{AOOT}^{-1} \mathbin{;} \mathsf{OT} \mathbin{;} \mathsf{OT}^{-1} \mathbin{;} \mathsf{AOOT} \\ \quad \equiv \{ \ OT \text{ is total and injective } \} \\ \mathsf{AOOT}^{-1} \mathbin{;} \mathsf{AOOT} \\ \quad \subseteq \{ \ AOOT \text{ assumed to be functional } \} \\ \mathsf{id} \end{array}$$

Functionality of original output transformers now appears to be a derived property more than a requirement, as we would start with an identity function (on the original output type), and repeatedly compose this with inverses of injective relations, i.e., functions.

This choice of *COOT* also trivially satisfies the applicability condition of finalisations.

10.4.3 Applicability

For applicability and correctness, observe that the embedding of the operations is the same as in Sect. 4.4, and thus the analysis of the domain of an operation there is still valid.

$$\text{ran}(\text{dom}\ \mathsf{AOp_i} \lhd \mathsf{R}) \subseteq \text{dom}\ \mathsf{COp_i}$$

$$
\begin{array}{l}
\quad \equiv \{ \text{ definitions } \} \\
\forall c, a, cis, cos, ais, aos \bullet \\
\quad\quad ((ais, aos, a) \mapsto (cis, cos, c) \in \mathsf{R} \land (ais, aos, a) \in \operatorname{dom} \mathsf{AOp_i}) \\
\quad\quad \Rightarrow (cis, cos, c) \in \operatorname{dom} \mathsf{COp_i} \\
\quad \equiv \{ \text{ embedding of } R\text{; analysis of dom } \} \\
\forall c, a, cis, cos, ais, aos \bullet \\
\quad\quad (ais, cis) \in \mathsf{IT}* \land (aos, cos) \in \mathsf{OT}* \land ais \neq \langle\rangle \\
\quad\quad \land\ (a, c) \in \{R \mid \theta AState \mapsto \theta CState\} \\
\quad\quad \land\ (\exists \operatorname{pre} AOp_i \bullet a = \theta AState \land head\, ais = \theta ?AOp_i) \\
\quad\quad \Rightarrow \\
\quad\quad cis \neq \langle\rangle \land (\exists \operatorname{pre} COp_i \bullet c = \theta CState \land head\, cis = \theta ?COp_i) \\
\quad \equiv \{ \text{ calculus: } cis \neq \langle\rangle \text{ follows, replace } cis \text{ by } cin = head\, cis, \\
\quad\quad \text{remove } aos, cos, \text{ replace } ais \text{ by } ain = head\, ais \ \} \\
\forall c, a, cin, ain \bullet (ain, cin) \in \mathsf{IT} \land (a, c) \in \{R \mid \theta AState \mapsto \theta CState\} \\
\quad\quad\quad\quad \land\ (\exists \operatorname{pre} AOp_i \bullet a = \theta AState \land ain = \theta ?AOp_i) \\
\quad\quad\quad\quad \Rightarrow \\
\quad\quad\quad\quad \exists \operatorname{pre} COp_i \bullet c = \theta CState \land cin = \theta ?COp_i \\
\quad \equiv \{ \text{ predicate calculus } \} \\
\forall c, a, cin \bullet (a, c) \in \{R \mid \theta AState \mapsto \theta CState\} \\
\quad\quad\quad \land \exists \operatorname{pre} AOp_i \bullet a = \theta AState \land ain = \theta ?AOp_i \\
\quad\quad\quad\quad\quad\quad \land \exists ain \bullet (cin, ain) \in \overline{\mathsf{IT}} \\
\quad\quad\quad \Rightarrow \\
\quad\quad\quad \exists \operatorname{pre} COp_i \bullet c = \theta CState \land cin = \theta ?COp_i \\
\quad \equiv \{ \text{ introduce schema quantification } \} \\
\forall CState; AState; ?COp_i \bullet \operatorname{pre}(\overline{IT} \gg AOp_i) \land R \Rightarrow \operatorname{pre} COp_i
\end{array}
$$

Thus, the applicability condition is mostly unchanged, except that we need to consider the abstract operation with the “new” inputs rather than its original inputs. It is no surprise that the output transformer is of no relevance here, as $\operatorname{pre}(Op \gg OT) = \operatorname{pre} Op$ for any total output transformer OT.

10.4.4 Correctness

The derivation for the correctness condition is as follows.

$$
\begin{array}{l}
(\operatorname{dom} \mathsf{AOp_i} \lhd \mathsf{R}) \fatsemi \mathsf{COp_i} \subseteq \mathsf{AOp_i}; \mathsf{R} \\
\quad \equiv \{ \text{ definitions } \} \\
\forall a, ais, aos, c, cis, cos, c', cis', cos' \bullet \\
\quad\quad ((ais, aos, a) \mapsto (cis, cos, c) \in \mathsf{R} \land (ais, aos, a) \in \operatorname{dom} \mathsf{AOp_i}
\end{array}
$$

$$\land\ (cis, cos, c) \mapsto (cis', cos', c') \in \mathsf{COp_i})$$
$$\Rightarrow (ais, aos, a) \mapsto (cis', cos', c) \in \mathsf{AOp_i} \mathbin{;} \mathsf{R}$$
$\equiv$ { embeddings }

$$\forall\, a, ais, aos, c, cis, cos, c', cis', cos' \bullet$$
$$(ais, cis) \in \mathsf{IT}{*} \land (aos, cos) \in \mathsf{OT}{*}$$
$$\land\ (a, c) \in \{R \bullet \theta AState \mapsto \theta CState\} \land ais \neq \langle\rangle$$
$$\land (\exists\, \mathrm{pre}\, AOp_i \bullet a = \theta AState \land head\, ais = \theta ?AOp_i)$$
$$\land\ cis \neq \langle\rangle \land cis' = tail\, cis$$
$$\land (\exists\, COp_i \bullet c = \theta CState \land c' = \theta CState'$$
$$\land\, head\, cis = \theta ?COp_i \land cos' = cos \frown \langle \theta !COp_i \rangle)$$
$$\Rightarrow$$
$$ais \neq \langle\rangle \land (tail\, ais, cis') \in \mathsf{IT}{*}$$
$$\land (\exists\, AOp_i;\ R' \bullet head\, ais = \theta ?AOp_i$$
$$\land\ a = \theta AState \land c' = \theta CState'$$
$$\land\ aos \frown \langle \theta !AOp_i \rangle \mapsto cos' \in \mathsf{OT}{*})$$

$\equiv$ { one point rule for $cis' = tail\, cis$; $ais \neq \langle\rangle$ follows; replace ais by $ain = head\, ais$ }

$$\forall\, a, ain, aos, c, cis, cos, c', cos' \bullet$$
$$(ain, head\, cis) \in \mathsf{IT} \land (aos, cos) \in \mathsf{OT}{*}$$
$$\land\ (a, c) \in \{R \bullet \theta AState \mapsto \theta CState\}$$
$$\land (\exists\, \mathrm{pre}\, AOp_i \bullet a = \theta AState \land ain = \theta ?AOp_i) \land cis \neq \langle\rangle$$
$$\land (\exists\, COp_i \bullet c = \theta CState \land c' = \theta CState'$$
$$\land\, head\, cis = \theta ?COp_i \land cos' = cos \frown \langle \theta !COp_i \rangle)$$
$$\Rightarrow$$
$$\exists\, AOp_i;\ R' \bullet ain = \theta ?AOp_i \land a = \theta AState$$
$$\land\ c' = \theta CState' \land \theta !AOp_i \mapsto last\, cos' \in \mathsf{OT}$$

$\equiv$ { replace cis by $cin = head\, cis$, cos' by $cout = last\, cos'$; aos and cos unconstrained }

$$\forall\, a, ain, c, cin, c', cout \bullet$$
$$(a, c) \in \{R \bullet \theta AState \mapsto \theta CState\} \land (ain, cin) \in \mathsf{IT}$$
$$\land (\exists\, \mathrm{pre}\, AOp_i \bullet a = \theta AState \land ain = \theta ?AOp_i)$$
$$\land (\exists\, COp_i \bullet c = \theta CState \land c' = \theta CState'$$
$$\land\ cin = \theta ?COp_i \land cout = \theta !COp_i)$$
$$\Rightarrow$$
$$\exists\, AOp_i;\ R' \bullet ain = \theta ?AOp_i \land a = \theta AState$$
$$\land\ c' = \theta CState' \land \theta !AOp_i \mapsto cout \in \mathsf{OT}$$

$\equiv$ { calculus }

$$\begin{array}{l}
\forall a, ain, c, c', cout \bullet \\
\quad (a, c) \in \{R \bullet \theta AState \mapsto \theta CState\} \\
\quad \wedge (\exists \operatorname{pre} AOp_i \bullet a = \theta AState \wedge ain = \theta ?AOp_i) \\
\quad \wedge (\exists COp_i;\ cin \bullet (ain, cin) \in \mathsf{IT} \wedge c = \theta CState \wedge c' = \theta CState' \\
\qquad\qquad \wedge\ cin = \theta ?COp_i \wedge cout = \theta !COp_i) \\
\quad \Rightarrow \\
\quad \exists AOp_i;\ R';\ aout \bullet ain = \theta ?AOp_i \wedge a = \theta AState \\
\qquad\qquad \wedge\ c' = \theta CState' \wedge aout = \theta !AOp_i \\
\qquad\qquad \wedge\ aout \mapsto cout \in \mathsf{OT} \\
\equiv \{ \text{ introduce schema quantification } \}
\end{array}$$

$$\begin{array}{l}
\forall AState;\ ?AOp_i;\ CState;\ CState';\ !COp_i \bullet \\
\quad R \wedge \operatorname{pre} AOp_i \wedge (IT \gg COp_i) \\
\quad \Rightarrow \\
\quad \exists AState' \bullet R' \wedge (AOp_i \gg OT)
\end{array}$$

Thus, whereas applicability needs to be checked for operations $\overline{IT} \gg AOp_i$ and COp_i, the correctness condition for IO refinement resembles the normal correctness condition for $AOp_i \gg OT$ and $IT \gg COp_i$.

10.5 Conditions of IO Refinement

Summarising the results of the preceding derivation, we have the following.

Definition 10.10 (IO downward simulation) Given conformal IO-mutable ADTs $A = (AState, AInit, \{AOp_i\}_{i \in I}, AOIT, AOOT)$ and $C = (CState, CInit, \{COp_i\}_{i \in I}, COIT, COOT)$. Let for $i \in I$, IT_i be an input transformer for COp_i which is total on $?AOp_i$. Let OT_i be a total injective output transformer for AOp_i. (IT denotes the disjoint sum of IT_i, and similar for OT.) The retrieve relation R defines a downward simulation between A and C if:

$$\begin{array}{l}
COIT = AOIT \gg IT \\
COOT = \overline{OT} \gg AOOT \\
\forall CState' \bullet CInit \Rightarrow \exists AState' \bullet R' \wedge AInit
\end{array}$$

and for all $i \in I$

$$\begin{array}{l}
\forall CState;\ AState;\ ?COp_i \bullet \operatorname{pre}(\overline{IT_i} \gg AOp_i) \wedge R \Rightarrow \operatorname{pre} COp_i \\
\forall AState;\ ?AOp_i;\ CState;\ CState';\ !COp_i \bullet \\
\qquad R \wedge \operatorname{pre} AOp_i \wedge (IT_i \gg COp_i) \Rightarrow \\
\qquad\quad \exists AState' \bullet R' \wedge (AOp_i \gg OT_i)
\end{array}$$

If this is the case, we also say that the *underlying ADTs* (or any of their matching operations) are related by an *IO downward simulation* or by *IO refinement*, and in that case omit the first two conditions. □

Theorem 10.1 *IO downward simulation generalises downward simulation on the underlying ADTs.*

Proof It could be argued that this already follows from the derivation of IO downward simulation being a generalisation of the standard downward simulation. Apart from that, by taking all input and output transformers to be the appropriate identities, the rules above reduce to the normal downward simulation rules.

The IO downward simulation rule has a large number of variables: one to three more, per operation, than the standard downward simulation rule. For that reason, it is somewhat unwieldy in practice. For ordinary data refinement, we can separate out the change of data representation from reduction of non-determinism, by first calculating the most general data refinement (see Chap. 5), and then using operation refinement (see Chap. 2). Similarly, many IO refinements can be decomposed into simple IO refinement steps, followed by ordinary data refinement. For that purpose, we present a number of simplified IO refinement rules. These do not refer to the original IO transformers, but the rules can be extended to include those components, too. □

Theorem 10.2 (Simple output refinement) *For any ADT, adding a declaration of a new output (from a non-empty set) to one of the operations constitutes a valid IO downward simulation.*

Proof Take the input and output transformers for all unaffected operations to be the appropriate identities. Assume we want to add output $x! : S$ to operation AOp, then the output transformer for AOp is given by

$$OT == \mathrm{OId}\ AOp \wedge [\, x! : S \,]$$

This output transformer is clearly total and injective (its converse just removes the new output $x!$). Applicability is trivial, as the new, unconstrained, output does not strengthen the precondition when it can be taken from a non-empty set. Correctness is simple as well, because $AOp \gg OT$ is exactly the concrete operation. □

Example 10.9 The birthday book transformation (Example 10.6) is an example of the application of this theorem, taking the added output as $[result! : \{ok\}]$. As there were no other outputs, this generator is also the relevant output transformer. Indeed it is the case that $AddBirthday \wedge Success = AddBirthday \gg Success$. Because the precondition of $AlreadyKnown$ is disjoint from that of $AddBirthday \wedge Success$, the complete operation $RAddBirthday$ is indeed an (IO) refinement of $AddBirthday$. □

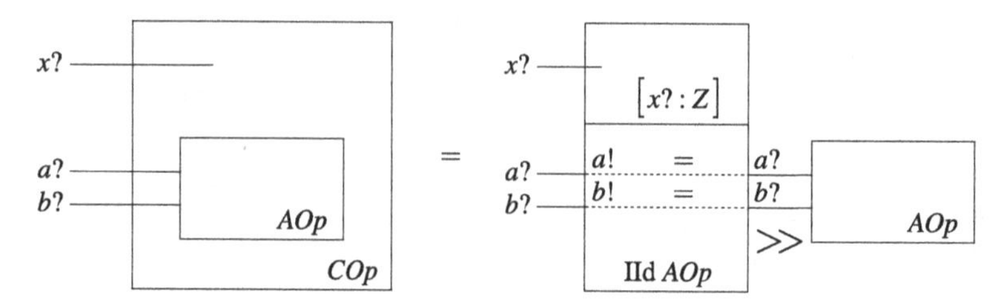

Fig. 10.2 Simple input refinement

Theorem 10.3 (Simple input refinement) *For any ADT, adding a declaration of a new input (from a non-empty set) to one of the operations constitutes a valid IO downward simulation.*

Proof Let the new input be $x? : Z$. Take the input and output transformers for all unaffected operations to be the appropriate identities. The input transformer for AOp is given by

$$IT == \mathrm{IId}\, AOp \wedge [x! : Z]$$

(see Fig. 10.2). The required concrete operation is then

$$COp == AOp \wedge [x? : Z] = \overline{IT} >> AOp$$

Using the identity relation as the retrieve relation, applicability is trivial. Correctness relies on the fact that $IT >> \overline{IT} >> AOp = AOp$. □

Example 10.10 The IO refinement on the *Travel* operation on the London Underground which uses a sequence of lines to control the trip (see Example 10.4) can be proved using simple input refinement, followed by ordinary operation refinement.

First IO refine *Travel* to

```
TravelL
ΔLU
h! : STATION
l? : seq LINE
─────────────
lu′ = lu
here ↦ here′ ∈ connected lu
h! = here′
```

This is allowed according to the simple input refinement rule above.

The operation *Itinerary* is an operation refinement of *TravelL*, and thus (using Theorem 10.1) an IO refinement of *Travel*.

This IO refinement could also be proved directly from the full IO downward simulation rule, but the number of variables appearing in such a proof would be high. □

Theorem 10.4 (Constructing IO refinement) *The operation AOp is IO refined by $COp = \overline{IT} \gg AOp \gg OT$ when the following conditions hold:*

- *OT is a total injective output transformer for AOp;*
- *IT is an* injective *input transformer for COp.*

Proof Take R to be the (identity on the) state space for AOp (and COp). Applicability is:

$$
\begin{array}{l}
\text{pre}(\overline{IT} \gg AOp) \\
\qquad \equiv \{ \; OT \text{ is total } \} \\
\text{pre}(\overline{IT} \gg AOp \gg OT) \\
\qquad \equiv \{ \text{ definition of } COp \; \} \\
\text{pre}\, COp
\end{array}
$$

For correctness, we have:

$$
\begin{array}{l}
\text{pre}\, AOp \wedge IT \gg COp \\
\qquad \equiv \{ \text{ definition of } COp \; \} \\
\text{pre}\, AOp \wedge IT \gg \overline{IT} \gg AOp \gg OT \\
\qquad \Rightarrow \{ \; IT \text{ is injective } \} \\
\text{pre}\, AOp \wedge AOp \gg OT \\
\qquad \equiv \{ \; X \Rightarrow \text{pre}\, X \; \} \\
AOp \gg OT
\end{array}
$$

□

Theorem 10.5 (Deconstructing IO refinement) *The operation AOp is IO refined by COp when the following conditions hold:*

- *$AOp = IT \gg COp \gg \overline{OT}$;*
- *OT is a total bijective output transformer for AOp;*
- *OT is surjective onto COp's possible output values, i.e.,*

$$\forall\, COp \bullet \exists_1\, AOp \bullet OT$$

- *IT is a* functional *input transformer for COp.*

Proof Take R to be the (identity on the) state space for AOp (and COp). Applicability is:

$$
\begin{array}{l}
\text{pre}(\overline{IT} \gg AOp) \\
\qquad \equiv \{ \text{ equation for } AOp \; \} \\
\text{pre}(\overline{IT} \gg IT \gg COp \gg \overline{OT}) \\
\qquad \Rightarrow \{ \; IT \text{ is functional } \} \\
\text{pre}(COp \gg \overline{OT})
\end{array}
$$

$$\begin{array}{l} \quad \equiv \{ \ OT \text{ is surjective, so } \overline{OT} \text{ is total } \} \\ \text{pre}\, COp \end{array}$$

Correctness is:

$$\begin{array}{l} \text{pre}\, AOp \wedge IT \gg COp \\ \quad \equiv \{ \text{ equation for } AOp \ \} \\ \text{pre}(IT \gg COp \gg \overline{OT}) \wedge IT \gg COp \\ \quad \Rightarrow \{ \text{ pre}\, Y \wedge X \Rightarrow X \text{ if } Y \Rightarrow X \ \} \\ IT \gg COp \\ \quad \equiv \{ \ OT \text{ functional and surjective on } COp \text{ outputs } \} \\ IT \gg COp \gg \overline{OT} \gg OT \\ \quad \equiv \{ \text{ equation for } AOp \ \} \\ AOp \gg OT \end{array}$$

There is, of course, also an upward simulation variant of IO refinement, which is derived similarly to the downward simulation rule. Note that this is only applicable for IO transformers whose base names of inputs and outputs are disjoint.[2] □

Definition 10.11 (IO upward simulation) Given conformal IO-mutable ADTs $A = (AState, AInit, \{AOp_i\}_{i \in I}, AOIT, AOOT)$ and $C = (CState, CInit, \{COp_i\}_{i \in I}, COIT, COOT)$. Let for $i \in I$, IT_i be an injective input transformer for COp_i which is total on $?AOp_i$ and total on $?COp_i$. Let OT_i be an injective output transformer for AOp_i which is total on $!AOp_i$ and total on $!COp_i$. (IT denotes the disjoint sum of IT_i, and similar for OT.) The retrieve relation T defines a upward simulation between A and C if:

$$\begin{array}{l} COIT = AOIT \gg IT \\ COOT = \overline{OT} \gg AOOT \\ \forall\, CState \bullet \exists\, AState \bullet T \\ \forall\, AState';\ CState' \bullet CInit \wedge T' \Rightarrow AInit \end{array}$$

and for all $i \in I$

$$\begin{array}{l} \forall\, CState;\ ?COp_i \bullet \\ \quad (\forall\, AState;\ ?AOp_i \bullet T \wedge ??IT_i \Rightarrow \text{pre}\, AOp_i) \Rightarrow \text{pre}\, COp_i \\ \forall\, CState;\ ?COp_i;\ CState';\ !COp_i;\ AState';\ !AOp_i \bullet \\ \quad COp_i \wedge T' \wedge !!OT_i \Rightarrow \\ \qquad \exists\, AState;\ ?AOp_i \bullet T \wedge ??IT_i \wedge (\text{pre}\, AOp_i \Rightarrow AOp_i) \end{array}$$

[2] This is a consequence of transformed inputs and outputs appearing in universal rather than existential quantifications in the derivation—existentially quantified transformed IO can be hidden using $\gg$.

If this is the case, we also say that the *underlying ADTs* (or any of their matching operations) are related by an *IO upward simulation* or by *IO refinement*, and omit the first two conditions. □

10.6 Further Examples

Example 10.11 **(Deleting an unused input)** In Example 5.3 (p. 136) it was claimed that deleting an unused input is a correct refinement step, in particular it was OK to delete input $i?$ in

$$\begin{array}{|l}
\textit{push}_C \\
\hline
\Delta CState \\
m? : \mathbb{N} \\
i? : \{0, 1\} \\
\hline
\exists t, s : \mathrm{seq}\,\mathbb{N} \bullet u_{merge}(t, s) \land u'_{merge}(t \frown \langle m? \rangle, s) \\
\hline
\end{array}$$

Taking $push_C$ as the abstract operation AOp, and COp the same operation without the declaration of $i?$, the required input transformer is

$$\begin{array}{|l}
IT \\
\hline
i? : \{0, 1\} \\
\hline
\end{array}$$

Although it is the case that $COp = \overline{IT} \gg AOp$, we cannot use Theorem 10.4 because IT is not injective. However, we can use Theorem 10.5, because $AOp = IT \gg COp$, and IT is functional; the identity output transformer is bijective and surjective. Thus, it is indeed a correct IO refinement to remove this (and, in general, any) unused input. □

Example 10.12 Let us verify that the desired refinement in Example 10.1, replacing an abstract amount of money by a sequence of coins in the *Withdraw* operation, can be verified.

The required output transformer is

$$\begin{array}{|l}
AmtToCoins \\
\hline
money? : \mathbb{N} \\
money! : \mathrm{seq}\, COINS \\
\hline
money? = +/money! \\
\hline
\end{array}$$

First, we need to check whether this output transformer is *total* on the abstract output, i.e., the input of the transformer. The resulting condition, a restriction on *COINS*, is that

$$\forall money? : \mathbb{N} \bullet \exists money! : \mathrm{seq}\, COINS \bullet money? = +/money!$$

which is satisfied iff $1 \in COINS$.

Since we have that $WithdrawC = Withdraw \gg AmtToCoins$, we will use Theorem 10.4 to verify the IO refinement. The input transformer is the identity, which is injective as required. The output transformer *AmtToCoins* is injective: every sequence of coins represents a unique amount of money. The injectivity of *AmtToCoins* means that, in the full IO-mutable ADT, the inverse of *AmtToCoins* can be used in the finalisation to reconstruct the original output (the amount of money).

As a side note, observe that *WithdrawC* cannot be IO refined to *Withdraw*, as *AmtToCoins* is *not* injective. However, if the IO-mutable ADT which contains *WithdrawC* has an original output transformer which turns the sequence of coins into the amount of money, replacing *WithdrawC* by *Withdraw* (with a corresponding change to the original output transformer) would, semantically, be correct. This demonstrates the "incompleteness" of IO refinement: refinement steps that depend on properties of the *original* input/output transformers are precluded. □

Example 10.13 Consider the definition of a simple stack-based calculator: the ADT $(Calc, Init, \{Digit, AllClear, Clear, Plus, Minus\})$.

Calc
$$stack : \mathrm{seq}_1 \mathbb{N}$$
$$edittop : \mathbb{B}$$

Init
$$Calc'$$

$$stack' = \langle 0 \rangle \land edittop'$$

The state consists of a stack of numbers, and a Boolean which indicates whether the top of the stack can be extended by entering more digits. (If not, entering a new digit starts a new stack top value.) Initially the stack contains a single 0, which can be edited. Entering a new digit has different results depending on the value of *edittop*:

Digit
$$\Delta Calc$$
$$d? : 0..9$$
$$top! : \mathbb{N}$$

$$edittop \Rightarrow top! = 10*(head\,stack) + d? \land stack' = \langle top! \rangle \frown tail\,stack$$
$$\neg edittop \Rightarrow top! = d? \land stack' = \langle top! \rangle \frown stack$$
$$edittop'$$

There are two "clearing" operations: *AllClear* empties (re-initialises) the stack, *Clear* only resets the top.

AllClear
$$\Delta Calc$$
$$top! : \mathbb{N}$$

$$Init$$
$$top! = 0$$

Clear
$$\Delta Calc$$
$$top! : \mathbb{N}$$

$$top! = 0 \land edittop'$$
$$stack' = \langle top! \rangle \frown (tail\,stack)$$

We provide two binary operations: addition and subtraction.

$$
\begin{array}{|l}
\multicolumn{1}{l}{\_\ Plus\ \_\_\_\_\_\_\_\_\_\_\_\_\_\_\_\_\_\_\_\_\_\_\_\_} \\
\Delta Calc \\
top! : \mathbb{N} \\
\hline
\#stack \geq 2 \land \neg edittop' \\
top! = (head(tail\ stack)) + (head\ stack) \\
stack' = \langle top! \rangle \frown (tail(tail\ stack)) \\
\hline
\end{array}
$$

$$
\begin{array}{|l}
\multicolumn{1}{l}{\_\ Minus\ \_\_\_\_\_\_\_\_\_\_\_\_\_\_\_\_\_\_\_\_\_\_\_\_} \\
\Delta Calc \\
top! : \mathbb{N} \\
\hline
\#stack \geq 2 \land \neg edittop' \\
top! = (head(tail\ stack)) - (head\ stack) \\
stack' = \langle top! \rangle \frown (tail(tail\ stack)) \\
\hline
\end{array}
$$

We can think of at least three textually simpler specifications for this: a display box data type which always provides $top! = head\ stack$; auxiliary operations *Push* and *Pop* such that the assignments to $stack'$ become $Pop \mathbin{;} Push$ or $Pop \mathbin{;} Pop \mathbin{;} Push$; or an auxiliary operation that abstracts the commonality between *Plus* and *Minus*. Using IO refinement, we can actually combine the latter two operations into a single operation, removing duplication of text.

We first define the type of binary operations, and their effects.

$$BINOP ::= plus \mid minus$$

$$
\begin{array}{|l}
effect : BINOP \to (\mathbb{N} \times \mathbb{N}) \to \mathbb{N} \\
\hline
effect\ plus(x, y) = x + y \\
effect\ minus(x, y) = x - y
\end{array}
$$

Now consider the operation *BinOp* below.

$$
\begin{array}{|l}
\multicolumn{1}{l}{\_\ BinOp\ \_\_\_\_\_\_\_\_\_\_\_\_\_\_\_\_\_\_\_\_\_\_\_\_} \\
\Delta Calc \\
top! : \mathbb{N} \\
op? : BINOP \\
\hline
\#stack \geq 2 \land \neg edittop' \\
top! = effect\ op?\ (head(tail\ stack), head\ stack)) \\
stack' = \langle top! \rangle \frown (tail(tail\ stack)) \\
\hline
\end{array}
$$

The operation *BinOp* is an IO refinement of both *Plus* and *Minus*, using the following generators as input transformers:

$$\begin{array}{l} \underline{DoPlus} \\ op! : \{plus\} \end{array} \qquad \begin{array}{l} \underline{DoMinus} \\ op! : \{minus\} \end{array}$$

The operation set of the underlying concrete ADT now contains two identical operations (which is fine, it is an indexed set). However, in the full IO-mutable ADT the two operations are still distinguished by different input transformations (*DoPlus* vs. *DoMinus*) having been added to the original input transformation. □

10.7 Bibliographical Notes

Hayes and Sanders [9] advocate the use of schema piping to abstract details of interfaces from the core of the specification. Their work is complementary to that described here, concentrating mostly on specification issues. Their "representation schemas" are very similar to our original IO transformers, but they do not consider changes to these as related to changes of inputs and outputs in operations.

Mikhajlova and Sekerinski [11] investigated class refinement and interface refinement for object-oriented languages. Their results have very reassuring similarities to ours. First, their "interface refinement" condition generalises "class refinement" in much the same way as our IO refinement generalises data refinement. Also, they conclude that sensible input transformers should be surjective, and output transformers functional, which is exactly what we require given that their transformers operate in the opposite direction. Also, the need to administrate the transformers that have been applied so far appears in their notion of "wrappers".

Work by Stepney, Cooper and Woodcock [18, 20] gives a generalisation of standard Z data refinement very similar to that presented here. The main differences are that global state is made explicit in initialisation and finalisation, and the assumption of a notion of global input and output which is separate from the global state itself. The necessary relations between local and global IO correspond to our original IO transformers.

They used their generalisation in a verification of the *Mondex* case study. The case study was a reduced version of a real development by the NatWest Development Team of a Smartcard product for electronic commerce. The case study and the derivation of the generalised refinement rules are described in two technical reports: [6, 19]. The Mondex case study has proved to be a fertile source of challenging issues in refinement, and has led to much work including workshops and a journal special issue [10] devoted to some of the challenges. This has included mechanisation of some of the parts of the case study and the generalised refinement conditions, see for example, [15].

IO refinement was also considered by Schneider and Treharne [16] in the context of $CSP \| B$, a combination of the process algebra CSP with the state based notation B (see Chap. 18 for a more detailed discussion of this approach).

Banach et al. defined *retrenchment* [1] as a generalisation of refinement (see Sect. 14.5), which aims to address a wider range of formal-based program development by allowing virtually arbitrary weakening of the conditions involved. One of the dimensions in which retrenchment allows weakening of refinement conditions is in changing inputs and outputs; however, unlike in our work, they do not view IO refinement as a liberalisation that is justified on the same semantic basis as refinement itself.

The ideas presented in this chapter were first published in [3]. The relation between IO-refinement and existential quantification (i.e., "hiding") of inputs and outputs in Z is considered in [2].

References

1. Banach, R., & Poppleton, M. (2000) Retrenchment, refinement and simulation. In Bowen et al. [5] (pp. 304–323).
2. Boiten, E. A. (2004). *Input/output abstraction of state based systems* (Technical Report 12-04). University of Kent, Computing Laboratory, June 2004.
3. Boiten, E. A., & Derrick, J. (1998). IO-refinement in Z. In A. Evans, D. J. Duke, & T. Clark (Eds.), *3rd BCS-FACS Northern Formal Methods Workshop*. Berlin: Springer. http://www.ewic.org.uk/.
4. Bowen, J. P., Fett, A., & Hinchey, M. G. (Eds.) (1998). *ZUM'98: the Z Formal Specification Notation*. *Lecture Notes in Computer Science: Vol. 1493*. Berlin: Springer.
5. Bowen, J. P., Dunne, S., Galloway, A., & King, S. (Eds.) (2000). *ZB2000: Formal Specification and Development in Z and B*. *Lecture Notes in Computer Science: Vol. 1878*. Berlin: Springer.
6. Cooper, D., Stepney, S., & Woodcock, J. C. P. (2002). *Derivation of Z refinement proof rules: forwards and backwards rules incorporating input/output refinement* (Technical Report YCS-2002-347). Department of Computer Science, University of York.
7. Fitzgerald, J. A., Jones, C. B., & Lucas, P. (Eds.) (1997). *FME'97: Industrial Application and Strengthened Foundations of Formal Methods*. *Lecture Notes in Computer Science: Vol. 1313*. Berlin: Springer.
8. Grundy, J., Schwenke, M., & Vickers, T. (Eds.) (1998). *International Refinement Workshop & Formal Methods Pacific '98*, Canberra, September 1998. *Discrete Mathematics and Theoretical Computer Science*. Berlin: Springer.
9. Hayes, I. J., & Sanders, J. W. (1995). Specification by interface separation. *Formal Aspects of Computing*, *7*(4), 430–439.
10. Jones, C. B., & Woodcock, J. C. P. (2008). Editorial: special issue on the Mondex challenge. *Formal Aspects of Computing*, *20*(1), 1–3.
11. Mikhajlova, A., & Sekerinski, E. (1997) Class refinement and interface refinement in object-oriented programs. In Fitzgerald et al. [7] (pp. 82–101).
12. Milner, R. (1989). *Communication and Concurrency*. New York: Prentice Hall.
13. Monty Python (1983). *The meaning of life*.
14. Roscoe, A. W., Woodcock, J. C. P., & Wulf, L. (1996). Non-interference through determinism. *Journal of Computer Security*, *4*(1), 27–54.
15. Schellhorn, G., Grandy, H., Haneberg, D., & Reif, W. (2006). The Mondex challenge: machine checked proofs for an electronic purse. In J. Misra, T. Nipkow, & E. Sekerinski (Eds.), *Formal Methods 2006*. *Lecture Notes in Computer Science: Vol. 4085* (pp. 16–31). Berlin: Springer.
16. Schneider, S. A., & Treharne, H. (2011). Changing system interfaces consistently: a new refinement strategy for CSP||B. *Science of Computer Programming*, *76*(10), 837–860.

17. Spivey, J. M. (1989). *The Z Notation: A Reference Manual. International Series in Computer Science*. New York: Prentice Hall.
18. Stepney, S., Cooper, D., & Woodcock, J. C. P. (1998) More powerful data refinement in Z. In Bowen et al. [4] (pp. 284–307).
19. Stepney, S., Cooper, D., & Woodcock, J. C. P. (2000). *An Electronic Purse: Specification, Refinement, and Proof* (Technical Monograph PRG-126). Oxford University Computing Laboratory.
20. Woodcock, J. C. P. (1998) Industrial-strength refinement. In Grundy et al. [8] (pp. 33–44).
21. Woodcock, J. C. P., & Davies, J. (1996). *Using Z: Specification, Refinement, and Proof*. New York: Prentice Hall.

Chapter 11
Weak Refinement

In the refinements we have looked at so far the abstract and concrete data types have been *conformal*, i.e., there has been a 1–1 correspondence between the abstract and concrete operations. Thus given an abstract program, we know exactly the concrete program that will simulate it.

We are now going to relax that assumption and look at ways in which the available operations differ in the abstract and concrete data types. In Chap. 12 we will see how we can split one abstract operation into a sequence of concrete operations but first, in this chapter, we are going to focus on the inclusion of *internal* operations in either of the abstract or concrete data types.

Internal operations are interpreted as being internal because they are not under the control of the environment of the ADT. In some way they represent operations that the system can invoke whenever their preconditions hold. For that reason they cannot have inputs. They are also not directly observable, so they have no outputs either.

They arise naturally in a number of settings. For example, in distributed systems they appear as a result of modelling concurrency or the non-determinism that is inherent in a model of such a system. In some notations, internal operations are used to model communication, and non-determinism arises as a by-product of such an approach. Internal operations are also central to obtaining abstract specification through hiding, a particularly important example of this is to enable communication between distributed components to be internalised—a central facet in the design of distributed systems.

The explicit modelling of internal operations is not strictly necessary in Z, since we can mimic their effect within the observable operations in a specification (more

---

J. Derrick, E.A. Boiten, *Refinement in Z and Object-Z*,
DOI 10.1007/978-1-4471-5355-9_11, © Springer-Verlag London 2014

on this later). However, it is often more convenient and more readable to describe an internal operation separately in the schema calculus, either with a distinguished name or with some informal commentary in the text telling us it is not part of the environmental interface.

Once we have explicit internal operations we need a way to refine specifications containing them, and we could either adapt or generalise the existing rules. We can adapt the existing rules by using so-called stuttering steps in a verification of a refinement. A stuttering step is an operation which does not change the state, and we look at this approach in Sect. 11.1.

However, this is not universally successful and the bulk of this chapter is concerned with a generalisation called *weak refinement*, which draws upon the experience of refining internal operations in process algebras. Section 11.2 motivates and defines the generalisation with subsequent sections discussing its properties and considering a number of examples.

Speaking of which, in order to motivate our discussion let us look at a very simple example in which the refinement introduces an internal operation.

Example 11.1 A two-dimensional world, in which some object moves about and its movement and position can be observed, is represented by the Z ADT $(2D, Init, \{Move, Where\})$ where

$2D$
$x, y : \mathbb{Z}$

$Init$
$2D'$

$x' = 0 \land y' = 0$

$Move$
$\Delta 2D$

$Where$
$\Xi 2D$
$x!, y! : \mathbb{Z}$

$x! = x' \land y! = y'$

We might consider refining this to a system which introduces an internal clock, which is left unchanged by *Move* and *Where* but incremented by an internal operation *Tick*1, e.g.,

$Timed2D1$
$x, y, clock : \mathbb{Z}$

$TInit1$
$Timed2D1'$

$x' = y' \land x' = 0 \land clock' = 0$

$TMove1$
$\Delta Timed2D1$

$clock' = clock$

$TWhere1$
$\Xi Timed2D1$
$x!, y! : \mathbb{Z}$

$x! = x' \land y! = y'$

$Tick1$
$\Delta Timed2D1$

$clock' = clock + 1$
$x' = x \land y' = y$

□

11.1 Using Stuttering Steps

Since internal operations are unobservable, one approach to refining them is to require that all internal operations in an ADT refine *skip*; in Z this is defined as $skip == [\Xi State]$ where *State* is the state space. Since a stuttering step is total the requirements for *skip* refining to a concrete internal operation *CIOp* are:

$$\forall AState; CState \bullet R \Rightarrow \text{pre } CIOp$$
$$\forall AState; CState; CState' \bullet R \land CIOp \Rightarrow \exists AState' \bullet \Xi AState \land R'$$

where R is the retrieve relation between *AState* and *CState*.

Example 11.2 In Example 11.1 the *Timed2D1* specification is a refinement of the *2D* ADT if we use stuttering steps. Considering the observable operations we see that *TMove1* refines *Move* and *TWhere1* refines *Where*. In addition, the internal operation *Tick1* refines $skip == [\Xi 2D]$ where the retrieve relation is the identity on the common components of the two ADTs (i.e., x and y). □

Another similar example is the following.

Example 11.3 Consider an abstraction of a plane flight from Leeds/Bradford airport to Manchester airport. The position of the plane is modelled by two coordinates; x the distance from Leeds and z the altitude (to the nearest metre). The initial specification has three operations: *Fly* to move the plane forwards together with two operations to raise and lower its altitude: *Higher* and *Lower*.

```
┌─ Plane0 ──────────────
│ x, z : ℕ
└───────────────────────

┌─ Init0 ───────────────
│ Plane0'
├───────────────────────
│ x' = 0 ∧ z' = 0
└───────────────────────

┌─ Fly0 ────────────────
│ ΔPlane0
├───────────────────────
│ x' = x + 1 ∧ z' = z
└───────────────────────

┌─ Higher0 ─────────────
│ ΔPlane0
├───────────────────────
│ x' = x ∧ z' = z + 1
└───────────────────────

┌─ Lower0 ──────────────
│ ΔPlane0
├───────────────────────
│ x' = x ∧ z' = z − 1
└───────────────────────
```

The second specification adds a third dimension y, and two new operations to describe the process of moving *Left* or *Right*.

$$3D == [y : \mathbb{Z}]$$
$$Plane_1 == Plane_0 \wedge 3D$$
$$Init_1 == Init_0 \wedge [3D' \mid y' = 0]$$
$$Fly_1 == Fly_0 \wedge \Xi 3D$$
$$Higher_1 == Higher_0 \wedge \Xi 3D$$
$$Lower_1 == Lower_0 \wedge \Xi 3D$$
$$Left == \Xi Plane_0 \wedge [\Delta 3D \mid y' = y - 1]$$
$$Right == \Xi Plane_0 \wedge [\Delta 3D \mid y' = y + 1]$$

The retrieve relation between the two specifications is the identity on x and z. Then both the internal operations *Left* and *Right* refine the abstract *skip* operation. □

There are a number of problems with this approach though. The more trivial objection is that we have had to break conformity and introduce stuttering steps in the original ADT before we can verify the refinement. This seems rather artificial. More importantly is the interpretation of internal operations and the potential for divergence through their introduction.

Internal operations are interpreted as being internal because they are not under the control of the environment of the ADT, and in some way they represent operations that the system can invoke whenever their preconditions hold. Where our programs so far were finite sequences of operations invoked by the environment of the ADT, we now get a more complex situation. In particular, internal operations getting invoked infinitely often by the ADT itself, called "livelock", is not desirable. We view this as a source of divergence. However, refinements of *skip* are (due to the applicability condition) always enabled, and thus have precisely this undesirable property. Consider the *Tick*1 operation in the example above. Since it is an internal operation there is nothing to stop it being repeatedly invoked by the system, this causes livelock preventing any other operation from occurring.

There is an issue here about what is observable. The *effect* of an internal operation really is unobservable. For example, we cannot observe the value of y in $Plane_1$ since there is no output related to it. Introducing such an output through output refinement is always possible, see Theorem 10.2. However, that would normally involve a change of the boundary of the system: "observations" of y would still be discarded by the finalisation. Nevertheless the internal operation does cause an effect on the rest of the system, as repeated application of this unobservable operation causes livelock. We could attempt to overcome this as follows.

Example 11.4 Let us introduce a variant of *Timed2D*1 from Example 11.1 where the clock counts down only a finite number of times and is then reset by every *Move* operation:

$$\begin{array}{l} \underline{\textit{Timed2D2}} \\ x, y : \mathbb{Z};\ clock : \mathbb{N} \end{array}$$

$$\begin{array}{l} \underline{\textit{TInit2}} \\ \textit{Timed2D2}' \\ \hline x' = 0 \wedge y' = 0 \end{array}$$

$$\begin{array}{l} \underline{\textit{TMove2}} \\ \Delta \textit{Timed2D2} \end{array}$$

$$\begin{array}{l} \underline{\textit{TWhere2}} \\ \Xi \textit{Timed2D2} \\ x!, y! : \mathbb{Z} \\ \hline x! = x' \wedge y! = y' \end{array}$$

$$\begin{array}{l} \underline{\textit{Tick2}} \\ \Delta \textit{Timed2D2} \\ \hline clock' = clock - 1 \\ x' = x \wedge y' = y \end{array}$$

In this specification we have attempted to solve the divergence of *Tick*2 by only allowing it to be invoked a finite number of times. However, this ADT is not a refinement of 2*D* even with stuttering steps. In particular, the applicability condition

$$R \Rightarrow \text{pre}\, Tick2$$

fails since the precondition of *Tick*2 requires that $clock \geq 1$ which we cannot guarantee from R. □

At issue really is the meaning of a precondition of an internal operation. For internal operations we need to take a *behavioural* interpretation of their precondition. That is, we do not want internal operations to be total, nor allow arbitrary weakenings of their preconditions, because we wish them to be applicable only at certain values in the state space. To see this consider the example of a protocol timeout being modelled as an internal operation. Clearly we wish the timeout to be invoked only at certain points within the protocol's behaviour. Incorrect refinements would result if we could weaken its precondition and timeout at arbitrary points.

Therefore we wish to view internal operations in such a way that they may not be applied outside their precondition, and we will encode this in our weak refinement rules.

11.2 Weak Refinement

In this section we define a set of weak refinement rules which overcome these problems. We begin with the definitions given in a relational context before expressing the requirements in the Z schema calculus.

11.2.1 The Relational Context

We still view refinement in terms of the external behaviour, and we define a set of rules that ensure the observable behaviour of the refined ADT is a valid refinement of the observable behaviour of the original ADT.

As before we take the sets of observable operations in a refinement to be conformal, but we extend the notion of a data type to include additionally a set of internal operations $\{IOp_j\}_{j \in J}$ for some index set J. The definition of conformity is adjusted to require that the indexing sets of *observable* operations coincide. However, we make no requirements on the conformity of the internal operations, and indeed weak refinement will allow the introduction or removal of internal operations during a refinement.

The definition of weak refinement is motivated by the approach taken to internal actions in a process algebra. In particular, the system might evolve under application of any of its internal operations between any two observable operations. Thus we adapt our notion of a complete program (see Definition 3.4) to involve any finite number of internal operations before and after any observable one.

Data refinement now requires that for any given program (i.e., choice of observable operations) the meaning of every complete program over C possibly involving internal operations is contained in the meaning of some complete program over A (possibly involving internal operations but using the same choice of observable operations). Formally, this leads to the following generalisation of Definition 3.5.

Definition 11.1 (Weak data refinement) Let $\mathsf{A} = (\mathsf{AState}, \mathsf{AInit}, \{\mathsf{AOp_i}\}_{\mathsf{i} \in \mathsf{I} \cup \mathsf{J}}, \mathsf{AFin})$ and $\mathsf{C} = (\mathsf{CState}, \mathsf{CInit}, \{\mathsf{COp_i}\}_{\mathsf{i} \in \mathsf{I} \cup \mathsf{K}}, \mathsf{CFin})$ be data types. Suppose further that J and K are both disjoint from I, and that the visible (external) operations in both data types are those from I only. (J and K need not be disjoint.) Projection of a program p on an index set L is denoted by $\mathsf{p} \upharpoonright \mathsf{L}$, and yields the sequence of all elements in sequence p that are in set L. For a program p over I, recall that the meaning in ADT A of its corresponding complete program is denoted $\mathsf{p_A}$. Define the meanings of programs with internal actions in A as follows (and similarly for C):

$$\hat{\mathsf{p}}_{\mathsf{A}} = \{\mathsf{q_A} \mid \mathsf{q} \in \mathsf{seq}(\mathsf{I} \cup \mathsf{J}) \land \mathsf{q} \upharpoonright \mathsf{I} = \mathsf{p}\}$$

Now C *weakly refines* A iff for each finite sequence p over I,

$$\bigcup \hat{\mathsf{p}}_{\mathsf{C}} \subseteq \bigcup \hat{\mathsf{p}}_{\mathsf{A}}$$

□

Observe that this is a correct generalisation: when J and K are empty, $\hat{\mathsf{p}}_{\mathsf{A}}$ and $\hat{\mathsf{p}}_{\mathsf{C}}$ are singleton sets containing only $\mathsf{p_A}$ and $\mathsf{p_C}$, respectively. Notice also that the stronger requirement that

$$\forall \mathsf{x} : \hat{\mathsf{p}}_{\mathsf{C}} \bullet \forall \mathsf{y} : \hat{\mathsf{p}}_{\mathsf{A}} \bullet \mathsf{x} \subseteq \mathsf{y}$$

is the stuttering steps approach.

In order to verify weak refinements we adapt the existing simulation rules, allowing finite numbers of internal operations between every observable operation. To do so we define relations $\mathsf{Int_A}$ and $\mathsf{Int_C}$ which include all possible finite evolutions as follows.

$$\mathsf{Int_A} = \left(\bigcup_{\mathsf{i} \in \mathsf{J}} \mathsf{AOp_i}\right)^*$$
$$\mathsf{Int_C} = \left(\bigcup_{\mathsf{i} \in \mathsf{K}} \mathsf{COp_i}\right)^*$$

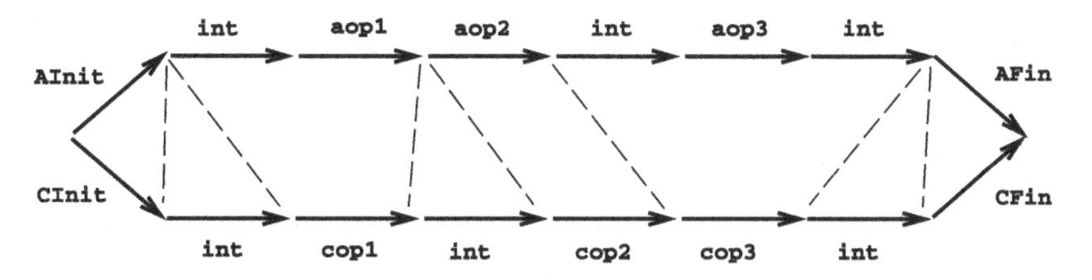

Fig. 11.1 A weak simulation between an abstract and concrete data type

We will still, of course, require a retrieve relation to be defined between the state spaces of the data types, and the definition of weak data refinement ensures that the observable operations are conformal.

Definition 11.2 (Weak downward simulation) Let $\mathsf{A} = (\mathsf{AState}, \mathsf{AInit}, \{\mathsf{AOp_i}\}_{\mathsf{i} \in \mathsf{I} \cup \mathsf{J}}, \mathsf{AFin})$ and $\mathsf{C} = (\mathsf{CState}, \mathsf{CInit}, \{\mathsf{COp_i}\}_{\mathsf{i} \in \mathsf{I} \cup \mathsf{K}}, \mathsf{CFin})$ be data types. Suppose further that J and K are both disjoint from I, and that the visible (external) operations in both data types are those from I only. The relation R on $\mathsf{AState} \wedge \mathsf{CState}$ is a *weak downward simulation* from AState to CState if

$$(\mathsf{CInit} \mathbin{⨾} \mathsf{Int_C}) \subseteq (\mathsf{AInit} \mathbin{⨾} \mathsf{Int_A}) \mathbin{⨾} \mathsf{R}$$
$$\mathsf{R} \mathbin{⨾} (\mathsf{Int_C} \mathbin{⨾} \mathsf{CFin}) \subseteq (\mathsf{Int_A} \mathbin{⨾} \mathsf{AFin})$$
$$\forall \mathsf{i} : \mathsf{I} \bullet \mathsf{R} \mathbin{⨾} (\mathsf{Int_C} \mathbin{⨾} \mathsf{COp_i} \mathbin{⨾} \mathsf{Int_C}) \subseteq (\mathsf{Int_A} \mathbin{⨾} \mathsf{AOp_i} \mathbin{⨾} \mathsf{Int_A}) \mathbin{⨾} \mathsf{R}$$ □

Figure 11.1 illustrates one possible retrieve relation (the dotted lines) used to verify a weak simulation.

We have not considered conditions to prevent the introduction of livelock in this definition, however, we will do so when we express the rules in the Z schema calculus.

11.2.2 The Schema Calculus Context

To express weak downward simulations in a more practical form we encode the idea of a finite number of internal operations before and after an observable operation in the Z schema calculus. In order to avoid quantifications over sequences of internal operations in the definition of weak refinement, we encode all possible finite internal evolution for a specification as a *single* operation *Int* (allowing us to write $Int \mathbin{⨾} Op \mathbin{⨾} Int$).

Let *Internals* be the set of all internal operations in the specification; this set can be typed as $\mathbb{P}\, StateOp$ for some schema type *StateOp*. Let $IntSeq == \mathrm{seq}\, Internals$, representing all *finite* sequences of internal operations. The *effect* of such a sequence is obtained using the operator $\circ : IntSeq \to StateOp$ defined, using distributed schema composition, by

$$\overset{\circ}{\langle\rangle} = \Xi State$$
$$\overset{\circ}{ops} = {}_{9}^{\circ}/ops \quad \text{for } ops \neq \langle\rangle$$

Every possible finite internal evolution is now described by the schema disjunction of the effects of all possible finite sequences of internal operations, i.e.,

$$Int == \exists x : IntSeq \bullet \overset{\circ}{x}$$

Or in other words, two states are related by *Int* iff there exists a series of internal operations x such that the combined effect $\overset{\circ}{x}$ of these operations relates the states.

We distinguish between internal operations in the concrete and abstract specifications by using the subscripts C and A on *Int*. Remember that our internal operations do not have input or output.

Example 11.5 The internal operations in the second specification in Example 11.3 were *Left* and *Right*. Thus here *Int* will be all possible sequences of these, i.e.,

$$Int == \Xi Plane_1 \vee Left \vee Right \vee (Left \mathbin{{}_{9}^{\circ}} Left) \vee (Left \mathbin{{}_{9}^{\circ}} Right) \vee \cdots$$

and this evaluates to $\Xi Plane_0 \wedge [\Delta 3D]$. □

The definition in Z of a weak downward simulation can now be given as follows. To prevent the introduction of livelock we use two additional conditions D1 and D2.

Definition 11.3 (Weak downward simulation) Given Z data types $A = (AState, AInit, \{AOp_i\}_{i \in I \cup J})$ and $C = (CState, CInit, \{COp_i\}_{i \in I \cup K})$, where J and K (the index sets denoting internal operations) are both disjoint from I. Then the relation R on $AState \wedge CState$ is a *weak downward simulation* from A to C if

$$\forall CState' \bullet (CInit \mathbin{{}_{9}^{\circ}} Int_C) \Rightarrow \exists AState' \bullet (AInit \mathbin{{}_{9}^{\circ}} Int_A) \wedge R'$$

and $\forall i : I$

$$\forall AState; CState \bullet \text{pre}(Int_A \mathbin{{}_{9}^{\circ}} AOp_i) \wedge R \Rightarrow \text{pre}(Int_C \mathbin{{}_{9}^{\circ}} COp_i)$$

$$\begin{aligned}&\forall AState; CState; CState' \bullet\\&\quad \text{pre}(Int_A \mathbin{{}_{9}^{\circ}} AOp_i) \wedge R \wedge (Int_C \mathbin{{}_{9}^{\circ}} COp_i \mathbin{{}_{9}^{\circ}} Int_C) \Rightarrow\\&\qquad \exists AState' \bullet R' \wedge (Int_A \mathbin{{}_{9}^{\circ}} AOp_i \mathbin{{}_{9}^{\circ}} Int_A)\end{aligned}$$

where we have elided quantification over inputs and outputs as before.

In addition, we require the existence of a well-founded set *WF* with partial order $<$, and a variant E which is an expression in the state variables satisfying the following conditions.

D1. $R \Rightarrow E \in WF$
D2. $\forall i : K \bullet R \wedge COp_i \Rightarrow E' < E$ □

Before we illustrate the definition with examples let us explain each of the conditions. Each possible initial state of the concrete specification now includes all possible evolutions of the initial state under internal operations. Therefore initialisation ensures that *every* initial concrete path (including all possible internal operations) can be matched by *some* initial abstract path (possibly involving internal operations). Notice that because *Int* encodes in a single schema the disjunction of all possible internal evolutions, explicit quantification over internal evolutions is not necessary.

In the presence of internal operations the applicability condition requires that if an abstract and concrete state are related by R, then whenever the abstract operation terminates possibly after any internal evolution then the concrete operation terminates after some internal evolution. Finally, in correctness every possible state after the concrete operation must be related by R' to a possible state after the abstract operation, except that now 'after' means an arbitrary number of internal operations may occur before and after the abstract operation.

Although we do not have to check conditions for internal operations, it is a useful sanity check to verify that every concrete internal operation when applied simulates zero or more abstract internal operations, i.e.,

$$R \land COp_i \Rightarrow \exists AState' \bullet R' \land Int_A$$

for all concrete internal operations (i.e., for $i \in K$), as this turns out to be true in all examples except for a few contrived ones.

The absence of any requirements on internal operations does not mean that the rules allow arbitrary changes of their preconditions. The circumstances when preconditions can be changed are governed by what observable operations are present in the abstract specification, and the correctness rules for *observable* operations prevent the arbitrary change of preconditions of internal operations.

Turning to the condition that prevents divergence, note that although internal operations decrease the variant E, there are no constraints on observable operations, which are allowed to increase the variant. This means that an internal operation can be invoked an infinite number of times, but not in an infinite sequence between observable operations. The conditions *D1* and *D2* are stronger than necessary: we could relax them to require only that divergence was reduced (rather than eliminated) in the concrete specification. We discuss this in Sect. 11.6 below.

Defining the abbreviation (called the weak version of an operation *Op*) Op_w to mean $Int \circ Op \circ Int$ for an observable *Op*, we can rewrite the three refinement conditions as

$$\forall CState' \bullet CInit_w \Rightarrow \exists AState' \bullet AInit_w \land R'$$
$$\forall AState; CState \bullet \text{pre}\, AOp \land R \Rightarrow \text{pre}\, COp$$
$$\forall AState; CState; CState' \bullet \text{pre}\, AOp \land R \land COp_w \Rightarrow \exists AState' \bullet R' \land AOp_w$$

To see the rules in practice let us revisit the examples that we introduced earlier.

Example 11.6 If we look at *Timed2D*1 from Example 11.1 we see that it is not a weak refinement of $2D$ because the internal operation *Tick*1 can introduce divergence, i.e., it breaks the divergence criteria *D1* and *D2*.

In a similar way the *Plane* specifications from Example 11.3 are not related by weak refinement because the internal operations introduce divergence.

However, *Timed2D2* is a valid weak refinement of $2D$. To see this note first that there are no internal operations in $2D$. Secondly, we note that *Tick*2 does not change the effect of any other variable apart from *clock*, and therefore does not alter the effect of the observable operations. This means that the conditions on the observable operations hold. Although we do not have to verify the correctness condition of *Tick*2 it does indeed hold since $R \wedge Tick2 \Rightarrow \exists 2D' \bullet \Xi 2D \wedge R'$. The divergence criteria in this case are also trivial ($\mathbb{N}$ being the set *WF*). □

Example 11.7 Another variant of the two-dimensional world $2D$ is the following specification where the effect of *Move* is non-deterministic to the observer, but determined elsewhere by the internal operation *SetNext*. The Boolean *nextset* is manipulated in such a way that *Move* and *SetNext* can only happen alternately.

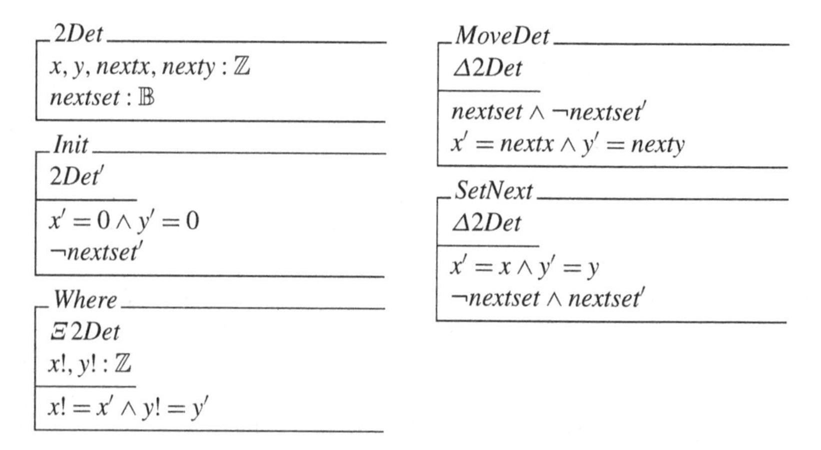

Here the internal operation *SetNext* when applied behaves as *skip* when the obvious simulation is used (however, due to its partiality it is not a refinement of *skip*). However, *MoveDet* (a partial operation) can only be a refinement of the total operation *Move* when the simulation does not relate concrete states outside pre *MoveDet* to any abstract states, i.e., those concrete states (where $\neg nextset$) are unreachable. From the fact that the concrete initialisation requires $\neg nextset$ we must then conclude that this description is not a standard refinement of $2D$.

However, this specification is a weak refinement of the original $2D$ system. To see this first calculate *Int*. Doing so we find it evaluates to $\Xi 2Det \vee SetNext$. Initialisation then requires (using the identity retrieve on $2D$)

$$\forall 2Det' \bullet (x' = 0 \wedge y' = 0) \Rightarrow \exists 2D' \bullet (x' = 0 \wedge y' = 0)$$

which is trivial, as are the conditions upon *Where*. For the *Move* operation applicability requires that

$$R \Rightarrow \text{pre}(Int \fatsemi MoveDet)$$

This is again trivial since the predicate of $\text{pre}(Int \fatsemi MoveDet)$ evaluates to true. For correctness we need

$$Int \fatsemi MoveDet \fatsemi Int \Rightarrow Move$$

Finally, we note there is no divergence since any two occurrences of *SetNext* have to be separated by an observable *MoveDet*. □

11.2.3 Further Examples

Here are two further examples which are more complicated than the ones we have looked at so far.

Example 11.8 We describe a small aspect of a time service mechanism which is concerned with the synchronisation between clocks in a distributed system.

We are given a finite set of *NODES* where each node has a type (either slave or master), and has an associated time given by its clock. In the fragment we look at here we are interested in the setting of the clocks when the master node has counted down to 1. When this happens the clock on the master node is set to a special value *sync_freq* and the clocks on the other nodes are set to a *watchdog* value. The *watchdog* value is the time-out period used by slave nodes, and the *sync_freq* value is the time-out period used by the master node. The system is constrained to ensure that *watchdog* is greater than *sync_freq*.

In our initial description the process is specified as a single observable *Tick* operation. *Tick* is composed of two parts: *Dec* decreases the timer if the clock has not reached 1, and if it has *Set* sets the timers and the types.

$$[NODES]$$
$$NTYPE ::= master \mid slave$$
$$PHASE_A ::= idle \mid completed$$

$$sync\_freq, watchdog : \mathbb{N}$$
$$\overline{\quad}$$
$$sync\_freq < watchdog$$

ATS

$$type_A : NODES \to NTYPE$$
$$timer_A : NODES \to \mathbb{N}_1$$
$$phase_A : PHASE_A$$
$$\overline{\quad}$$
$$phase_A = completed \Rightarrow \exists_1\, n : NODES \bullet type_A(n) = master$$

$$
\begin{array}{|l}
\hline
Init_A \\
ATS' \\
\hline
phase'_A = idle \\
\hline
\end{array}
$$

$$
\begin{array}{|l}
\hline
Dec_A \\
\Delta ATS \\
n? : NODES \\
\hline
phase_A = idle \\
timer_A(n?) > 1 \\
timer'_A = timer_A \oplus \{n? \mapsto timer_A(n?) - 1\} \\
type'_A = type_A \\
phase'_A = phase_A \\
\hline
\end{array}
$$

$$
\begin{array}{|l}
\hline
Set_A \\
\Delta ATS \\
n? : NODES \\
\hline
phase_A = idle \\
type_A(n?) = master \\
timer_A(n?) = 1 \\
timer'_A = \{(n, m) \mid (n = n? \wedge m = sync\_freq) \vee \\
\qquad\qquad\qquad (n \neq n? \wedge m = watchdog)\} \\
type'_A = \{(n, s) \mid (n = n? \wedge s = master) \vee (n \neq n? \wedge s = slave)\} \\
phase'_A = completed \\
\hline
\end{array}
$$

$$Tick_A == Dec_A \vee Set_A$$

We refine this description to one in which the observable $Tick_A$ operation is decomposed into an observable $Tick_C$ operation and an internal *Update* operation. The $Tick_C$ operation deals with the settings of the master node and the *Update* completes the process by visiting each remaining node in turn. To describe this we introduce a new *updating* state, and record the set of nodes that we have *visited*.

$$PHASE_C ::= idle \mid updating \mid completed$$

$$
\begin{array}{|l}
\hline
CTS \\
type_C : NODES \rightarrow NTYPE \\
timer_C : NODES \rightarrow \mathbb{N}_1 \\
visited : \mathbb{P}\, NODES \\
phase_C : PHASE_C \\
\hline
phase_C = completed \Rightarrow visited = NODES \\
\hline
\end{array}
$$

$$
\begin{array}{|l}
\hline
\multicolumn{1}{l}{Init_C} \\
CTS' \\
\hline
visited' = \varnothing \\
phase'_C = idle \\
\hline
\end{array}
$$

The concrete *Dec* part of the operation is essentially the same as the abstract version, but in *Set* we mark the node as *visited* and enter an *updating* phase.

$$
\begin{array}{|l}
\hline
\multicolumn{1}{l}{Dec_C} \\
\Delta CTS \\
n? : NODES \\
\hline
phase_C = idle \\
timer_C(n?) > 1 \\
timer'_C = timer_C \oplus \{n? \mapsto timer_C(n?) - 1\} \\
type'_C = type_C \\
phase'_C = phase_C \\
visited' = visited \\
\hline
\end{array}
$$

$$
\begin{array}{|l}
\hline
\multicolumn{1}{l}{Set_C} \\
\Delta CTS \\
n? : NODES \\
\hline
phase_C = idle \\
type_C(n?) = master \\
timer_C(n?) = 1 \\
timer'_C = timer_C \oplus \{n? \mapsto sync\_freq\} \\
type'_C = type_C \\
phase'_C = updating \\
visited' = \{n?\} \\
\hline
\end{array}
$$

$$Tick_C == Dec_C \vee Set_C$$

In an *updating* phase the internal *Update* operation can be applied. This visits one more node and is applicable until all the nodes have been visited.

$$
\begin{array}{|l}
\hline
\multicolumn{1}{l}{Update} \\
\Delta CTS \\
\hline
phase_C = updating \\
\exists x : NODES \bullet \\
\quad (x \notin visited \wedge visited' = visited \cup \{x\} \\
\quad \wedge\ timer'_C = timer_C \oplus \{x \mapsto watchdog\} \\
\quad \wedge\ type'_C = type_C \oplus \{x \mapsto slave\}) \\
visited' = NODES \Rightarrow phase'_C = completed \\
visited' \neq NODES \Rightarrow phase'_C = phase_C \\
\hline
\end{array}
$$

The concrete description is a weak refinement of the abstract. To see this first note that the variant that guarantees no divergence is $E = \#(NODES\text{-}visited)$. Thus the correctness of the refinement depends on the eventual implementation of the given type *NODES* by a *finite* type. Next note that *Update* is only applicable in the *updating* phase, this means that the internal operation does not play a *rôle* in the initialisation or applicability conditions. To verify these we need to write down the retrieve relation—the following will suffice.

$$
\begin{array}{|l}
\multicolumn{1}{l}{R} \\
\hline
ATS \\
CTS \\
\hline
phase_A = idle \Leftrightarrow phase_C = idle \\
phase_C \in \{updating, completed\} \Rightarrow phase_A = completed \\
phase_C \in \{idle, completed\} \Rightarrow type_C = type_A \land timer_C = timer_A \\
phase_C = updating \Rightarrow visited \lhd type_C = visited \lhd type_A \\
phase_C = updating \Rightarrow visited \lhd timer_C = visited \lhd timer_A \\
\hline
\end{array}
$$

The retrieve relation expresses the idea that in an *updating* phase $type_C$ and $type_A$ will be equal on *visited* (and similarly for *timer*). Since *visited* increases and *NODES* is finite, the approximation eventually terminates. With this retrieve in place initialisation and applicability follow. For correctness we note that *Update* plays no part until the clock reaches 1, when it has it is easy to see that $Tick_C \mathbin{;} Int$ has the same effect as $Tick_A$. □

Example 11.9 At an abstract level the process of delivering email can be considered as a direct communication between the sender and recipient. However, an implementation will need to consider the route that emails take through the nodes comprising the network being used. Then, to the user of email, the operations which control the passing of messages through the network are internal to the system, and all the user is interested in is the process of sending and receiving email.

We can consider an abstract description of such a mail service as follows. Assume we are given a set of users and messages. We describe a $Send_0$ operation which transmits message $m?$ from user $u?$ to user $v?$, and a $Receive_0$ operation in which the oldest message from a user is read. Which user $u!$ the $Receive_0$ operation reads messages from is selected non-deterministically.

$$[USER, MESSAGE]$$

$$
\begin{array}{|l}
\multicolumn{1}{l}{Email} \\
\hline
mail : USER \times USER \rightarrow \operatorname{seq} MESSAGE \\
\hline
\end{array}
$$

$$
\begin{array}{|l}
\multicolumn{1}{l}{Init_0} \\
\hline
Email' \\
\hline
\forall u, v : USER \bullet mail'(u, v) = \langle\, \rangle \\
\hline
\end{array}
$$

$$
\begin{array}{|l}
\hline
\textit{Send}_0 \\
\Delta \textit{Email} \\
u?, v? : \textit{USER} \\
m? : \textit{MESSAGE} \\
\hline
\textit{mail}' = \textit{mail} \oplus \{(u?, v?) \mapsto \textit{mail}(u?, v?) \frown \langle m? \rangle\} \\
\hline
\end{array}
$$

$$
\begin{array}{|l}
\hline
\textit{Receive}_0 \\
\Delta \textit{Email} \\
u!, v? : \textit{USER} \\
m! : \textit{MESSAGE} \\
\hline
\exists\, u! : \textit{USER};\ m! : \textit{MESSAGE} \bullet \\
\quad \textit{mail}(u!, v?) \neq \langle\, \rangle \\
\quad m! = \textit{head}\,\textit{mail}(u!, v?) \\
\quad \textit{mail}' = \textit{mail} \oplus \{(u!, v?) \mapsto \textit{tail}\,\textit{mail}(u!, v?)\} \\
\hline
\end{array}
$$

In addition, a *Multicast* operation is provided which sends out a single message to a number of users.

$$
\begin{array}{|l}
\hline
\textit{Multicast}_0 \\
\Delta \textit{Email} \\
u? : \textit{USER} \\
g? : \mathbb{P}\,\textit{USER} \\
m? : \textit{MESSAGE} \\
\hline
\textit{mail}' = \textit{mail} \oplus \{(u?, g) \mapsto \textit{mail}(u?, g) \frown \langle m? \rangle \mid g \in g?\} \\
\hline
\end{array}
$$

The implementation of this email service uses an underlying network (*net*) consisting of a number of nodes (taken from the set *NODE*). Each user is given a node as its address (by *addr*), and any two nodes in the network are connected together by a *path*, which is a sequence of nodes through the network. The variable *net* describes the connectivity of the network: every pair of nodes in *net* is connected by a *path*.

$$
\begin{array}{|l}
\textit{net} : \textit{NODE} \leftrightarrow \textit{NODE} \\
\textit{addr} : \textit{USER} \rightarrow \textit{NODE} \\
\textit{path} : \textit{NODE} \times \textit{NODE} \rightarrow \operatorname{seq} \textit{NODE} \\
\hline
\forall\, u, v : \textit{USER} \bullet (\textit{addr}(u), \textit{addr}(v)) \in \textit{net} \\
\forall\, a, b : \textit{NODE} \bullet (a, b) \in \textit{net} \Rightarrow \\
\quad (a, b) \in \operatorname{dom} \textit{path} \\
\quad a = \textit{head}\,\textit{path}(a, b) \\
\quad b = \textit{last}\,\textit{path}(a, b) \\
\quad \forall\, i, j : \operatorname{dom} \textit{path}(a, b) \bullet (\textit{path}(a, b)(i), \textit{path}(a, b)(j)) \in \textit{net} \\
\quad \#\operatorname{ran} \textit{path}(a, b) = \#\textit{path}(a, b)
\end{array}
$$

The first predicate says that all the users are connected via the network. The second constraint says that there is a path between each two points on the network. The final line in the definition constrains the paths so they contain no cycles (by forcing all nodes to be distinct).

To implement the functionality of the abstract specification messages will be passed along the network from the source to the final destination down paths. Each node has a sequence of messages which it currently *store*s and a sequence of messages which are ready for final delivery (*dest*). Between each node in the network is a communication *link* (one in each direction) which holds the messages currently in transit between those two nodes. The messages stored and in transit are tagged with the sender and the recipient user's identity.

$$PKT == USER \times USER \times MESSAGE$$

Network

$store, dest : NODE \rightarrow \operatorname{seq} PKT$
$link : NODE \times NODE \rightarrow \operatorname{seq} PKT$

$Init_1$

$Network'$

$\forall n, m : NODE \bullet store'(n) = \langle\rangle \wedge dest'(n) = \langle\rangle \wedge link'(n, m) = \langle\rangle$

$Send_1$ and $Receive_1$ then add and access the local stores at a given user's address.

$Send_1$

$\Delta Network$
$u?, v? : USER$
$m? : MESSAGE$

$link' = link$
$dest' = dest$
$store' = store \oplus \{addr(u?) \mapsto store(addr(u?)) \frown \langle(u?, v?, m?)\rangle\}$

$Receive_1$

$\Delta Network$
$u!, v? : USER$
$m! : MESSAGE$

$dest(addr(v?)) \neq \langle\rangle$
$store' = store$
$link' = link$
$dest' = dest \oplus \{addr(v?) \mapsto tail\, dest(addr(v?))\}$
$\exists u! : USER; m! : MESSAGE \bullet$
$\quad \langle u!, v?, m!\rangle = head\, dest(addr(v?))$

$$
\begin{array}{l}
\textit{Multicast}_1 \\
\hline
\Delta \textit{Network} \\
u? : \textit{USER} \\
g? : \mathbb{P}\,\textit{USER} \\
m? : \textit{MESSAGE} \\
\hline
\textit{link}' = \textit{link} \\
\textit{dest}' = \textit{dest} \\
\exists\, s : \operatorname{iseq} \textit{PKT} \bullet \\
\qquad \operatorname{ran} s = \{(u?, g, m?) \mid g \in g?\} \\
\qquad \textit{store}' = \textit{store} \oplus \{\textit{addr}(u?) \mapsto \textit{store}(\textit{addr}(u?)) \frown s\}
\end{array}
$$

Messages are then moved through the network by two internal operations *Forward* and *Store*. *Forward* takes the next message in a node's *store* and passes it either to *dest* (if it is destined for that node) or onto a *link* that forms part of the path for that particular message. *Store* takes the next message in a *link* and passes it onto the *store* at the end of the *link*. Which nodes *Forward* and *Store* decide to update is chosen non-deterministically at this level of abstraction.

$$
\begin{array}{l}
\textit{Forward} \\
\hline
\Delta \textit{Network} \\
\hline
\exists\, x : \textit{NODE};\ p : \textit{PKT} \bullet \textit{store}(x) \neq \langle\,\rangle \\
\qquad p = \textit{head}\,\textit{store}(x) \\
\qquad p.2 = x \Rightarrow (\textit{link}' = \textit{link} \land \\
\qquad\qquad \textit{dest}' = \textit{dest} \oplus \{x \mapsto \textit{dest}(x) \frown p\}) \\
\qquad p.2 \neq x \Rightarrow (\textit{dest}' = \textit{dest} \land \\
\qquad\qquad \exists\, y : \textit{NODE} \bullet \textit{link}' = \textit{link} \oplus \{(x, y) \mapsto \textit{link}(x, y) \frown p\} \land \\
\qquad\qquad \exists\, s, t : \operatorname{seq} \textit{NODE} \bullet s \frown \langle x, y \rangle \frown t = \textit{path}(p.1, p.2)) \\
\qquad\qquad \textit{store}' = \textit{store} \oplus \{x \mapsto \textit{tail}\,\textit{store}(x)\}
\end{array}
$$

$$
\begin{array}{l}
\textit{Store} \\
\hline
\Delta \textit{Network} \\
\hline
\textit{dest}' = \textit{dest} \\
\exists\, x, y : \textit{NODE};\ p : \textit{PKT} \bullet \textit{link}(x, y) \neq \langle\,\rangle \\
\qquad p = \textit{head}\,\textit{link}(x, y) \\
\qquad \textit{link}' = \textit{link} \oplus \{(x, y) \mapsto \textit{tail}\,\textit{link}(x, y)\} \\
\qquad \textit{store}' = \textit{store} \oplus \{y \mapsto \textit{store}(y) \frown p\}
\end{array}
$$

The *Network* specification is a weak refinement of the *Email* specification. The retrieve relation which links the two specifications maps the email messages between two users to the concatenation of their messages in the *store*, *link* and *dest* that lie on the path between the addresses of the two users.

The internal operations do not introduce livelock because *Forward* is applicable only when *store* is non-empty, and *Store* is applicable only when *link* is non-empty. Since paths do not contain cycles it is not possible to perform an infinite chain of $Forward \fatsemi Store \fatsemi Forward \fatsemi \cdots$ internal operations: eventually all messages will reach their *dest*.

The implementation of the abstract multicast operation as $Multicast_0$ together with simple *Forward* and *Store* operations is inefficient. An interesting exercise would be to think about how it could be more realistically implemented with appropriate internal operations in the network. □

11.3 Properties

Refinement is based upon the reduction of non-determinism in a specification. Adding internal operations has introduced an additional form of non-determinism into the language. Although the examples we have looked at so far have introduced internal operations in the refinement it is no surprise that weak refinement allows us to reduce this type of non-determinism by removing internal operations.

Example 11.10 The first specification contains three observable operations, *Coin*, *Coffee* and *Tea*, together with an internal operation *Fault* which represents the machine breaking down. Clearly such an event is not under the control of the environment and therefore must be considered as an internal operation. After a fault occurs the machine gives out only tea instead of only coffee.

$$SELECT ::= nul \mid tea \mid coffee \mid empty$$

```
┌─ VM ──────────────────
│ select : SELECT
└───────────────────────

┌─ Init ────────────────
│ VM'
├───────────────────────
│ select' = nul
└───────────────────────

┌─ Coin ────────────────
│ ΔVM
│ coin? : ℕ
├───────────────────────
│ select = nul
│ select' = coffee
└───────────────────────

┌─ Coffee ──────────────
│ ΔVM
│ cup! : SELECT
├───────────────────────
│ cup! = coffee
│ select = coffee
│ select' = empty
└───────────────────────
```

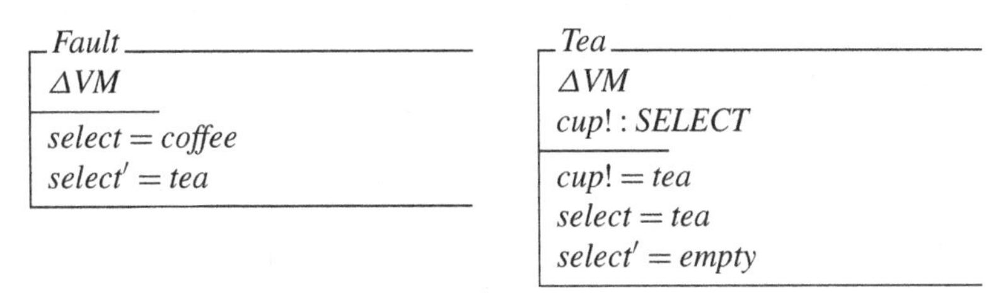

We can refine out the internal operation to a second specification which just contains two operations *Coin* and *Tea* (i.e., where the fault does occur initially) as follows,

VM

$select : SELECT$

Init

VM'

$select' = nul$

Coin

ΔVM
$coin? : \mathbb{N}$

$select = nul$
$select' = coffee$

Tea

ΔVM
$cup! : SELECT$

$cup! = tea$
$select = coffee$
$select' = empty$

The retrieve relation between the two is the identity, and with this it is easy to see that the second is a weak refinement of the first. □

Example 11.11 The abstract *VM* could also be correctly weak refined to one in which *Coffee* and *Tea* were offered after a *Coin*. The choice of which drink is taken is now external as opposed to the internal non-determinism due to the internal *Fault* operation. □

In these examples the reverse refinements do not hold, i.e., we cannot introduce non-determinism, and the abstract *VM* is not a weak refinement of either of the concrete variants. Weak refinement also allows us to change the granularity of the internal operations as the following example shows.

Example 11.12 Both specifications below specify a clock where the *Minute* operation is observable, but the *Tick* operation (which increments the seconds) is internal. However, the seconds counter of the abstract specification is incremented by 10 at a time, whereas the seconds in the concrete specification are incremented one at a time.

Clock

$m, s : \mathbb{N}$

$0 \leq m \leq 59$

Init

$Clock'$

$m' = 0 \land s' = 0$

$Tick_A$
$\Delta Clock$

$s \neq 0 \bmod 60$
$s' = s + 10$
$m' = m$

$Minute_A$
$\Delta Clock$
$m! : \mathbb{N}$

$s = 0 \bmod 60$
$s' = s + 10$
$m' = m + 1 \bmod 60$
$m! = m'$

The second specification uses the same state space and has exactly the same initialisation. The only difference is in the two operations.

$Tick_C$
$\Delta Clock$

$s \neq 0 \bmod 60$
$s' = s + 1$
$m' = m$

$Minute_C$
$\Delta Clock$
$m! : \mathbb{N}$

$s = 0 \bmod 60$
$s' = s + 1$
$m' = m + 1 \bmod 60$
$m! = m'$

The retrieve relation between the two is the identity, and with this it is easy to see that they are weak refinements of each other. □

11.3.1 Weak Refinement Is Not a Pre-congruence

One desirable property for a refinement relation to have is that of being a pre-congruence. That is, if specification S is refined by S', then in any context $C[.]$, $C[S']$ refines $C[S]$. By a context we mean that C enlarges the specification S with additional operations. Thus pre-congruence says that if only one bit of the specification is altered in a refinement then the refinement is compositional. Due to the presence of internal operations, weak refinement is not a pre-congruence. To see this consider the following example.

Example 11.13 Consider our vending machines again. The concrete specification consisting of *Init*, *Coin* and *Tea* and that containing *Init*, *Coin*, *Fault* and *Tea* (but note not *Coffee*) are equivalent under weak refinement. However, suppose we add the following *Coffee* operation to each specification.

$Coffee$
ΔVM
$cup! : SELECT$

$cup! = coffee$
$select = coffee$
$select' = empty$

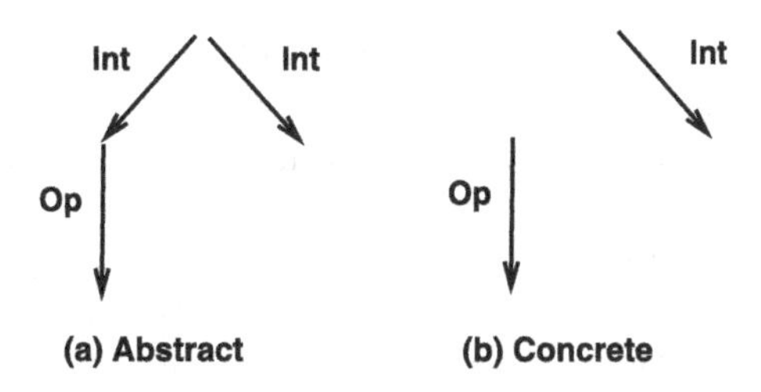

Fig. 11.2 Refining the internal operations separately

Then, as observed above, the specification containing *Coin*, *Fault*, *Coffee* and *Tea* is not a weak refinement of the specification containing *Coin*, *Coffee* and *Tea*. Although we added an *observable* operation *Coffee* the *VM* specification now contains some non-determinism that was not present without it. □

So pre-congruence is lost with weak refinement. Incidentally, this counter-example is the same example that shows weak bisimulation is not a congruence in a process algebra, so the result here is not surprising and the ability to find observational relations which *are* (pre-)congruences can be non-trivial. It is an open question as to whether similar solutions to this problem exist in the state-based world as they do for process algebras.

In addition to the issue of pre-congruence, it is also not the case that individual refinements of the internal operations will give rise to a weak refinement of the whole specification. To see this consider Fig. 11.2 which sketches an example with an observable operation *Op* and an internal operation *Int*. Under the "refinement" the observable operation is unchanged, and normal refinement rules are applied to *Int*, and here we just reduce some non-determinism. However, having done this we see that the specification in (b) is not a weak refinement of the specification in (a). It is the reduction of non-determinism in the internal operation that has caused the problem, and indeed with no reduction of non-determinism in internal operations the result holds.

11.3.2 Internal Operations with Output

Clearly internal operations cannot have inputs, but also in the model we have considered here internal operations cannot have outputs. This is because if an internal operation did have an output then it would not have an unobservable effect.

This is the usual approach to modelling reactive systems, but an alternative approach is to consider *active* systems. In an active system events can be under the control of the system but not the environment, such events are internal but can have observable effects. This differs from the notion of internal we have adopted here (which coincides with that used in a process algebra), which equates internal with no observable transition or effect, including output.

However, some applications are naturally modelled as an active system where it is desirable to be able to specify an internal event which does have data associated with it.

For example, an alarm notification in a managed object is internal in the sense that it is under the control of the system and not the environment. The notification occurs as an output which clearly needs to be visible, however the operation is atomic and is best represented as an individual internal operation, with output, in the specification.

To be applicable to such specifications the weak refinement rules would have to be adapted further to allow internal operations to contain outputs, and this would require appropriate finalisations to be used.

11.4 Upward Simulations

Of course not every weak refinement can be verified by a weak downward simulation, and sometimes we need to use an upward simulation, for which we give a weak version now. The relational characterisation is a simple modification of Definition 11.2, and gives rise to the following schema calculus version:

Definition 11.4 (Weak upward simulation) Given Z data types $A = (AState, AInit, \{AOp_i\}_{i \in I \cup J})$ and $C = (CState, CInit, \{COp_i\}_{i \in I \cup K})$, where J and K (the index sets denoting internal operations) are both disjoint from I. Then the relation T on $AState \wedge CState$ is a *weak upward simulation* from A to C if for all $i \in I$

$$\forall AState'; CState' \bullet (CInit \fatsemi Int_C) \wedge T' \Rightarrow (AInit \fatsemi Int_A)$$

$$\forall CState \bullet (\forall AState \bullet T \Rightarrow \text{pre}(Int_A \fatsemi AOp_i)) \Rightarrow \text{pre}(Int_C \fatsemi COp_i)$$

$$\begin{array}{l} \forall AState'; CState; CState' \bullet \\ \quad (\forall AState \bullet T \Rightarrow \text{pre}(Int_A \fatsemi AOp_i)) \Rightarrow \\ \quad ((Int_C \fatsemi COp_i \fatsemi Int_C) \wedge T' \Rightarrow \exists AState \bullet T \wedge (Int_A \fatsemi AOp_i \fatsemi Int_A)) \end{array}$$

where we have elided quantification over inputs and outputs as before.

In addition, we require the existence of a well-founded set WF with partial order $<$, and a variant E which is an expression in the state variables satisfying the following conditions.

D1. $T \Rightarrow E \in WF$
D2. $\forall i : K \bullet T \wedge COp_i \Rightarrow E' < E$ □

Example 11.14 A simple unreliable protocol is modelled by a sequence a denoting the messages currently inside the protocol. *Send* and *Receive* operations add and remove messages in the system. The system can also *Crash* and when it does it goes into a recovery mode (modelled by *rec*) from which it can *Recover*. However, during this phase it can *Lose* messages. The *Lose* operation is internal and it non-deterministically deletes an arbitrary number of elements from a. In its description $del(I, a)$ deletes elements with indices in I from a.

$[M]$

$Protocol_A$

$a : \mathrm{seq}\, M$
$rec : \mathbb{B}$

$Init_A$

$Protocol'_A$

$a' = \langle \rangle$
$\neg rec'$

$Send_A$

$\Delta Protocol_A$
$m? : M$

$a' = a \frown \langle m? \rangle$
$rec' = rec$

$Receive_A$

$\Delta Protocol_A$
$m! : M$

$a = \langle m! \rangle \frown a'$
$rec' = rec$

$Crash_A$

$\Delta Protocol_A$

$a' = a$
rec'

$Recover_A$

$\Delta Protocol_A$

$a' = a$
$rec \wedge \neg rec'$

$Lose_A$

$\Delta Protocol_A$

$\exists I \bullet I \subseteq \mathrm{dom}\, a \mid a' = del(I, a)$
$rec \wedge rec'$

In an implementation the choice of which messages to lose in a crash may be postponed until after recovery—the choice being dependent on race conditions between network channels for example. Thus instead of losing messages after a crash we merely *mark* elements for potential deletion using an internal operation *Mark*. After recovery some of the marked messages may be lost, the latter being modelled by the internal operation *Drop*.

To represent the marking each element in the protocol is tagged with a Boolean where *true* denotes that it has been marked. To mark messages we use the function $mark(I, c)$ which marks all elements in c with indices in I.

$Protocol_C$

$c : \mathrm{seq}(M \times \mathbb{B})$
$rec : \mathbb{B}$

$Init_C$

$Protocol'_C$

$c' = \langle \rangle$
$\neg rec'$

$$
\begin{array}{|l}
\hline
Send_C \\
\Delta Protocol_C \\
m? : M \\
\hline
c' = c \frown \langle (m?, false) \rangle \\
rec' = rec \\
\hline
\end{array}
$$

$$
\begin{array}{|l}
\hline
Receive_C \\
\Delta Protocol_C \\
m! : M \\
\hline
c' = tail\ c \\
m! = first(head\ c) \\
rec' = rec \\
\hline
\end{array}
$$

$$
\begin{array}{|l}
\hline
Crash_C \\
\Delta Protocol_C \\
\hline
c' = c \\
rec' \\
\hline
\end{array}
$$

$$
\begin{array}{|l}
\hline
Recover_C \\
\Delta Protocol_C \\
\hline
c' = c \\
rec \wedge \neg rec' \\
\hline
\end{array}
$$

$$
\begin{array}{|l}
\hline
Mark_C \\
\Delta Protocol_C \\
\hline
\exists I \bullet I \subseteq \mathrm{dom}\, c \mid c' = mark(I, c) \\
rec \wedge rec' \\
\hline
\end{array}
$$

$$
\begin{array}{|l}
\hline
Drop_C \\
\Delta Protocol_C \\
\hline
\exists I \bullet I \subseteq \{i \mid i \in \mathrm{dom}\, c \wedge c(i).2\} \wedge c' = del(I, c) \\
\hline
\end{array}
$$

The retrieve relation between the two specifications is

$$
\begin{array}{|l}
\hline
T \\
Protocol_A \\
Protocol_C \\
\hline
a \in del\_marked(c) \\
\hline
\end{array}
$$

where $del\_marked(c)$ deletes some of the marked elements in c and removes the tags from the remaining elements. That is, given a c we relate this to many possible a's where a is the same as c but the tags have been removed, and some of the marked elements have been deleted. Clearly T is not functional, for example if $a_1, a_2 \in M$ then both $\langle a_1 \rangle$ and $\langle a_2 \rangle$ in $Protocol_A$ are related to $\langle (a_1, true), (a_2, true) \rangle$ in $Protocol_C$.

The concrete specification is a refinement of the abstract, but cannot be verified using a weak downward simulation. To see why consider Fig. 11.3. From this we can see that the correctness of $Recover_C$ will fail to hold since $T \wedge (Recover_C \fatsemi Drop_C)$ will relate $\langle a_1 \rangle$ to $\langle (a_2, true) \rangle$ which cannot be simulated by $Recover_A$ followed by any abstract internal operations. Thus we need a weak upward simulation to verify this refinement.

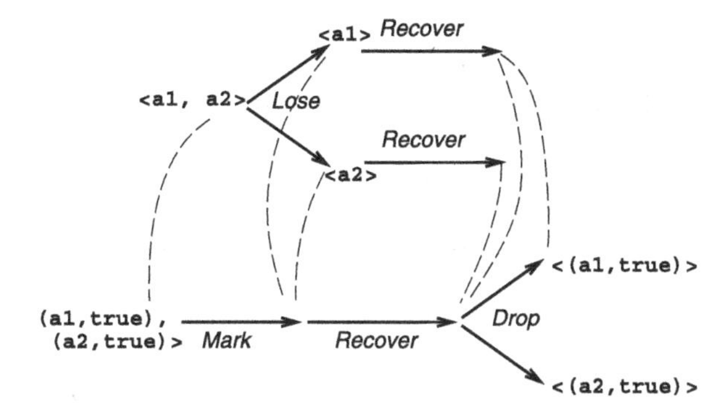

Fig. 11.3 Non-determinism in the protocol specifications

With an upward simulation we can see that the correctness of $Recover_C$ now holds. For example, $(Recover_C \mathbin{⨾} Drop_C) \land T'$ can be matched by $(Recover_A \land T)$ as the diagram in Fig. 11.3 illustrates. □

11.5 Removing Internal Operations

Although many of the examples we have looked at *introduced* internal operations in the process of refinement, that is by no means always the case. We saw in Example 11.12 how the granularity of internal operations could be changed, and Example 11.10 contained a small example where we could remove an internal *Fault* operation. A further example is provided in the following, where we remove an internal operation in a refinement, transferring its functionality to an observable operation and reducing the non-determinism to deterministic behaviour.

Example 11.15 The new low cost airline *AlwaysLate* allows bookings to be made via its online booking web site. To make a booking customers fill in an online form and receive an acknowledgement. At the initial level of abstraction we use the following state space, where the values *nnul*, *fnul* and *tnul* represent empty fields in the web site form.

$$[FID, NAME]$$
$$RESPONSE ::= confirmed \mid full \mid processing$$
$$SCREEN ::= tickets\_sent \mid flight\_full \mid try\_later$$

$$nnul : NAME;\ fnul : FID;\ tnul, timeout : \mathbb{N}$$
$$tnul = 0 \land timeout > 0$$

$$
\begin{array}{|l}
\hline
\textit{AlwaysLate} \\
name : NAME \\
flight : FID \\
tickets : \mathbb{N} \\
response : RESPONSE \\
timer : \mathbb{N} \\
\hline
\end{array}
$$

$$
\begin{array}{|l}
\hline
\textit{Init} \\
AlwaysLate' \\
\hline
name' = nnul \wedge flight' = fnul \wedge tickets' = tnul \\
\hline
\end{array}
$$

The operation *Form* allows users to enter the relevant details for buying any number of tickets in their own name on a flight. After sending off the form the system does a certain amount of processing before an *Acknowledgement* is delivered. As with all web sites, sometimes the booking is successful, at other times we find the flight is full and, most annoyingly, occasionally the system does not respond and we are asked to try again later.

At the customer's level of abstraction, which outcome we get is non-deterministic. When we specify the system we could embed this non-determinism as part of the description of the *Acknowledgement* operation. In actual fact, the uncertainty is due to some internal processing and we choose to describe it more explicitly by describing the appropriate internal operations.

We will use two internal operations *Process* and *Tick*. *Process* represents the processing that decides if the flight is full or not. *Tick* is our clock, and if too much time occurs before an *Acknowledgement* we will timeout.

$$
\begin{array}{|l}
\hline
\textit{Form} \\
\Delta AlwaysLate \\
name? : NAME \\
flight? : FID \\
tickets? : \mathbb{N} \\
\hline
name = nnul \wedge flight = fnul \wedge tickets = tnul \\
name' = name? \wedge flight' = flight? \wedge tickets' = tickets? \\
timer' = timeout \\
response' = processing \\
\hline
\end{array}
$$

$$
\begin{array}{|l}
\hline
\textit{Tick} \\
\Delta AlwaysLate \\
\hline
timer' = timer - 1 \\
name' = name \wedge flight' = flight \\
tickets' = tickets \wedge response' = response \\
\hline
\end{array}
$$

Process

$$\Delta AlwaysLate$$

$$\begin{aligned}
&response = processing\\
&name \neq nnul \wedge flight \neq fnul \wedge tickets \neq tnul\\
&timer' = timer\\
&name' = name \wedge flight' = flight\\
&tickets' = tickets \wedge response' \in \{confirmed, full\}
\end{aligned}$$

Acknowledgement

$$\begin{aligned}
&\Delta AlwaysLate\\
&screen! : SCREEN
\end{aligned}$$

$$\begin{aligned}
&timer = 0 \Rightarrow screen! = try_later\\
&response = confirmed \wedge timer \neq 0 \Rightarrow screen! = tickets_sent\\
&response = full \wedge timer \neq 0 \Rightarrow screen! = flight_full\\
&name' = nnul \wedge flight' = fnul \wedge tickets' = tnul\\
&timer' = 0
\end{aligned}$$

We now refine this description to one where we add details of how the booking is made, and in particular we specify the circumstances when the flight is full. The *Process* operation is no longer needed and the non-determinism now becomes deterministic functionality within *Acknowledgement*. In the concrete specification we assign seats to passengers, so we declare:

$$[SEAT]$$

Booking

$$booking : FID \rightarrow SEAT \nrightarrow NAME$$

$Init_1$

$$Booking'$$

$$\forall fid : FID \bullet booking'(fid) = \varnothing$$

Thus for a given flight identifier *fid*, *booking*(*fid*) assigns seats to passengers. The state space *AlwaysLate* is exactly as before, except we have no need for the *response* variable. The process of booking, i.e., *Form*, is unchanged, and we may still timeout, so the *Tick* operation remains identical to above. The *Process* operation disappears, and *Acknowledgement* has a new description. *Timeout* determines whether we have waited too long. *Clear* clears the fields in the booking form. *BookOK* determines whether there are enough unused seats on the given *flight* and, if so, allocates them to *name*. *Full* determines whether the flight is full (i.e., cannot satisfy this particular request).

Timeout

$\Delta AlwaysLate$
$\Xi Booking$
$screen! : SCREEN$

$timer = 0 \wedge screen! = try\_later$

Clear

$\Delta AlwaysLate$
$screen! : SCREEN$

$name' = nnul \wedge flight' = fnul \wedge tickets' = tnul$
$timer' = 0$

BookOK

$\Delta AlwaysLate$
$\Delta Booking$
$screen! : SCREEN$

$timer \neq 0$
$\exists S : \mathbb{P}\, SEAT \bullet \#S = tickets$
$\quad \forall s : S \bullet s \notin \mathrm{dom}\, booking(flight)$
$\quad \{flight\} \lhd booking' = \{flight\} \lhd booking$
$\quad booking'(flight) = booking(flight) \oplus \{(s, name) \mid s \in S\}$
$screen! = tickets\_sent$

Full

$\Delta AlwaysLate$
$\Xi Booking$
$screen! : SCREEN$

$timer \neq 0$
$\#(SEAT \setminus \mathrm{dom}\, booking(flight)) < tickets$
$screen! = flight\_full$

$$Acknowledgement_1 == (Timeout \vee BookOK \vee Full) \wedge Clear$$

The concrete description is a weak refinement of the abstract, and what was non-determinism due to internal operations has become deterministic behaviour in the observable operations. It is fairly easy to verify the conditions for weak refinement once you spot that the retrieve relation should link the abstract states that occur before and after the *Process* operation to the same concrete state, as in Fig. 11.4.

Divergence is easily dealt with, since it is clear that *Tick* can only occur a finite number of times. Furthermore each concrete *Tick* is matched exactly by an abstract

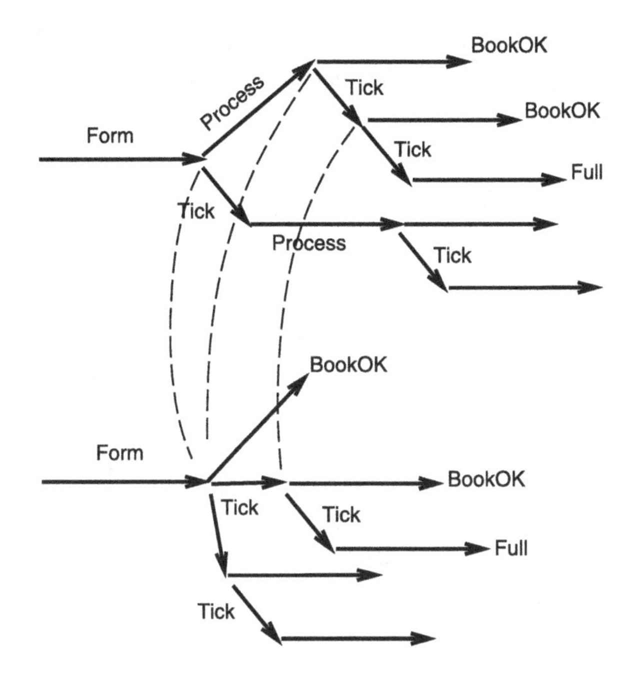

Fig. 11.4 A small fragment of the airline specification and retrieve relation

Tick, therefore the only conditions we must check are those connected with the *Acknowledgement* operation. □

In fact we can go further than this example suggests, and dispense with internal operations altogether. This is a consequence of the following: if specification S_2 is a weak refinement of specification S_1, then there exists equivalent specifications to S_1 and S_2, denoted T_1, T_2 respectively, not containing internal operations such that T_2 is a standard Z refinement of specification T_1.

This is demonstrated in [23] by showing that given a Z specification with internal operations explicitly specified, we can dispense with such operations by adding their non-determinism to the observable operations present. Then it is possible to prove that the weak refinement of a specification with internal operations implies the normal Z refinement if the internal operations are absorbed into the observable ones.

This is discussed in the context of divergence-free specifications, i.e., specifications without divergence due to livelock of internal operations. The notion of equivalence is failures-divergences equivalence (see Chap. 19, Sect. 19.1), which is the same as testing equivalence [11] since we have no divergence.

The first step is to derive equivalent specifications with only observable operations for every specification containing internal operations. This is done by redefining each observable operation AOp to an operation AOp_S, where $AOp_S == Int_A \mathbin{;} AOp \mathbin{;} Int_A$. The initialisation schema $INIT$ is also redefined as $INIT_S$. The transformed Z specification is obtained by replacing each observable operation with its redefinition and removing all the internal operations.

Example 11.16 Consider the *2Det* specification from Example 11.7. The equivalent specification without internal operations is thus

```
┌─ 2Det ──────────────────
│ x, y, nextx, nexty : ℤ
│ nextset : bool
└─────────────────────────
```

```
┌─ Init_S ────────────────
│ 2Det'
├─────────────
│ x' = 0 ∧ y' = 0
│ ¬nextset'
└─────────────────────────
```

```
┌─ Where_S ───────────────
│ Δ2Det
│ x!, y! : ℤ
├─────────────
│ x! = x' ∧ y! = y'
│ x' = x ∧ y' = y
│ ((nextx' = nextx ∧ nexty' = nexty
│   ∧ nextset' = nextset)∨
│ (¬nextset ∧ nextset'))
└─────────────────────────
```

```
┌─ MoveDet_S ─────────────
│ Δ2Det
├─────────────
│ x' = nextx ∧ y' = nexty
└─────────────────────────
```

□

With this transformation in place it is possible to prove the following.

Theorem 11.1 *Let S_1 and S_2 be Z specifications possibly containing internal operations. Let S_2 be a weak refinement of S_1. Then there exists equivalent specifications to S_1 and S_2, denoted T_1, T_2 respectively, not containing internal operations such that T_2 is a standard Z refinement of the specification T_1.*

11.6 Divergence

The simulation rules defined in Sect. 11.2.2 used two conditions *D1* and *D2* which ensure the concrete specification contains no livelock. These simulation rules are sound, but can be weakened further to require just that any divergence is potentially reduced and that we cannot introduce new divergence where there was none before.

To derive these we define a *process data type* which models both divergence *and* blocking in a single embedding, essentially by having separate "bottom" values for both of those. This allows one to derive refinement rules directly for any relational formalism which gives rise to both kinds of errors once the areas of divergence and blocking have been made explicit. In fact, it covers *any* relational formalism modelling at most two kinds of errors, one of which is chaotic (i.e., "anything can happen, including something bad" after this kind of error), and whose combination satisfies some simple constraints. The details of how to use this to derive simulation rules for a variety of semantic interpretations of Z, including the construction of the relational embeddings from Z ADTs, are given in [10]. Here we discuss the basics of this approach at the relational level only.

To do this we need to think about the relative ordering of the two kinds of erroneous behaviours. In particular, we need to decide what observations should be possible when the semantics leads to a non-deterministic choice between any combination of the three behaviours: “normal”, “divergent” and “blocking”.

As before we will view “divergent” behaviour as chaotic like in CSP, this means that a choice between divergent and normal behaviour should appear as divergence, and also that the choice between divergence and blocking should result in divergence. All in all, this means that there is no observable difference between possible and certain divergence, and that divergence is a zero of non-deterministic choice. The remaining issue is the choice between normal and blocking behaviour. It would be possible, using a model of partial relations (as in Sect. 3.3.3), to take deadlock as a unit of choice, and therefore to not observe *possible* blocking. However, consistent with usual semantics for Z and for CSP, we will distinguish possible blocking in our model.

We can now define a process data type with partial operations allowing for both divergence and blocking. Its reduction is the basic data type obtained by removing all blocking and divergence information.

Definition 11.5 (Process data type; reduction) A *process data type* is a quadruple

$$(\mathsf{State}, \mathsf{Inits}, \{\mathsf{Op}_i\}_{i \in I}, \mathsf{Fin})$$

where Inits is a subset of State; every operation Op_i is a triple $(\mathsf{N}, \mathsf{B}, \mathsf{D})$ where N is a relation on State such that $\mathrm{dom}\,\mathsf{N}$ and sets D and B form a partition of State; Fin is a relation from State to G.

Its *reduction* is the basic data type $(\mathsf{State}, \mathsf{G} \times \mathsf{Inits}, \{\mathsf{N}_i\}_{i \in I}, \mathsf{Fin})$ □

In an operation $\mathsf{Op} = (\mathsf{N}, \mathsf{B}, \mathsf{D})$ the relation N represents the operation's normal effect; the sets B and D represent states where the operation would lead to blocking and divergence, respectively. The three sets forming a partition excludes certain situations, such as miracles and possible (as opposed to certain) deadlock from a given state, and ensures that it can be represented by a *total* data type. Possible deadlock still occurs, however, whenever a program leads to multiple states, some but not all of which are deadlocked.

The blocking and non-blocking approaches to operations are, of course, special cases of process data types, in particular:

- the blocking operation Op is represented by $(\mathsf{Op}, \overline{\mathrm{dom}\,\mathsf{Op}}, \varnothing)$, i.e., it never diverges, and blocks in the complement of the operation's domain;
- the non-blocking operation Op is represented by $(\mathsf{Op}, \varnothing, \overline{\mathrm{dom}\,\mathsf{Op}})$, i.e., it diverges in the complement of the operation's domain, but never blocks.

However, in representing ADTs with internal operations, livelock adds to the divergence area.

We view this as an intermediate formalism that most partial relation frameworks can be embedded into. In turn, it is embedded into the total relations framework in

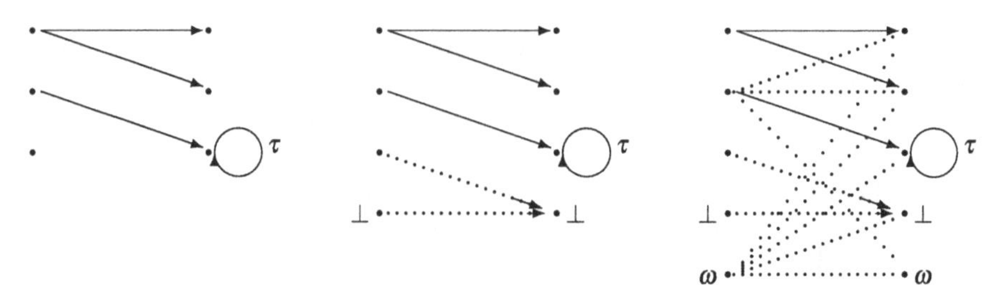

Fig. 11.5 The original Op, and a divergent after-state; with $\mathsf{B}_\perp \times \{\perp\}$ added; finally also with $\mathsf{D}_\omega \times \mathsf{State}_\omega \cup \{(\omega, \perp)\}$

order to define its simulation rules. The embedding is given below. In effect, it fixes the semantics of a process data type.

In the embeddings we will use state spaces enhanced with special values $\perp$, ω and no, that are different values not already contained in any local or global state space of interest. Blocking is represented by $\perp$ and ω represents divergence. The impossibility of making an observation in a final state is encoded in no, and this is added to the global state only. The embedding (and semantics) of a process data type is then defined as follows.

Definition 11.6 (Enhanced state; embedding of a process data type) For any set State, let

$$\mathsf{State}_{\perp,\omega,\mathsf{no}} == \mathsf{State} \cup \{\perp, \omega, \mathsf{no}\}$$

and similarly for sets subscripted with subsets of these special values.

A process data type (State, Inits, $\{(\mathsf{N}_i, \mathsf{B}_i, \mathsf{D}_i)\}_{i \in I}$, Fin) with global state G is embedded into the total data type ($\mathsf{State}_{\perp,\omega}$, Init, $\{[\![\mathsf{Op}]\!]_\mathsf{i}\}_{i \in I}$, $[\![\mathsf{Fin}]\!]$) where

$$\begin{aligned}
&\mathsf{Init} == \mathsf{G}_{\perp,\omega,\mathsf{no}} \times \mathsf{Inits} \\
&[\![(\mathsf{N}, \mathsf{B}, \mathsf{D})]\!] == \mathsf{N} \cup (\mathsf{B}_{\perp,\omega} \times \{\perp\}) \cup (\mathsf{D}_\omega \times \mathsf{State}_\omega) \\
&[\![\mathsf{Fin}]\!] == \mathsf{Fin} \cup (\overline{\mathrm{dom}\,\mathsf{Fin}} \times \{\mathsf{no}\}) \cup \{(\perp, \perp)\} \cup \{\omega\} \times \mathsf{G}_{\omega,\mathsf{no}}
\end{aligned}$$ □

As a process data type has a set of initial states rather than an initialisation relation, the embedding's initialisation relates every global state to all such states. The normal effect of an operation is part of the embedded one. In addition, every blocking state including $\perp$ is related to $\perp$, every state where the operation diverges to every state including ω, and divergent state ω is linked to all states even including $\perp$. Finalisation makes both blocking and divergence visible globally. Figure 11.5 illustrates this operation embedding.

Refinement of process data types is derived as usual by embedding them into total data types, applying the simulation rules for total data types, and then eliminating $\perp$, ω, and no from the resulting rules—i.e., expressing them in terms of the process data types only. This results in the following derivation.

Theorem 11.2 (Downward simulation for process data types) *The relation* R *between* AState *and* CState *is a downward simulation between the process data types* (AState, AInits, $\{\mathsf{AOp}_i\}_{i \in I}$, AFin) *and* (CState, CInits, $\{\mathsf{COp}_i\}_{i \in I}$, CFin), *iff*

$$
\begin{array}{l}
\mathsf{CInits} \subseteq \mathrm{ran}(\mathsf{AInits} \lhd \mathsf{R}) \\
\mathsf{R} \fatsemi \mathsf{CFin} \subseteq \mathsf{AFin} \\
(\mathrm{dom}\,\mathsf{AFin}) \lhd \mathsf{R} = \mathsf{R} \rhd (\mathrm{dom}\,\mathsf{CFin})
\end{array}
$$

and $\forall\, i : I,$ *for* $\mathsf{AOp}_i = (\mathsf{AN}, \mathsf{AB}, \mathsf{AD})$, $\mathsf{COp}_i = (\mathsf{CN}, \mathsf{CB}, \mathsf{CD})$

$$
\begin{array}{l}
\mathsf{AD} ⩤ \mathsf{R} \fatsemi \mathsf{CN} \subseteq \mathsf{AN} \fatsemi \mathsf{R} \\
\mathrm{dom}(\mathsf{R} \rhd \mathsf{CB}) \subseteq \mathsf{AB} \\
\mathrm{dom}(\mathsf{R} \rhd \mathsf{CD}) \subseteq \mathsf{AD}
\end{array}
$$

□

The rules for initialisation and finalisation are identical to the usual rules. The three rules for operations are the expected generalisations. The first is correctness, ensuring correct after-states, provided the abstract system does not diverge; in the blocking approach, this proviso is immaterial. The second and third rules both relate to applicability. When $(\mathsf{B}, \mathsf{D}) = (\varnothing, \overline{\mathrm{dom}\,\mathsf{Op}_i})$ (i.e., the non-blocking interpretation) the second is vacuously true and the third reduces to the usual $\mathrm{ran}(\mathrm{dom}\,\mathsf{AOp}_i \lhd \mathsf{R}) \subseteq \mathrm{dom}\,\mathsf{COp}_i$; in the opposite (blocking) case, the second reduces to the same familiar condition and the third is trivially true.

Similarly we get the following for upward simulations.

Theorem 11.3 (Upward simulation for process data types) *The relation* T *between* CState *and* AState *is an upward simulation between the process data types* (AState, AInits, $\{\mathsf{AOp}_i\}_{i \in I}$, AFin) *and* (CState, CInits, $\{\mathsf{COp}_i\}_{i \in I}$, CFin), *iff*

$$
\begin{array}{l}
\mathrm{ran}(\mathsf{CInits} \lhd \mathsf{T}) \subseteq \mathsf{AInits} \\
\mathsf{CFin} \subseteq \mathsf{T} \fatsemi \mathsf{AFin} \\
\overline{\mathrm{dom}\,\mathsf{CFin}} \subseteq \mathrm{dom}(\mathsf{T} ⩥ \mathrm{dom}\,\mathsf{AFin})
\end{array}
$$

and $\forall\, i : I,$ *for* $\mathsf{AOp}_i = (\mathsf{AN}, \mathsf{AB}, \mathsf{AD})$, $\mathsf{COp}_i = (\mathsf{CN}, \mathsf{CB}, \mathsf{CD})$

$$
\begin{array}{l}
\mathrm{dom}(\mathsf{T} \rhd \mathsf{AD}) ⩤ \mathsf{CN} \fatsemi \mathsf{T} \subseteq \mathsf{T} \fatsemi \mathsf{AN} \\
\mathsf{CB} \subseteq \mathrm{dom}(\mathsf{T} \rhd \mathsf{AB}) \\
\mathsf{CD} \subseteq \mathrm{dom}(\mathsf{T} \rhd \mathsf{AD})
\end{array}
$$

□

Using these as a basis [10] derives simulations for Z data types containing internal operations defined through an embedding into process data types. The semantic basis there is a failures-divergences model, which is subtly different from the blocking model for Z we use here, see Chap. 19. The paper also contains closure properties of weak simulations, e.g., weak downwards simulations are closed under post-composition with concrete internal behaviour that does not lead to divergence. This allows some simplifications of the derived simulation rules.

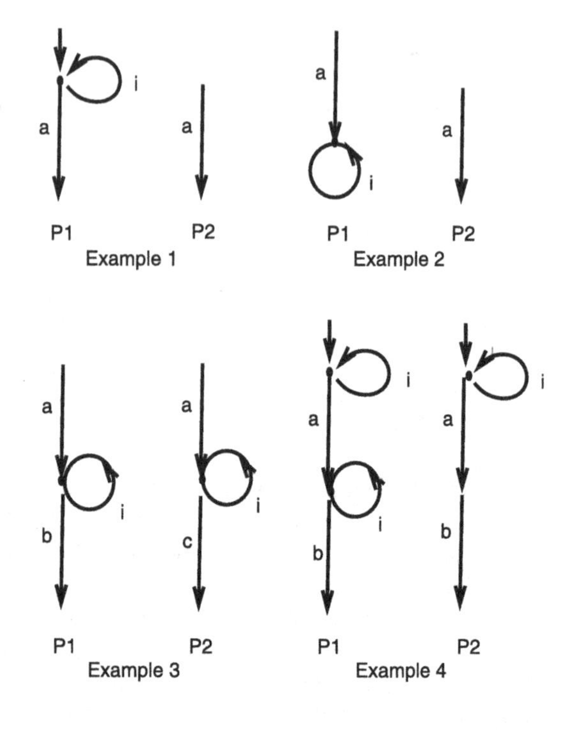

Fig. 11.6 Simple labelled transition systems. Reprinted from [23] kind permission of Springer Science and Business Media

So far, we (also in [10]) have used a catastrophic view of divergence, i.e., after a divergence anything might happen. There is an alternative, non-catastrophic, view of divergence. This is adopted by the process algebra LOTOS, where testing equivalence ignores divergence or treats it in a fair manner [27].

Formally the catastrophic view is based upon the idea that a process diverges after the trace σ if any of its subtraces diverge. The non-catastrophic view says that a system P diverges after σ iff there is a state reachable from P by σ such that in that state it is possible to engage in an infinite sequence of internal events. For example, consider the pairs of systems given as simple labelled transition systems in Fig. 11.6.

In Examples 1 and 2, P_1 and P_2 are testing equivalent (P_1 and P_2 have the same traces and refusals) but P_1 and P_2 are not equivalent in the failures-divergences model (because in both cases P_1 diverges whereas P_2 does not). However, in Example 3, P_1 and P_2 are not testing equivalent (they have different traces), yet they are failures-divergences equivalent (the traces only differ after a point of divergence). Finally, Example 4 exhibits two systems which are both testing and failures-divergences equivalent (they have the same traces and refusals and both diverge initially).

A non-catastrophic view of livelock has the distinct advantage of simpler refinement rules in Z. We can drop requirements *D1* and *D2*, and merely use the initialisation, applicability and correctness criteria to verify weak refinements, resulting in simpler rules than those needed for the process data types.

11.7 Bibliographical Notes

The use of internal operations in Z is not new, representative examples include [20, 21, 30, 36–38]. For example, Strulo [36] considers the use of Z in network management and describes the need for both observable and internal operations in this application area. A particular example is described of a network manager's view of a router within a network. There, alarm notifications are a typical example of internal events which are specified as usual but with informal commentary describing which operations are observable and which are internal. A similar approach and application area is described by Wezeman and Judge in [37] and Rafsanjani in [30].

Cusack and Wezeman in [20] adopt a number of conventions for the use of Z for the specification of OSI network management standards. In particular, they make the distinction between internal and observable operations according to whether an operation has input/output, all operations without input or output variables are considered internal, all others are observable. Their work is placed in an object-oriented setting and they consider notions of subtyping based upon conformance instead of refinement.

Evans in [24] considers the use of Z for the specification of parallel systems, and in particular discusses issues of liveness and fairness in dynamic specifications. Informal commentary partitions the specification into internal and observable operations, and Evans considers the refinement relations needed for Z specifications of concurrent systems.

Similar work to that described in this chapter has appeared in other state-based formalisms. For example, Butler [16] considers the specification and refinement of internal actions in the B method. The email specification in Example 11.9 above is adapted from a B specification given in [17]. Additional work in this area also includes the work of Lano, e.g., [25], and the time service example has been specified in B in [26].

The work described in this chapter first appeared in [10, 22, 23]. Specifically, [22, 23] derive the simulations presented in Sects. 11.2 and 11.4 where explicit conditions preventing livelock are used. In [10] we defined the idea of process data types that provided a general flexible scheme for incorporating the two main "erroneous" concurrent behaviours: deadlock and divergence, into relational refinement. This was shown to subsume previous characterisations. The paper then derived fully general simulation conditions for specifications containing both internal operations and outputs that corresponds to CSP failures-divergences refinement. Moreover, the theory has been formally specified and verified using the interactive theorem prover KIV [31] by Gerhard Schellhorn. Such mechanisation provides additional assurance that the derivation of theorems and simulation definitions are indeed correct, and that derived rules are sound.

Related theories verified previously using KIV include the preliminary work on weak refinement ([23], in [32]) and Cooper, Stepney and Woodcock's generalised Z refinement rules ([19], in [34]). The mechanisation for the theory in [10] required three weeks of work: one for setting up specifications, one for proving theorems and one for adapting to changes. Doing the mechanisation uncovered a few missing assumptions, and helped to shorten some proofs. A Web presentation automatically generated from KIV specifications and proofs is available [33].

The treatment of divergence in [10] corresponds to CSP failures-divergences refinement, however, as noted there are other less catastrophic views of divergence, and [9] discusses how these can be modelled in a relational framework, and in doing so derives relational simulation conditions for process algebraic refinement incorporating differing notions of divergence.

Other formalisms, such as TLA and automata, have a long history of using stuttering steps to verify refinements involving internal actions, see, for example, [1, 35]. These languages take a different approach to the introduction of livelock by using fairness criteria on the observable operations. For example, in [35] liveness conditions are used to describe which executions of the automaton are considered to represent a properly working system. The use of such liveness conditions or fairness criteria provide a language which is strictly more expressive than Z, since we cannot write a Z specification with an infinitely often enabled internal operation that does not involve livelock.

In [35] Sogaard-Andersen, Lynch and Lampson demonstrate the correctness of reliable at-most-once message delivery protocols using timed I/O automata. In particular, two protocol refinements are verified: a clock-based protocol and a five-packet handshake protocol. The verification of these refinements is split into a number of stages, with the two implementations being downward simulations of an intermediate description, which itself is an upward simulation of the original problem specification.

In addition to its size, the case study is interesting because it is necessary to use an upward simulation in the verification. It also verifies a timed model (the clock protocol) against the untimed problem statement by using various embedding results to move between the models. The specifications in Example 11.14 are simplified versions of parts of this case study.

Stuttering steps have also been used in action systems [4, 5] to verify refinements with internal actions. Back and von Wright in [6] define refinement in terms of the execution traces of an action system. Forward and backward simulation methods are then described which explicitly deal with internal (i.e., stuttering) actions. Event-B [2] is a specification method evolved from B and inspired by action systems. Butler has explored internal actions in Event-B [18]. The issues of internal actions, and action refinement (see Chap. 12) through refining skip are closely intertwined in Event-B. This issue is explored in detail by Boiten in [7].

References

1. Abadi, M., & Lamport, L. (1991). The existence of refinement mappings. *Theoretical Computer Science*, *2*(82), 253–284.
2. Abrial, J.-R. (2010). *Modelling in Event-B*. Cambridge: Cambridge University Press.
3. Araki, K., Galloway, A., & Taguchi, K. (Eds.) (1999). *International Conference on Integrated Formal Methods 1999 (IFM'99)*. York: Springer.

4. Back, R. J. R., & Kurki-Suonio, R. (1988). Distributed cooperation with action systems. *ACM Transactions on Programming Languages and Systems*, *10*(4), 513–554.
5. Back, R. J. R., & Kurki-Suonio, R. (1989). Decentralization of process nets with centralized control. *Distributed Computing*, *3*(2), 73–87.
6. Back, R. J. R., & von Wright, J. (1994). Trace refinement of action systems. In B. Jonsson & J. Parrow (Eds.), *CONCUR'94: Concurrency Theory*, Uppsala, Sweden, August 1994. *Lecture Notes in Computer Science: Vol. 836* (pp. 367–384). Berlin: Springer.
7. Boiten, E. A. (2012). Introducing extra operations in refinement. *Formal Aspects of Computing*. doi:10.1007/s00165-012-0266-z.
8. Boiten, E. A., & Derrick, J. (2000). Liberating data refinement. In R. C. Backhouse & J. N. Oliveira (Eds.), *Mathematics of Program Construction. Lecture Notes in Computer Science: Vol. 1837* (pp. 144–166). Berlin: Springer.
9. Boiten, E. A., & Derrick, J. (2009) Modelling divergence in relational concurrent refinement. In Leuschel and Wehrheim [28] (pp. 183–199).
10. Boiten, E. A., Derrick, J., & Schellhorn, G. (2009). Relational concurrent refinement II: internal operations and outputs. *Formal Aspects of Computing*, *21*(1–2), 65–102.
11. Bolognesi, T., & Brinksma, E. (1988). Introduction to the ISO specification language LOTOS. *Computer Networks and ISDN Systems*, *14*(1), 25–59.
12. Bowen, J. P. & Nicholls, J. E. (Eds.) (1992). *Seventh Annual Z User Workshop*. London: Springer.
13. Bowen, J. P., & Hall, J. A. (Eds.) (1994). *ZUM'94, Z User Workshop. Workshops in Computing*. Cambridge: Springer.
14. Bowen, J. P. & Hinchey, M. G. (Eds.) (1995). *ZUM'95: the Z Formal Specification Notation. Lecture Notes in Computer Science: Vol. 967*. Limerick: Springer.
15. Bowen, J. P., Hinchey, M. G., & Till, D. (Eds.) (1997). *ZUM'97: the Z Formal Specification Notation. Lecture Notes in Computer Science: Vol. 1212*. Berlin: Springer.
16. Butler, M. (1997) An approach to the design of distributed systems with B AMN. In Bowen et al. [15] (pp. 223–241).
17. Butler, M. (1999). Distributed electronic mail system. In E. Sekerinski & K. Sere (Eds.), *Program Development by Refinement—Case Studies Using the B Method. FACIT*. Berlin: Springer.
18. Butler, M. J. (2009) Decomposition structures for Event-B. In Leuschel and Wehrheim [28] (pp. 20–38).
19. Cooper, D., Stepney, S., & Woodcock, J. C. P. (2002). *Derivation of Z refinement proof rules: forwards and backwards rules incorporating input/output refinement* (Technical Report YCS-2002-347). Department of Computer Science, University of York.
20. Cusack, E., & Wezeman, C. (1992) Deriving tests for objects specified in Z. In Bowen and Nicholls [12] (pp. 180–195).
21. Derrick, J., Boiten, E. A., Bowman, H., & Steen, M. W. A. (1996) Supporting ODP-translating LOTOS to Z. In Najm and Stefani [29] (pp. 399–406).
22. Derrick, J., Boiten, E. A., Bowman, H., & Steen, M. W. A. (1997) Weak refinement in Z. In Bowen et al. [15] (pp. 369–388).
23. Derrick, J., Boiten, E. A., Bowman, H., & Steen, M. W. A. (1998). Specifying and refining internal operations in Z. *Formal Aspects of Computing*, *10*, 125–159.
24. Evans, A. S. (1997) An improved recipe for specifying reactive systems in Z. In Bowen et al. [15] (pp. 275–294).
25. Lano, K. (1997) Specifying reactive systems in B AMN. In Bowen et al. [15] (pp. 242–274).
26. Lano, K., & Androutsopoulos, K. (1999) Reactive system refinement of distributed systems in B. In Araki et al. [3] (pp. 415–434).
27. Leduc, G. (1991). *On the role of implementation relations in the design of distributed systems using LOTOS*. PhD thesis, University of Liège, Liège, Belgium.
28. Leuschel, M. & Wehrheim, H. (Eds.) (2009). *Integrated Formal Methods, 7th International Conference, IFM 2009. Lecture Notes in Computer Science: Vol. 5423*. Berlin: Springer.

29. Najm, E. & Stefani, J. B. (Eds.) (1996). *First IFIP International Workshop on Formal Methods for Open Object-Based Distributed Systems*, Paris, March 1996. London: Chapman & Hall.
30. Rafsanjani, G. H. B. (1994). *ZEST—Z extended with structuring: a users's guide* (Technical report). BT.
31. Reif, W., Schellhorn, G., Stenzel, K., & Balser, M. (1998). Structured specifications and interactive proofs with KIV. In W. Bibel & P. Schmitt (Eds.), *Automated Deduction—A Basis for Applications, Volume II: Systems and Implementation Techniques* (pp. 13–39). Dordrecht: Kluwer Academic.
32. Schellhorn, G. (2005). ASM refinement and generalizations of forward simulation in data refinement: a comparison. *Theoretical Computer Science*, *336*(2–3), 403–435.
33. Schellhorn, G. (2006). Web presentation of the KIV proofs of 'Relational Concurrent Refinement with Internal Operations and Output'. http://www.informatik.uni-augsburg.de/swt/projects/Refinement/Web/CSPRef/.
34. Schellhorn, G., Grandy, H., Haneberg, D., & Reif, W. (2006). The Mondex challenge: machine checked proofs for an electronic purse. In J. Misra, T. Nipkow, & E. Sekerinski (Eds.), *Formal Methods 2006. Lecture Notes in Computer Science: Vol. 4085* (pp. 16–31). Berlin: Springer.
35. Sogaard-Andersen, J. F., Lynch, N. A., & Lampson, B. W. (1993). *Correctness of communication protocols—a case study* (Technical report). Laboratory for Computer Science, MIT.
36. Strulo, B. (1995) How firing conditions help inheritance. In Bowen and Hinchey [14] (pp. 264–275).
37. Wezeman, C., & Judge, A. J. (1994) Z for managed objects. In Bowen and Hall [13] (pp. 108–119).
38. Woodcock, J. C. P., & Davies, J. (1996). *Using Z: Specification, Refinement, and Proof*. New York: Prentice Hall.

Chapter 12
Non-atomic Refinement

In the last chapter we started to relax the assumption of conformity, that is the idea that the abstract and concrete data types should have corresponding sets of operations, by introducing internal operations into our specifications. In this chapter we go one step further and consider the idea of *non-atomic* refinements where conformity is abandoned completely and the abstract and concrete specifications have different sets of observable operations. The particular case we will be interested in is when an abstract operation is refined by not one, but by a sequence of concrete operations thus allowing a change of granularity when we develop a specification.

The need for such refinements arises in a variety of situations. For example, a protocol might be specified abstractly as a single operation, but in a later development refined into a sequence of operations describing the structure of how the protocol works in more detail. So an abstract *send* operation might be decomposed into a sequence of concrete operations *prepare_to_send*; *get_permission*; *transmit*; *ack*.

Non-atomic refinements are useful in this situation because they allow the initial specification to be described independently of the structure of the eventual implementation. The desired final structure needed in the implementation can then be introduced by non-atomic refinements which support a change of operation granularity absent if we require conformity. Methods in Java, for example, only allow a single result parameter. Forcing such a constraint on a Z specification of a system all the way through to development would be unwieldy and there are clear benefits in freeing the initial designs from such language-dependent constraints.

Non-atomic refinements also enable considerations of efficiency to be gradually introduced. For example, in a protocol we might describe sending a file as one operation, whereas in an implementation we wish to transmit the file byte by byte, where these transmissions are interleaved with the other actions of the protocol.

Excerpts from pp. 1483–1493 of [14] are reprinted within this chapter with kind permission of Springer Science and Business Media.
Excerpts from pp. 11–12, 15–16 of [17] are reprinted within Sect. 12.7.4 with kind permission of Springer Science and Business Media.

J. Derrick, E.A. Boiten, *Refinement in Z and Object-Z*,
DOI 10.1007/978-1-4471-5355-9_12, © Springer-Verlag London 2014

As with internal operations, one approach is to use abstract skip operations to model the change of granularity when we introduce sequences of concrete operations. In Sect. 12.1 we derive the relational basis for refinements of this kind and give a Z formulation for the appropriate simulation conditions.

However, not all non-atomic refinements can be verified in such a manner. In particular, we are going to look at refinements where we need to transform the input and output in the concrete decomposition. One example we will look at will be a coffee machine in which a 3-digit code is entered in order to select a drink. Initially this is described as a single atomic operation, but in a subsequent refinement we split the input of digits across several concrete operations. Since we are transforming the inputs in this manner, the refinement cannot be verified using abstract steps of *skips*.

Sections 12.2–12.7 of this chapter look at support for this scenario in more detail and we use the technique of IO refinement from Chap. 10 in doing so.

12.1 Non-atomic Refinement via Stuttering Steps

In this section we look at how non-atomic refinement can be supported within the standard framework by introducing *skip* operations into the abstract specification. Consider the following example.

Example 12.1 A simple protocol transmits messages between two users using a *Send* operation.

$[M]$

State
$q : \mathrm{seq}\, M$

$Init_0$
$State'$

$q' = \langle\, \rangle$

Send
$\Delta State$
$m? : M$

$q' = q \frown \langle m? \rangle$

Our refinement introduces the idea that there is a local *store* for each user and a communication *link* between them. The abstract *Send* operation is now decomposed into an operation which adds a message to the local *store* and an operation which transmits this onto the communication *link*.

Network
$store, link : \mathrm{seq}\, M$

$Init_1$
$Network'$

$store' = \langle\, \rangle \wedge link' = \langle\, \rangle$

$$\begin{array}{|l} \textit{PrepareToSend} \\ \hline \Delta Network \\ m? : M \\ \hline link' = link \\ store' = store \frown \langle m? \rangle \\ \hline \end{array}$$

$$\begin{array}{|l} \textit{Transmit} \\ \hline \Delta Network \\ \hline store \neq \langle\rangle \Rightarrow \\ \quad (store' = tail\,store \wedge link' = link \frown \langle head\,store \rangle) \\ store = \langle\rangle \Rightarrow \Xi Network \\ \hline \end{array}$$

We wish to consider *PrepareToSend* $\fatsemi$ *Transmit* as being a refinement of *Send*, where the retrieve relation between the specifications has clearly to be the following.

$$\begin{array}{|l} R \\ \hline State \\ Network \\ \hline q = link \frown store \\ \hline \end{array}$$

In the case of the *Send* operation we can see that *PrepareToSend* refines *Send* and *Transmit* refines *skip*, thus *PrepareToSend* $\fatsemi$ *Transmit* is a concrete representation of *Send* as required. □

However, in considering *PrepareToSend* and *Transmit* to be observable we have broken the conformity of the observable operations and we need to consider the consequences of this on the definition of refinement. Before we look at the appropriate rules in Z we look at the definitions in a relational context.

12.1.1 Semantic Considerations

Given data types A and C we are now interested in a situation where they have different indexing sets ($\mathsf{I_A}$ and $\mathsf{I_C}$ respectively) for their observable operations. To give a definition of refinement between such data types the relationship between $\mathsf{I_A}$ and $\mathsf{I_C}$ needs to be described. Whereas for conformal data types the mapping between them is the identity, in a general refinement this could be an arbitrary relation. However, here we will restrict ourselves to a mapping $\rho : \mathsf{I_A} \rightarrow \operatorname{seq} \mathsf{I_C}$, so that ρ describes the concrete counterparts for each operation in A. For example, if $\rho(3) = \langle 6, 4 \rangle$ then this means that operation $\mathsf{AOp_3}$ is refined by the sequence $\mathsf{COp_6} \fatsemi \mathsf{COp_4}$. To write

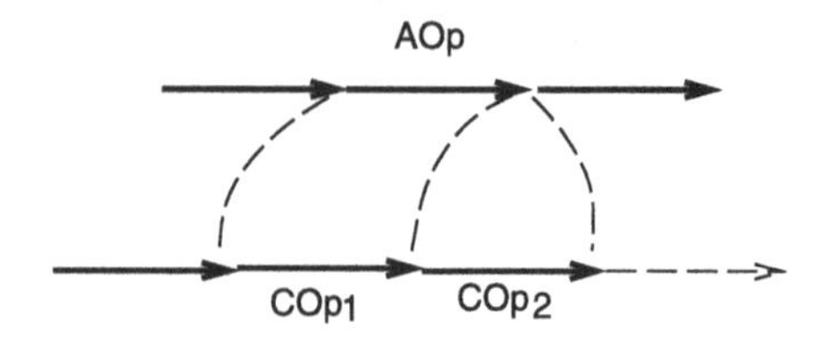

Fig. 12.1 Non-atomic refinement of the abstract operation AOp

down the definition formally we note that ρ has to be applied to every concrete index, and the resulting sequence of sequences has to be flattened (using $\frown/$) to obtain a concrete program.

We can now define non-atomic refinement together with a safety property as follows.

Definition 12.1 (Non-atomic refinement) C non-atomically refines A with respect to ρ iff for each finite sequence p over I_A,

$$(\frown/(\rho \circ p))_C \subseteq p_A$$

In addition, we say that C safely non-atomically refines A if C non-atomically refines A and $\forall q : \text{seq}\, I_C \bullet \exists q : \text{seq}\, I_A \bullet q_C \subseteq p_A$. □

For example, if $I_A = \{1\}$, $I_C = \{1, 2\}$ and $\rho(1) = \langle 1, 2\rangle$ then non-atomic refinement requires that, amongst others, $\mathsf{CInit} \mathbin{;} \mathsf{COp}_1 \mathbin{;} \mathsf{COp}_2 \mathbin{;} \mathsf{CFin} \subseteq \mathsf{AInit} \mathbin{;} \mathsf{AOp}_1 \mathbin{;} \mathsf{AFin}$. This is a sort of liveness condition: to every abstract program there is an equivalent concrete one.

The second, safety, property requires that to every concrete program there must be some abstract equivalent that it refines, e.g., there will be an abstract equivalent to $\mathsf{CInit} \mathbin{;} \mathsf{COp}_2 \mathbin{;} \mathsf{CFin}$ in addition to one equivalent to $\mathsf{CInit} \mathbin{;} \mathsf{COp}_1 \mathbin{;} \mathsf{COp}_2 \mathbin{;} \mathsf{CFin}$.

Let us consider the case when in addition to non-atomic refinement, safety is required. We would like to use simulations to make step-by-step comparisons as before. To ease the presentation let us suppose that in the two data types the indices coincide except that one of the abstract operations AOp is refined by the sequence $\mathsf{COp}_1 \mathbin{;} \mathsf{COp}_2$.

For the remainder of this chapter, unless otherwise stated, we shall consider the requirements arising from downward simulations. In the partial relation setting in order that $(\frown/(\rho \circ p))_C \subseteq p_A$ we will require that

$$\widetilde{R} \mathbin{;} \widehat{\mathsf{COp}_1} \mathbin{;} \widehat{\mathsf{COp}_2} \subseteq \widehat{\mathsf{AOp}} \mathbin{;} \widetilde{R} \tag{12.1}$$

whilst the safety requirement means we need to find abstract counterparts to COp_1 and COp_2 which we denote p^1_A and p^2_A such that

$$\widetilde{R} \mathbin{;} \widehat{\mathsf{COp}_1} \subseteq p^1_A \mathbin{;} \widetilde{R} \quad \text{and} \quad \widetilde{R} \mathbin{;} \widehat{\mathsf{COp}_2} \subseteq p^2_A \mathbin{;} \widetilde{R}$$

The obvious choice for p^1_A and p^2_A is for one to be the original abstract operation and for the other to be a stuttering step (see Fig. 12.1). These choices are sufficient, but not necessary, however whilst it is possible to construct examples where the concrete operations are refining different abstract operations it is difficult to construct realistic examples.

Proposition 12.1 *With the usual notation, suppose that*

$$\widetilde{\mathsf{R}} \mathbin{;} \widehat{\mathsf{COp_1}} \subseteq \widehat{\mathsf{AOp}} \mathbin{;} \widetilde{\mathsf{R}} \quad \textit{and} \quad \widetilde{\mathsf{R}} \mathbin{;} \widehat{\mathsf{COp_2}} \subseteq \mathsf{skip} \mathbin{;} \widetilde{\mathsf{R}} \qquad (12.2)$$

Then the requirements for non-atomic refinement of AOp *into* $\mathsf{COp_1} \mathbin{;} \mathsf{COp_2}$ *with safety are met.* □

The proof is trivial since

$$\widetilde{\mathsf{R}} \mathbin{;} \widehat{\mathsf{COp_1}} \mathbin{;} \widehat{\mathsf{COp_2}} \subseteq \widehat{\mathsf{AOp}} \mathbin{;} \widetilde{\mathsf{R}} \mathbin{;} \widehat{\mathsf{COp_2}} \subseteq \widehat{\mathsf{AOp}} \mathbin{;} \mathsf{skip} \mathbin{;} \widetilde{\mathsf{R}} = \widehat{\mathsf{AOp}} \mathbin{;} \widetilde{\mathsf{R}}$$

The underlying conditions on the partial relations follow easily. The first is the standard condition for refining AOp by $\mathsf{COp_1}$, namely that (we elide the identities over input and output streams for the moment):

$$(\mathrm{dom}\,\mathsf{AOp} \lhd \mathsf{R} \mathbin{;} \mathsf{COp_1}) \subseteq \mathsf{AOp} \mathbin{;} \mathsf{R}$$
$$\mathrm{ran}((\mathrm{dom}\,\mathsf{AOp}) \lhd \mathsf{R}) \subseteq \mathrm{dom}\,\mathsf{COp_1}$$

The second condition is the requirement that $\mathsf{COp_2}$ refines skip.

$$\mathsf{R} \mathbin{;} \mathsf{COp_2} \subseteq \mathsf{R}$$
$$\mathrm{ran}\,\mathsf{R} \subseteq \mathrm{dom}\,\mathsf{COp_2}$$

This compares with the stuttering step approach to weak refinement discussed in Sect. 11.1, but made into visible operations.

12.1.2 Using the Schema Calculus

The conditions for a simple non-atomic downward simulation can now easily be written down. Again we consider just one abstract operation being refined to the sequence $COp_1 \mathbin{;} COp_2$. Without loss of generality assume that COp_1 refines AOp and COp_2 refines a *skip*. The requirements on COp_1 refining AOp are the standard ones whilst those on COp_2 are that:

$$\forall AState;\ CState \bullet R \Rightarrow \mathrm{pre}\, COp_2$$
$$\forall AState;\ CState;\ CState' \bullet R \wedge COp_2 \Rightarrow \exists AState' \bullet \Xi AState \wedge R'$$

In the protocol example above these requirements are met and thus we can consider *Send* to be refined by the concrete sequence *PrepareToSend* $\mathbin{;}$ *Transmit*.

However, the requirements that one of the concrete operations refines the abstract operation and the other refines *skip* are rather strong, and there are valid non-atomic refinements which cannot be satisfied in such a manner. Here is an example.

Example 12.2 This example looks at a load balancing algorithm. To consider the load balancing aspects the communication network is described as a connected graph (V, E), where the nodes V represent processes and edges E the communication paths between processes.

The individual load for each node is found by the function *load*, and the constant *threshold* is used to determine which nodes are overloaded. When a node has passed the threshold the operation *BalLoad* can be called to reduce it and to pass its load to a neighbouring, more lightly loaded, node.

$[V]$

$$E : \mathbb{F}(V \times V)$$

$$threshold : \mathbb{N}$$
$$threshold > 0$$

LBA
$$load : V \to \mathbb{N}$$

Init
$$LBA'$$

BalLoad
$$\Delta LBA$$
$$i?, j? : V$$
$$(i?, j?) \in E$$
$$load(i?) < threshold$$
$$load(j?) \geq threshold$$
$$load' = load \oplus \{i? \mapsto load(i?) + 1, j? \mapsto load(j?) - 1\}$$

Let us refine this specification to one where the change of loads is achieved in two phases. In the first phase a node chooses which neighbouring node to change loads with, and in the second phase the actual loads are altered. The first phase will be controlled by an operation *Select*, and the second phase by the operation *Change*. In this description the function Q represents an active communication over a particular edge, the actual communication being hidden at this level of abstraction.

CLBA
$$load : V \to \mathbb{N}$$
$$Q : E \to \mathbb{B}$$
$$\forall (i,j) : E \bullet Q(i,j) \Rightarrow (\neg Q(j,i) \land (\neg \exists k : V \mid k \notin \{i,j\} \bullet Q(i,k) \lor Q(j,k) \lor Q(k,i) \lor Q(k,j)))$$

CInit

$CLBA'$

$\forall e : E \bullet \neg Q'(e)$

The invariant in the state space ensures that only one link between two nodes can be active at any one time, and that no other communication involving those nodes is currently taking place. *Select* picks a node $i?$ to transfer some of its load to, and as long as the link between $i?$ and $j?$ is currently unused it then opens this link for communication.

Select

$\Delta CLBA$
$i?, j? : V$

$(i?, j?) \in E$
$load(i?) < threshold$
$load(j?) \geq threshold$
$\neg Q(i?, j?)$
$load' = load$
$Q' = Q \oplus \{(i?, j?) \mapsto true\}$

Change then transfers the load for an open link.

Change

$\Delta CLBA$

$\exists i, j \bullet (Q(i, j) \wedge$
$\quad Q' = Q \oplus \{(i, j) \mapsto false\} \wedge$
$\quad load' = load \oplus \{i \mapsto load(i) + 1, j \mapsto load(j) - 1\})$

Notice that neither *Select* or *Change* refine an abstract *skip* operation since *Select* has inputs and *Change* alters the *load*. Thus we cannot verify this refinement via a stuttering step. However, we could still consider it to be a valid non-atomic refinement since $Select \fatsemi Change$ produces the same effect as *BalLoad*. What we have lost though is that *Select* and *Change*, by themselves, have no abstract counterpart. □

The next section develops the necessary framework to verify such refinements.

12.2 Non-atomic Refinement Without Stuttering

In this section we drop the safety requirement from Definition 12.1 and thus weaken our requirements to just that every abstract program has an equivalent concrete program. This allows us to decompose an abstract operation into a sequence of concrete operations without requiring that any of the concrete operations correspond to abstract stuttering steps.

We discuss the basis of the approach here and later on, in Sect. 12.3, look at how to decompose inputs and outputs in the abstract operation across the sequence of concrete operations.

Working for the moment at the relational level, the single requirement we have to fulfil is:[1]

$$\widetilde{R} \mathbin{⨾} \widehat{COp_1} \mathbin{⨾} \widehat{COp_2} \subseteq \widehat{AOp} \mathbin{⨾} \widetilde{R} \tag{12.3}$$

As before we can extract the underlying conditions on the partial relations, and find the condition above is equivalent to the following:

$$(\mathrm{dom}\, AOp \lhd R \mathbin{⨾} COp_1 \mathbin{⨾} COp_2) \subseteq AOp \mathbin{⨾} R \tag{12.4}$$

$$\mathrm{ran}((\mathrm{dom}\, AOp) \lhd R) \subseteq \mathrm{dom}\, COp_1 \tag{12.5}$$

$$\mathrm{ran}((\mathrm{dom}\, AOp) \lhd R \mathbin{⨾} COp_1) \subseteq \mathrm{dom}\, COp_2 \tag{12.6}$$

Furthermore, if the first concrete operation COp_1 is deterministic we can replace the last two (applicability) conditions by a single condition.

Proposition 12.2 *If* COp_1 *is deterministic then*

$$\mathrm{ran}((\mathrm{dom}\, AOp) \lhd R) \subseteq \mathrm{dom}\, COp_1 \land$$
$$\mathrm{ran}((\mathrm{dom}\, AOp) \lhd R \mathbin{⨾} COp_1) \subseteq \mathrm{dom}\, COp_2$$

is equivalent to the condition $\mathrm{ran}((\mathrm{dom}\, AOp) \lhd R) \subseteq \mathrm{dom}(COp_1 \mathbin{⨾} COp_2)$. □

The requirement on COp_1 being deterministic is necessary to ensure that the resultant condition implies $\mathrm{ran}((\mathrm{dom}\, AOp) \lhd R \mathbin{⨾} COp_1) \subseteq \mathrm{dom}\, COp_2$, the other implications always hold.

The conditions characterised in (12.3) define a simulation that is transitive. In particular, further non-atomic refinements of standard (atomic) refinements give rise overall to a non-atomic refinement.

Theorem 12.1 *Non-atomic refinement is transitive.*

Proof There are four cases to consider. Firstly when the initial refinement is subject to a further atomic refinement. Secondly, two cases when one of the operations has a further non-atomic refinement, and lastly when both have further non-atomic refinements. These are pictured in Fig. 12.2.

Without loss of generality consider the first case. Thus suppose that $\widetilde{R} \mathbin{⨾} \widehat{COp_1} \mathbin{⨾} \widehat{COp_2} \subseteq \widehat{AOp} \mathbin{⨾} \widetilde{R}$ for some retrieve relation R. Suppose in addition that COp_i is

[1] If you think we should require $\widetilde{R} \mathbin{⨾} \widehat{COp_1 \mathbin{⨾} COp_2} \subseteq \widehat{AOp} \mathbin{⨾} \widetilde{R}$ note that $\widehat{COp_1 \mathbin{⨾} COp_2} \subseteq \widehat{COp_1} \mathbin{⨾} \widehat{COp_2}$.

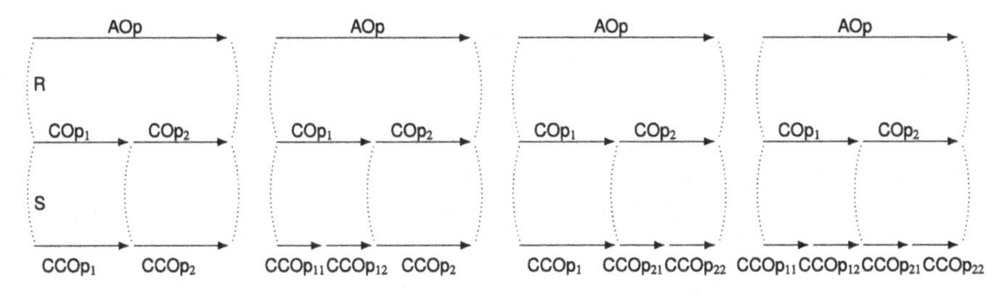

Fig. 12.2 Non-atomic refinement is transitive. Reprinted from [14] with kind permission of Springer Science and Business Media

further refined by $\mathsf{CCOp_i}$ using a retrieve relation $\widetilde{\mathsf{S}}$, i.e., we have: $\widetilde{\mathsf{S}} \fatsemi \widehat{\mathsf{CCOp_i}} \subseteq \widehat{\mathsf{COp_i}} \fatsemi \widetilde{\mathsf{S}}$, for $\mathsf{i} = 1, 2$. Then

$$\begin{array}{l} \widetilde{\mathsf{R}} \fatsemi \widetilde{\mathsf{S}} \fatsemi \widehat{\mathsf{CCOp_1}} \fatsemi \widehat{\mathsf{CCOp_2}} \\ \quad\subseteq \\ \widetilde{\mathsf{R}} \fatsemi \widehat{\mathsf{COp_1}} \fatsemi \widetilde{\mathsf{S}} \fatsemi \widehat{\mathsf{CCOp_2}} \\ \quad\subseteq \\ \widetilde{\mathsf{R}} \fatsemi \widehat{\mathsf{COp_1}} \fatsemi \widehat{\mathsf{COp_2}} \fatsemi \widetilde{\mathsf{S}} \\ \quad\subseteq \\ \widehat{\mathsf{AOp}} \fatsemi \widetilde{\mathsf{R}} \fatsemi \widetilde{\mathsf{S}} \end{array}$$

Hence, using retrieve relation $\widetilde{\mathsf{R}} \fatsemi \widetilde{\mathsf{S}}$ we find that $\widehat{\mathsf{CCOp_1}} \fatsemi \widehat{\mathsf{CCOp_2}}$ is a non-atomic refinement of $\widehat{\mathsf{AOp}}$. The other cases are similar. □

We can now give a characterisation of these rules in the Z schema calculus.

Definition 12.2 (Non-atomic downward simulation without IO transformations) Let R be a retrieve relation between data types $(AState, AInit, \{AOp_1\})$ and $(CState, CInit, \{COp_1, COp_2\})$, and $\rho(1) = \langle 1, 2 \rangle$.

Then R is a non-atomic downward simulation with respect to ρ if, in addition to the initialisation, the following hold.

$$\begin{array}{l} \forall AState; CState; CState' \bullet \operatorname{pre} AOp_1 \land (COp_1 \fatsemi COp_2) \land R \Rightarrow \\ \qquad\qquad\qquad\qquad \exists AState' \bullet R' \land AOp_1 \\ \forall AState; CState \bullet \operatorname{pre} AOp_1 \land R \Rightarrow \operatorname{pre} COp_1 \\ \forall AState; CState; CState' \bullet \operatorname{pre} AOp_1 \land R \land COp_1 \Rightarrow (\operatorname{pre} COp_2)' \end{array}$$
□

The conditions in this definition express the following requirements. The first says that the effect of $COp_1 \fatsemi COp_2$ is consistent with that of AOp_1 (but can of course reduce any non-determinism in AOp_1). The second says that COp_1 can be invoked whenever AOp_1 can, and the third says that when COp_1 has been completed COp_2 can be invoked (we write $(\operatorname{pre} COp_2)'$ so the before-state matches the after-state of COp_1). Informally these are clearly the correct conditions for a refinement of AOp_1 into $COp_1 \fatsemi COp_2$.

These conditions generalise to a non-atomic refinement with an arbitrary number of concrete operations in the obvious manner.

Example 12.3 With this, weaker, requirement for non-atomic refinement we can verify that the concrete load balancing specification is a refinement of the abstract. We use a retrieve relation which maps an abstract value to the same concrete value but with Q set to false.

Then to verify the refinement we have to demonstrate three conditions:

$$\begin{array}{l}
\forall LBA;\ CLBA;\ CLBA' \bullet \\
\quad \text{pre}\, BalLoad \wedge (Select \mathbin{;} Change) \wedge R \Rightarrow \exists LBA' \bullet R' \wedge BalLoad \\
\forall LBA;\ CLBA \bullet \text{pre}\, BalLoad \wedge R \Rightarrow \text{pre}\, Select \\
\forall LBA;\ CLBA;\ CLBA' \bullet \text{pre}\, BalLoad \wedge R \wedge Select \Rightarrow (\text{pre}\, Change)'
\end{array}$$

and these are easily seen to be true. □

The conditions embodied in Definition 12.2 assume that, although the functionality of *AOp* has been decomposed across the sequence of concrete operations, the input parameters are preserved within one of the concrete operations, and similarly for the output parameters. However, in an arbitrary change of granularity we may wish to duplicate or spread such parameters throughout the sequence of concrete operations.

Example 12.4 Consider again the load balancing specification. Instead of refining the *BalLoad* into *Select* followed by *Change*, the *Change* operation is replaced by an *Alter* operation which requires as input, the nodes whose loads are being altered. With the same state space *CLBA* as above *Alter* is:

$$\begin{array}{|l}
\textit{Alter} \\
\hline
\Delta CLBA \\
i?, j? : V \\
\hline
Q(i?, j?) \\
Q' = Q \oplus \{(i?, j?) \mapsto \mathit{false}\} \\
load' = load \oplus \{i? \mapsto load(i?) + 1, j? \mapsto load(j?) - 1\} \\
\hline
\end{array}$$

□

In this example the inputs from the abstract *BalLoad* operation have been replicated across the two concrete operations and we need additional machinery in order to be able to verify this as a valid refinement. The ideas from Chap. 10 will provide the framework that we need.

12.3 Using IO Transformations

In this section we consider the input and output transformations that are needed to support quite general non-atomic refinements. In doing so we will use IO refinement to produce a set of conditions that allow inputs and outputs to be replicated or distributed throughout a concrete decomposition. We have just seen an example which replicates the inputs, to see the need for distribution consider the following example.

Example 12.5 Returning to our two-dimensional world (cf. Example 11.1) this time with a translate operation.

```
┌─ 2D ──────────────────
│ x, y : ℤ
└───────────────────────

┌─ Init ────────────────
│ 2D′
├──────────
│ x′ = 0 ∧ y′ = 0
└───────────────────────

┌─ Translate ───────────
│ Δ2D
│ x?, y? : ℤ
├───────────────────────
│ x′ = x + x? ∧ y′ = y + y?
└───────────────────────
```

We now wish to decompose *Translate* into two operations *TranslateX* and *TranslateY*, each of which moves a single coordinate.

```
┌─ TranslateX ──────────
│ Δ2D
│ x? : ℤ
├──────────
│ x′ = x + x?
│ y′ = y
└───────────────────────

┌─ TranslateY ──────────
│ Δ2D
│ y? : ℤ
├──────────
│ x′ = x
│ y′ = y + y?
└───────────────────────
```

In order that $TranslateX \mathbin{;} TranslateY$ non-atomically refines *Translate*, we need a suitable mapping between the inputs of the abstract operation and the inputs of the concrete operations. □

Notice that we really do need a transformation of inputs, for suppose that *TranslateY* is replaced by the following, equivalent, schema definition

```
┌─ TranslateY₁ ─────────────────────────
│ Δ2D
│ x? : ℤ
├──────────
│ x′ = x
│ y′ = y + x?
└───────────────────────────────────────
```

Due to the double use of input name $x?$, the composition $TranslateX \mathbin{;} TranslateY_1$ does not bear comparison with *Translate*. Clearly we need a properly defined mechanism when considering inputs and outputs and the IO refinement framework provides one. Thus in a refinement we will dispense with the identity mappings between inputs and outputs and we will use input and output transformers whose structure is identical to that of the mapping $\rho : I_A \rightarrow \operatorname{seq} I_C$. Before we derive the necessary characterisation, we will illustrate the ideas with an example.

Example 12.6 To decompose *Translate* we use an input transformer *IT* which takes the inputs of *Translate* and transforms them to a sequence which will be used as inputs (one by one) in the concrete sequence $TranslateX \mathbin{⨟} TranslateY$. We can do this by using the following.

$$
\begin{array}{|l}
\hline
\textit{IT} \\
x?, y? : \mathbb{N} \\
d! : \operatorname{seq} \mathbb{N} \\
\hline
d! = \langle x?, y? \rangle \\
\hline
\end{array}
$$

The conditions we verify then use the input transformation conditions from Definition 10.10 together with the Definition 12.2. Thus we will verify requirements such as

$$
\begin{array}{l}
\forall 2D;\ 2D' \bullet (\operatorname{pre} Translate \land (IT \gg \\
TranslateX[d?(1)/x?] \mathbin{⨟} TranslateY[d?(2)/y?]) \land R \land R') \Rightarrow Translate
\end{array}
$$

which upon expansion is seen to be true. The purpose of the input transformer is to take in inputs for *Translate* and turn them into a sequence $d!$ to be used as inputs for *TranslateX* and *TranslateY*. The substitutions, e.g., $TranslateX[d?(1)/x?]$, then ensure that the inputs are used, one by one, in the correct sequence. Because we are combining the schemas using piping, the $d!$ in *IT* will be identified with the $d?$ in $TranslateX[d?(1)/x?]$ which is why the substitution appears as $[d?(1)/x?]$ rather than $[d!(1)/x?]$. □

12.3.1 Semantic Considerations Again

IO refinement, as defined in Chap. 10, is a mechanism to refine the inputs/outputs of one abstract operation into one concrete operation. To apply it fully to non-atomic refinements we need to be able to spread the inputs/outputs throughout a sequence of concrete operations as we did in the *Translate* example. This means that there will not necessarily be a 1–1 mapping between the number of abstract inputs/outputs and the number of concrete inputs/outputs. Furthermore, in general (cf. Example 12.11), the input transformations might not be between two operations, but a whole sequence of concrete operations, the length of which is only determined at the time of applicability by the abstract inputs.

Therefore we will use mappings from an abstract input to a sequence of concrete inputs representing the inputs needed in the decomposition. That is, we replace the maps IT and OT by $\mathsf{R_{in}}$ and $\mathsf{R_{out}}$ where

$$
\begin{array}{l}
\mathsf{R_{in}} : AInput \leftrightarrow \operatorname{seq} CInput \\
\mathsf{R_{out}} : AOutput \leftrightarrow \operatorname{seq} COutput
\end{array}
$$

and $\mathsf{R_{in}}$ is total on *AInput*, and $\mathsf{R_{out}}$ is total on *AOutput* and functional on $\operatorname{seq} COutput$.

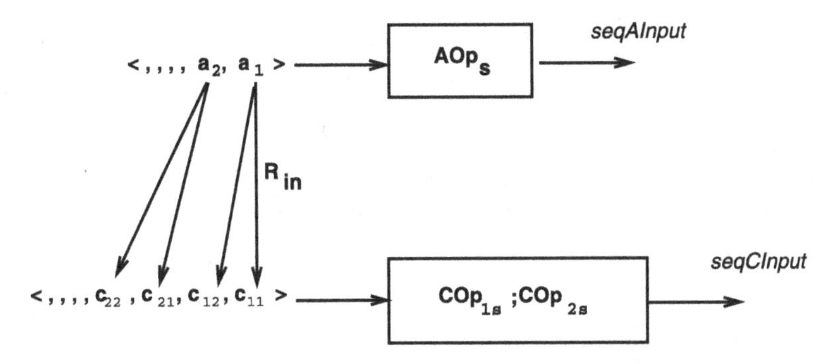

Fig. 12.3 Input transformers in non-atomic refinement. Reprinted from [14] with kind permission of Springer Science and Business Media

In the following discussion, we will use two different embeddings of operations. First, we will use an embedding into *heterogeneous* relations (cf. Fig. 4.1, p. 102) where, for example, the embedding of *AOp* has the type:

$$\mathsf{AOp} : (AState \times AInput) \leftrightarrow (AState \times AOutput)$$

and similarly for concrete operations. In order to derive the refinement conditions we also need to embed operations as *homogeneous* relations, where the ("augmented") state space contains sequences of inputs and outputs, exactly as in Chaps. 4 and 10. The homogeneous embedding of *AOp* will be denoted $\mathsf{AOp_s}$, and will be a relation on type $\mathrm{seq}\,AInput \times \mathrm{seq}\,AOutput \times AState$ as in Definition 4.5, p. 105.

We consider, as usual, the decomposition of AOp into $\mathsf{COp_1} \fatsemi \mathsf{COp_2}$. With mappings $\mathsf{R_{in}}$ and $\mathsf{R_{out}}$ describing how the inputs and outputs of AOp are turned into those for $\mathsf{COp_1}$ and $\mathsf{COp_2}$, and a retrieve relation R between the state spaces, the retrieve relation $\mathsf{R_s}$ on the augmented state will be given by

$$\mathsf{R_s} = {}^\frown/\,\mathsf{R_{in}}* \parallel {}^\frown/\,\mathsf{R_{out}}* \parallel \mathsf{R}$$

Thus, states, inputs and outputs are transformed fully independently. The relation ${}^\frown/\,\mathsf{R_{in}}*$ takes an input sequence $\mathrm{seq}\,AInput$ and creates one concrete input sequence by concatenating together the effect of $\mathsf{R_{in}}$ for each item in $\mathrm{seq}\,AInput$. If there are two concrete operations in the refinement, then $\mathsf{R_{in}}$ maps each abstract input into a two element sequence of concrete inputs, the first for consumption by $\mathsf{COp_1}$, the second for $\mathsf{COp_2}$ (see Fig. 12.3).

For example, the input for a single *Translate* operation is mapped to a pair of concrete inputs, the first for use by *TranslateX*, and the second for use by *TranslateY*.

Similarly in the load balancing specifications, the mapping replicates the input from *BalLoad* to provide inputs for *Select* and inputs for *Alter*.

We can now take the three non-atomic refinement conditions described in terms of an augmented state. For the particular case of decomposing AOp into two concrete operations, we will conveniently assume that the results of $\mathsf{R_{in}}$ and $\mathsf{R_{out}}$ are pairs rather than two-element sequences.

$$(\mathrm{dom}\, \mathsf{AOp_S} \lhd \mathsf{R_S} \,_9^o\, \mathsf{COp_{1S}} \,_9^o\, \mathsf{COp_{2S}}) \subseteq \mathsf{AOp_S} \,_9^o\, \mathsf{R_S} \tag{12.7}$$

$$\mathrm{ran}((\mathrm{dom}\, \mathsf{AOp_S}) \lhd \mathsf{R_S}) \subseteq \mathrm{dom}\, \mathsf{COp_{1S}} \tag{12.8}$$

$$\mathrm{ran}((\mathrm{dom}\, \mathsf{AOp_S}) \lhd \mathsf{R_S} \,_9^o\, \mathsf{COp_{1S}}) \subseteq \mathrm{dom}\, \mathsf{COp_{2S}} \tag{12.9}$$

We turn these into equivalent conditions on the operations with input and output at each step: AOp, $\mathsf{COp_1}$ and $\mathsf{COp_2}$ in the usual way. They become the following:

$$\mathrm{dom}\, \mathsf{AOp} \lhd (\mathsf{R} \| \mathsf{R_{in}}) \,_9^o\, (\mathsf{id} \| \mathsf{COp_1}) \,_9^o\, (\mathsf{COp_2} \| \mathsf{id}) \subseteq \mathsf{AOp} \,_9^o\, (\mathsf{R} \| \mathsf{R_{out}}) \tag{12.10}$$

$$\mathrm{ran}(\mathrm{dom}\, \mathsf{AOp} \lhd (\mathsf{R} \| \mathsf{R_{in}})) \subseteq \mathrm{dom}\, \mathsf{COp_1} \tag{12.11}$$

$$\mathrm{ran}(\mathrm{dom}\, \mathsf{AOp} \lhd (\mathsf{R} \| \mathsf{R_{in}}) \,_9^o\, (\mathsf{id} \| \mathsf{COp_1})) \subseteq \mathrm{dom}\, \mathsf{COp_2} \tag{12.12}$$

where again we require that $\mathsf{R_{out}}$, like OT, is injective.

In the formalisation of these conditions we need to write $(\mathsf{id} \| \mathsf{COp_1})$ and $(\mathsf{COp_2} \| \mathsf{id})$ because a single abstract input has become a pair of concrete inputs, one for $\mathsf{COp_1}$ and one for $\mathsf{COp_2}$, and the expressions $(\mathsf{id} \| \mathsf{COp_1})$ and $(\mathsf{COp_2} \| \mathsf{id})$ select the correct inputs.

12.3.2 The Z Characterisation

To represent the relational characterisation in Z we need to represent the information expressed in $(\mathsf{id} \| \mathsf{COp_1})$ etc., which says the concrete operations take the transformed input one at a time. This restores the distinction between the individual inputs, and for this reason we add substitutions, such as $TranslateX[d?(1)/x?]$, to describe the consumption process explicitly in the operations.

The following definition expresses the refinement of AOp into a fixed sequence $COp_1 \,_9^o\, COp_2$. In the definition we assume, without loss of generality, that the operations AOp, COp_1 and COp_2 have the following form.

AOp
$z? : Z$
$w! : W$
$\ldots$

COp_1
$x? : X$
$u! : U$
$\ldots$

COp_2
$y? : Y$
$V! : V$
$\ldots$

Furthermore, our input and output transformers have the following form.

IT
$z? : Z$
$d! : X \times Y$
$\ldots$

OT
$w? : W$
$e! : U \times V$
$\ldots$

Definition 12.3 (Non-atomic downward simulation with IO transformations) Let R be a retrieve relation between data types $(AState, AInit, \{AOp\})$ and $(CState, CInit, \{COp_1, COp_2\})$. Let IT be an input transformer for $COp_1 \,_9^o\, COp_2$ which is

total on the abstract inputs. Let OT be a total injective output transformer for AOp. Then R defines a non-atomic IO downward simulation if, in addition to the initialisation:

$$
\begin{array}{l}
\forall z? : Z;\ e! : U \times V;\ AState;\ CState;\ CState' \bullet \\
\quad \operatorname{pre} AOp \land R \land \\
\quad (IT \gg COp_1[d?.1/x?, e!.1/u!] \fatsemi COp_2[d?.2/y?, e!.2/v!]) \\
\qquad \Rightarrow \exists AState' \bullet R' \land (AOp \gg OT) \\
\forall d? : X \times Y;\ AState;\ CState \bullet \\
\quad \operatorname{pre}(\overline{IT} \gg AOp) \land R \Rightarrow \operatorname{pre} COp_1[d?.1/x?] \\
\forall d? : X \times Y;\ u! : U;\ AState;\ CState \bullet \\
\quad \operatorname{pre}(\overline{IT} \gg AOp) \land R \land COp_1[d?.1/x?] \Rightarrow (\operatorname{pre} COp_2[d?.2/y?])'
\end{array}
$$

□

In fact explicit substitutions (as in *TranslateX*) are only necessary when the decomposition of AOp involves more than one occurrence of the same input or output parameter names in the concrete operations. If COp_1 and COp_2 are distinct operations with distinct parameter names then the formalisation is simplified by the omission of the substitutions. Dealing with an arbitrary number of inputs and outputs follows in the obvious manner.

The syntactic form of the substitutions depends on the exact form of the input and output transformers. If IT contains a output which is a sequence (as in *TranslateX*) then component selection is, for example, written as $d!(1)$, whereas if IT contains a output which is a tuple then component selection is via the dot notation, as in $d!.1$.

We have seen how the definition verifies the refinement from Example 12.6 and the load balancing description is similar.

Example 12.7 The IT we use to verify that *BalLoad* is refined by $Select \fatsemi Alter$ is one that replicates the inputs.

$$
\begin{array}{|l}
\textbf{IT} \\ \hline
i?, j? : V \\
d! : \operatorname{seq}(V \times V) \\ \hline
d! = \langle (i?, j?), (i?, j?) \rangle
\end{array}
$$

and we can verify that

$$
\begin{array}{l}
\operatorname{pre} BalLoad \land (IT \gg Select[d?(1)/(i?, j?)] \fatsemi Alter[d?(2)/(i?, j?)]) \land R \Rightarrow \\
\qquad \exists CLBA' \bullet R' \land BalLoad
\end{array}
$$

as required.

□

12.4 Further Examples

We look at three more examples here. The first uses simple input transformers, the second includes the use of an output transformer, and the last is the coffee machine example mentioned in the introduction.

Example 12.8 Consider the simple addressable memory specification given in Example 4.7 (p. 115). The operations $Read_0$ and $Write_0$ have natural non-atomic refinements. For example, the $Write_0$ operation has a refinement

$$CheckCache_1 \fatsemi WrCache_1$$

We can consider this as a non-atomic refinement using the input transformer

$$\begin{array}{|l}
\hline IT \\
a? : A \\
d? : D \\
d! : A \times (A \times D) \\
\hline
d! = (a?, (a?, d?)) \\
\hline
\end{array}$$

In a similar fashion $Read_0$ is non-atomically refined by

$$CheckCache_1 \fatsemi RdCache_1 \fatsemi [\Xi Mem_1;\ a? : A;\ d! : D \mid d! = c_1(a?)]$$

Here the input $a?$ in $Read_0$ is replicated across all three concrete operations whilst the output $d!$ is used only in the last operation after the cache has been updated. □

Example 12.9 A library stores books which its users can borrow, return, reserve, etc. We describe such a system as follows. Every copy of a *TITLE* has an identity denoted by *ITEM* (so we can have multiple copies of a given book), and we keep a record of whether the book is on loan (*active*) or not. The users of the library will be taken from the set *MEMBER*.

$$[ITEM, MEMBER, TITLE]$$
$$ACTIVE ::= active \mid nonactive$$
$$NOTICE ::= return\langle\langle ITEM\rangle\rangle$$

The state of the library is modelled as a partial function *items* which records the books potentially available for loan, *loans* on the other hand records which items have been lent out to which user.

$$\begin{array}{|l}
\hline Library \\
items : ITEM \pfun TITLE \times ACTIVE \\
loans : ITEM \pfun MEMBER \\
\hline
\operatorname{dom} loans \subseteq \operatorname{dom} items \\
\hline
\end{array}$$

We describe two operations: $Borrow_0$ and $Reserve_0$.

$$
\begin{array}{|l}
\multicolumn{1}{l}{\underline{\quad Borrow_0 \quad}} \\
\Delta Library \\
i? : ITEM \\
m? : MEMBER \\
\hline
i? \in \operatorname{dom} items \\
i? \notin \operatorname{dom} loans \\
loans' = loans \oplus \{i? \mapsto m?\} \\
\{i?\} \ntriangleleft items' = \{i?\} \ntriangleleft items \\
items'(i?) = (first\ items(i?), active) \\
\hline
\end{array}
$$

$$
\begin{array}{|l}
\multicolumn{1}{l}{\underline{\quad Reserve_0 \quad}} \\
\Xi Library \\
b? : TITLE \\
m? : MEMBER \\
recall! : MEMBER \times NOTICE \\
\hline
\exists i : ITEM;\ m : MEMBER \bullet \\
\qquad items(i) = (b?, active) \\
\qquad loans(i) = m \\
\qquad m \neq m? \\
\qquad recall! = (m, return(i)) \\
\hline
\end{array}
$$

The $Borrow_0$ operation is pretty obvious. The $Reserve_0$ operation will issue a recall notice to a member which has a copy of the book out. If there is a choice one member is picked at random.

When one actually borrows a book, the process is usually completed in two stages. First the borrowing details are entered into the machine (e.g., by the bar code on the book and your library card), and secondly the book is then swiped to deactivate the metal strip which stops the alarm going off when you leave the library with the book. We can represent this as a non atomic refinement of $Borrow_0$ into $Borrow_1 \mathbin{;} Active$ where the new operations are:

$$
\begin{array}{|l}
\multicolumn{1}{l}{\underline{\quad Borrow_1 \quad}} \\
\Delta Library \\
i? : ITEM \\
m? : MEMBER \\
\hline
i? \in \operatorname{dom} items \\
i? \notin \operatorname{dom} loans \\
loans' = loans \oplus \{i? \mapsto m?\} \\
items' = items \\
\hline
\end{array}
$$

$$
\begin{array}{|l}
\hline
\textit{Active} \\
\Delta \textit{Library} \\
i? : \textit{ITEM} \\
\hline
\textit{loans}' = \textit{loans} \\
\{i?\} \ntriangleleft \textit{items}' = \{i?\} \ntriangleleft \textit{items} \\
\textit{items}'(i?) = (\textit{first items}(i?), \textit{active}) \\
\hline
\end{array}
$$

We also change the granularity of $Reserve_0$, and describe it as a *CheckOnLoan* operation followed by a *Recall* operation. To do this we augment the state space by *RecState*.

$$
\begin{array}{|l}
\hline
\textit{RecState} \\
\textit{recall} : \textit{MEMBER} \times \textit{ITEM} \\
\hline
\end{array}
$$

$$
\begin{array}{|l}
\hline
\textit{CheckOnLoan} \\
\Xi \textit{Library} \\
\Delta \textit{RecState} \\
b? : \textit{TITLE} \\
\textit{onloanto}! : \textit{MEMBER} \\
\hline
\exists\, i : \textit{ITEM} \bullet \\
\quad \textit{items}(i) = (b?, \textit{active}) \\
\quad \textit{loans}(i) = \textit{onloanto}! \\
\quad \textit{recall}' = (\textit{onloanto}!, i) \\
\hline
\end{array}
$$

$$
\begin{array}{|l}
\hline
\textit{Recall} \\
\Xi \textit{Library} \\
\Xi \textit{RecState} \\
m? : \textit{MEMBER} \\
\textit{recall}! : \textit{MEMBER} \times \textit{NOTICE} \\
\hline
m? \neq \textit{first recall} \\
\textit{recall}! = (\textit{first recall}, \\
\qquad\qquad \textit{return}(\textit{second recall})) \\
\hline
\end{array}
$$

The refinement of $Borrow_0$ is easy to verify with an input transformer that simply duplicates the input $i?$. The refinement of $Reserve_0$ allows the librarian to check who the recall notice is being sent to before it is delivered (after all the library might not want too many recall notices going to the Vice-Chancellor). The $Reserve_0$ decomposition uses an output transformer:

$$
\begin{array}{|l}
\hline
\textit{OT} \\
\textit{recall}? : \textit{MEMBER} \times \textit{NOTICE} \\
\textit{onloanto}! : \textit{MEMBER} \\
\textit{recall}! : \textit{MEMBER} \times \textit{NOTICE} \\
\hline
\textit{recall}! = \textit{recall}? \\
\textit{onloanto}! = \textit{first recall}? \\
\hline
\end{array}
$$

together with an input transformer which spreads the inputs $b?$ and $m?$ across the two concrete operations.

We can then verify

$$
\begin{array}{l}
\text{pre}\, Reserve_0 \wedge (IT \gg CheckOnLoan \fatsemi Recall) \wedge R \wedge R' \\
\qquad\qquad \Rightarrow (Reserve_0 \gg OT)
\end{array}
$$

where the retrieve relation R is the identity on *Library*. □

Example 12.10 A simple vending machine allows a user to input a 3-digit code to *Select* a drink of their choice, which is delivered by the *Dispense* operation. (The input of money is considered an irrelevant implementation detail here!) In the initial description the code to be entered is input within a single operation *Select*.

$DRINK ::= black\_coffee \mid white\_coffee$

AState

$drink : \mathrm{seq}\,\mathbb{N}$

AInit

$AState'$

$drink' = \langle\,\rangle$

Select

$\Delta AState$
$i? : \mathrm{seq}\,\mathbb{N}$

$\#i? = 3$
$drink = \langle\,\rangle \wedge drink' = i?$

Dispense

$\Delta AState$
$cup! : DRINK$

$drink = \langle 1, 2, 3\rangle \Rightarrow cup! = black\_coffee$
$drink = \langle 4, 5, 6\rangle \Rightarrow cup! = white\_coffee$
$drink' = \langle\,\rangle$

In the implementation we accept the 3-digit code digit by digit, with each digit being entered via a separate operation. The single abstract *Select* operation will thus be split into a sequence consisting of three concrete operations: *FirstDigit*, *SecondDigit* and *ThirdDigit*.

CState

$cdrink : \mathrm{seq}\,\mathbb{N}$

CInit

$CState'$

$cdrink' = \langle\,\rangle$

FirstDigit

$\Delta CState$
$c? : \mathbb{N}$

$cdrink = \langle\,\rangle \wedge cdrink' = \langle c?\rangle$

SecondDigit

$\Delta CState$
$c? : \mathbb{N}$

$\#cdrink = 1 \Rightarrow cdrink' = cdrink \frown \langle c?\rangle$
$\#cdrink \neq 1 \Rightarrow cdrink' = cdrink$

ThirdDigit

$$\Delta CState$$
$$c? : \mathbb{N}$$

$$\#cdrink = 2 \Rightarrow cdrink' = cdrink \frown \langle c? \rangle$$
$$\#cdrink \neq 2 \Rightarrow cdrink' = cdrink$$

Dispense

$$\Delta CState$$
$$cup! : DRINK$$

$$cdrink = \langle 1, 2, 3 \rangle \Rightarrow cup! = black\_coffee$$
$$cdrink = \langle 4, 5, 6 \rangle \Rightarrow cup! = white\_coffee$$
$$cdrink' = \langle \rangle$$

The retrieve relation R will identify *drink* and *cdrink* and the input transformer IT we use is the following:

IT

$$i? : \mathrm{seq}\,\mathbb{N}$$
$$d! : \mathrm{seq}\,\mathbb{N}$$

$$i? = d!$$

With this in place we can express the refinement conditions that have to be verified. For example the correctness criteria require that

$$\begin{array}{l} \mathrm{pre}\, Select \wedge R \wedge (IT >> \\ (FirstDigit[d?(1)/c?] \fatsemi SecondDigit[d?(2)/c?] \fatsemi ThirdDigit[d?(3)/c?])) \\ \qquad \Rightarrow \exists AState' \bullet R' \wedge Select \end{array}$$

which is easily checked, as are the other conditions. □

12.5 Varying the Length of the Decomposition

Definition 12.3 deals with the case when the length of the concrete decomposition is fixed. We can also verify refinements when the length of decomposition is only known when the input is provided.

Example 12.11 Consider the *Multicast* operation from Example 11.9 (p. 278). We could decompose this to a sequence of concrete operations which send the message $m?$ to one user in the group $g?$ at a time. In fact we can use $Send_0$ for this and refine $Multicast_0$ as $Send_0 \fatsemi \cdots \fatsemi Send_0$. The length of this decomposition is not fixed, that is, the number of $Send_0$ operations varies depending on the size of the group $\#g?$. The appropriate input transformer is

$$
\begin{array}{|l}
\multicolumn{1}{l}{\_IT\_\_\_\_} \\
g? : \mathbb{P}\,USER \\
d! : \mathrm{seq}\,USER \\
\hline
\mathrm{ran}\,d! = g? \\
\hline
\end{array}
$$

This input transformer allows us to transmit the message to the users via $Store_0$ in any order. All that remains is to work out the expression representing the decomposition, and we consider this in general now. □

Let us suppose that an abstract operation

$$
\begin{array}{|l}
\multicolumn{1}{l}{\_AOp\_\_\_\_} \\
a? : A \\
\vdots \\
\hline
\vdots \\
\hline
\end{array}
$$

is decomposed into a sequence consisting of concrete operations $COp \mathbin{;} \cdots \mathbin{;} COp$ where COp is given by

$$
\begin{array}{|l}
\multicolumn{1}{l}{\_COp\_\_\_\_} \\
c? : C \\
\vdots \\
\hline
\vdots \\
\hline
\end{array}
$$

Suppose that the input $a?$ determines the length in the concrete decomposition. To verify such a refinement we use an input transformer

$$
\begin{array}{|l}
\multicolumn{1}{l}{\_IT\_\_\_\_} \\
a? : A \\
d! : \mathrm{seq}\,C \\
\hline
a? = mapping(d!) \\
\hline
\end{array}
$$

where *mapping* describes how one large input is broken down into individual pieces. Then the correct concrete decomposition for use in the refinement rules above is given by

$$
\mathbin{;}/\{(i, COp[d?(i)/c?]) \mid i \in \mathrm{dom}\,d?\}
$$

In this expression distributed schema composition ($\fatsemi/$) is being used along the sequence $\langle COp[d?(1)/c?], \ldots, COp[d?(m)/c?]\rangle$ where $m = \#d?$. This expression produces a schema composition of the correct number of COp operations according to the size of $d?$ (the concrete representation of $a?$) as required.

So, for example, the correctness criterion we have to verify for the multicast operation is

$$\begin{array}{l} \mathrm{pre}\, Multicast_0 \wedge (IT \gg \fatsemi/\{(i, Send_0[d?(i)/v?]) \mid i \in \mathrm{dom}\, d?\}) \\ \qquad \Rightarrow Multicast_0 \end{array}$$

since OT and R are both the identity. The applicability conditions are (for $i \in 0..(\#d? - 1)$)

$$\begin{array}{l} \mathrm{pre}(\overline{IT} \gg Multicast_0) \Rightarrow \mathrm{pre}\, Send_0[d?(1)/v?] \\ \mathrm{pre}(\overline{IT} \gg Multicast_0) \wedge Send_0[d?(i)/v?] \Rightarrow (\mathrm{pre}\, Send_0[d?(i+1)/v?])' \end{array}$$

Example 12.12 We specify a bank consisting of a number of electronic booths where users may deposit money and check their balances. In the initial description an account is represented by a mapping from account names to amounts. The specification includes operations allowing accounts to be opened, money to be deposited and balances checked.

$$\begin{array}{l} [NAME] \\ MONEY == \mathbb{N} \\ RECEIPT ::= balance\langle\langle MONEY\rangle\rangle \end{array}$$

$$\begin{array}{l} ABank == [act : NAME \nrightarrow MONEY] \\ ABankInit == [ABank' \mid act' = \varnothing] \\ AOpenAcct == [\Delta ABank;\ n? : NAME \mid act' = act \oplus \{n? \mapsto 0\}] \end{array}$$

Deposit

$$\begin{array}{|l} \Delta ABank \\ n? : NAME \\ p? : MONEY \\ \hline n? \in \mathrm{dom}\, act \\ act' = act \oplus \{n? \mapsto act(n?) + p?\} \end{array}$$

Balance

$$\begin{array}{|l} \Xi ABank \\ n? : NAME \\ b! : MONEY \\ r! : RECEIPT \\ \hline n? \in \mathrm{dom}\, act \\ b! = act\ n? \\ r! = balance(b!) \end{array}$$

In an implementation an atomic *Deposit* operation is unrealistic and we will transfer the amounts coin by coin at every booth. This will allow interleaving of these operations with actions at other booths. To specify this we use a collection of temporary accounts *tct* and split the *Deposit* operation into a transaction consisting of a *Begin*, a succession of *Next* operations transferring the amount coin by coin with a *End* operation ending the process. A temporary account is now represented by sequences of coins. The *End* operation takes this sequence and sums the coins entered, updating the concrete account with the result of this calculation (where $+/$ represents distributed summation over a sequence). The concrete specification is as follows.

$$COINS == \{1, 2, 5, 10\}$$

CBank

$$cct : NAME \pfun MONEY$$
$$tct : NAME \pfun \operatorname{seq} COINS$$

$$\operatorname{dom} tct \subseteq \operatorname{dom} cct$$

$$CBankInit == [CBank' \mid cct' = \varnothing \land tct' = \varnothing]$$

Begin

$$\Delta CBank$$
$$n? : NAME$$

$$n? \in \operatorname{dom} cct$$
$$tct' = tct \oplus \{n? \mapsto \langle\,\rangle\}$$
$$cct' = cct$$

Next

$$\Delta CBank$$
$$n? : NAME$$
$$c? : COINS$$

$$n? \in \operatorname{dom} tct$$
$$tct' = tct \oplus \{n? \mapsto (tct\ n?) \frown \langle c? \rangle\}$$
$$cct' = cct$$

End

$$\Delta CBank$$
$$n? : NAME$$

$$n? \in \operatorname{dom} tct$$
$$tct' = \{n?\} \ndres tct$$
$$cct' = cct \oplus \{n? \mapsto cct(n?) + (+/(tct\ n?))\}$$

The retrieve relation *R* identifies *act* and *cct*. The number of concrete operations needed is only determined by the input *p*? (in fact the number of *Next* operations needed is determined by the coins used as long as they sum to the correct amount $p?$), and this can continually vary. To verify the refinement we use input transformer

$$\begin{array}{|l} \hline IT \\ p? : MONEY \\ d! : \mathrm{seq}\ COINS \\ n?, n! : NAME \\ \hline +/(d!) = p? \wedge n! = n? \\ \hline \end{array}$$

and decompose *Deposit* into the sequence

$$Begin \mathbin{;} (\mathbin{;}/\{(i, Next[d?(i)/c?]) \mid i \in \mathrm{dom}\, d?\}) \mathbin{;} End$$

This expression produces a schema composition of the correct number of *Next* operations according to the size of $d?$. Upon calculating this composition it is easy to see that all the conditions for a non-atomic refinement are met.

The *Balance* operation could also be refined to two operations, one which outputs the balance to the screen, and the other which prints a paper receipt:

$$\begin{array}{|l} \hline B_1 \\ \Xi CBank \\ n? : NAME \\ b! : MONEY \\ \hline n? \in \mathrm{dom}\, cct \\ b! = cct\ n? \\ \hline \end{array} \qquad \begin{array}{|l} \hline B_2 \\ \Xi CBank \\ n? : NAME \\ r! : RECEIPT \\ \hline n? \in \mathrm{dom}\, cct \\ r! = balance(cct\ n?) \\ \hline \end{array}$$

Then using an identity input transformer and appropriate output transformer *Balance* is refined by $B_1 \mathbin{;} B_2$. □

12.6 Upward Simulations

In a fashion similar to the derivation in Sect. 12.2 we can derive conditions for refinements due to non-atomic upward simulations. In the relational setting the requirement we have to fulfil is (for a retrieve relation T):

$$\widehat{\mathsf{COp_1}} \mathbin{;} \widehat{\mathsf{COp_2}} \mathbin{;} \widetilde{\mathsf{T}} \subseteq \widetilde{\mathsf{T}} \mathbin{;} \widehat{\mathsf{AOp}} \tag{12.13}$$

To extract the underlying conditions on the partial relations we expand out the expression as follows.

$$\begin{aligned} \widehat{\mathsf{COp_1}} \mathbin{;} \widehat{\mathsf{COp_2}} \mathbin{;} \widetilde{\mathsf{T}} = {} & (\mathsf{COp_1} \mathbin{;} \mathsf{COp_2} \mathbin{;} \mathsf{T}) \cup (\overline{\mathrm{dom}\,\mathsf{COp_1}} \times \mathsf{AState}_\perp) \cup \\ & (\mathrm{dom}(\mathsf{COp_1} \mathbin{\rhd\!\!\!-} \mathrm{dom}\,\mathsf{COp_2}) \times \mathsf{AState}_\perp) \\ \widetilde{\mathsf{T}} \mathbin{;} \widehat{\mathsf{AOp}} = {} & (\mathsf{T} \mathbin{;} \mathsf{AOp}) \cup (\mathrm{dom}(\mathsf{T} \mathbin{\rhd\!\!\!-} \mathrm{dom}\,\mathsf{AOp}) \times \mathsf{AState}_\perp) \cup \\ & (\{\perp_{\mathsf{CState}}\} \times \mathsf{AState}_\perp) \end{aligned}$$

where the totalisation of the retrieve relation T used in an upward simulation is

$$\widetilde{T} = T \cup (\{\bot_{CState}\} \times AState_{\bot})$$

Doing a case-by-case analysis on the domains we find that $\widehat{COp_1} \mathbin{;} \widehat{COp_2} \mathbin{;} \widetilde{T} \subseteq \widetilde{T} \mathbin{;} \widehat{AOp}$ if and only if the following three conditions hold.

$$\overline{\operatorname{dom} COp_1} \subseteq \operatorname{dom}(T \rhd \operatorname{dom} AOp)$$
$$\operatorname{dom}(COp_1 \rhd \operatorname{dom} COp_2) \subseteq \operatorname{dom}(T \rhd \operatorname{dom} AOp)$$
$$\operatorname{dom}(T \rhd \operatorname{dom} AOp) \lhd (COp_1 \mathbin{;} COp_2 \mathbin{;} T) \subseteq T \mathbin{;} AOp$$

These can be rephrased in the Z schema calculus as the following definition.

Definition 12.4 (Non-atomic upward simulation without IO transformations) Let T be a retrieve relation between data types $(AState, AInit, \{AOp_1\})$ and $(CState, CInit, \{COp_1, COp_2\})$, and $\rho(1) = \langle 1, 2 \rangle$.

Then T is a non-atomic upward simulation with respect to ρ if, in addition to the initialisation, the following hold.

$$\forall CState \bullet (\forall AState \bullet T \Rightarrow \operatorname{pre} AOp) \Rightarrow \operatorname{pre} COp_1$$
$$\forall CState \bullet (\forall AState \bullet T \Rightarrow \operatorname{pre} AOp) \Rightarrow$$
$$\qquad \forall CState' \bullet (COp_1 \Rightarrow (\operatorname{pre} COp_2)')$$
$$\forall AState'; CState; CState' \bullet (\forall AState \bullet T \Rightarrow \operatorname{pre} AOp) \Rightarrow$$
$$\qquad (COp_1 \mathbin{;} COp_2 \wedge T' \Rightarrow \exists AState \bullet T \wedge AOp)$$

□

12.7 Properties

Although we have dropped the requirement that concrete operations refine *skip* if they do not refine the whole abstract operation, there are a number of other safety-like properties that one can require of a non-atomic refinement. We briefly discuss some of these now.

12.7.1 No Effect Elsewhere

Consider a refinement of abstract AOp into $COp_1 \mathbin{;} COp_2$. If COp_2 happens straight after COp_1, then the composition refines AOp. But what happens if we are in a concrete state that was not a result of a COp_1 invocation? We could now require that COp_2 had no effect. Thus outside the range of COp_1, COp_2 should be skip, and we can formalise this, viewing it through the retrieve relation naturally, as

$$R \mathbin{;} (\operatorname{ran} COp_1 \lhd COp_2) \subseteq skip \mathbin{;} R = R$$

Example 12.13 Looking at the bank account example we see that the range of the *Begin* operation is $n? \in \mathrm{dom}\, tct$, so for this property to hold we would require that outside this region any application of *Next* has no effect. If this were desirable then it could be achieved by adding a disjunction of $n? \notin \mathrm{dom}\, tct \wedge tct' = tct$ to the *Next* operation. We could amend the *End* operation in the same way for similar reasons if needed. □

Example 12.14 In a similar fashion in our vending machine decomposition (Example 12.10) *SecondDigit* and *ThirdDigit* only have an effect if they occur as part of the sequence $FirstDigit \fatsemi SecondDigit \fatsemi ThirdDigit$. □

Likewise the operations *Change* and *Alter* from Examples 12.2 and 12.3 above could be rewritten to have this property if needed. This, of course, has the effect of making the operations total, preventing weakening of their preconditions under any subsequent refinement.

12.7.2 Non-interference

Non-interference is the property that if COp_1 has occurred, then the only operation to have an effect should be COp_2, i.e., other operations cannot interfere with a concrete transaction yet to be completed. This is modelled by saying that COp_1 followed by any other operation (apart from COp_2) is the same as COp_1:

$$\mathsf{COp}_1 \fatsemi \mathsf{Op} \subseteq \mathsf{COp}_1 \quad \text{for all } \mathsf{Op} \neq \mathsf{COp}_2$$

Example 12.15 Consider the coffee machine. Perhaps it would make sense for the *Dispense* operation to have no effect until all inputs have been entered. Non-interference with *FirstDigit* is thus required, i.e.,

$$FirstDigit \fatsemi Dispense \subseteq FirstDigit$$

and we could amend *Dispense* to ensure that it behaves like this. □

Example 12.16 In the bank account example a concrete *Withdraw* operation is required not to interfere with a transaction once it has started, i.e., we require that $Begin \fatsemi Withdraw \subseteq Begin$. To do this, any *Withdraw* operation must contain the predicate $n? \in \mathrm{dom}\, tct \Rightarrow \Xi CBank$. □

Non-interference is connected with locking a resource at the start of a non-atomic refinement, which stops that resource being altered unintentionally until the concrete decomposition has been completed.

Example 12.17 Consider the reading and writing of a single queue in a memory.

$$\begin{array}{|l}\hline \textit{Mem}_0 \\ q : \operatorname{seq} M \\ \hline \end{array}$$

$$\begin{array}{|l}\hline \textit{Store}_0 \\ \Delta \textit{Mem}_0 \\ m? : M \\ \hline q' = \langle m? \rangle \frown q \\ \hline \end{array} \qquad \begin{array}{|l}\hline \textit{Load}_0 \\ \Delta \textit{Mem}_0 \\ m! : M \\ \hline q = q' \frown \langle m! \rangle \\ \hline \end{array}$$

In a shared variable model of concurrent execution a number of threads have access to the memory. To control the access we introduce operations to *Lock* and *Unlock* ownership of the shared memory. The threads are given identifiers from $\mathbb{N}_1$.

$$\textit{THREAD} == \mathbb{N}_1$$

$$\begin{array}{|l}\hline \textit{Mem}_1 \\ q : \operatorname{seq} M \\ \textit{key} : \mathbb{N} \\ \hline \end{array}$$

$$\begin{array}{|l}\hline \textit{Lock}_1 \\ \Delta \textit{Mem}_1 \\ u? : \textit{THREAD} \\ \hline q' = q \\ \textit{key} = 0 \Rightarrow \textit{key}' = u? \\ \textit{key} \neq 0 \Rightarrow \textit{key}' = \textit{key} \\ \hline \end{array} \qquad \begin{array}{|l}\hline \textit{Unlock}_1 \\ \Delta \textit{Mem}_1 \\ u? : \textit{THREAD} \\ \hline q' = q \\ \textit{key} = u? \Rightarrow \textit{key}' = 0 \\ \textit{key} \neq u? \Rightarrow \textit{key}' = \textit{key} \\ \hline \end{array}$$

$$\begin{array}{|l}\hline \textit{Store}_1 \\ \Delta \textit{Mem}_1 \\ m? : M \\ u? : \textit{THREAD} \\ \hline \textit{key} = u? \Rightarrow q' = \langle m? \rangle \frown q \\ \textit{key} \neq u? \Rightarrow q' = q \\ \textit{key}' = \textit{key} \\ \hline \end{array} \qquad \begin{array}{|l}\hline \textit{Load}_1 \\ \Delta \textit{Mem}_1 \\ m! : M \\ u? : \textit{THREAD} \\ \hline \textit{key} = u? \Rightarrow q = q' \frown \langle m! \rangle \\ \textit{key} \neq u? \Rightarrow q' = q \\ \textit{key}' = \textit{key} \\ \hline \end{array}$$

This refinement is interesting because it uses both IO refinement and a non-atomic refinement. The IO refinement comes in because we are adding an input $u?$ which was not there before, however the refinement is correct because the changes to the memory are identical in both specifications. That is, it is irrelevant which threads load or store from the memory, the change to it is the same. We also have a non-atomic refinement of $\textit{Load}_0$ into $\textit{Lock}_1 \mathbin{;} \textit{Load}_1 \mathbin{;} \textit{Unlock}_1$, and similarly for $\textit{Store}_0$.

The locking works because once any thread has locked the resource, any further concrete operation has no effect until that thread has completed the cycle of operations, that is, from any state linked to an abstract one:

$$Lock_1 \mathbin{⨾} Load_1 = Lock_1$$

as long as the thread in $Lock_1$ and $Load_1$ are different. □

This idea can be extended further to describe a shared filestore.

Example 12.18 At the abstract level a single file is represented as a mapping from a name to that file, and we are equipped with a *Write* operation which overwrites the contents of an existing named file.

$[NAME, FILE, USER]$

AF

$file_A : NAME \pfun FILE$

$Write_A$

ΔAF
$n? : NAME$
$f? : FILE$

$n? \in \mathrm{dom}\, afs$
$file'_A = file_A \oplus \{n? \mapsto f?\}$

In a more concrete version we wish to introduce a number of users. Each user can work on a copy of a file by *Lock*-ing it and then *Unlock*-ing it when they have finished. The effect of a *Lock* is to create their own working copy (wc) of the file and only at the *Unlock* stage is the main file overwritten with the updated working copy. This will allow operations of different users to be interleaved.

CF

$file_C : NAME \pfun FILE$
$wc : USER \rightarrow NAME \pfun FILE$

$Write_C$

ΔCF
$u? : USER$
$n? : NAME$
$f? : FILE$

$file'_C = file_C$
$n? \notin \mathrm{dom}\, wc(u?) \Rightarrow \Xi CF$
$n? \in \mathrm{dom}\, wc(u?) \Rightarrow wc' = wc \oplus \{u? \mapsto wc(u?) \oplus \{n? \mapsto f?\}\}$

$$
\begin{array}{|l}
\textit{Lock} \\
\hline
\Delta CF \\
u? : USER \\
n? : NAME \\
\hline
file'_C = file_C \\
n? \in \mathrm{dom}\, wc(u?) \Rightarrow \Xi CF \\
n? \notin \mathrm{dom}\, wc(u?) \Rightarrow wc' = wc \oplus \{u? \mapsto wc(u?) \oplus \{n? \mapsto file_C(n?)\}\} \\
\hline
\end{array}
$$

$$
\begin{array}{|l}
\textit{Unlock} \\
\hline
\Delta CF \\
u? : USER \\
n? : NAME \\
\hline
wc' = wc \\
n? \notin \mathrm{dom}\, wc(u?) \Rightarrow \Xi CF \\
n? \in \mathrm{dom}\, wc(u?) \Rightarrow file'_C = file_C \oplus \{n? \mapsto wc(u?)(n?)\} \\
\hline
\end{array}
$$

Then, as expected, $Write_A$ is non-atomically refined by $Lock \mathbin{;} Write_C \mathbin{;} Unlock$. Furthermore, the behaviour of the locking mechanism has been arranged with a number of properties in mind. In particular, $Write_C$ and *Unlock* have been defined so that they have no effect unless an appropriate *Lock* has already taken place. The behaviour also allows multiple locks and writes to be taking place, however, the effect of any concrete sequence will be determined by the last *Unlock* to occur, that is, $file_C$ will reflect the final change to be committed irrespective of when the actual concrete write took place. □

12.7.3 Interruptive

Interruptive is the opposite effect to non-interference, and models a cancellation of a half-completed sequence of concrete operations. An alternative design of a coffee machine might model the dispense operation so that it cancels a sequence of inputs unless the sequence is complete. This requirement is then $FirstDigit \mathbin{;} Dispense \subseteq Dispense$, and in general

$$\mathsf{COp}_1 \mathbin{;} \mathsf{Op} \subseteq \mathsf{Op} \quad \text{for all } \mathsf{Op} \neq \mathsf{COp}_2$$

Example 12.19 Another example of this is the effect of a *Begin* operation in our bank account after any incomplete deposit. If after any sequence of a *Begin* and a succession of *Next* operations we perform another *Begin* (with the same input name $n?$), then the effect is to interrupt the transaction and to ignore the half-completed sequence, that is, we have:

$$Begin \mathbin{;} Begin = Begin \qquad Next \mathbin{;} Begin = Begin$$

□

The requirements of non-interference and interruptiveness are alternative choices that might be required of a non-atomic refinement. Which, if any, is chosen very much depends on the system being modelled. They are also properties of the resultant concrete specification as opposed to properties inherited from the abstract specification through the refinement.

Note that these properties are not necessarily preserved under refinement, however, they are preserved if we take the weakest refinement. Consider the safety condition for the second concrete operation in a decomposition:

$$\mathsf{R} \mathbin{⨾} (\operatorname{ran} \mathsf{COp}_1 \lhd \mathsf{COp}_2) \subseteq \mathsf{R}$$

It is easy to construct simple examples to show that if COp_1 and COp_2 are both refined further to, say, CCOp_1 and CCOp_2 respectively, then $\mathsf{R} \mathbin{⨾} (\operatorname{ran} \mathsf{CCOp}_1 \lhd \mathsf{CCOp}_2) \not\subseteq \mathsf{R}$. The problem arises because non-determinism in COp_1 can be reduced in an arbitrary refinement, therefore $\operatorname{ran} \mathsf{CCOp}_1$ is smaller than $\operatorname{ran} \mathsf{COp}_1$, and thus CCOp_2 may no longer be skip outside this range. However, taking the weakest refinement ensures that non-determinism is not reduced more than necessary, and that therefore $\mathsf{R} \mathbin{⨾} (\operatorname{ran} \mathsf{CCOp}_1 \lhd \mathsf{CCOp}_2) \subseteq \mathsf{R}$ holds.

The situation is the same for the non-interference and interruptive properties.

12.7.4 Using Coupled Simulations

An alternative to outlining specific properties that a non-atomic refinement should possess (see Sects. 12.7.2 and 12.7.3) is to strengthen the definition of non-atomic refinement so that these hold by construction (i.e., the simulation conditions guarantee that a certain property holds). The use of *coupled simulations* is one approach to doing this.

Specifically it tackles the following two issues: the current definition of non-atomic refinement given above allows for the sequence $COp_1 \mathbin{⨾} COp_2$ to be started "in the middle" (i.e., COp_2 can be executed without COp_1 having occurred) and it allows it to remain unfinished (i.e., there isn't a COp_2 operation after a COp_1).

For example, consider the non-atomic refinement specified in Example 12.5, which met the conditions in Definition 12.3. Although this is an acceptable refinement, the concrete components *TranslateX* and *TranslateY* can be invoked an arbitrary number of times in any order. In other words, we have failed to capture the requirement that we cannot do *TranslateY* unless *TranslateX* has already happened at some point in the past. And it allows the concrete sequence to remain unfinished. The definition we derive below rectifies these deficiencies. Further, the definition we give holds for both the non-blocking and the blocking model, as these are additional conditions to the standard definition of non-atomic refinement given, e.g., in Definition 12.3.

Already, there are other potential issues though. For example, suppose the abstract specification had an *Inverse* operation, which was decomposed into *InvX* and *InvY* (we denote concrete variables as x_C and y_C here):

```
┌─ Inverse ──────────────────────────
│ Δ2D
├──────────────
│ x' = −x ∧ y' = −y
└────────────────────────────────────
```

```
┌─ InvX ─────────────────     ┌─ InvY ─────────────────
│ Δ2D_C                       │ Δ2D_C
├──────────                   ├──────────
│ x'_C = −x_C ∧ y'_C = y_C    │ x'_C = x_C ∧ y'_C = −y_C
│ ...                         │ ...
└────────────────────────     └────────────────────────
```

We then have the additional problem: the concrete non-atomic operations can be interleaved in such a way that they do not match any sequence of abstract operations. For example, both $TranslateX \mathbin{⨾} InvX \mathbin{⨾} TranslateY \mathbin{⨾} InvY$ and $TranslateX \mathbin{⨾} InvX \mathbin{⨾} InvY \mathbin{⨾} TranslateY$ are valid sequences in the concrete specification. However, although the first sequence refines $Translate \mathbin{⨾} Inverse$ the second sequence has no abstract counterpart (using a retrieve relation R which relates appropriate concrete and abstract variables and is the identity on inputs).

This problem has arisen because the concrete specification has allowed *interleavings* which were not possible in the abstract specification. *Translate* and *Inverse* are not independent since they modify the same variables. Thus a 'concurrent' execution of their refinements, i.e., an interleaving of the concrete operations, may in principle lead to states with no matching abstract counterpart.

The basic idea of a *coupled simulation* is to complement the existing retrieve relation for non-atomic refinement with a second simulation relation which excludes these undesired behaviours. These two relations have to be *coupled* (i.e., agree) at specific points, thus ensuring a tight correspondence of concrete and abstract specification.

Thus this new definition tackles the following additional issues over and above non-atomic refinement as defined above:

1. After COp_1 it is always possible to do COp_2 (the *completion* of refinement).
2. It is not possible to do COp_2 unless the beginning of the non-atomic operation has already started, i.e., COp_1 has already happened at some previous point (we cannot start "in the middle").
3. At the concrete level refinements of two (or more) abstract operations may only be interleaved if the interleaving matches a sequence of abstract operations.

Constraint We do not consider *auto-concurrency* here: once a refined instance of an abstract operation has been started, we may not start another refinement unless the first is completed.

We use a coupled simulation to record whether the first part of a non-atomic sequence has commenced, by being indexed with a sequence of concrete operations—those which have been started but not finished.

For example, if we have two abstract operations AOp and BOp, being refined into $COp_1 \mathbin{;} COp_2$ and $DOp_1 \mathbin{;} DOp_2$, respectively, a coupled simulation relation $R^{\langle DOp_1, COp_1 \rangle}$ records that sequences have commenced (first concrete component of BOp and then that of AOp) but neither has finished. When a concrete sequence has been completed, the first concrete component operation is removed from the index.

The coupling condition requires that the indexed simulation relation agrees with the standard retrieve relation when the sequence is empty, which is exactly the case when refinements that have been started have finished.

For a sequence of concrete operations S we write the coupled simulations as R^S. We also use the notation $s \in S$ (for a sequence S) to stand for $s \in \operatorname{ran} S$ and $S \setminus s$ to stand for the sequence S with the first occurrence of s removed. The coupling condition is that (and this guarantees the final condition to hold):

$$R^{\langle\rangle} = R$$

We also record which non-atomic sequences have started but not yet finished. This requires that if an operation COp_1 (which begins a non-atomic decomposition) is executed the resulting concrete state should be related to the same abstract state as before, however, not using R^S but instead $R^{S \frown \langle COp_1 \rangle}$. This is captured as:

$$\begin{array}{l} \forall AState, CState, CState' \bullet \\ \qquad R^S \land COp_1 \Rightarrow \exists AState' \bullet \Xi AState \land (R^{S \frown \langle COp_1 \rangle})' \end{array}$$

The first additional requirement is then that if we have started the decomposition then we can carry on:

$$\forall AState, CState \bullet R^S \land COp_1 \in S \Rightarrow \operatorname{pre} COp_2$$

The requirement that one cannot do COp_2 unless the beginning of the non-atomic operation has already commenced, can be checked by inspecting the current index S ($COp_1 \in S$). This is given by the following

$$\begin{array}{l} \forall AState, CState, CState' \bullet \\ \qquad R^S \land COp_2 \Rightarrow COp_1 \in S \land \exists AState' \bullet AOp \land (R^{S \setminus \langle COp_1 \rangle})' \end{array}$$

IO transformers will, in general, be necessary since the inputs and outputs might well be spread throughout the concrete decomposition, and when we add them in, this leads to the following definition of non-atomic coupled simulation:

Definition 12.5 (Non-atomic coupled downward simulation with IO transformation) Given data types A and C, C is a non-atomic coupled downward simulation of A if there is a retrieve relation R and input and output transformers IT and OT showing that C is a non-atomic downward simulation of A, and there is a family of simulation relations R^S such that the following hold.

C $R^{\langle\rangle} = R$

S1 $\forall AState, CState, CState' \bullet R^S \wedge COp_1 \Rightarrow \exists AState' \bullet \Xi AState \wedge (R^{S \frown \langle COp_1 \rangle})'$

S2 $\forall AState, CState \bullet R^S \wedge COp_1 \in S \Rightarrow \text{pre } COp_2$

S3 $\forall AState, CState, CState' \bullet IT \gg (R^S \wedge COp_2) \Rightarrow$

$$COp_1 \in S \wedge \exists AState' \bullet (AOp \gg OT) \wedge (R^{S \setminus \langle COp_1 \rangle})' \qquad \square$$

As an example, consider the translate example given above. R is the identity on abstract and concrete inputs (modulo renaming). The coupled retrieve relation R^S records the effects of part of the concrete operation. With inputs in the concrete operation this necessitates the input being part of the coupled simulation. We set $R^{\langle\rangle} = R$, so that C is fulfilled. The other coupled relations can be calculated as

$R^{\langle TranslateX \rangle}$
$2D$
$2D_C$
$x_C? : \mathbb{Z}$

$x = x_C - x_C?$
$y = y_C$

It then should be easy to check that the conditions required in Definition 12.5 hold. For example, calculating $(R^{\langle TranslateX \rangle} \wedge TranslateY)$ we find it simplifies to

$2D$
$\Delta 2D_C$
$x_C?, y_C? : \mathbb{Z}$

$x = x_C - x?$
$y = y_C + y_C?$

which matches with $Translate \wedge R'$ upon application of appropriate input transformers (i.e., meets condition $S3$).

12.7.5 Further Non-atomic Refinements

In this chapter we have only considered non-atomic refinements where an operation gets decomposed into a sequence of operations whose length is fixed or determined by the inputs. More general non-atomic refinements should be possible, but approached with some caution. First, we might consider the situation where an operation gets decomposed into a sequence of operations whose length depends on the state. This is clearly useful, and should be possible under certain circumstances. However, effectively such a decomposition would represent the situation where an abstract program (a sequence of instructions) is linked to a concrete program which is based on a richer notion of program: namely one which allows while-loops in addition to sequences. As a consequence, our refinement theory would have to become much more complicated, in order to avoid the problems caused by the combination of loops, partiality, and unbounded non-determinism. We will see an example of where we might need such a refinement in Chap. 13 below.

Another avenue of further investigation is to consider decomposition operations whose source is not a single operation, but a sequence of operations. In particular, under which circumstances is it acceptable to replace $AOp_1 \mathbin{;} AOp_2$ by $COp_1 \mathbin{;} COp_2$, and will refinements like that remove the need for having both upward and downward simulation?

12.8 Bibliographical Notes

Ideas on refining the atomicity of actions go back to Lipton [22]. Such atomicity refinement has also been studied in TLA by Lamport and Schneider [21] and Cohen and Lamport [10].

Within the action systems formalism [4, 5] there has also been considerable work on this subject. Ralph Back's initial work is described in [2] and [3]. This has been extended by Sere and Walden [27] (for action systems with procedures) and Büchi and Sekerinski [8] (for concurrent object oriented action systems).

Non-atomic refinement has been studied extensively in the context of process algebras, usually under the term action refinement [1, 19, 31]. De Bakker and de Vink in [12] provide a survey. There has also been work in a Petri net setting [7]. The issue in a process algebra is the use of an interleaving model of concurrency, which is incompatible with action refinement. In particular, if $(a|||b)$ is expanded as $(a;\ b) + (b;\ a)$ then an action refinement of b into $(b_1;\ b_2)$ breaks the requirement that semantic equivalences should be congruences with respect to refinement.

However, in Z (and Object-Z) there are no such constraints because there are no global behavioural constructors such as $\|$ in a process algebra. In fact, although there is a $\|$ primitive in Object-Z as we shall see later, this is a schema calculus operator which builds a single new operation rather than defining a temporal constraint over existing behaviours.

The issue of non-atomic refinement in a state-based language was raised in [35], and the results in Sects. 12.1–12.6 of this chapter were first formulated in [14].

An alternative idea to that of interruptiveness (Sect. 12.7.3) is to be able to undo an action. This idea has been called *compensation* by Butler and Ferreira [9, 32], and has been explored in the context of process algebras.

The work on coupled simulations in Sect. 12.7.4 was done in conjunction with Heike Wehrheim for the blocking model of Object-Z in [17], and a combined Object-Z and CSP notation in [18].

Refinement has been used to verify particular *correctness* properties of concurrent algorithms as well as for their development. The most pertinent notion is that of linearizability, i.e., concrete runs should have abstract counterparts (see

Sect. 12.7.4). More specifically, linearizability as introduced by Herlihy and Wing in [20] permits one to view operations on concurrent objects as though they occur in some sequential order:

> Linearizability provides the illusion that each operation applied by concurrent processes takes effect instantaneously at some point between its invocation and its response.

There have been many approaches to verifying linearizability, model checking [33], separation logic [24], combinations of logics [29, 30] etc., but refinement as also been applied, e.g., by Colvin et al. in [11] as well as Derrick, Schellhorn and Wehrheim in [15]. The essential idea in [15] is to set up a particular type of non-atomic refinement to show that if A is refined by C then C is linearizable with respect to the abstract specification A.

In [16] a formal theory is derived that *relates* refinement theory and linearizability, and which has been fully mechanised using the interactive theorem prover KIV [25] (see Sect. 11.7). Specifically, proof obligations are derived which are shown to imply linearizability, by constructing an upward simulation between extended concrete and abstract data types out of the proof obligations. All definitions and proof steps for this generic theory have been formalised and mechanically verified using KIV, as has their application to a number of case studies (e.g., a lock-free, non-blocking algorithm from [23] and an implementation of a set from [29]).

Linearizable algorithms can be very subtle, and the techniques developed for their verification are usually only sound, and not complete, but the refinement approach has been taken further in [26] where a sound and complete method is presented based upon constructing a particular type of simulation relation.

References

1. Aceto, L. (1992). *Action Refinement in Process Algebras*. London: Cambridge University Press.
2. Back, R. J. R. (1989). A method for refining atomicity in parallel algorithms. In *PARLE'89 Parallel Architectures and Languages Europe*, Eindhoven, Netherlands, June 1989. *Lecture Notes in Computer Science: Vol. 366* (pp. 199–216). Berlin: Springer.
3. Back, R. J. R. (1993). *Atomicity refinement in a refinement calculus framework* (Reports on computer science and mathematics 142). Åbo Akademi University.
4. Back, R. J. R., & Kurki-Suonio, R. (1988). Distributed cooperation with action systems. *ACM Transactions on Programming Languages and Systems*, *10*(4), 513–554.
5. Back, R. J. R., & Kurki-Suonio, R. (1989). Decentralization of process nets with centralized control. *Distributed Computing*, *3*(2), 73–87.
6. Bloomfield, R., Marshall, L., & Jones, R. (Eds.) (1988). *VDM'88*. *Lecture Notes in Computer Science: Vol. 328*. Berlin: Springer.
7. Brauer, W., Gold, R., & Vogler, W. (1991). A survey of behaviour and equivalence preserving refinements of Petri nets. In G. Rozenberg (Ed.), *Advances in Petri Nets*. *Lecture Notes in Computer Science: Vol. 483* (pp. 1–46). Berlin: Springer.
8. Büchi, M., & Sekerinski, E. (1999). *Refining concurrent objects* (TUCS Technical Report 298). Abo Akademi.
9. Butler, M. J., Ferreira, C., & Ng, M. Y. (2005). Precise modelling of compensating business transactions and its application to BPEL. *Journal of Universal Computer Science*, *11*(5), 712–743.

10. Cohen, E., & Lamport, L. (1998). Reduction in TLA. In D. Sangiorgi & R. de Simone (Eds.), *CONCUR'98. Lecture Notes in Computer Science: Vol. 1466* (pp. 317–331). Berlin: Springer.
11. Colvin, R., Doherty, S., & Groves, L. (2005). Verifying concurrent data structures by simulation. *Electronic Notes in Theoretical Computer Science, 137*, 93–110.
12. de Bakker, J. W., & de Vink, E. P. (1994). Bisimulation semantics for concurrency with atomicity and action refinement. *Fundamenta Informaticae, 20*, 3–34.
13. Davies, J. & Gibbons, J. (Eds.) (2007). *Integrated Formal Methods, 6th International Conference, IFM 2007. Lecture Notes in Computer Science: Vol. 4591*. Berlin: Springer.
14. Derrick, J., & Boiten, E. A. (1999) Non-atomic refinement in Z. In Wing et al. [34] (pp. 1477–1496).
15. Derrick, J., Schellhorn, G., & Wehrheim, H. (2007) Proving linearizability via non-atomic refinement. In Davies and Gibbons [13] (pp. 195–214).
16. Derrick, J., Schellhorn, G., & Wehrheim, H. (2011). Mechanically verified proof obligations for linearizability. *ACM Transactions on Programming Languages and Systems, 33*(1), 4.
17. Derrick, J., & Wehrheim, H. (2003). Using coupled simulations in non-atomic refinement. In D. Bert, J. P. Bowen, S. King, & M. Waldén (Eds.), *ZB 2003: Formal Specification and Development in Z and B. Lecture Notes in Computer Science: Vol. 2651* (pp. 127–147). Berlin: Springer.
18. Derrick, J., & Wehrheim, H. (2005) Non-atomic refinement in Z and CSP. In Treharne et al. [28] (pp. 24–44).
19. Gorrieri, R. (1991). *Refinement, atomicity and transactions for process description languages*. PhD thesis, University of Pisa, Pisa.
20. Herlihy, M., & Wing, J. M. (1990). Linearizability: a correctness condition for concurrent objects. *ACM Transactions on Programming Languages and Systems, 12*(3), 463–492.
21. Lamport, L., & Schneider, F. B. (1989). *Pretending atomicity* (Technical Report 44). Compaq Systems Research Center.
22. Lipton, R. J. (1975). Reduction: a method of proving properties of parallel programs. *Communications of the ACM, 18*(12), 717–721.
23. Michael, M. M., & Scott, M. L. (1998). Nonblocking algorithms and preemption-safe locking on multiprogrammed shared-memory multiprocessors. *Journal of Parallel and Distributed Computing, 51*(1), 1–26.
24. Parkinson, M., Bornat, S., & O'Hearn, P. (2007). Modular verification of a non-blocking stack. In *ACM SIGPLAN-SIGACT Symposium on Principles of Programming Languages (POPL '07)* (pp. 297–302). New York: ACM.
25. Reif, W., Schellhorn, G., Stenzel, K., & Balser, M. (1998). Structured specifications and interactive proofs with KIV. In W. Bibel & P. Schmitt (Eds.), *Automated Deduction—A Basis for Applications, Volume II: Systems and Implementation Techniques* (pp. 13–39). Dordrecht: Kluwer Academic.
26. Schellhorn, G., Wehrheim, H., & Derrick, J. (2012). How to prove algorithms linearisable. In *Proceedings of the 24th International Conference on Computer Aided Verification, CAV'12* (pp. 243–259). Berlin: Springer.
27. Sere, K., & Waldén, M. (1997). Data refinement of remote procedures. In *International Symposium on Theoretical Aspects of Computer Software (TACS'97). Lecture Notes in Computer Science: Vol. 1281* (pp. 267–294). Berlin: Springer.
28. Treharne, H., King, S., Henson, M. C., & Schneider, S. A. (Eds.) (2005). *ZB 2005: Formal Specification and Development in Z and B, 4th International Conference of B and Z Users. Lecture Notes in Computer Science: Vol. 3455*. Berlin: Springer.
29. Vafeiadis, V., Herlihy, M., Hoare, C. A. R., & Shapiro, M. (2006). Proving correctness of highly-concurrent linearisable objects. In *ACM SIGPLAN Symposium on Principles and Practice of Parallel Programming (PPoPP '06)* (pp. 129–136). New York: ACM.
30. Vafeiadis, V., & Parkinson, M. J. (2007). A marriage of rely/guarantee and separation logic. In L. Caires & V. Thudichum Vasconcelos (Eds.), *International Conference on Concurrency Theory (CONCUR 2007). Lecture Notes in Computer Science: Vol. 4703* (pp. 256–271). Berlin: Springer.

31. van Glabbeek, R. J. (1990). *Comparative concurrency semantics and refinement of actions*. PhD thesis, Vrije Universiteit, Amsterdam.
32. Vaz, C., & Ferreira, C. (2012). On the analysis of compensation correctness. *Journal of Logic and Algebraic Programming*, *81*(5), 585–605.
33. Vechev, M., & Yahav, E. (2008). Deriving linearizable fine-grained concurrent objects. In *ACM SIGPLAN Conference on Programming Language Design and Implementation (PLDI '08)*, New York, NY, USA (pp. 125–135). New York: ACM.
34. Wing, J. M., Woodcock, J. C. P., & Davies, J. (Eds.) (1999). *FM'99 World Congress on Formal Methods in the Development of Computing Systems*. *Lecture Notes in Computer Science: Vol. 1708*. Berlin: Springer.
35. Woodcock, J. C. P., & Dickinson, B. (1988) Using VDM with rely and guarantee-conditions: experiences of a real project. In Bloomfield et al. [6] (pp. 434–458).

Chapter 13
Case Study: A Digital and Analogue Watch

The Managing Director of our local watch manufacturer has decided to use Z to describe the functionality of the latest watches in production. Given an initial abstract specification there are a number of possible implementations, and in this chapter we use a collection of different refinement techniques to verify that the implementations are correct with respect to the initial design.

13.1 The Abstract Design

The basic requirements of a watch are that it keeps track of the hours and minutes, and for our model that it is equipped with an alarm. The state space and initialisation for the initial design are as follows.

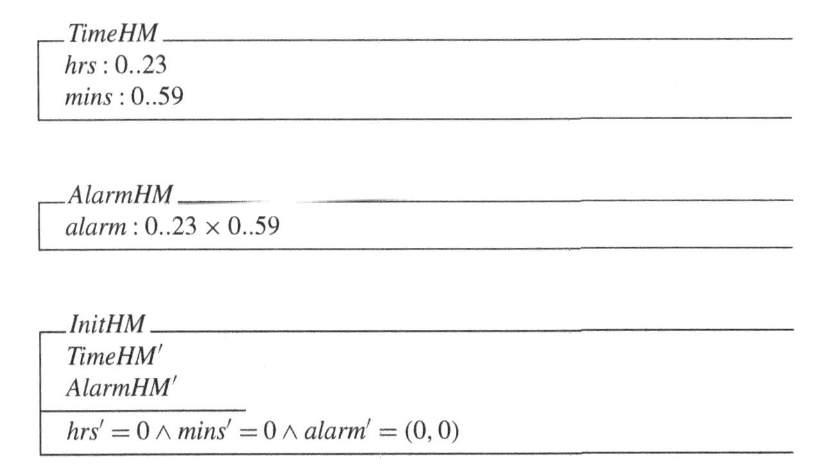

The full state is given by *TimeHM* $\wedge$ *AlarmHM*, but it will turn out to be convenient to separate the actual timekeeping and the alarm in further developments. The Z schema calculus allows us to do so, and a realistic larger-scale Z specification

J. Derrick, E.A. Boiten, *Refinement in Z and Object-Z*,
DOI 10.1007/978-1-4471-5355-9_13, © Springer-Verlag London 2014

will typically have the specification of its state split across a number of schemas like above. An additional advantage in this case are that there are no invariants relating the two subareas of the state, and for that reason we can often refine operations that affect only one of the subareas without explicitly considering the other subarea.[1]

A collection of operations describe the functionality of the watch. One aspect of the required functionality is that at any point we should be able to reset the current time and the alarm.

$$\begin{array}{|l}
\hline
\textit{ResetTime} \\
\Delta \textit{TimeHM} \\
hrs? : 0..23 \\
mins? : 0..59 \\
\hline
hrs' = hrs? \\
mins' = mins? \\
\hline
\end{array}$$

$$\begin{array}{|l}
\hline
\textit{ResetAlarm} \\
\Delta \textit{AlarmHM} \\
hrs? : 0..23 \\
mins? : 0..59 \\
\hline
alarm' = (hrs?, mins?) \\
\hline
\end{array}$$

The full specifications of these operations are $\textit{ResetTime} \land \Xi \textit{AlarmHM}$ and $\textit{ResetAlarm} \land \Xi \textit{TimeHM}$.

We also clearly need to be able to display both the current time, and also determine when the alarm is likely to go off.

$$\begin{array}{|l}
\hline
\textit{ShowTime} \\
\Xi \textit{TimeHM} \\
hrs! : 0..23 \\
mins! : 0..59 \\
\hline
hrs = hrs! \\
mins = mins! \\
\hline
\end{array}$$

$$\begin{array}{|l}
\hline
\textit{ShowAlarm} \\
\Xi \textit{AlarmHM} \\
alarm! : 0..23 \times 0..59 \\
\hline
alarm! = alarm \\
\hline
\end{array}$$

Note that *ShowTime* only displays the time when the operation is evoked. This is the functionality of displays of the early LED watches, which, to save power, were blank unless the appropriate button was pressed. With the demise of LED displays we really require that the display be permanent, and we discuss the correct way to represent this in Sect. 13.2 below. However, for the moment we stick with *ShowTime* as described above.

[1]The construction here appears to be some kind of *product* of ADTs, which has monotonicity properties with respect to refinement.

We also need an operation to make time pass which we describe as *AddMin* (which, again, may be conjoined with $\Xi AlarmHM$).

$$
\begin{array}{|l}
\textit{AddMin} \\
\hline
\Delta TimeHM \\
\hline
mins' = (mins + 1) \bmod 60 \\
\mathbf{let}\, \alpha == (mins + 1) \operatorname{div} 60 \bullet \\
\quad hrs' = (hrs + \alpha) \bmod 24 \\
\hline
\end{array}
$$

We could consider *AddMin* to be internal (certainly the user should not have to press a button to make time pass!), but in fact it does not really matter to our specification either way. However, if we made it an internal operation we would need to consider issues of divergence and fairness with respect to the other operations. In particular, if it were internal under the current semantic interpretation it could diverge (since it is always applicable), whereas what we really need to say is that it is internal but all other operations can be evoked between its occurrences. To describe this properly we would need to use a time-based notation such as timed Object-Z [4], a timed process algebra [3] or timed automata [2].

Other operations to display and reset the date might also be described. There would also be an operation to describe when the alarm should sound, probably specified as (where *SOUND* is given an appropriate definition)

$$
\begin{array}{|l}
\textit{Alarm} \\
\hline
\Xi TimeHM \\
\Xi AlarmHM \\
alarm! : SOUND \\
\hline
alarm = (hrs, mins) \\
alarm! = ring \\
\hline
\end{array}
$$

Note that this should be considered as an internal (active) operation as discussed in Sect. 11.3.2, that is, an internal operation which has an output. However, we do not pursue this aspect further in this chapter.

13.2 Grey Box Specification and Refinement

As we pointed out above, the specification of the *ShowTime* operation was not entirely satisfactory. Unless we were really describing an LED display we would like to consider the display of the time to be continuous. Also, the *ResetTime* operation has a commonly occurring form, stating only that the components *hrs* and *mins* may be modified by replacing them with inputs with the same base names.

Grey box specification (cf. Chap. 9) provides a more appropriate way of modelling such operations by allowing state components to be observable or modifiable.

Thus a grey box specification of the system above will be:

$$(TimeHM \wedge AlarmHM, AlarmHM, TimeHM, InitHM, \{ResetAlarm, AddMin, Alarm\})$$

By including state components in the second and third components of the grey box data type, operations to observe and modify are made implicit. The second component of the grey box denotes the schema of all observable, but not directly modifiable, components. Due to the type of *alarm* not matching that of the inputs of *ResetAlarm*, the latter operation still needs to be explicitly included and *AlarmHM* is considered observable but not modifiable. The third component of the grey box represents the modifiability and observability of *TimeHM*, replacing operations *ResetTime* and *ShowTime*.

However, as explained in Chap. 9, grey box specifications allow only a limited range of refinements, due to the observable state components' having to remain part of the state in further refinement. In other words, they are good for implementation specifications, but not for abstract specifications. The same disadvantage does not hold for *display box* specifications, also defined in Chap. 9. Informally, they are ADTs some of whose "operations" (*displays*) do not need to mention the after-state because it is implicitly assumed to be equal to the before-state. The two observation operations could be defined as displays, for example *ShowTime* would be

$$\begin{array}{|l}
\multicolumn{1}{l}{DispTime_A} \\
\hline
TimeHM \wedge AlarmHM \\
hrs! : 0 \,..\, 23 \\
mins! : 0 \,..\, 59 \\
\hline
hrs = hrs! \\
mins = mins! \\
\hline
\end{array}$$

We could now replace our "circular" timekeeping by an eternal clock,[2] i.e.,

$$\begin{array}{|l}
\multicolumn{1}{l}{Eternal} \\
\hline
mins_0 : \mathbb{N} \\
\hline
\end{array}$$

using the obvious retrieve relation

$$\begin{array}{|l}
\multicolumn{1}{l}{GroundHog} \\
\hline
Eternal \\
TimeHM \\
\hline
hrs = (mins_0 \text{ div } 60) \text{ mod } 24 \\
mins = mins_0 \text{ mod } 60 \\
\hline
\end{array}$$

[2]Specifications with unbounded numbers in them are not directly implementable on real computers which have bounded storage. Different approaches to addressing this are given in Sect. 14.4 and the bibliographical notes of Chap. 14.

The weakest refinement of $DispTime_A$ using this retrieve relation can be calculated using the simple rule, cf. Chap. 5: the retrieve relation is functional from *Eternal* to *TimeHM*. This will result in:

$$
\begin{array}{|l}
\hline
\multicolumn{1}{l}{\textit{DispTime}_C} \\
\textit{Eternal} \land \textit{AlarmHM} \\
\textit{hrs}! : 0 \mathinner{\ldotp\ldotp} 23 \\
\textit{mins}! : 0 \mathinner{\ldotp\ldotp} 59 \\
\hline
\textit{hrs}! = (\textit{mins}_0 \operatorname{div} 60) \bmod 24 \\
\textit{mins}! = \textit{mins}_0 \bmod 60 \\
\hline
\end{array}
$$

As this operation is total, it is a display. Moreover, it is deterministic, so any display box refinement using *GroundHog* as the retrieve relation will refine display $DispTime_A$ to $DispTime_C$.

Aside Unlike for $DispTime_A$, the weakest refinement of *AddTime* using *GroundHog* is *not* the "obvious" operation: instead of requiring $min_0' = min_0 + 1$, it allows addition of any n such that $n \bmod 1440 = 1$.

13.3 An Analogue Watch: Using IO Refinement

The watch we have defined so far has a digital display. However, this manufacturer also makes analogue watches and we should be able to implement the above specification as an analogue watch. To do so let

$$ANGLE == 0..359$$

which represents the angle between a hand and the positive y-axis. We define a function to convert between the two displays

$$
\begin{array}{|l}
Angle : 0..23 \times 0..59 \rightarrow ANGLE \times ANGLE \\
\hline
Angle(x, y).1 = 30 * (x \bmod 12) + y \operatorname{div} 2 \\
Angle(x, y).2 = 6 * y
\end{array}
$$

To model the analogue watch, we now replace the original *ShowTime* by

$$
\begin{array}{|l}
\hline
\multicolumn{1}{l}{\textit{ShowTimeHands}} \\
\Xi \textit{TimeHM} \\
\textit{hrs}! : \textit{ANGLE} \\
\textit{mins}! : \textit{ANGLE} \\
\hline
(\textit{hrs}!, \textit{mins}!) = \textit{Angle}(\textit{hrs}, \textit{mins}) \\
\hline
\end{array}
$$

As this refinement changes the type of outputs, it can only be valid as an IO refinement. We can prove this using Theorem 10.4 (constructing IO refinement):

$ShowTimeHands = ShowTime \gg OT$, where OT essentially applies the function *Angle* to its inputs, and outputs the result. OT is total (defined at every "time" instant) and injective (every hour and minute has a unique representation in terms of the hands). Observe that the representation using just the big hand is insufficient: the function div 2 is not injective.

Similarly, injectivity of *Angle* is also sufficient to prove that an analogous equivalent of *ResetTime* can be given:

ResetTimeHands

$$\Delta TimeHM$$
$$hrs? : ANGLE$$
$$mins? : ANGLE$$

$$(hrs?, mins?) \in \operatorname{ran} Angle$$
$$(hrs?, mins?) = Angle(hrs', mins')$$

The restriction on $(hrs?, mins?)$ reflects the fact that some combinations of angles are impossible. However, this also includes angles that are possible on real analogue watches, but which would not reflect an exact number of minutes on the digital watch, e.g., $(0, 3)$ might be said to represent half a minute past noon. We might want to assume the behavioural interpretation of this operation in order to prevent $(0, 3)$ getting linked to an arbitrary time in refinement.

13.4 Adding Seconds: Weak Refinement

In the face of overseas competition it has become necessary to increase the precision of the timekeeping of the watches, and the decision has been taken to implement a component that will count in seconds. To do so we use the new state space

TimeHMS

$$TimeHM$$
$$secs : 0..59$$

InitHMS

$$TimeHMS'$$
$$InitHM$$

$$secs' = 0$$

With this state space we will only change the *AddMin* operation, and add an internal *Tick* operation to increment the seconds. All other operations remain the same. The two operations of interest are described by

$$
\begin{array}{|l}
\hline
\textit{Tick} \\
\Delta \textit{TimeHMS} \\
\Xi \textit{TimeHM} \\
\hline
secs' = secs + 1 \\
secs \neq 59 \\
\hline
\end{array}
$$

$$
\begin{array}{|l}
\hline
\textit{AddMinS} \\
\Delta \textit{TimeHMS} \\
\hline
secs = 59 \\
secs' = 0 \\
mins' = (mins + 1) \bmod 60 \\
\mathbf{let}\, \alpha == (mins + 1) \operatorname{div} 60 \bullet \\
\qquad hrs' = hrs + \alpha \\
\hline
\end{array}
$$

This new specification is a weak refinement of the original *TimeHM* specification. To see this, first note that an appropriate retrieve relation to use would be one which is the identity on *TimeHM*. With this in place the necessary conditions are easily verified.

For example, note that *Tick* can clearly only be applied a finite (e.g., less than 60) number of times. Thus divergence is not a problem. In addition, the *AddMin* operation is effectively refined by *AddMinS* and a number of internal *Tick* operations. Whenever the precondition of *AddMin* holds then either the precondition of *AddMinS* holds or *Tick*s will lead to a state where the precondition holds.

We can also add the seconds to the *ShowTime* operation. To do this, we first use Simple Output Refinement (Theorem 10.2) to add an unconstrained output $secs!$, and then operation refinement to add $secs! = secs$. However, we need to be aware that this refinement, like any IO refinement, changes the "boundaries" of the system. The full IO-mutable ADT concerned would have an original output transformer which would remove the information about seconds.

13.5 Resetting the Time: Using Non-atomic Refinement

The original descriptions of the two operations *ResetTime* and *ResetAlarm* were hopelessly unrealistic. Digital watches do not have an interface that allows you to enter a new arbitrary time to which it is instantly and atomically set. Rather, there is a painfully tedious process of incrementing the hours and minutes of the watch by pressing a selection of buttons repeatedly. However, specifying that process at the initial level of abstraction would be wrong; different watches will want to use different mechanisms (e.g., resetting an analogue watch is completely different), and such a description does not capture the essential process in a clear way. Although the granularity of this type of description is very different from that given above, we can use non-atomic refinement to verify correctness of such a specification against the original abstract description.

Our original description is going to be refined to a watch with a simple display together with two buttons *ButA* and *ButB*. In order to perform some of the functions of the watch, the instructions describe how the two buttons need to be pressed. The original interface

$$(TimeHM \wedge AlarmHM, InitHM, \{ShowTime, ShowAlarm, ResetTime, ResetAlarm, AddMin, Alarm\})$$

will thus be refined to

$$(TimeDG \wedge AlarmHM, InitDG, \{Display, ButA, ButB, AddMin, Alarm\})$$

The watch we implement has a number of *MODE*s, the first two concern whether the time or the alarm setting should be displayed, and the others determine which settings can be currently changed.

$$MODE ::= display_t \mid display_a \mid change_h \mid change_m \mid change_a$$

The state space *AlarmHM* is unchanged, and we add a single *mode* to *TimeHM*:

$$\begin{array}{|l}
\hline
\textit{TimeDG} \\
TimeHM \\
mode : MODE \\
\hline
\end{array}$$

$$\begin{array}{|l}
\hline
\textit{InitDG} \\
TimeDG' \\
InitHM \\
\hline
\end{array}$$

The *AddMin* operation is unchanged apart from requiring that it does not change the *mode* of the watch. The watch in question has a single display which can either show the time or the alarm (but not both) at any particular moment. So we will refine *ShowTime* and *ShowAlarm* to a single display operation.

$$\begin{array}{|l}
\hline
\textit{Display} \\
\Xi TimeDG \\
\Xi AlarmHM \\
display! : 0..23 \times 0..59 \\
\hline
(mode \in \{display_t, change_h, change_m\} \wedge display! = (hrs, mins)) \\
\vee \\
(mode \in \{display_a, change_a\} \wedge display! = alarm) \\
\hline
\end{array}$$

The final two operations are the buttons *ButA* and *ButB*. The purpose of button *A* is just to switch between the modes, thus allowing different functionality to be achieved with the second button.

$$
\begin{array}{|l}
\hline
\textit{ButA} \\
\Delta \textit{TimeDG} \\
\Xi \textit{TimeHM} \\
\Xi \textit{AlarmHM} \\
\hline
mode = display_t \Rightarrow mode' = display_a \\
mode = display_a \Rightarrow mode' = change_h \\
mode = change_h \Rightarrow mode' = change_m \\
mode = change_m \Rightarrow mode' = change_a \\
mode = change_a \Rightarrow mode' = display_t \\
\hline
\end{array}
$$

ButB is slightly more complex, and we describe it bit by bit. If *B* is pressed when in either of the display modes, nothing happens. However, if the watch is in $change_h$ mode, then the act of pressing *B* will increment the hours by one, and similarly for $change_m$ mode. Finally to reset the alarm we have to move through the minutes and hours by pressing *B* in $change_a$ mode.

$$
\begin{array}{|l}
\hline
\textit{ButBDisp} \\
\Xi \textit{TimeDG} \\
\Xi \textit{AlarmHM} \\
\hline
mode \in \{display_t, display_a\} \\
\hline
\end{array}
$$

$$
\begin{array}{|l}
\hline
\textit{ButBHrs} \\
\Delta \textit{TimeDG} \\
\Xi \textit{AlarmHM} \\
\hline
mode = change_h \\
mode' = mode \\
mins' = mins \\
hrs' = (hrs + 1) \bmod 24 \\
\hline
\end{array}
$$

$$
\begin{array}{|l}
\hline
\textit{ButBMins} \\
\Delta \textit{TimeDG} \\
\Xi \textit{AlarmHM} \\
\hline
mode = change_m \\
mode' = mode \\
hrs' = hrs \\
mins' = (mins + 1) \bmod 60 \\
\hline
\end{array}
$$

$$
\begin{array}{|l}
\hline
\textit{ButBAlm} \\
\Xi \textit{TimeDG} \\
\Delta \textit{AlarmHM} \\
\hline
\textit{mode} = \textit{change}_a \\
\textit{mode}' = \textit{mode} \\
\textbf{let}\ \beta == (\textit{mins} + 1) \bmod 60;\ \gamma == (\textit{mins} + 1) \operatorname{div} 60; \\
\quad \alpha == (\textit{hrs} + \gamma) \bmod 24 \bullet \\
\quad\quad \textit{alarm}' = (\alpha, \beta) \\
\hline
\end{array}
$$

$$\textit{ButB} == \textit{ButBDisp} \vee \textit{ButBHrs} \vee \textit{ButBMins} \vee \textit{ButBAlm}$$

This completes our description.

As claimed, *TimeDG* is a non-atomic refinement of *TimeHM*. The retrieve relation we use to verify this is the identity on $\textit{TimeHM} \wedge \textit{AlarmHM}$ and the *mode* set to $\textit{display}_t$ (there are other possibilities, for example mapping to $\textit{display}_a$ is equally valid).

We then need to describe the relationship between the abstract operations and appropriate sequences of concrete operations, and to do so we abbreviate n schema compositions of an operation *Op* to Op^n.

We claim that *ShowTime* is refined by *Display*, and *ShowAlarm* by $\textit{ButA} \fatsemi \textit{Display} \fatsemi \textit{ButA}^4$, and a little thought will clearly show this to be the case. The IO transformations needed for these two refinements are relatively simple, but still not the identity. *ButA* does not have any input/output, so the output of *ShowTime* is represented completely in *Display*. Here, in fact, we need a simple IO refinement because the two outputs *hrs*! and *mins*! have become a single *display*!, and, because of this, standard refinement would not be sufficient here. Thus we need to use an output transformer OT (which satisfies all the right conditions):

$$
\begin{array}{|l}
\hline
\textit{OT} \\
\textit{hrs}? : 0..23 \\
\textit{mins}? : 0..59 \\
\textit{display}! : 0..23 \times 0..59 \\
\hline
\textit{display}! = (\textit{hrs}?, \textit{mins}?) \\
\hline
\end{array}
$$

The non-atomic refinements of *ResetAlarm* and *ResetTime* are more complicated, and, in fact, under the refinement all the inputs in these two operations disappear completely. What happens is that the inputs representing the new time settings are implemented as a series of pushes to button B, and the number of pushes equals the increments needed to reach the new time from the current time. Thus *ResetTime* gets refined by

$$\textit{ButA}^2 \fatsemi \textit{ButB}^\alpha \fatsemi \textit{ButA} \fatsemi \textit{ButB}^\beta \fatsemi \textit{ButA}^2$$

where α and β are the difference between the current and desired hours and minutes respectively. That is,

$$\alpha = (hrs? - hrs) \bmod 24$$
$$\beta = (mins? - mins) \bmod 60$$

The length of the concrete decomposition needed thus varies according to the current time and to the new time to which we wish to reset our watch. Indeed experimenting with JD's watch we find that this mirrors the actual behaviour exactly.

Observe that this seems to involve a kind of IO transformer that we have not previously considered: one which depends not only on the input value, but also on the current state (see comments in Sect. 12.7.5). We had good reasons not to consider such input transformers, because they break abstraction, by depending on the actual representation of the state. However, it is not really the state which is being used here. The only way in which the user of the watch knows how often to press button B is by looking at the display, i.e., by observing previous *output*. The kind of input transformer involved, which uses previous outputs to determine new inputs, is abstract alright, but would take us firmly into the area of *interactive systems*, which we do not pursue further in this book.

References

1. Araki, K., Galloway, A., & Taguchi, K. (Eds.) (1999). *International Conference on Integrated Formal Methods 1999 (IFM'99)*. York: Springer.
2. Lynch, N. A., & Vaandrager, F. (1992). Forward and backward simulations for timing-based systems. In J. W. de Bakker, W.-P. de Roever, C. Huizing, & G. Rozenberg (Eds.), *Real-Time: Theory in Practice, REX Workshop*, Mook, The Netherlands, June 1991. *Lecture Notes in Computer Science: Vol. 600* (pp. 397–446). Berlin: Springer.
3. Schneider, S. (2000). *Concurrent and Real-Time Systems: the CSP Approach*. New York: Wiley.
4. Smith, G., & Hayes, I. J. (1999) Towards real-time Object-Z. In Araki et al. [1] (pp. 49–65).

Chapter 14
Further Generalisations

This book considers systems whose specifications are centred around abstract data types of the form $(State, Init, \{Op_i\}_{i \in I})$. Formally developing such systems amounts to changing (some of) the components of these data types. In this book we have slowly expanded the extent of changes allowed in refinement. In Chap. 2 we considered operation refinement, where only the operations are allowed to change. Data refinement as described in Chaps. 3 and 4 also allows the initialisation and the state to change. In IO refinement, Chap. 10, we also allowed the inputs and outputs of operations to change. The last two chapters have dealt with refinement relations where the index set I is allowed to change in refinement, to model operations not under the control of the environment (Chap. 11), or to model a change of focus and granularity (Chap. 12).

In this chapter, we consider two more ways in which the operation index set may evolve during formal development: *alphabet extension* and *alphabet translation*, together with a generalisation which looks at *approximate refinement* between abstract and concrete systems.

14.1 Alphabet Extension

Ideas from process algebra served in Chap. 11 to motivate a generalisation of Definition 3.5 that allows internal actions. From the fact that most refinement relations in process algebra do not allow addition of new *observable* actions ("extension of the alphabet") we might then conclude that refinement of ADTs should not lead to extended sets of visible operations either.

In the behavioural approach this appears to be a defensible position. When our specification is some usual drinks machine using actions *coin*, *coffee*, and *tea*, we

Excerpts from pp. 159–160 of [8] are reprinted within Sect. 14.1 with kind permission of Springer Science and Business Media.
Excerpts from pp. 377–381 of [9] are reprinted within Sect. 14.4 with kind permission of Springer Science and Business Media.

J. Derrick, E.A. Boiten, *Refinement in Z and Object-Z*,
DOI 10.1007/978-1-4471-5355-9_14, © Springer-Verlag London 2014

would be distinctly unhappy if the implementing machine came with a button that allowed the removal of all the money from the machine. Such an operation, which clearly disrupts the system state, should not be possible.

However, in the "contract" interpretation it is less clear that allowing extra operations should be unsound. The canonical view of an ADT in this interpretation is of a software component whose operations are only required to function as specified when their preconditions hold. There being additional operations available in the component besides the ones the customer intends to use does not appear to be a problem. Indeed, it would require some disingenuity to distinguish between an operation that "does not exist" and one whose precondition is universally *false*. The consequence of identifying non-existent operations with disabled ones is obvious: in the behavioural approach, no extra operations may be added; in the contract approach, they may be added at will.

The corresponding generalisation of Definition 3.5 is given below. Implicitly it generalises the notion of conformity to the concrete index set *containing* the abstract index set, and as before it quantifies over all *abstract* programs only, not making any restrictions on those concrete programs which contain indices *not* in the abstract index set. This is in contrast to weak refinement, where *all concrete* behaviours are considered.

Definition 14.1 (Data refinement with alphabet extension) Let $\mathsf{A} = (\mathsf{AState}, \mathsf{AInit}, \{\mathsf{AOp_i}\}_{\mathsf{i} \in \mathsf{I}}, \mathsf{AFin})$ and $\mathsf{C} = (\mathsf{CState}, \mathsf{CInit}, \{\mathsf{COp_i}\}_{\mathsf{i} \in \mathsf{I} \cup \mathsf{J}}, \mathsf{CFin})$ be relational data types. The data type C *refines* A *with alphabet extension* iff for each finite sequence p over I

$$\mathsf{p_C} \subseteq \mathsf{p_A}$$ □

Observe that this is a correct generalisation: when J is empty, this reduces to Definition 3.5. The informal justification of this definition is that it ensures that the concrete system behaves like the abstract one when executing any sequence of instructions that the abstract system allowed.

The translation of this generalised relational refinement rule to Z ADTs is obvious: when we assume the contractual approach, we can freely add operations to the Z ADT as (part of) a refinement step.

Example 14.1 The specification of a simple stack-based calculator in Example 10.13 (p. 260) did not state what happens when the calculator is dropped on the floor. As far as the implementor is concerned, this is mostly because the specifier would not even wish to consider this happening. For that reason, the implementor feels that the following behaviour should be fully acceptable when the calculator is dropped:

```
┌─ Drop ──────────────────────────────
│ ΔCalc
│ top! : ℕ
├──────────────
│ stack' = ⟨top!⟩
│ top! = 88888888
│ ¬edittop'
└─────────────────────────────────────
```

Thus, the calculator which displays all the behaviour listed in its specification, and which displays 88888888 when dropped, should be considered an acceptable implementation of the specification given in Example 10.13 under the alphabet extension view. □

14.2 Alphabet Translation

In Chap. 12 we considered the situation where one abstract operation was represented by a number of concrete operations, whose combined effects related to the abstract behaviour. Although the combination uses schema composition rather than schema conjunction (which is not always well-behaved with respect to refinement, due to the *rôle* of preconditions), it can still be viewed as a kind of conjunctive combination.

One might also consider the situation where an abstract operation abstracts from multiple concrete operations in a *disjunctive* way.

Example 14.2 Consider again our two-dimensional world, with an operation to move one step in any direction along any axis.

```
┌─ 2D ─────────────────
│ x, y : ℤ
└──────────────────────

┌─ Init ───────────────
│ 2D'
├──────────
│ x' = 0 ∧ y' = 0
└──────────────────────

┌─ Step ───────────────
│ Δ2D'
├──────────
│ | x − x' | + | y − y' | = 1
└──────────────────────
```

A more concrete view of this might be to replace *Step* by four operations (say *North*, *South*, *East*, *West*) which represent the four directions we could move in.

With the standard refinement rules, we have two ways of representing this. We could observe that $Step = North \lor South \lor East \lor West$ and leave it at that. This has the disadvantage that we end up with a single concrete operation rather than four. The alternative is to include four copies of *Step* in the abstract specification (with distinct indices), and refine these to the four different concrete operations. This may be unsatisfactory because it implies that the abstract specification should contain information about the number of concrete views of a single operation that exist. □

A more faithful representation of the sort of situation described above can be obtained using what we call *alphabet translation*. The definition given below actually subsumes the possibility of alphabet extension described in the previous section.

Definition 14.2 (Data refinement with alphabet translation) Let $\mathsf{A} = (\mathsf{AState}, \mathsf{AInit}, \{\mathsf{AOp_i}\}_{\mathsf{i} \in \mathsf{I}}, \mathsf{AFin})$ and $\mathsf{C} = (\mathsf{CState}, \mathsf{CInit}, \{\mathsf{COp_i}\}_{\mathsf{i} \in \mathsf{J}}, \mathsf{CFin})$ be relational data types. Let α be a total and injective mapping from I to J. Then C *refines* A *with alphabet translation* α iff

$$\forall \mathsf{p} : \mathsf{seq}\, \mathsf{I} \bullet \exists \mathsf{q} : \mathsf{seq}\, \mathsf{J} \bullet (\mathsf{p}, \mathsf{q}) \in \alpha* \wedge \mathsf{q_C} \subseteq \mathsf{p_A}$$ □

Observe that this is a correct generalisation of Definition 14.1: the identity relation on I is a total and injective mapping from I to $\mathsf{I} \cup \mathsf{J}$, and its extension to sequences is the identity over $\mathsf{seq}\, \mathsf{I}$.

The Z analogue of this is obvious: it includes the normal conditions between any pair of operations linked by α (and none for concrete operations whose index is not in $\operatorname{ran} \alpha$). This requirement is *stronger* than requiring that AOp_i is refined by the disjunction of the concrete operations it represents $\exists j \bullet (i, j) \in \alpha \wedge COp_j$, as the precondition of each linked COp_j should match that of AOp_i.

Example 14.3 We can verify the replacement of *Step* by *North*, *South*, etc. in Example 14.2 as a data refinement with alphabet translation: let the abstract data type be $(2D, Init, \{AOp_1 = Step\})$ and the concrete one $(2D, Init, \{COp_1 = North, COp_2 = East, COp_3 = South, COp_4 = West\})$. The required alphabet translation is then $\{(1, 1), (1, 2), (1, 3), (1, 4)\}$. □

14.3 Compatibility of Generalisations

The generalisations proposed above and in previous chapters are all compatible with each other. First, IO refinement can be combined with the other generalisations, as it (like traditional refinement) derives from Definition 3.5, which is generalised by the other definitions. Weak refinement and alphabet translation are orthogonal, as in Definition 14.2 it is possible to generalise by taking the closures of p and q under composition with internal behaviour, with the appropriate extra quantification. The same sort of construction could be applied to Definition 12.1 to combine non-atomic refinement with weak refinement. The combination of alphabet translation and non-atomic refinement is obtained by taking the decomposition mapping ρ to be an *arbitrary* total relation between abstract indices and sequences of concrete indices. Injectivity of alphabet translation may need to be weakened or reinterpreted in that situation. The combination of all three generalisations seems feasible theoretically, although mind-boggling in practice. As is already the case with operation refinement and data refinement, and as was demonstrated for IO refinement, it is probably advisable to perform refinement in "one dimension at a time".

14.4 Approximate Refinement

Refinement, as we have presented it so far, asks for the concrete system to be *exactly* like the abstract system—subject to our resolving some of the non-determinism in the abstract. However, in many (most?) practical situations development is not so idealistic, and indeed we might find we can only approximate the behaviour of the abstract system.

For example, consider the specification of a simple buffer:

```
┌─ Buffer ──────────────
│ cont : seq ℤ
└───────────────────────

┌─ EmptyBuf ────────────
│ Buffer'
├───────────
│ cont' = ⟨ ⟩
└───────────────────────

┌─ Remove ──────────────
│ ΔBuffer
│ out! : ℤ
├───────────
│ cont = ⟨out!⟩ ⁀ cont'
└───────────────────────

┌─ Insert∞ ─────────────
│ ΔBuffer
│ in? : ℤ
├───────────
│ cont' = cont ⁀ ⟨in?⟩
└───────────────────────
```

In reality, this might be implemented by a bounded buffer of a particular size, e.g., $n = 256$. In this implementation, $Insert_\infty$ is replaced by $Insert_n$, where this is defined as:

```
┌─ Insertₙ ──────────────────────────────────────────────────
│ ΔBuffer
│ in? : ℤ
├───────────
│ (#cont < n ∧ cont' = cont ⁀ ⟨in?⟩) ∨ (#cont >= n ∧ cont' = cont)
└────────────────────────────────────────────────────────────
```

This bounded buffer is only an approximate refinement (in some sense) of the abstract infinite buffer—it certainly does not meet the requirements of Definition 4.3.

To model this type of approximate refinements we can consider chains of specifications (S_n), where we will identify a chain with its limit:

Definition 14.3 A sequence of specifications $(S_n)_{n \in \mathbb{N}}$ is considered equivalent to its limit. □

Different approximations give different notions of a limit, and we can consider to what extent the limit S is close to any given element in the chain. Data refinement asks for consistency of observations for *all* programs. One natural metric we can consider is one defined in terms of program length, where we assign a distance to specifications which agree on observations up to a *certain length*. This is easiest if phrased in terms of equivalence (i.e., data refinement in both directions). Thus we define

Definition 14.4 (Program length metric) We define the metric d_l on specifications as follows:

$$d_l(A, C) = \begin{cases} 0 & \text{if } A =_{data} C \\ 2^{-n} & \text{if } n = \min\{m : \mathbb{N} \mid \exists \mathsf{p} \bullet \mathsf{p}_C \neq \mathsf{p}_A \wedge \#\mathsf{p} = m\} \end{cases}$$

where the length of a program (i.e., $\#\mathsf{p}$) is the number of operations plus one (for the initialisation). □

This defines a metric on the set of equivalence classes (with respect to $=_{data}$) of specifications. The key idea here is that two specifications are considered close if it takes a long time to tell them apart, where a 'long time' is the length of the shortest program which can observe the difference.

Limits with respect to this metric are characterised as follows. The sequence S_n converges to S whenever, $d_l(S_n, S) \to 0$, i.e., $2^{-n} \to 0$ where n is the minimum length of program needed to distinguish S_n from S.

As an example, to calculate the metric for the buffer example we note the following. The shortest program that can observe that Buf_n, a buffer of size n, does not have infinite capacity has size $2n + 2$: first $n + 1$ elements are inserted (the last of which is the first one to be ignored), then n *Remove* operations are successful, and the next *Remove* operation fails.[1]

Thus $d_l(Buf_n, Buf) = 2^{-(2n+3)}$ and so $Buf_n \to Buf$.

So the metric has formalised our intuition that Buf_n gets closer to its idealised behaviour as n gets larger. Furthermore, the metric quantifies this closeness numerically.

Notice that the definition of the metric is, as one would hope, not sensitive to small changes. For example, if we consider finite and infinite stacks (i.e., inserting and removing from the same end) we can observe the difference more quickly than in the buffer example (since we do not have to remove all the elements first). However, the distance is $2^{-(n+1)}$, and thus we still get convergence to the infinite stack as one would expect.

Although its definition seems to assume observations being characterised by outputs, it, in fact, also works for specifications that use different finalisations. This is because there will always be a minimum length program where any difference can be observed—whether that be due to an output of an operation, or a finalisation after the last operation.

Approximate refinement gives a way of making the transition from mathematics-based specifications to ones using bounded resources, such as limited storage in the example above, or using numbers of bounded size or precision instead of mathematical numbers. It can also be used in contexts where the natural notion of correctness is an approximate one, e.g. in cryptographic constructs, using probabilistic algorithms.

---

[1]The standard semantics of applying *Remove* outside its precondition allows for any result including the "correct" one and a completely undefined one; however, we are looking for *equality* of semantics of programs rather than the *inclusion*.

14.5 Bibliographical Notes

The ideas in the first two sections of this chapter were first presented in *Liberating Data Refinement* [8]. The consequences of allowing extra operations in classes for behavioural subtyping are also studied by Liskov and Wing [16]. Their "constraint rule" is closely related to our alphabet extension rule. This issue is discussed in an Object-Z like context by Helke and Santen [15]. Event-B [1] allows alphabet translation under the name of "splitting events". The paper [7] explores this and other ways of extending ADT alphabets.

Section 14.2 above briefly touched on conjunctive composition of operations, and the relation between refinement and the Z schema calculus. A thorough study of this topic was made by Groves [13, 14].

The work presented in Sect. 14.4 first appeared in [9]. The program length metric described here was introduced by De Bakker and Meyer [11].

The observation that idealised specifications correspond to "realistic" specifications with resource bounds tending to infinity has been discussed in several contexts. For example, in [17], Neilson defined ∞-refinement $\sqsubseteq_\infty$ in terms of ordinary refinement $\sqsubseteq$ as follows:

$$A \sqsubseteq_\infty B \Leftrightarrow \exists c_1, c_2, \ldots, c_n : \mathit{ResourceLimit} \bullet \lim_{c_1, c_2, \ldots, c_n \to \infty, \infty, \ldots, \infty} (A \sqsubseteq B)$$

where the resource constraints c_i appear free in B and not at all in A. Such a refinement step establishes resource limits as constants in the specification; Neilson implicitly indicates that subsequent refinement may fix the values of these constants.

Banach and Poppleton have defined and investigated a generalisation of refinement called retrenchment [3]. They add "within" and "concedes" relations to every refinement step, indicating where preconditions are strengthened and postconditions weakened. These allow for developments which are not quite correctness preserving to be documented. However, this documentation refers to the internals of a specification at any given point in the development trace, and is thus hard to relate to external behaviour. Clearly, by taking a strong enough "within" relation, retrenchment holds between any pair of specifications—its value is in the documentation of where and how refinement has been relaxed. For applications of retrenchment to a larger case study, the Mondex purse (see Sect. 10.7), see [2, 4–6].

Similar ideas are explored by Smith [18] for real-time specification—importantly, this work concentrates on the properties that are preserved by development steps which are not quite refinement steps but so-called "realisations".

References

1. Abrial, J.-R. (2010). *Modelling in Event-B*. Cambridge: Cambridge University Press.
2. Banach, R., Jeske, C., Poppleton, M., & Stepney, S. (2007). Retrenching the purse: the balance enquiry quandary, and generalised and (1, 1) forward refinements. *Fundamenta Informaticae*, *77*(1–2), 29–69.

3. Banach, R., & Poppleton, M. (2000) Retrenchment, refinement and simulation. In Bowen et al. [10] (pp. 304–323).
4. Banach, R., Poppleton, M., Jeske, C., & Stepney, S. (2005) Retrenching the purse: Finite sequence numbers, and the tower pattern. In Fitzgerald et al. [12] (pp. 382–398).
5. Banach, R., Poppleton, M., Jeske, C., & Stepney, S. (2006). Retrenching the purse: hashing injective CLEAR codes, and security properties. In *Leveraging Applications of Formal Methods, Second International Symposium, ISoLA*, Cyprus (pp. 82–90). New York: IEEE Press.
6. Banach, R., Poppleton, M., & Stepney, S. (2006). Retrenching the purse: finite exception logs, and validating the small. In *30th Annual IEEE/NASA Software Engineering Workshop (SEW-30)* (pp. 234–248). Los Alamitos: IEEE Comput. Soc.
7. Boiten, E. A. (2012). Introducing extra operations in refinement. *Formal Aspects of Computing*. doi:10.1007/s00165-012-0266-z.
8. Boiten, E. A., & Derrick, J. (2000). Liberating data refinement. In R. C. Backhouse & J. N. Oliveira (Eds.), *Mathematics of Program Construction. Lecture Notes in Computer Science: Vol. 1837* (pp. 144–166). Berlin: Springer.
9. Boiten, E. A., & Derrick, J. (2005) Formal program development with approximations. In Treharne et al. [19] (pp. 374–392).
10. Bowen, J. P., Dunne, S., Galloway, A., & King, S. (Eds.) (2000). *ZB2000: Formal Specification and Development in Z and B. Lecture Notes in Computer Science: Vol. 1878*. Berlin: Springer.
11. de Bakker, J. W., & Meyer, J.-J. C. (1992). Metric semantics for concurrency. In J. W. de Bakker & J. J. M. Rutten (Eds.), *Ten Years of Concurrency Semantics: Selected Papers of the Amsterdam Concurrency Group* (pp. 104–130). Singapore: World Scientific.
12. Fitzgerald, J. A., Hayes, I. J., & Tarlecki, A. (Eds.) (2005). *FM 2005: Formal Methods, International Symposium of Formal Methods Europe. Lecture Notes in Computer Science: Vol. 3582*. Berlin: Springer.
13. Groves, L. (2002). Refinement and the Z schema calculus. *Electronic Notes in Theoretical Computer Science*, *70*(3), 70–93.
14. Groves, L. (2005) Practical data refinement for the Z schema calculus. In Treharne et al. [19] (pp. 393–413).
15. Helke, S., & Santen, T. (2001). Mechanized analysis of behavioral conformance in the Eiffel base libraries. In J. N. Oliveira & P. Zave (Eds.), *FME'01—International Symposium of Formal Methods Europe. Lecture Notes in Computer Science: Vol. 2021* (pp. 20–42). Berlin: Springer.
16. Liskov, B., & Wing, J. M. (1994). A behavioural notion of subtyping. *ACM Transactions on Programming Languages and Systems*, *16*(6), 1811–1841.
17. Neilson, D. S. (1990). *From Z to C: illustration of a rigorous development method*. PhD thesis, Oxford University Computing Laboratory.
18. Smith, G. (1999). From ideal to realisable real-time specifications. In N. Leslie (Ed.), *Fifth New Zealand Formal Program Development Colloquium*. Number 99-1 in IIMS Technical Report. Institute of Information and Mathematical Sciences, Massey University at Albany.
19. Treharne, H., King, S., Henson, M. C., & Schneider, S. A. (Eds.) (2005). *ZB 2005: Formal Specification and Development in Z and B, 4th International Conference of B and Z Users. Lecture Notes in Computer Science: Vol. 3455*. Berlin: Springer.

Part III
Object-Oriented Refinement

In this part, we will look at how the theory of refinement can be used in the context of object orientation. To do so we introduce the Object-Z specification language as a canonical example, and show how refinement between classes and collections of classes can be verified.

Chapter 15 provides an introduction to Object-Z. In doing so it concentrates on the additional features in Object-Z and the differences between the languages that impact on the theory of refinement.

In Chap. 16 we adapt the refinement rules derived for Z to Object-Z specifications consisting of a single class, taking into account the consequences of the different interpretation of preconditions. We also consider how weak refinement and non-atomic refinement may be applied to Object-Z. Finally, we discuss the relation between refinement, subtyping and inheritance in Object-Z.

The final chapter of this part, Chap. 17, considers more complex Object-Z specifications, in particular ones that involve objects and promotions of their operations. Specifications consisting of multiple classes enable a wide range of more complex refinements, by allowing (for example) classes to be split or merged. This chapter investigates how such refinements can be verified, and concludes with an investigation of compositionality of refinement in Object-Z.

Chapter 15
An Introduction to Object-Z

Object-Z is an object-oriented extension of the specification language Z, which has been developed over a number of years and is one of the most mature of all the proposals to extend Z in an object-oriented fashion.

The arguments for and against object orientation are well rehearsed and we will not repeat them here, however an object-oriented extension of Z does offer a number of enhancements over Z. In particular it offers support for encapsulation, inheritance and polymorphism.

As we have seen, schemas and the schema calculus are very useful techniques for structuring the description of individual operations, however there are only weak mechanisms in Z for structuring the whole specification into a number of related parts. The 2002 ISO standard for Z introduced the idea of sections, which can explicitly import previously defined sections, and this gives some large scale structuring abilities. However, this encapsulation really only gives a very limited form of modularisation allowing a series of toolkits and libraries to be built up.

Object-oriented extensions to Z offer a more powerful mechanism by splitting the specification into a number of interacting classes and objects. In Object-Z a specification consists of a collection of classes, and within a class the state and operations are written using schemas. Each class consists of a state space and an initialisation together with a collection of operations which change the state (and thus effectively defines an abstract data type).

The classes in a specification can be related in a number of ways. For example, an Object-Z class may inherit another class. Such inheritance allows complex classes and specifications to be built from simpler components in an iterative fashion in a similar way to the use of schema inclusion at an operation level. Instantiation is also supported because a class is viewed as a template for objects of that class, enabling classes to refer to objects of other classes as state variables. For each object instantiated from a class, its state and operations conform to the state and operations of that class. Instantiation also enables a limited form of polymorphism to be supported, by allowing an object of a subclass to be substituted when an object of a superclass was expected.

J. Derrick, E.A. Boiten, *Refinement in Z and Object-Z*,
DOI 10.1007/978-1-4471-5355-9_15,

15.1 Classes

A complete Object-Z specification consists of a number of interacting class definitions. In addition to the classes, which define the behaviour of the specification, any number of global type definitions, abbreviations and declarations may be introduced as in a typical Z specification.

The classes are related by inheritance or instantiation. Inheritance allows a limited amount of reuse to be supported by using the definitions of one class in another. A class can also refer to instances of other classes as state variables, and operations can be performed on these instances changing their internal state without changing the reference to them. Although a specification consists of a number of classes, each specification is considered to have one main *system class* through which the behaviour of the specification is viewed.

A class is represented as a named box with zero or more generic parameters. The class schema may include local type or constant definitions, at most one state schema and initial state schema together with zero or more operation schemas. The operations define the behaviour of the class by specifying any input and output together with a description of how the state variables change.

A class schema has the following form.

ClassName[generic parameters]
Visibility list
inherited classes
local definitions
state schema
initial state schema
operation schemas

The Visibility List The visibility list, which is optional, restricts external access to the listed operations or attributes (i.e., constants or variables defined in the state schema). In a Z or Object-Z specification some operations may be defined solely to build up more complex operations via the schema calculus. One of the weaknesses of Z is that there is no way to say formally that these are not part of the intended interface in any implementation, and the purpose of the visibility list in Object-Z is to enable the interface of a specification to be made precise.

If no visibility list is given then no attributes are visible, but all operations (including *INIT*) are visible.[1] The symbol $\upharpoonright$ is used to denote visibility.

[1]This convention differs from the reference manual [20], which also has *attributes* visible by default.

Inherited Classes The inherited classes are names of other classes which are inherited by this class. We discuss inheritance in Sect. 15.2 below.

Local Definitions The local type and constant definitions are as in Z, and their scope extends to the end of the class in which they are defined. Constants are associated with fixed values unchanged by any of the operations in a class. However, the value of a constant may differ for different object instantiations of the class. Of course any state variable or constant can only take values in accordance with the class invariant, i.e., with the predicate defined in the state schema.

The State Schema The state schema defines the state variables of a class, and along with any local axiomatic definitions it defines the possible states of the class. The state schema is unnamed, and the variables declared are local to the class (except for those in the visibility list).

The Initial State Schema The initial state schema *INIT* defines the initial states of a class. The declaration available to *INIT* is implicitly the state schema of the class, because the state variables and constants of the class are available in the environment in which it is interpreted. The predicate in *INIT* implicitly includes the state schema's predicate (and any predicates associated with constants). The form of *INIT* differs from Z in that it consists of a predicate (as opposed to being an operation), and is therefore given in terms of unprimed variables as opposed to their primed form.

Operation Schemas As in Z each operation consists of a declaration together with a predicate, and the usual conventions apply so that names ending in a ? denote input, and those ending in a ! denote output. In the predicate primes ($'$) are used to denote the value of a state variable after an operation has occurred. Operation schemas differ from Z in that each operation has a Δ-list which contains those state variables which may change when the operation is applied to an object of that class. An operation does not change the state variables that are not listed in its Δ-list. By convention the absence of a Δ-list is equivalent to an empty Δ-list.

Operations are interpreted in an environment which includes the state variables and constants of a class. The environment is enriched with the state variables in primed form, allowing the state schema's predicate in primed and unprimed form to be implicitly included in each operation definition.

When defining a schema in horizontal form in Object-Z the symbol $\widehat{=}$ is used, as in $Op \mathrel{\widehat{=}} [\ldots]$, as opposed to the symbol $==$ now used in standard Z.

Example 15.1 Consider the following specification of an addressable memory (cf. p. 115).

$[A]$

$Memory[D]$

$\upharpoonright(INIT, Read, Write)$

$m : A \rightarrow\!\!\!\!\!\!+\, D$

$INIT$

$m = \varnothing$

$Read$

$a? : A$
$d! : D$

$a? \in \operatorname{dom} m$
$d! = m(a?)$

$Write$

$\Delta(m)$
$a? : A$
$d? : D$

$m' = m \oplus \{a? \mapsto d?\}$

This class is generic because the type D is unspecified, i.e., D is its generic parameter. The scope of this parameter is the whole class, and therefore it is available in each schema definition. Externally only the operations $INIT$, $Read$ and $Write$ are visible, this means, for example, that the variable m cannot be read except by use of the visible operations.

The state schema here has one state variable m, a partial function from A to D, which is initially empty. In this example the operation $Read$ outputs the data value $d!$ associated with a particular address $a?$. Because $Read$ has no Δ-list, m is unchanged by this operation. $Write$ on the other hand alters m, updating it according to its inputs $a?$ and $d?$. □

Generic Parameters and Renaming Each formal generic parameter essentially introduces a basic type with scope extending to the end of the class. Such generic parameters must be replaced by actual types when a class is used. For example, the following replaces D by the type $\mathbb{Z}$ in the class $Memory$:

$$Memory[\mathbb{Z}]$$

In addition to replacing generic parameters by actual types, attributes and variables can be renamed. The scope of the renaming is the whole class, with simultaneous substitution being indicated by list pairs. For example, the following renames m to mem, $d!$ to $int!$ and $d?$ to $int?$.

$$Memory[\mathbb{Z}][mem/m, int!/d!, int?/d?]$$

Renaming is primarily used in inheritance allowing attributes and variables to be renamed to avoid name clashes.

Secondary Variables An Object-Z state schema is identical in purpose to a Z state schema. However, in addition to the usual variable declarations (known as primary variables in Object-Z) the state schema can also introduce *secondary* variables. These are listed after a Δ-symbol in the state schema and they are usually defined in terms of the primary variables, and used as a means to represent derived information. Secondary variables may be changed by any operation. Although useful in many circumstances, we shall not consider them further here.

15.1.1 The Object-Z Schema Calculus

Like Z, Object-Z has a schema calculus which enables complex operations to be built in an incremental fashion. In Object-Z the schema calculus operators are known as operation operators. These are restricted to seven operators: conjunction $\wedge$, choice $[]$, sequential composition $\fatsemi$, two parallel composition operators $\|$ and $\|_!$, enrichment $\bullet$ and hiding. So, for example, piping and the operator pre are not included in the Object-Z schema calculus.

When using the binary operators to combine two operations the usual rules of type compatibility apply when merging declarations. In addition the Δ-lists of the combined operations are merged so that the new operation can change any variable which either of its constituent operations could have changed.

Conjunction and Hiding Conjunction and hiding have the same effect as their Z counterparts.

Non-deterministic Choice Non-deterministic choice is denoted [] and it behaves as disjunction. When combining schemas with choice, additional predicates are explicitly introduced to model the variables that are unchanged. For example, given the following two operations:

*Op*1
$\Delta(x)$

$x' = x + 1$

*Op*2
$\Delta(y)$

$y' = y - 1$

Then the choice, $Op1[]Op2$, is equivalent to

Op
$\Delta(x, y)$

$(x' = x + 1 \wedge y' = y) \vee (y' = y - 1 \wedge x' = x)$

Parallel Composition and Scope Enrichment The two types of parallel composition are discussed in Sect. 15.4 below and the scope enrichment operator is discussed in Sect. 15.3.1.

Sequential Composition and Piping Sequential composition in Object-Z is slightly different from the same named operator in Z in that it additionally defines communication between outputs of the first operation and the inputs of the second operation.

For consistency between the Z and Object-Z specifications, and to support non-atomic refinement in a uniform way, we shall use the Z sequential composition operator and, for convenience, we shall also use piping $\gg$ as defined for Z.

Distributed Operators Object-Z also has three distributed versions of the operators $\wedge$, $[]$ and $\fatsemi$, for details see Smith [20]. Smith [20] also discusses recursive operation definitions which we will not have use for here.

15.1.2 Late Binding of Operations

There is a subtlety concerning the use of Δ-lists which is worth mentioning, and this concerns the so-called late binding of operations.

Δ-lists were originally introduced as a way to extend operations when using inheritance, and are used to postpone fixing which variables are unchanged until the actual application of an operation. This means that two operations might be identical when they are actually applied to an object, but have slightly different effects when combined with other operations. To understand this we need to see an example. Consider the following two operations.

*Op*1
$\Delta(x)$
$x' = x + y$

*Op*2
$\Delta(x, y)$
$x' = x + y$ $y' = y$

When they are applied, in the system class or to other objects, their effect is identical, in particular the value of y does not change.

However, if we conjoin these two operations with a third, say,

*Op*3
$\Delta(y)$
$y' = y + 1$

then we get the result that $Op2 \wedge Op3$ has predicate false, whereas $Op1 \wedge Op3$ has predicate $x' = x + y \wedge y' = y + 1$. Thus the actual effect of an operation is determined when it is applied (hence the late binding), and this means that we are not free to replace an operation by a seemingly equivalent one (e.g., $Op2$ for $Op1$) unless we know how it is used throughout the specification.

15.1.3 Preconditions

Although preconditions are not included in the Object-Z schema calculus we shall of course need them in our theory of refinement. The calculation of a precondition of an operation in a class is the same as in Z (quantifying over the class' state space), however the interpretation of a precondition is different in Object-Z to that in Z. As we have seen in Z the interpretation of an operation specified as a partial relation is that it behaves as specified when used within its precondition, but that outside its precondition, anything may happen, including divergence.

However, in Object-Z the interpretation is that an operation cannot occur at all outside its precondition, that is, we adopt the *behavioural* interpretation discussed in Chap. 3. Thus in the class *Memory* above the precondition of *Read* is $a? \in \mathrm{dom}\, m$ and the operation is not applicable outside this region.

This differing interpretation will, of course, affect the refinement rules when expressed in Object-Z because the totalisation will be different as we discuss below.

15.2 Inheritance

The purpose of inheritance is to allow incremental specification by building complex classes from simple components.

Inheritance works as follows. Everything (attributes, operations, etc.) is inherited apart from the visibility list. The type and constant definitions of the inherited class and those declared explicitly in the derived class are merged. The state schemas of the inherited and derived class are also merged, and the two predicates in the initialisations are conjoined.

The operation schemas of the inherited and derived class are also merged. Schemas with the same name are conjoined, and therefore their declarations must be compatible for the inheritance to well defined. An inherited operation with a name distinct from those in the new class is implicitly included in the inheritance.

Multiple inheritance is defined, as usual, as a sequence of single inheritance steps. The result is the same as inheriting all the classes individually in any order.

15.2.1 Example: A Bank Account

As an example of inheritance consider the following specification of a bank account given as a single class which allows money to be deposited and withdrawn and balances checked.

$[NAME]$

BankAcct

$\upharpoonright(INIT, Balance, Withdraw, Deposit, name, money)$

$name : NAME$

$money : \mathbb{N}$

INIT

$money = 0$

Balance

$bal! : \mathbb{Z}$

$bal! = money$

Withdraw

$\Delta(money)$

$amount! : \mathbb{N}$

$money' = money - amount!$

Deposit

$\Delta(money)$

$amount? : \mathbb{N}$

$money' = money + amount?$

Using inheritance we can specify a savings account (which pays interest and requires a non-negative balance) as follows.

InterestAcct

$\upharpoonright(INIT, Balance, Withdraw, Deposit, SetRate, PayInterest)$

BankAcct

$interestrate : \mathbb{R}$

INIT

$interestrate = 0$

SetRate

$\Delta(interestrate)$

$rate? : \mathbb{R}$

$interestrate' = rate?$

PayInterest

$\Delta(money)$

$money' = money * (1 + interestrate)$

Withdraw

$$\Delta(money)$$
$$amount! : \mathbb{N}$$
$$money \geq amount! + 10$$

InterestAcct inherits the state schema, initialisation and operations from *BankAcct*. *InterestAcct* has defined additional operations *SetRate* and *PayInterest*, and placed further constraints on *Withdraw* (any *Withdraw* must now keep a minimum balance of 10 in the account). (We assume that values are rounded up or down to integers in *PayInterest*.) Visibility lists are not inherited thus this class defines a new visibility list which includes operations from the inherited class. We can give an equivalent definition of *InterestAcct* as follows.

InterestAcct

$$\upharpoonright(INIT, Balance, Withdraw, Deposit, SetRate, PayInterest)$$
$$name : NAME$$

$$money : \mathbb{N}$$
$$interestrate : \mathbb{R}$$

INIT

$$money = 0 \wedge interestrate = 0$$

Balance

$$bal! : \mathbb{Z}$$
$$bal! = money$$

Withdraw

$$\Delta(money)$$
$$amount! : \mathbb{N}$$
$$money \geq amount! + 10$$
$$money' = money - amount!$$

Deposit

$$\Delta(money)$$
$$amount? : \mathbb{N}$$
$$money' = money + amount?$$

SetRate

$$\Delta(interestrate)$$
$$rate? : \mathbb{R}$$
$$interestrate' = rate?$$

PayInterest

$$\Delta(money)$$
$$money' = money * (1 + interestrate)$$

We can see the purpose of late binding now. If we inherited a class where operation $Op1$ was defined then we can extend this definition by describing how the operation now changes y. However, if we had replaced $Op1$ by $Op2$, since $Op2$ constrains y not to change we cannot extend the definition further in the inheritance.

15.3 Object Instantiation

In addition to inheritance the encapsulation offered by classes allows Object-Z to exploit the idea of object instantiation. That is, a class may include variables which are references to objects. This is achieved by declarations of the form $c : C$ which declares c to be a reference to an object of the class described by C.

Informally we can view references to objects as follows. If C is a class the declaration $c : C$ declares c to be a variable whose value is a reference to an object of class C. In order to use such objects we use attributes or operations that are in the visibility list of c. For example, $c.INIT$ is a predicate which denotes whether the object c conforms to C's initial state schema. Since $c.INIT$ is a predicate it can appear in an initialisation or predicate part of an operation, provided it is visible. Similarly $c.Op$ is an operation which transforms the object referenced by c according to the definition of the operation Op defined in the class C.

Example 15.2 As a very simple example consider the following description of a class *Count*.

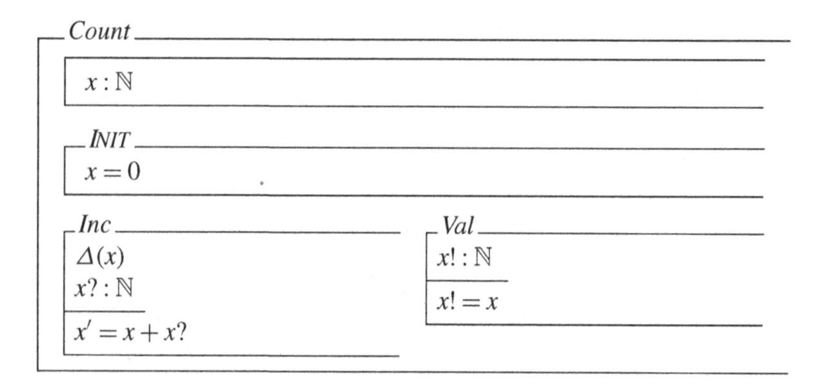

together with the class *CountAgain* specified as:

CountAgain

$ct : Count$

INIT

$ct.INIT$

$Inc \mathrel{\widehat{=}} ct.Inc$

$Val \mathrel{\widehat{=}} ct.Val$

The class *CountAgain* contains a reference *ct* to the class *Count*. This reference is initialised according to the initialisation in *Count*.

The operations *Inc* and *Val* have been *promoted* to operations on the reference *ct*. Here we have given the operations identical names, although of course they could have been called anything. When these operations are applied *ct* evolves according to the definition of the operations in the class *Count*. The operations *Inc* and *Val* therefore consume input and produce output as defined by the promoted operations *Inc* and *Val* from *Count*. An operation promotion such as *ct.Inc* has a Δ-list, however, as it does not change any variable in *CountAgain* its Δ-list in this example is empty. □

When an object reference declaration $c : C$ is made it is not implicitly initialised. If initialisation is desired, then the predicate $c.INIT$ needs to be included in the initialisation of the class where the declaration is made.

Similarly multiple object references to a class are not necessarily distinct. Thus the declaration $c, d : C$ does not imply that c and d reference distinct objects, and the additional predicate $c \neq d$ is needed to ensure that.

15.3.1 Modelling Aggregates

In Chap. 6 we looked at the technique of promotion which enabled us to model a collection of indexed components in a uniform way, promoting local operations to global ones. The use of object references in Object-Z allows a similar effect to be achieved without the need for a complex promotion or framing schema to be used when promoting the local operations. This is due to the reference semantics for Object-Z, which means that with a declaration $c : C$ the object referenced by c may evolve without changing the value of c. In line with this, the predicate $c' = c$ in an operation means that c refers to the same object after the operation, but it does not imply that its attribute values also remain unchanged. In order to express this, we assume that every class defines an implicit operation Ξ, which states that all object attributes are unchanged as well as recursively applying Ξ where needed: to attributes which are objects themselves, and to the elements of attributes which are sets, sequences, or indexed maps of objects.

Convention 15.1 (Ξ in Object-Z) We assume every class defines an implicit Ξ operation which keeps all attributes identical, as well as doing the same for all objects which are attributes or contained in attributes which are sets, sequences, and indexed maps of objects.

In addition, we use the following shorthand:

$$\Xi(S) \mathrel{\widehat{=}} \forall s : S \bullet s.\Xi$$ □

Thus in *Count*, we would have the implicit

$$\begin{array}{|l}\multicolumn{1}{l}{\Xi} \\ \hline \Delta(x) \\ \hline x' = x \\ \hline\end{array}$$

In order to define Ξ also for *CountAgain*, we need an additional Object-Z operator, viz. enrichment.

Enrichment Enrichment of one operation by another, denoted

$$schema_1 \bullet schema_2$$

means that the declarations in $schema_1$ are accessible when interpreting $schema_2$. Semantically this means that $schema_1 \bullet schema_2$ is the same as the conjunction of the two schemas, since the scope of any variable described in $schema_1$ now extends to the end of $schema_2$. Enrichment is necessary because a promotion such as $c.Op$ is defined in the language as an (operation) expression rather than a predicate, and therefore cannot directly appear in the predicate part of a schema definition. With enrichment, we can give the implicit Ξ operator for *CountAgain* as

$$\Xi \mathrel{\widehat{=}} [\Delta(ct) \mid ct' = ct] \bullet ct.\Xi$$

We have already seen this effect in the class *CountAgain* above where *Val* is promoted by simply writing $ct.Val$. This changes the state of the object referenced by *ct* but does not change *ct* itself since *ct* is not in the Δ-list of *Val*. Therefore although the reference after the application of an operation is denoted ct' we can deduce that $ct = ct'$.

This aspect is particularly effective when modelling a collection of objects. For example, suppose that in a particular class we declare $counts : \mathbb{P}\,Count$. Then in that class our promotion schema (in fact a selection schema) will be created in two steps:

$$\begin{array}{|l}\multicolumn{1}{l}{Select} \\ \hline c? : Count \\ \hline c? \in counts \\ \hline\end{array}$$

selects one element of *counts*, and then any change is constrained to that element by defining, using enrichment

$$SelectOne \mathrel{\widehat{=}} Select \bullet \Xi(counts \setminus \{c?\})$$

Enrichment is used frequently in promotion in Object-Z. For example, we can now apply *Val* to our selected object by writing

$$OneVal \mathrel{\widehat{=}} SelectOne \bullet c?.Val$$

Multiple selection environments are also often useful. For example, to select two objects we might define

$$\begin{array}{|l}
\hline
\textit{SelTwo} \\
c?, d? : Count \\
\hline
c? \neq d? \\
\{c?, d?\} \subseteq counts \\
\hline
\end{array}$$

$$SelectTwo \mathrel{\widehat=} SelTwo \bullet \Xi(counts \setminus \{c?, d?\})$$

allowing the operation *Val* to be applied to both objects by writing

$$TwoVal \mathrel{\widehat=} SelectTwo \bullet c?.Val \wedge d?.Val[y!/x!]$$

where we rename one of the outputs to prevent the two outputs being identified in the conjunction.

In all these cases the object reference does not change, and therefore does not appear in any Δ-list even though the object referenced by it does evolve. Object references do appear in a Δ-list when the references themselves are altered. For example, an operation to add a new object to the set *counts* could be specified as follows:

$$\begin{array}{|l}
\hline
\textit{New} \\
\Delta(counts) \\
c? : Count \\
\hline
c? \notin counts \\
counts' = counts \cup \{c?\} \\
\hline
\end{array}$$

If $c?$ was to be initialised then the predicate in this operation would have to include $c?.\mathit{INIT}$ as well, and Ξ could be used to ensure the rest of *counts* does not change with this operation.

15.3.2 Example: A Bank

We can now use some of these techniques to describe a *Bank* class consisting of a set of bank accounts. Initially there are no accounts, and an operation *Open* allows new accounts to be opened; the predicate $o?.\mathit{INIT}$ in this operation ensures that any newly opened account will be in its initial state. Promoted operations are also provided to access the individual accounts. For example, *Balance* will check the balance of any account $o?$.

$Bank$

$\upharpoonright(\mathit{INIT}, Open, Balance, Withdraw, Deposit)$

$accts : \mathbb{P}\, BankAcct$

INIT

$accts = \varnothing$

$Open1$

$\Delta(accts)$
$o? : BankAcct$

$o? \notin accts$
$accts' = accts \cup \{o?\}$
$o?.\mathit{INIT}$

$Open \mathrel{\widehat=} Open1 \bullet \Xi(accts)$
$Select \mathrel{\widehat=} [\, o? : accts \,] \bullet \Xi(accts \setminus \{o?\})$
$Balance \mathrel{\widehat=} Select \bullet o?.Balance$
$Withdraw \mathrel{\widehat=} Select \bullet o?.Withdraw$
$Deposit \mathrel{\widehat=} Select \bullet o?.Deposit$

15.3.3 Object Containment

When modelling aggregates it is often useful to be able to model object containment in a simple fashion and Object-Z has a syntactic abbreviation to do so which restricts external referencing to an object.

The basic property of object containment is that an object cannot be contained by two distinct objects at the same time. For example, a national bank can be thought of as a collection of individual branches, where each branch contains its own collection of accounts. Clearly accounts at different branches are distinct, and so we write something like

$NationalBank$

$banks : \mathbb{P}\, Bank$

$\forall a, b : Bank \bullet a \neq b \Rightarrow a.accts \cap b.accts = \varnothing$

$\vdots$

$Balance \mathrel{\widehat=} [b? : banks] \bullet b?.Balance \land \Xi(accts \setminus \{b?\})$

$\vdots$

(note that this would actually be illegal due to *accts* not being visible).

The point of the predicate in the state schema is to ensure an account cannot be contained in two different banks. To avoid explicitly writing out such predicates we use a containment notation on the objects we wish to contain, e.g., accounts. Thus in this example we amend the state schema of *Bank* to

$$accts : \mathbb{P}\, BankAcct_{©}$$

and we can then simply write

$$banks : \mathbb{P}\, Bank$$

for the state schema of *NationalBank*.

Notice that object containment does not imply exclusive control, for example, accounts can be referenced by any object, but only contained in one bank. Clearly an object can be indirectly contained by another object, but direct containment is unique and the syntax allows us to capture this relationship in a concise way.

15.4 Communicating Objects

In addition to modelling aggregates of objects, Object-Z has the facility to model the communication between objects, and it provides additional schema operators, denoted $\|$ and $\|_!$, to allow such inter-object communication.

Parallel Composition The two operators $\|$ and $\|_!$ model the parallel composition of two operations. The operator $\|$ used as in

$$schema_1 \parallel schema_2$$

conjoins the two schemas but identifies and hides inputs and outputs having identical types and base names. The base name of a declaration is defined to be the name with any decoration of ? or ! removed. The effect of this definition is to communicate output values of one schema to input values of another. Communication is in both directions.

For example, we could compose operations *Read* and *Write* from the *Memory* class. The composition *Read* $\|$ *Write* is semantically equivalent to

$$\Delta(m)$$
$$a? : A$$

$$a? \in \operatorname{dom} m$$
$$\exists d \bullet d = m(a?) \wedge m' = m \oplus \{a? \mapsto d\}$$

which simplifies to

$$\begin{array}{|l}
a? : A \\
\hline
a? \in \operatorname{dom} m
\end{array}$$

Notice how the $\|$ operator equates base names regardless of whether they were originally decorated with ? or !. Furthermore, notice that only the pairs of matching input and output variables which have been identified are hidden, thus in the composition of *Read* and *Write* the input $a?$ is not hidden, making it available for further synchronisation.

As another example, the definition

$$Swap \mathrel{\widehat=} SelectTwo \bullet (c?.Val \parallel d?.Inc)$$

has the effect of communicating the value $x!$ in $c?.Val$ to $x?$ in $d?.Inc$. Thus $d?.x$ is incremented by the amount $c?.x$.

The operator $\|$ also allows two-way communication. For example, an operation with declaration $x?$ and $y!$ can be composed in parallel with an operation with declaration $x!$ and $y?$. The resultant effect is the communication of $x!$ and $y!$ between the two schemas.

We can use this idea to specify a money transfer operation at the bank. We do so by creating a new class *BankTransfer* which inherits *Bank* and defines one new visible operation *Transfer* to transfer money from $o?$ to $u?$. The communication synchronises on the base name (i.e., *amount*) and thus the value *amount*? is transferred between the two accounts.

$$\begin{array}{|l}
\textit{BankTransfer} \\
\hline
\upharpoonright(\mathrm{INIT}, Open, Balance, Withdraw, Deposit, Transfer) \\
Bank \\
SelectTwo \mathrel{\widehat=} [o?, u? : accts \mid o? \neq u?] \bullet \Xi(accts \setminus \{o?, u?\}) \\
Transfer \mathrel{\widehat=} SelectTwo \bullet (o?.Withdraw \parallel u?.Deposit)
\end{array}$$

Associative Parallel Composition The associative parallel composition operator $\|_!$ is similar to $\|$ except that the output variables are not hidden, this allows them to synchronise with other input (and output) variables in further parallel compositions.

For example, the composition *Read* $\|_!$ *Write* is semantically equivalent to

$$\begin{array}{|l}
\Delta(m) \\
a? : A \\
d! : D \\
\hline
a? \in \operatorname{dom} m \\
d! = m(a?)
\end{array}$$

The schema operator $\|_!$ is associative, however $\|$ is not.

15.5 Semantics

Object-Z extends Z with the notion of encapsulation provided by objects and classes, and uses a *reference semantics* to describe their meaning.

In a reference semantics, which is the semantics adopted by most object-oriented programming languages, the variable in an object instantiation does not directly represent an object but rather *refers* to an object. A reference semantics is convenient when specifying many data structures such as those involving recursive definitions, however, it does bring its own complications particularly with respect to several aspects of refinement.

In order to understand a specification involving object instantiation we have to give a meaning to declarations such as $a : A$ for a class A. To do so Object-Z extends the Z type system by regarding classes as types. Every class (apart from those with unsatisfiable initialisation) defines a set of objects and the declaration $a : A$ introduces a variable a whose value is the *identity* of one of the objects of type A. For each class C, there is an associated universe of all object identities, denoted $\mathbb{O}_C$, one for each possible instantiation of the class.

This emphasis on object identity means that objects are used in a different way to other variables declared in a specification. For example, given a declaration $n : \mathbb{N}$, n refers to a particular value and this value might change throughout the specification. However, an object a of type A refers to the *identity* of a object as opposed to its value, and it is this property that differentiates a reference from a value semantics. Although references to an object can change in a specification the objects themselves are persistent, and therefore creation and deletion of objects does not figure in Object-Z. However, including *INIT* in the visibility list of a class C allows other classes to treat C objects “as if they were new”.

The declaration $a : A$ introduces a reference a to an object. This object has a *state* corresponding to its defining class. If the class A defines a constant then each object of type A has a fixed value of that constant, however, different objects of type A may take different values as long as they conform to the definition of A. For example, given declarations $acct_1, acct_2 : BankAcct$ then $acct_1.name$ and $acct_2.name$ may be different, but $acct_1.name$ is constant in the sense that it cannot be altered by any operation.

The state of an object is accessed and altered by using the dot notation introduced in Sect. 15.3. So, for example, $ct.x$ denotes the value of the variable x in the state of object referenced by ct (which is of type *Count* defined earlier). Only attributes declared as visible may be accessed in this fashion.

The only means of changing an object's state is by applying a (visible) operation, as in $ct.Inc$. The use of a reference semantics is important here because although the state of ct has changed by application of *Inc* its identity has not, and the object is still referred to by ct after the operation. This means that the Δ-list of $ct.Inc$ is empty.

However, as we have already seen object references can appear in a Δ-list when, for example, introducing a new object reference. See, for example, the operation *New* specified in Sect. 15.3.1. We can also change the reference to an object, but this does not alter the state of any object. Thus we can change the reference to a bank account $acct : BankAcct$ by using the operation

$$\begin{array}{|l}
\hline
\textit{ChangeName} \\
\Delta(acct) \\
\hline
acct' \neq acct \\
acct'.name \neq acct.name \\
\hline
\end{array}$$

which causes *acct* to refer to a different object with a different *name*.

The promotion of an initialisation defines a predicate which is true if that object is in its initial state. The precondition of an operation is defined in a similar way. Given a promotion $a.Op$ this operation is applicable whenever the state of the object referenced by a satisfies the precondition of Op. We can thus make the following definition.

Definition 15.1 (Precondition) Let A be an Object-Z class containing an operation Op. Suppose that the state declaration $a : A$ appears in another class, then in that class $\text{pre}\, a.Op$ is defined to be $a.\text{pre}\, Op$. □

15.5.1 Polymorphism and Generic Parameters

When using a class as a type all generic parameters must be instantiated. Thus, for example, when we used the class *Memory* in *MemorySet* we used the declaration $mem : \mathbb{F}\, Memory[\mathbb{Z}]$—leaving D uninstantiated is not permitted.

The Object-Z class system also supports polymorphic declarations. A *class union* allows a declaration to belong to one of a set of classes, so that given classes C and D we can write $o : C \cup D$. Then the object o refers to an object of either class C or D. A particular form of class union is the polymorphic declaration $o : \downarrow C$. This means that o refers to either the class C or any of its *subclasses*, i.e., classes which are inherited from C.

For example, instead of the declaration

$$accts : \mathbb{P}\, BankAcct$$

in the class *Bank* given in Sect. 15.3.1 we could define its state as

$$accts : \mathbb{P}(\downarrow BankAcct)$$

allowing *accts* to contain interest accounts as well as normal bank accounts.

15.6 Example: A Football League

In the final example of this chapter we specify in Object-Z the football league described, in Z, in Sect. 6.2. There we described a single club as a local state together with a collection of operations, and then the league was specified as a global state indexed over some set I. Promotion was used to describe operations on the football league in terms of the local operations on the local state (i.e., single clubs).

In our Object-Z specification of the same system we can use encapsulation to good effect. A single club will be given as a single object, and a league will be a class containing a number of clubs. Operations on the football league can then be described using operation application on individual clubs, and this simplifies their description considerably.

Club

$\upharpoonright(points, INIT, Join, Buy)$

$members : ID \rightarrowtail\!\!\!\!\!\!\twoheadrightarrow PEOPLE$
$away : \mathbb{P}\, ID$
$points : \mathbb{N}$

$away \subseteq \operatorname{dom} members$

INIT

$members = \varnothing$
$points = 0$

Join

$\Delta(members)$
$app? : PEOPLE$
$id! : ID$

$app? \notin \operatorname{ran} members$
$id! \notin \operatorname{dom} members$
$members' = members \cup \{(id!, app?)\}$

Buy

$\Delta(away)$
$app? : PEOPLE$

$\exists x : ID \bullet ((x, app?) \in members \wedge away' = away \cup \{x\})$

League

$clubs, europe : \mathbb{P}\, Club$

$europe \subseteq clubs$

INIT

$clubs = \varnothing$

$Select \mathrel{\widehat=} [\, i? : clubs \,] \bullet \Xi(clubs \setminus \{i?\})$
$Buy \mathrel{\widehat=} Select \bullet i?.Buy$
$Join \mathrel{\widehat=} Select \bullet i?.Join$

GNew
$\Delta(clubs)$
$i? : Club$

$i? \notin clubs$
$i?.INIT$
$clubs' = clubs \cup \{i?\}$

$New \mathrel{\widehat=} GNew \bullet \Xi(clubs)$

GDel
$\Delta(clubs)$
$i? : Club$

$i? \in clubs$
$i?.INIT$
$clubs' = clubs \setminus \{i?\}$

$Del \mathrel{\widehat=} GDel \bullet \Xi(clubs)$

GRel
$\Delta(clubs)$
$i? : Club$

$i? \in clubs$
$clubs' = clubs \setminus \{i?\}$
$\forall j : clubs \bullet i?.points \leq j.points$

$Relegate \mathrel{\widehat=} GRel \bullet \Xi(clubs)$

A number of points are worth noting about this specification. Firstly, notice that since the object references provide a means to reference individual clubs, the use of an indexing set I is not necessary in the Object-Z specification. In addition, because *INIT* is a predicate on the unprimed state we can describe deletion of a club in terms of $i?.INIT$ without having to define an explicit termination state schema as we did in the version given in Z. Finally, *points* in *Club* had to be visible in order to specify *Relegate*.

15.7 Bibliographical Notes

There have been a number of proposals to extend Z with object-oriented features including MOOZ [16], ZEST [6, 23] and OOZE [1], and also proposals to use Z in an object-oriented style [13, 14]. Both [21] and [22] provide comparisons of some of these approaches, and a collection of object-oriented specification case studies has also been published [15].

Object-Z grew out of work at the SVRC, University of Queensland. Early definitions of the language include [8, 10], and these were accompanied by work on its semantics [7, 19]. The reference manual by Smith [20] provides a definitive account of the current syntax together with a discussion of the semantics. More recent work on the semantics includes [11] and [5]. There has also been work on a logic for Object-Z by Smith [18] and Griffiths [12]. An introductory textbook [9] by Duke and Rose was published in 2000.

An Object-Z home page is maintained on the WWW at:

http://itee.uq.edu.au/~smith/objectz.html

This contains a comprehensive list of references plus a description of the available tool support.

References

1. Alencar, A. J., & Goguen, J. A. (1991). OOZE: an object oriented Z environment. In P. America (Ed.), *ECOOP '91—Object-Oriented Programming. Lecture Notes in Computer Science: Vol. 512* (pp. 180–199). Berlin: Springer.
2. Bjørner, D., Hoare, C. A. R., & Langmaack, H. (Eds.) (1990). *VDM'90: VDM and Z!—Formal Methods in Software Development. Lecture Notes in Computer Science: Vol. 428.* Berlin: Springer.
3. Bowen, J. P., & Hall, J. A. (Eds.) (1994). *ZUM'94, Z User Workshop. Workshops in Computing*. Cambridge: Springer.
4. Bowen, J. P. & Hinchey, M. G. (Eds.) (1995). *ZUM'95: the Z Formal Specification Notation. Lecture Notes in Computer Science: Vol. 967.* Limerick: Springer.
5. Butler, S. (1998). *Behaviour in object-based systems: a formal approach.* PhD thesis, University of Queensland.
6. Cusack, E., & Rafsanjani, G. H. B. (1992). ZEST. In S. Stepney, R. Barden, & D. Cooper (Eds.), *Object Orientation in Z. Workshops in Computing* (pp. 113–126). Berlin: Springer.
7. Duke, D. J., & Duke, R. (1990) Towards a semantics for Object-Z. In Bjørner et al. [2] (pp. 244–261).
8. Duke, R., King, P., Rose, G. A., & Smith, G. (1991). *The Object-Z specification language version 1* (Technical Report 91-1). Software Verification Research Centre, Department of Computer Science, University of Queensland.
9. Duke, R., & Rose, G. A. (2000). *Formal Object-Oriented Specification Using Object-Z. Cornerstones of Computing*. New York: Macmillan.
10. Duke, R., Rose, G. A., & Smith, G. (1995). Object-Z: a specification language advocated for the description of standards. *Computer Standards & Interfaces*, *17*, 511–533.
11. Griffiths, A. (1997). *A formal semantics to support modular reasoning in Object-Z.* PhD thesis, University of Queensland.
12. Griffiths, A. (1997). Modular reasoning in Object-Z. In *Asia-Pacific Software Engineering Conference and International Computer Science Conference*. Los Alamitos: IEEE Comput. Soc.
13. Hall, J. A. (1990) Using Z as a specification calculus for object-oriented systems. In Bjørner et al. [2] (pp. 290–318).
14. Hall, J. A. (1990) Specifying and interpreting class hierarchies in Z. In Bowen and Hall [3] (pp. 120–138).

15. Lano, K. & Haughton, H. (Eds.) (1994). *Object Oriented Specification Case Studies*. New York: Prentice Hall.
16. Meira, S. L., & Cavalcanti, A. L. C. (1990) Modular object oriented Z specifications. In Nicholls [17] (pp. 173–192).
17. Nicholls, J. E. (Ed.) (1990). *Z User Workshop*, Oxford. *Workshops in Computing*. Berlin: Springer.
18. Smith, G. (1995) Extending W for Object-Z. In Bowen and Hinchey [4] (pp. 276–295).
19. Smith, G. (1995). A fully abstract semantics of classes for Object-Z. *Formal Aspects of Computing*, *7*(3), 289–313.
20. Smith, G. (2000). *The Object-Z Specification Language*. Dordrecht: Kluwer Academic.
21. Stepney, S., Barden, R., & Cooper, D. (Eds.) (1992). *Object Orientation in Z*. *Workshops in Computing*. Berlin: Springer.
22. Stepney, S., Barden, R., & Cooper, D. (1992). A survey of object orientation in Z. *Software Engineering Journal*, *7*(2), 150–160.
23. Zadeh, H. B. (1996). *Using ZEST for specifying managed objects* (Technical report). British Telecom.

Chapter 16
Refinement in Object-Z

We now turn our attention to applying the theory of refinement to Object-Z. In this chapter we adapt the existing refinement rules to Object-Z specifications consisting of a single class, and then in the next chapter we will consider issues of refinement concerning objects and their instantiation and use within classes.

To do so we first look at the underlying relational characterisation in Sect. 16.1 where we describe the simulation rules for single Object-Z classes. These have exactly the same form as in Z except that they do not allow the weakening of a precondition in a refinement. We subsequently look at generalisations of refinement; Sect. 16.2 discusses weak refinement in the context of Object-Z and Sect. 16.3 non-atomic refinement. We finish with some comments on the relationship between refinement, subtyping and inheritance.

16.1 The Simulation Rules in Object-Z

Although we are working in the context of an object-oriented language complete with ideas of class and object, our view of refinement as being the reduction of non-determinism in an ADT remains unchanged. To consider an Object-Z specification as an ADT we view its behaviour through its system class, and this defines the state, initialisation and operations in the ADT. Data refinement therefore can be defined in a similar way in Object-Z as in Z, and simulations form a sound and jointly complete refinement methodology.

Refinement in Object-Z only differs from that in Z in how partial relations are totalised. Section 3.3 described how different totalisations can be used in order to model "contractual" or "behavioural" interpretations of the precondition of a relation (and hence an operation). As pointed out in Sect. 15.1.3, Object-Z adopts a blocking model for preconditions, and we use a behavioural interpretation to describe this. That is, operations do not diverge outside their preconditions in Object-Z, and therefore the simulation rules for partial relations do not allow a widening of domain. Note that in this chapter we base the Object Z refinement rules on the

J. Derrick, E.A. Boiten, *Refinement in Z and Object-Z*,
DOI 10.1007/978-1-4471-5355-9_16, © Springer-Verlag London 2014

relational theory, rather than the alternative option of using the histories semantics for Object-Z [17, 18], see [3] and Chap. 19 for further discussion.

All this means that preconditions cannot be weakened in a refinement. The remaining aspects are the same, and so we can define downward and upward simulation in Object-Z in a fashion similar to Z. As usual we will denote the state schema in the class A by $A.STATE$, and $A.INIT$ will denote the initialisation.

Definition 16.1 (Downward simulation in Object-Z) An Object-Z class C is a downward simulation of the class A if there is a retrieve relation R on $A.STATE \wedge C.STATE$ such that every visible abstract operation AOp is recast into a visible concrete operation COp and the following hold.

1. $\forall C.STATE \bullet C.INIT \Rightarrow (\exists A.STATE \bullet A.INIT \wedge R)$
2. $\forall A.STATE; C.STATE \bullet R \Rightarrow (\mathrm{pre}\, AOp \Leftrightarrow \mathrm{pre}\, COp)$
3. $\forall A.STATE; C.STATE; C.STATE' \bullet R \wedge COp \Rightarrow \exists A.STATE' \bullet R' \wedge AOp$ □

Definition 16.2 (Upward simulation in Object-Z) An Object-Z class C is an upward simulation of the class A if there is a retrieve relation T on $A.STATE \wedge C.STATE$ such that every visible abstract operation AOp is recast into a visible concrete operation COp and the following hold.

1. $\forall A.STATE; C.STATE \bullet C.INIT \wedge T \Rightarrow A.INIT$
2. $\forall C.STATE \bullet \exists A.STATE \bullet T \wedge (\mathrm{pre}\, AOp \Rightarrow \mathrm{pre}\, COp)$
3. $\forall A.STATE'; C.STATE; C.STATE' \bullet T' \wedge COp \Rightarrow \exists A.STATE \bullet T \wedge AOp$ □

Note that these definitions are concerned only with the *visible* operations, any changes are permitted to the non-visible operations, as long as refinement holds for the external interface of the class. This is because the non-visible operations are only defined for the purposes of structuring the specification (see definition of visibility in Sect. 15.1).

Although we claimed that preconditions could not be weakened under a refinement, in the case of a downward simulation this is not strictly true. A more precise statement reads: preconditions of an operation cannot be weakened in the reachable states of a specification. That is, in condition 2 the preconditions are only required to be equivalent in the image of R. The relation R is existentially quantified, and it can be safely constrained to apply only to the states reachable from the initialisation, cf. Sect. 2.4.1. Therefore outside the reachable states the precondition of a concrete operation can be anything at all.

The second point to note is that in an upward simulation preconditions really cannot be weakened. Although condition 2 of an upward simulation looks as if it allows weakening, condition 3, along with the totality of T on $C.STATE$, prevents this. In addition, because T is total on $C.STATE$ (this is a consequence of condition 2) refinement cannot introduce new unreachable states in the sense that there can be with a downward simulation. This is really due to the nature of an upward simulation which picks an arbitrary point in the concrete execution and works backwards to see if it could be simulated from some abstract initialisation.

A couple of examples will suffice to illustrate the definitions. One aspect that impacts upon the verification of the conditions is the use of Δ-lists in Object-Z, and we discuss this issue in the first example.

Example 16.1 Our initial specification consists of a single class ATM_0 which describes a cashpoint machine. The behaviour of the class is defined by operations *Insert*, *Passwd*, *Withdraw*, etc. A user of the machine inserts a card which is modelled by reading the account number $account?$. A four-digit PIN is then given, and if this matches the account then the user is able to proceed and *Withdraw* money. Accounts are opened by the *OpenAcct* operation, where we assume that $max(\varnothing) = 0$. (A deposit operation is also assumed to be available but we elide its definition.)

The bank accounts are modelled as a partial function *accts* from account numbers to amounts. The PIN for a given account m is given by $pins(m)$. The card currently inside the machine is represented by *card* (0 represents no card present), and we use a Boolean *trans* to determine whether a transaction is allowed to proceed.

ATM_0

$accts : \mathbb{N} \rightarrow\!\!\!\!\!\!+\; \mathbb{N}$
$pins : \mathbb{N} \rightarrow\!\!\!\!\!\!+\; \mathbb{N} \times \mathbb{N} \times \mathbb{N} \times \mathbb{N}$
$card : \mathbb{N}$
$trans : \mathbb{B}$

$\mathrm{dom}\, accts = \mathrm{dom}\, pins$
$0 \notin \mathrm{dom}\, accts$

INIT

$accts = \varnothing \wedge pins = \varnothing \wedge card = 0 \wedge \neg trans$

Eject

$\Delta(card, trans)$

$card' = 0 \wedge \neg trans'$

OpenAcct

$\Delta(accts, pins)$
$account! : \mathbb{N}$
$pin? : \mathbb{N} \times \mathbb{N} \times \mathbb{N} \times \mathbb{N}$

$account! = (max\, \mathrm{dom}\, accts) + 1$
$accts' = accts \cup \{(account!, 0)\}$
$pins' = pins \cup \{(account!, pin?)\}$

$$\begin{array}{l}
\underline{\textit{Insert}} \\
\Delta(card) \\
card? : \mathbb{N} \\
\hline
card = 0 \land card? \in \mathrm{dom}\, pins \\
card' = card?
\end{array}$$

$$\begin{array}{l}
\underline{\textit{Passwd}} \\
\Delta(trans) \\
pin? : \mathbb{N} \times \mathbb{N} \times \mathbb{N} \times \mathbb{N} \\
\hline
card \neq 0 \land \neg trans \\
trans' = (pins(card) = pin?)
\end{array}$$

$$\begin{array}{l}
\underline{\textit{Withdraw}} \\
\Delta(accts, card, trans) \\
amount?, money! : \mathbb{N} \\
\hline
trans \\
accts' = accts \oplus \{card \mapsto accts(card) - money!\} \\
((card' = 0 \land money! = 0 \land \neg trans') \lor \\
(card' = card \land 0 \leq money! \leq amount? \land trans'))
\end{array}$$

The withdraw operation in this class is non-deterministic. Sometimes the bank gives out a sum of money, but at other times it just keeps the card, and to the user it is unclear why this happens.

In a data refinement we place ATM_0 by ATM_1 where we have modelled accounts as a sequence instead of a partial function, and changed the representation of *trans*. We have also reduced the non-determinism in *Withdraw* so that money is given out only if there are sufficient funds in the account. Of course because of the meaning of preconditions in Object-Z we cannot weaken the precondition of *Passwd* as we did in Example 7.5 (p. 182) but there is no need to since outside pre *Passwd* the operation is blocked.

$$\begin{array}{l}
\underline{ATM_1} \\
\quad \begin{array}{l}
accts : \mathrm{seq}\, \mathbb{N} \\
pins : \mathbb{N} \nrightarrow \mathbb{N} \times \mathbb{N} \times \mathbb{N} \times \mathbb{N} \\
card : \mathbb{N} \\
trans : \{0, 1\} \\
\hline
\mathrm{dom}\, accts = \mathrm{dom}\, pins \\
0 \notin \mathrm{dom}\, accts
\end{array} \\
\quad \begin{array}{l}
\underline{\textit{INIT}} \\
accts = \varnothing \land pins = \varnothing \land card = 0 \land trans = 0
\end{array} \\
\quad \begin{array}{l}
\underline{\textit{Eject}} \\
\Delta(card, trans) \\
\hline
card' = 0 \land trans' = 0
\end{array}
\end{array}$$

OpenAcct

$\Delta(accts, pins)$
$account! : \mathbb{N}$
$pin? : \mathbb{N} \times \mathbb{N} \times \mathbb{N} \times \mathbb{N}$

$account! = max \operatorname{dom} accts'$
$accts' = accts \frown \langle 0 \rangle$
$pins' = pins \cup \{(account!, pin?)\}$

Insert

$\Delta(card)$
$card? : \mathbb{N}$

$card = 0 \land card? \in \operatorname{dom} pins$
$card' = card?$

Passwd

$\Delta(trans)$
$pin? : \mathbb{N} \times \mathbb{N} \times \mathbb{N} \times \mathbb{N}$

$card \neq 0 \land trans = 0$
$pins(card) = pin? \Rightarrow trans' = 1$
$pins(card) \neq pin? \Rightarrow trans' = 0$

Withdraw

$\Delta(accts, card, trans)$
$amount?, money! : \mathbb{N}$

$trans = 1$
$accts' = accts \oplus \{card \mapsto accts(card) - money!\}$
$((card' = 0 \land money! = 0 \land trans' = 0 \land$
$\quad amount? \geq accts(card)) \lor$
$(card' = card \land money! = amount? \land$
$\quad amount? < accts(card) \land trans' = 1))$

The retrieve relation between the two classes is (remembering that a sequence is in fact a function):

R

$ATM_0.\text{STATE}$
$ATM_1.\text{STATE}$

$ATM_0.trans \Leftrightarrow (ATM_1.trans = 1)$
$ATM_0.pins = ATM_1.pins$
$ATM_0.accts = ATM_1.accts$
$ATM_0.card = ATM_1.card$

Notice how we prefix variables with the name of their class to avoid name clashes.

To illustrate verification of the refinement consider the correctness of *Withdraw*. This requires us to show that

$$\begin{array}{l}\forall ATM_0.\mathit{STATE};\ ATM_1.\mathit{STATE};\ ATM_1.\mathit{STATE}' \bullet \\ \quad R \wedge ATM_1.Withdraw \Rightarrow \exists ATM_0.\mathit{STATE}' \bullet R' \wedge ATM_0.Withdraw\end{array}$$

This is straightforward (cf. the similar proofs in Chap. 4) except that the Δ-list in an operation causes an additional proof obligation. To see this look at the Δ-list in *Withdraw*: $\Delta(accts, card, trans)$, there is actually a hidden predicate in there, namely that $pins' = pins$. Thus in addition to the explicit predicates, we have to prove that $ATM_0.pins' = ATM_0.pins$ as part of the correctness of $ATM_0.Withdraw$.

However, this is easy because we can use the corresponding predicate $ATM_1.pins' = ATM_1.pins$ which is part of $ATM_1.Withdraw$. The other conditions follow similarly. □

To summarise the last point: whenever a state variable x does not occur in the Δ-list of an operation Op, we have to *verify* that $x' = x$ whenever Op occurs on the right of an implication, and we may *assume* that $x' = x$ whenever Op occurs to the left of the implication in a refinement verification condition.

Another example of a simple refinement in Object-Z is the following.

Example 16.2 We can refine the *Memory* class from Example 15.1 (p. 365) into the class $Memory_1$ (see also Example 4.7) which uses a cache of fixed size, and a different flushing strategy (always when the buffer is full).

$$\begin{array}{l}
\textbf{Memory}_1[D] \\
\upharpoonright(\mathit{INIT}, Read, Write) \\
\quad n : \mathbb{N}_+ \\
\hline
\quad m_1 : A \nrightarrow D \\
\quad c_1 : A \nrightarrow D \\
\hline
\quad \mathit{INIT} \\
\quad m_1 = \varnothing \wedge c_1 = \varnothing \\
\hline
\quad Flush \\
\quad \Delta(m_1, c_1) \\
\quad \#\operatorname{dom} c_1 < n \Rightarrow m_1' = m_1 \wedge c_1' = c_1 \\
\quad \#\operatorname{dom} c_1 = n \Rightarrow \\
\qquad (\exists x : A \bullet x \in \operatorname{dom} c_1 \wedge \\
\qquad\quad m_1' = m_1 \oplus \{x \mapsto c_1(x)\} \\
\qquad\quad \{x\} \lhd c_1' = \{x\} \lhd c_1 \\
\qquad\quad \operatorname{dom} c_1' = \operatorname{dom} c_1 \setminus \{x\})
\end{array}$$

$$CheckCache \mathrel{\widehat=} [a? : A \mid a? \notin \operatorname{dom} c_1] \bullet Flush$$
$$[]$$
$$[a? : A \mid a? \in \operatorname{dom} c_1]$$

WrCache

$\Delta(c_1)$
$a? : A$
$d? : D$

$c_1' = c_1 \oplus \{a? \mapsto d?\}$

RdCache

$\Delta(c_1)$
$a? : A$

$a? \notin \operatorname{dom} c_1 \Rightarrow c_1' = c_1 \oplus \{a? \mapsto m_1(a?)\}$
$a? \in \operatorname{dom} c_1 \Rightarrow c_1' = c_1$

$Write \mathrel{\widehat=} CheckCache \mathbin{;} WrCache$
$Read \mathrel{\widehat=} CheckCache \mathbin{;} RdCache \mathbin{;} [a? : A;\ d! : D \mid d! = c_1(a?)]$

The verification is again straightforward. □

16.2 Weak Refinement in Object-Z

In a similar way to the standard refinement rules we can adapt generalisations of refinement to the Object-Z setting. For example it is straightforward to adapt the weak simulation rules, and in doing so we can formulate weak downward simulation in a similar fashion to before.

Remember (cf. Chap. 11) that the purpose of weak refinement is to give a proper treatment to *internal* operations. Internal operations are those which are under the control of the system as opposed to the environment, and they can be invoked (by the object itself) whenever their preconditions hold. To make clear which operations are internal in an Object-Z specification we list them after the visibility list by writing $Internals(Op_1, \ldots, Op_n)$. Internal operations should not be confused with other operations which are in a class but not in the visibility list, as these are used to structure the specification and cannot be invoked by the environment.

Weak refinement is a generalisation of refinement for specifications containing internal operations. The conditions encapsulated in weak refinement are the same as standard refinement except evolution due to internal operations is allowed before and after each visible operation.

To model this we again need to build the idea of the effect of all the internal operations in a class as *Int* (see Sect. 11.2.2). That is, *Int* is an operation which relates any two states that can be related by a finite sequence of internal operations. The definition of weak refinement is then essentially the same as in Z apart from the inability to widen preconditions of visible operations.

Definition 16.3 (Weak downward simulation) Let $A = (A.\mathit{STATE}, A.\mathit{INIT}, \{AOp_i\}_{i \in I \cup J})$ and $C = (C.\mathit{STATE}, C.\mathit{INIT}, \{COp_i\}_{i \in I \cup K})$ be Object-Z data types, where J and K (the index sets denoting internal operations) are both disjoint from I. Then the relation R on $A.\mathit{STATE} \wedge C.\mathit{STATE}$ is a *weak downward simulation* from A to C if

$$\forall C.\mathit{STATE} \bullet (C.\mathit{INIT} \, ⨾ \, Int_C) \Rightarrow \exists A.\mathit{STATE} \bullet (A.\mathit{INIT} \, ⨾ \, Int_A) \wedge R$$

and $\forall i : I$

$$\forall A.\mathit{STATE}; \; C.\mathit{STATE} \bullet R \Rightarrow (\mathrm{pre}(Int_A \, ⨾ \, AOp_i) \Leftrightarrow \mathrm{pre}(Int_C \, ⨾ \, COp_i))$$
$$\forall A.\mathit{STATE}; \; C.\mathit{STATE}; \; C.\mathit{STATE}' \bullet$$
$$R \wedge (Int_C \, ⨾ \, COp_i \, ⨾ \, Int_C) \Rightarrow \exists A.\mathit{STATE}' \bullet R' \wedge (Int_A \, ⨾ \, AOp_i \, ⨾ \, Int_A)$$

eliding quantification over inputs and outputs as before.

In addition, we require the existence of a well-founded set WF with partial order $<$, and a variant E which is an expression in the state variables satisfying the following conditions.

D1. $R \Rightarrow E \in WF$
D2. $\forall i : K \bullet R \wedge COp_i \Rightarrow E' < E$ □

Weak upward simulation is similar.

Example 16.3 As an example consider an alternative teller machine ATM_2.

ATM_2

$Internals(Reset)$
ATM_0

$limit : \mathbb{N}$

INIT
$limit = 0$

$Reset$
$\Delta(limit)$
$limit > 0$
$limit' = limit - 1$

$Withdraw$
$\Delta(limit)$
$limit < 4$
$limit' = limit + 1$

This teller only allows four withdraws in a row, and after the fourth it requires its *limit* to be *Reset*. This last operation is internal to the bank and represents the bank's control on how many times someone can withdraw.[1]

It is necessary to use weak downward simulation to verify the refinement. Standard refinement would fail because, for example,

$$\text{pre}\, ATM_0.Withdraw \not\Leftrightarrow \text{pre}\, ATM_2.Withdraw$$

as would be required.

However, because weak refinements allow an arbitrary number of internal operations before and after each external operation, ATM_2 is a weak downward simulation of ATM_0. For example, to verify applicability for *Withdraw* we can show that

$$R \Rightarrow (\text{pre}\, ATM_0.Withdraw \Leftrightarrow \text{pre}(\exists\, n \bullet Reset^n \mathbin{⨾} ATM_2.Withdraw))$$

where R is the projection onto $limit > 0$. The other conditions are similar. □

16.3 IO and Non-atomic Refinement in Object-Z

The definition of an ADT representing an Object-Z specification mentioned at the start of Sect. 16.1 assumed that the state of the system class was hidden. If this is not the case then the visible attributes can be considered as observable, and we have to apply the techniques of Grey Box refinement (see Chap. 9) to such a specification.

IO and non-atomic refinement can also be adapted for use in Object-Z. IO refinement (see Chap. 10) generalises refinement by allowing the inputs and outputs in an operation to be altered in a refinement. Normally the inputs and outputs cannot be changed, but in IO refinement we use input and output transformers *IT* and *OT*, which are essentially retrieve relations between the inputs and outputs.

This allows the types of inputs and outputs to be changed, and also for inputs and outputs to be added or removed if appropriate, for example, by adding a constant output or removing an input which is never used. Again we require that *IT* and *OT* are total on the abstract input and output types, and that *OT* is injective. For an IO refinement of one abstract Object-Z operation into a concrete version the following definition is used together with the normal initialisation condition.

Definition 16.4 (IO downward simulation for a single operation) Let *IT* be an input transformer for *COp* which is total on the abstract inputs. Let *OT* be a total injective output transformer for *AOp*. The retrieve relation *R* defines an IO refinement if (initialisation is as before):

---

[1]There is an implicit fairness assumption in the formulation of weak refinement in that if an internal operation is enabled then it is assumed that it will eventually occur. This is the same assumption underlying the use of internal events in many process algebras.

Applicability:

$$\forall A.\mathit{STATE};\ C.\mathit{STATE} \bullet R \Rightarrow (\mathrm{pre}(\overline{IT} \gg AOp) \Leftrightarrow \mathrm{pre}\, COp)$$

Correctness:

$$\begin{array}{l}\forall A.\mathit{STATE};\ C.\mathit{STATE};\ C.\mathit{STATE}' \bullet \\ \qquad R \wedge (IT \gg COp) \Rightarrow \exists A.\mathit{STATE}' \bullet R' \wedge (AOp \gg OT)\end{array}$$

The correctness criteria requires that COp with the input transformation should produce a result related by R and the output transformation to one that AOp could have produced. □

In fact we can consider IT and OT to be *class wrappers* acting on the whole class rather than on a per operation basis. If we rename any inputs and outputs in the operations so that they are all distinct we can define the effect of the transformers pointwise (i.e., operation by operation) and write

$$A \gg OT \sqsubseteq IT \gg C$$

whenever class A is IO refined into class C. Then IT and OT represent the wrappers that must be provided in any implementation.

16.3.1 Non-atomic Refinement

Non-atomic refinement looks at how the granularity of an operation can be altered in a refinement. The refinement rules given in Definitions 16.1 and 16.2 assume that the data types are conformal, i.e., that there is a 1–1 correspondence between abstract and concrete operations. However, this fixes the level of granularity early in the development life-cycle and does not allow implementation considerations to be introduced later (e.g., the chosen abstract operations might be inefficiently large to be atomic in reality).

Non-atomic refinement tackles these concerns by allowing an abstract operation to be decomposed into a sequence of concrete operations. To do so it uses some of the ideas from IO refinement. This is necessary because in a non-atomic refinement we often want to spread the inputs/outputs of the abstract operation throughout the concrete decomposition. Therefore we use input and output transformers IT and OT which map abstract inputs/outputs to a *sequence* of concrete inputs/outputs.

Non-atomic refinement has the same form as in Z, and the definition expresses the refinement of AOp into a fixed sequence $COp_1 \fatsemi COp_2$. The first condition is the requirement that the combined effect of $COp_1 \fatsemi COp_2$ is consistent with AOp. The next two conditions then require that the preconditions of the decomposition match up, and equal the precondition of AOp.

Definition 16.5 (Non-atomic downward simulation with IO transformations) Let *IT* be an input transformer for $COp_1 \mathbin{{}_{9}^{\circ}} COp_2$ which is total on the abstract inputs. Let *OT* be a total injective output transformer for *AOp*. The retrieve relation *R* defines a non-atomic downward simulation if:

$$\forall A.\mathit{STATE};\ C.\mathit{STATE} \bullet R \Rightarrow (\mathrm{pre}(\overline{IT} \gg AOp) \Leftrightarrow \mathrm{pre}\, COp_1)$$
$$\forall A.\mathit{STATE};\ C.\mathit{STATE};\ C.\mathit{STATE}' \bullet R \land COp_1 \Rightarrow (\mathrm{pre}\, COp_2)'$$
$$\forall A.\mathit{STATE};\ C.\mathit{STATE};\ C.\mathit{STATE}' \bullet$$
$$(IT \gg COp_1 \mathbin{{}_{9}^{\circ}} COp_2) \land R \Rightarrow \exists A.\mathit{STATE}' \bullet R' \land (AOp \gg OT) \qquad \square$$

As in Z we sometimes need some explicit substitutions to ensure that the inputs and outputs in the decomposition remain distinct. Again these are only necessary when the decomposition of *AOp* involves more than one occurrence of the same input or output parameter names in the concrete operations. If COp_1 and COp_2 are distinct operations with distinct parameter names then the formalisation is simplified by the omission of the substitutions.

When *IT* and *OT* are the identity the definition gives, as before, non-atomic refinement without IO transformations.

Example 16.4 In our ATM machines the *Passwd* operation accepted the PIN in one go, but in an implementation we wish to accept the 4 digits one at a time, with each digit being entered by a separate operation. We are therefore going to split this single operation into a sequence consisting of 4 operations: *First*, *Second*, *Third* and *Fourth*. Note that, as usual, the denial to proceed is only flagged after all the digits have been inserted. The definitions of the other operations are the same as before and are elided.

ATM$_3$

$accts : \mathbb{N} \twoheadrightarrow \mathbb{N}$
$pins : \mathbb{N} \twoheadrightarrow \mathbb{N} \times \mathbb{N} \times \mathbb{N} \times \mathbb{N}$
$card : \mathbb{N}$
$trans : \mathbb{B}$
$temp : \mathrm{seq}\, \mathbb{N}$

$\mathrm{dom}\, accts = \mathrm{dom}\, pins$
$0 \notin \mathrm{dom}\, accts$

$\mathit{INIT} \mathrel{\widehat=} \cdots$
$Eject \mathrel{\widehat=} \cdots$
$OpenAcct \mathrel{\widehat=} \cdots$
$Insert \mathrel{\widehat=} \cdots$
$Withdraw \mathrel{\widehat=} \cdots$

$$\begin{array}{|l}
\textit{First} \\
\Delta(temp) \\
c? : \mathbb{N} \\
\hline
card \neq 0 \land \neg trans \\
temp = \langle\rangle \land temp' = \langle c? \rangle
\end{array}$$

$$\begin{array}{|l}
\textit{Second} \\
\Delta(temp) \\
c? : \mathbb{N} \\
\hline
\#temp = 1 \land temp' = temp \frown \langle c? \rangle
\end{array}$$

$$\begin{array}{|l}
\textit{Third} \\
\Delta(temp) \\
c? : \mathbb{N} \\
\hline
\#temp = 2 \\
temp' = temp \frown \langle c? \rangle
\end{array}$$

$$\begin{array}{|l}
\textit{Fourth} \\
\Delta(trans) \\
c? : \mathbb{N} \\
\hline
\#temp = 3 \\
pins(card) = \langle temp(1), temp(2), temp(3), c? \rangle \Rightarrow trans' \\
pins(card) \neq \langle temp(1), temp(2), temp(3), c? \rangle \Rightarrow \neg trans'
\end{array}$$

Here *Passwd* is refined by the sequence $First \mathbin{;} Second \mathbin{;} Third \mathbin{;} Fourth$ modulo some renaming of inputs,[2] however, a single instance of *First* does not correspond on its own to any abstract level operation, but only a bit of *Passwd*'s functionality.

To verify the refinement between ATM_3 and ATM_1 we use an input transformer *IT* that maps an abstract input and a sequence of concrete inputs representing the inputs needed in the decomposition.

$$\begin{array}{|l}
\textit{IT} \\
pin? : \mathbb{N} \times \mathbb{N} \times \mathbb{N} \times \mathbb{N} \\
d! : \operatorname{seq} \mathbb{N} \\
\hline
pin? = (d!(1), d!(2), d!(3), d!(4))
\end{array}$$

This takes in a single input *pin*? for the abstract operation *Passwd* and maps it to a sequence for use by the concrete operations. The concrete operations then use this sequence one by one.

There are no outputs in this particular example so the output transformer is the identity. The net result is the following set of criteria that need to be verified.

[2] Failing to use renaming here would identify all four digits, limiting the specification to a very commonly used and thus very insecure subset of pin codes.

$$
\begin{array}{l}
R \wedge (IT \gg (First[d?(1)/c?] \fatsemi \\
\qquad\qquad Second[d?(2)/c?] \fatsemi Third[d?(3)/c?] \fatsemi Fourth[d?(4)/c?])) \\
\qquad \Rightarrow \exists ATM_0.\mathit{STATE}' \bullet R' \wedge Passwd \\
R \Rightarrow (\mathrm{pre}(\overline{IT} \gg Passwd) \Leftrightarrow \mathrm{pre}\, First[d?(1)/c?]) \\
R \wedge First[d?(1)/c?] \Rightarrow (\mathrm{pre}\, Second[d?(2)/c?])' \\
R \wedge (First[d?(1)/c?] \fatsemi Second[d?(2)/c?]) \Rightarrow (\mathrm{pre}\, Third[d?(3)/c?])' \\
R \wedge (First[d?(1)/c?] \fatsemi Second[d?(2)/c?] \fatsemi Third[d?(3)/c?]) \\
\qquad \Rightarrow (\mathrm{pre}\, Fourth[d?(4)/c?])'
\end{array}
$$

Remember that substitutions, e.g., $[d?(1)/c?]$, are used in order to restore the distinction between the individual inputs, and they allow the process of consuming the inputs one by one to be described explicitly in the refinement conditions. □

The next example shows that initialisation can also be decomposed in a non-atomic refinement.

Example 16.5 In a Monte Carlo simulation an approximation to the minimum of a function on an interval is obtained by computing the function at random values in the interval. We specify a class MC_0 which performs this approximation for a given function f on [0, 1], the set of reals between 0 and 1 inclusive.

The initialisation generates the random numbers, stored in a sequence r, and computes the first value. *Check* then computes the next approximation by using the head of r.

MC_0

$$
\begin{array}{l}
r : \mathrm{seq}\, \mathbb{R} \\
h : \mathbb{R}
\end{array}
$$

INIT

$$\#r = 1000 \wedge \#r = \#\mathrm{ran}\, r \wedge \mathrm{ran}\, r \subseteq [0, 1] \wedge h = f(0)$$

Check

$$
\begin{array}{l}
\Delta(r, h) \\
h! : \mathbb{R} \\
\hline
r \neq \langle\rangle \\
r' = tail\, r \\
h' = min(f(head\, r), h) \\
h! = h'
\end{array}
$$

We are going to refine the atomicity of the initialisation so that the generation of random numbers is done one by one rather than all at once. First of all we split the initialisation to define an explicit *Generate* operation.

MC_1

$r : \text{seq}\,\mathbb{R}$
$h : \mathbb{R}$
$b : \mathbb{B}$

INIT
$b \wedge h = f(0)$

Generate
$\Delta(r, b)$

$b \wedge \neg b'$
$\#r' = 1000 \wedge \#r' = \#\text{dom}\,r' \wedge \text{ran}\,r' \subseteq [0, 1]$

Check
$\Delta(r, h)$
$h! : \mathbb{R}$

$\neg b$
$r \neq \langle\rangle$
$r' = tail\,r$
$h' = min(f(head\,r), h)$
$h! = h'$

The retrieve relation between these two classes is the projection onto $\neg b$. Verifying the non-atomic refinement for the initialisation amounts to showing

$$\forall MC_1.\text{STATE} \bullet (MC_1.\text{INIT} \,{}_{9}^{\circ}\, MC_1.Generate) \Rightarrow \exists MC_0.\text{STATE} \bullet MC_0.\text{INIT} \wedge R$$
together with $\quad MC_1.\text{INIT} \Rightarrow \text{pre}\,MC_1.Generate$

We now perform the refinement in which the random numbers are generated one by one, and we also refine the atomicity of *Check*.

MC_2

$r : \text{seq}\,\mathbb{R}$
$random : \mathbb{P}\,\mathbb{R}$
$h, x, y : \mathbb{R}$
$b, c : \mathbb{B}$
$n : \mathbb{N}$

INIT
$b \wedge h = f(0) \wedge random = \varnothing \wedge c$

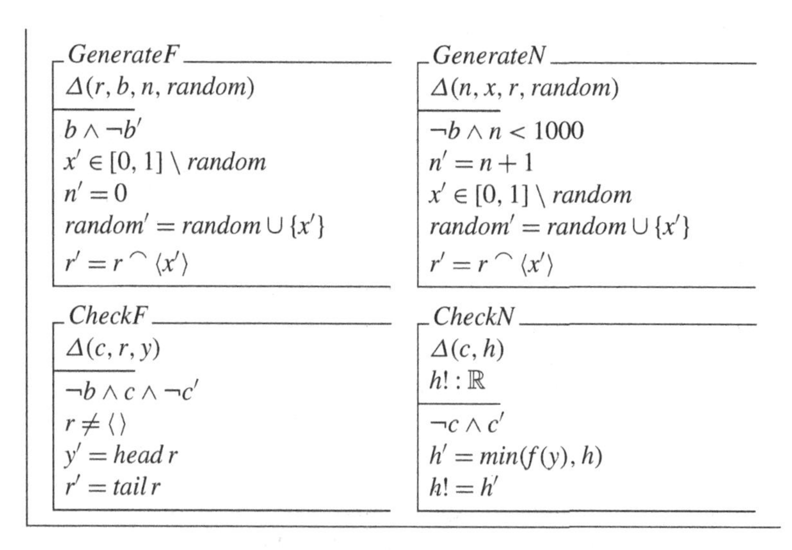

Then $MC_1.Generate$ is refined by $(MC_2.GenerateF \mathbin{\text{\textsf{;}}} MC_2.GenerateN^{999})$ and $MC_1.Check$ by $(MC_2.CheckF \mathbin{\text{\textsf{;}}} MC_2.CheckN)$. Furthermore, *GenerateN* and the *Check* operations can now be interleaved, opening the way for a parallel implementation. □

Coupled simulations can also be used in Object-Z. Indeed, their definition was first defined for that use, and the conditions given in Sect. 12.7.4 have to be used on top of the appropriate blocking definition of refinement as given in Definition 16.5.

16.4 Refinement, Subtyping and Inheritance

The notions of inheritance and subtyping are sometimes confused with refinement. As we have seen, inheritance in Object-Z has no relation to refinement. Inheritance is a syntactic device to allow a certain amount of reuse. However, since it allows additional constraints to be placed in the predicate of an operation, an inherited class is not necessarily a refinement of the base class.

Of course sometimes an inheritance can be a refinement; in Example 16.3 ATM_2 is a (weak) refinement of the class it inherits. Furthermore, any operation refinement can be given as an inheritance by simply inheriting the reduction of non-determinism that the refinement defines (compare Lemma 2.3). Clearly, however, this does not hold for data refinement.

The relation between subtyping and refinement is less clear-cut, and depends on the definitions of subtyping used. The ideas behind subtyping and refinement coincide when discussing the behaviour of individual operations, in that subtyping is usually phrased in terms of pre- and postconditions, although some definitions do allow widening of preconditions. However, there are two ways in which subtyping is sometimes extended:

- by considering non-conformal data types, i.e., the addition of new operations;
- by considering history invariants on the objects behaviour.

When allowing the addition of new operations in a subtype the behaviour of these operations is sometimes constrained to be derivable from the existing operation, so that the behaviour of the subtype is indistinguishable from that of the supertype. When this is the case the subtype is a non-atomic refinement of the supertype. If arbitrary new operations are allowed in the subtype, no correspondence exists with refinement.

The further aspect of subtyping that differs from refinement is the possibility of history invariants constraining the subtype's behaviour, which some definitions of subtyping allow. Such temporal constraints are not covered by the step-by-step simulation rules we use for refinement, although it should be noted that some languages (e.g., automata) include preservation of liveness properties as part of the definition of refinement.

Indeed early versions of Object-Z had a notion of history invariant which could be included in a class definition to express safety and liveness properties to which the objects of a class had to conform. Using such a notation, liveness constraints could be placed on the class ATM_2 to express the property that resets must happen infinitely often. This would stop the bank from perpetually refusing the *Withdraw* operation from being enabled. Such a history invariant appears at the end of a class definition, as the following example illustrates, where the temporal logic operators $\Box$ (always), $\Diamond$ (eventually) and $\bigcirc$ (next) can be used in the description.

ATM_4

ATM_2

$\Box(limit \geq 4) \Rightarrow \Diamond((\bigcirc limit) < limit)$

16.5 Bibliographical Notes

Definitions of the simulation rules for Object-Z were first given by Smith and Derrick [19] for a restricted subset of Object-Z where classes could not contain objects as state variables (which were unnecessary as the language was being combined with CSP). The definition of coupled simulations for Object-Z was first given in [9].

Refinement in Object-Z is also briefly discussed by Smith in [17] and Duke et al. in [10] in the context of behavioural compatibility. The central view of behavioural compatibility taken in [10] is the same notion that underlies refinement, i.e., an object ob_2 is compatible with an object ob_1 if ob_1 can be replaced by ob_2 without the substitution resulting in any detectable difference in behaviour. However, the blocking model defined in [10] differentiates between output messages controlled by the object and input messages which are under the control of the environment. It is also concerned with object substitutability whereas refinement is concerned with the behaviour and substitutability of the whole specification. Refinement in Object-Z as we have discussed here takes a simpler view in that operations are atomic and are under control of the environment irrespective of whether they contain input or output declarations.

Refinement methods have also been considered for a number of other object-oriented specification languages. In [4] a refinement methodology for OO action systems is presented, OO action systems being based upon the action systems formalism. Refinement for OO action systems is defined within the refinement calculus framework [2], and refinement is designed to preserve trace behaviour. There has also been work on refinement for object-oriented languages based around TLA [15] which includes, for example, DisCo [14] and TLO [6].

Work that is relevant to subtyping [1] in Object-Z includes that of [8, 16, 20]. Ideas on how to reconcile refinement with subtyping where new operations can be added are contained in [11, 21] which looks at these questions in the context of CSP and its failures-divergences semantics.

References

1. America, P. (1990). A parallel object-oriented language with inheritance and subtyping. In *Proceedings of the OOPSLA/ECOOP '90 Conference on Object-Oriented Programming Systems, Languages and Applications* (pp. 161–168). Published as ACM SIGPLAN Notices, volume 25, number 10.
2. Back, R. J. R., & von Wright, J. (1994). Trace refinement of action systems. In B. Jonsson & J. Parrow (Eds.), *CONCUR'94: Concurrency Theory*, Uppsala, Sweden, August 1994. *Lecture Notes in Computer Science: Vol. 836* (pp. 367–384). Berlin: Springer.
3. Bolton, C., & Davies, J. (2002). Refinement in Object-Z and CSP. In M. Butler, L. Petre, & K. Sere (Eds.), *Integrated Formal Methods (IFM 2002). Lecture Notes in Computer Science: Vol. 2335* (pp. 225–244). Berlin: Springer.
4. Bonsangue, M., Kok, J., & Sere, K. (1999) Developing object-based distributed systems. In Ciancarini et al. [7] (pp. 19–34).
5. Bowen, J. P. & Hinchey, M. G. (Eds.) (1995). *ZUM'95: the Z Formal Specification Notation. Lecture Notes in Computer Science: Vol. 967*. Limerick: Springer.
6. Canver, E., & von Henke, F. W. (1999) Formal development of object-based systems in a temporal logic setting. In Ciancarini et al. [7] (pp. 419–436).
7. Ciancarini, P., Fantechi, A., & Gorrieri, R. (Eds.) (1999). *Formal Methods for Open Object-Based Distributed Systems*. Dordrecht: Kluwer Academic.
8. Cusack, E. (1991). Inheritance in object oriented Z. In P. America (Ed.), *ECOOP '91—Object-Oriented Programming. Lecture Notes in Computer Science: Vol. 512* (pp. 167–179). Berlin: Springer.
9. Derrick, J., & Wehrheim, H. (2003). Using coupled simulations in non-atomic refinement. In D. Bert, J. P. Bowen, S. King, & M. Waldén (Eds.), *ZB 2003: Formal Specification and Development in Z and B. Lecture Notes in Computer Science: Vol. 2651* (pp. 127–147). Berlin: Springer.
10. Duke, R., Bailes, C., & Smith, G. (1996). A blocking model for reactive objects. *Formal Aspects of Computing*, *8*(3), 347–368.
11. Fischer, C., & Wehrheim, H. (2000). Behavioural subtyping relations for object-oriented formalisms. In T. Rus (Ed.), *Algebraic Methodology and Software Technology. Lecture Notes in Computer Science: Vol. 1816* (pp. 469–483). Berlin: Springer.
12. Fitzgerald, J. A., Jones, C. B., & Lucas, P. (Eds.) (1997). *FME'97: Industrial Application and Strengthened Foundations of Formal Methods. Lecture Notes in Computer Science: Vol. 1313*. Berlin: Springer.

13. Hinchey, M. G. & Liu, S. (Eds.) (1997). *First International Conference on Formal Engineering Methods (ICFEM'97)*. Hiroshima, Japan, November 1997. Los Alamitos: IEEE Comput. Soc.
14. Järvinen, H. M., & Kurki-Suonio, R. (1991). DisCo specification language: marriage of actions and objects. In *11th International Conference on Distributed Computing Systems*, Washington, DC, USA, May 1991 (pp. 142–151). Los Alamitos: IEEE Comput. Soc.
15. Lamport, L. (1994). The temporal logic of actions. *ACM Transactions on Programming Languages and Systems*, *16*(3), 872–923.
16. Liskov, B., & Wing, J. M. (1995) Specifications and their use in defining subtypes. In Bowen and Hinchey [5] (pp. 245–263).
17. Smith, G. (1995). A fully abstract semantics of classes for Object-Z. *Formal Aspects of Computing*, *7*(3), 289–313.
18. Smith, G. (1997) A semantic integration of Object-Z and CSP for the specification of concurrent systems. In Fitzgerald et al. [12] (pp. 62–81).
19. Smith, G., & Derrick, J. (1997) Refinement and verification of concurrent systems specified in Object-Z and CSP. In Hinchey and Liu [13] (pp. 293–302).
20. Strulo, B. (1995) How firing conditions help inheritance. In Bowen and Hinchey [5] (pp. 264–275).
21. Wehrheim, H. (2000). Behavioural subtyping and property preservation. In S. F. Smith & C. L. Talcott (Eds.), *Formal Methods for Open Object-Based Distributed Systems (FMOODS 2000)* (pp. 213–231). Dordrecht: Kluwer Academic.

Chapter 17
Class Refinement

Having looked at how a single Object-Z class is refined by another, we now consider more complex Object-Z specifications, and in particular ones that involve objects and promotions of their operations.

As soon as we have objects in a class, we have to deal with object references and look at how they should be treated in a refinement. The issue that arises is the following. In the refinements we have looked at so far a simulation compares each variable before and after an operation, that is, it compares x and x' for every variable x. However, if x is now a reference to an object and an operation is applied to x then the value of the object referenced by x may change, but its reference is still denoted by x. This has obvious consequences when we interpret the simulation rules in Object-Z, and Sect. 17.1 looks at this issue.

Following on from this the subsequent section looks at how the structure of the classes can be altered in a refinement. For example, clearly the classes *Count* and *CountAgain* from Example 15.2 have identical behaviour, and our simulation rules should be able to show that they are refinements of each other.

The final issue we look at in this chapter is that of compositionality. In Chap. 6 we discussed conditions when, in Z, a promotion of a refinement was a refinement of a promotion, and there is a similar issue in Object-Z. If a class A is refined to a class B, are classes that contain objects of type B refinements of those that contain objects of type A?

We begin first, though, by looking in general at how to treat object references in a refinement.

17.1 Objects and References

Our refinement methodology works by viewing a specification's behaviour through its system class. Therefore what we need to do now is to show how this can be used to verify refinements which involve objects and operations applied to them.

J. Derrick, E.A. Boiten, *Refinement in Z and Object-Z*,
DOI 10.1007/978-1-4471-5355-9_17, © Springer-Verlag London 2014

Because a class defines a type, objects may be used in a variety of ways in a specification. Consider the following example which uses the *BankAcct* class defined in Sect. 15.2.1 (p. 369).

$Bank_0$

$accts : \mathbb{P}\, BankAcct$
$account : BankAcct$

INIT
$accts = \varnothing$
$account.\text{INIT}$

Pick
$o! : BankAcct$
$o! \in accts$

Select
$o? : BankAcct$
$o? \in accts$

$Balance \mathrel{\widehat{=}} Select \bullet o?.Balance \wedge \Xi(accts) \wedge account.\Xi$

Open
$\Delta(accts)$
$o? : BankAcct$
$o? \notin accts$
$accts' = accts \cup \{o?\}$
$o?.\text{INIT}$

$OpenAcct \mathrel{\widehat{=}} Open \bullet \Xi(accts) \wedge account.\Xi$

OD
$\Delta(accts, account)$
$account' \notin accts$
$accts' = accts \cup \{account'\}$
$account'.\text{INIT}$

$OpenDeposit \mathrel{\widehat{=}} OD \bullet account'.Deposit \wedge \Xi(accts)$
$ChangeName \mathrel{\widehat{=}} [\Delta(account) \mid account' \neq account] \bullet \Xi(accts)$

This $Bank_0$ class uses object instantiations in a variety of ways. For example, there is a simple promotion of the *Balance* operation on a particular object $o?$. *Pick* non-deterministically selects an output reference. *Open* on the other hand adds to the existing object references contained in *accts*, whereas *ChangeName* and *OpenDeposit* alter a particular reference *account* and the latter even promotes an operation on this new reference.

A potential refinement of this class is the following, where we have replaced $\mathbb{P}\, BankAcct$ by $\operatorname{seq} BankAcct$ and reduced some of the non-determinism in *Pick* (assuming *money* is visible in *BankAcct*) and *ChangeName* (assuming *name* is visible).

$$
\begin{array}{l}
\textbf{Bank}_1 \\
\quad accts : \text{seq}\, BankAcct \\
\quad account : BankAcct \\[1ex]
\quad \textit{INIT} \\
\quad\quad accts = \langle\,\rangle \\
\quad\quad account.\textit{INIT} \\[1ex]
\quad Select \\
\quad\quad o? : BankAcct \\
\quad\quad \hline \\
\quad\quad o? \in \text{ran}\, accts \\[1ex]
\quad Balance \mathrel{\widehat{=}} Select \bullet o?.Balance \land \Xi(\text{ran}\, accts) \land account.\Xi \\[1ex]
\quad Pick \\
\quad\quad o! : BankAcct \\
\quad\quad \hline \\
\quad\quad o! \in \text{ran}\, accts \land \forall\, o : \text{ran}\, accts \bullet o!.money \leq o.money \\[1ex]
\quad Open \\
\quad\quad \Delta(accts) \\
\quad\quad o? : BankAcct \\
\quad\quad \hline \\
\quad\quad o? \notin \text{ran}\, accts \\
\quad\quad accts' = accts \frown \langle o? \rangle \\
\quad\quad o?.\textit{INIT} \\[1ex]
\quad OpenAcct \mathrel{\widehat{=}} Open \bullet \Xi(\text{ran}\, accts) \land account.\Xi \\[1ex]
\quad OD \\
\quad\quad \Delta(accts, account) \\
\quad\quad \hline \\
\quad\quad account' \notin \text{ran}\, accts \\
\quad\quad accts' = accts \frown \langle account' \rangle \\
\quad\quad account'.\textit{INIT} \\[1ex]
\quad OpenDeposit \mathrel{\widehat{=}} OD \bullet account'.Deposit \land \Xi(\text{ran}\, accts) \\[1ex]
\quad CN \\
\quad\quad \Delta(account) \\
\quad\quad \hline \\
\quad\quad account' \neq account \land account'.name \neq account.name \\[1ex]
\quad ChangeName \mathrel{\widehat{=}} CN \bullet \Xi(\text{ran}\, accts)
\end{array}
$$

To verify this refinement we need to understand the effect on both the references and the objects referenced in the class when an operation occurs. That is, we will determine the meaning of $\theta State$ when the state contains object references.

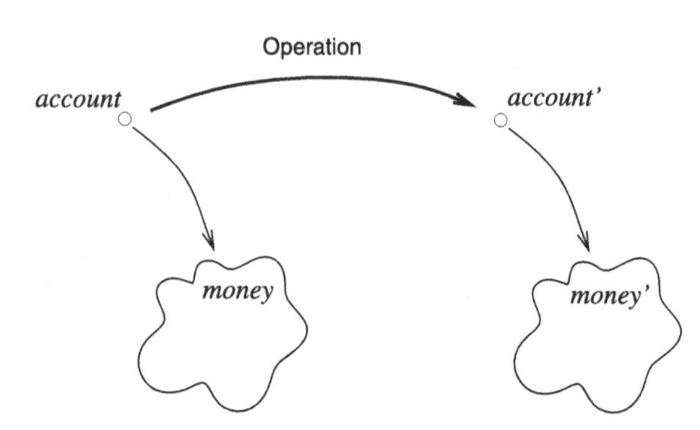

Fig. 17.1 The evolution of the *account* object

17.1.1 Understanding Object Evolution

When an object reference is declared, e.g., $account : BankAcct$, both the reference and the value of the object referenced may be changed by an operation. Thus *account* is replaced by $account'$ and $account.money$ evolves to $account'.money'$ (according to the predicate of the operation), see Fig. 17.1.

To understand the latter, note that $account'.money$ refers to the pre-state value of the variable *money* of $account'$, whereas we wish to consider the post-state $money'$ in the object referenced now by $account'$, i.e., $account'.money'$. Expressions such as $account'.money'$ are not part of the Object-Z language (i.e., we cannot write them within an Object-Z specification), however our usage here is consistent with their use in defining the language. In particular, this is the same mechanism used in the design of the language when defining $\Delta State$.

Because a simulation compares $A.\mathit{STATE}$ with $A.\mathit{STATE}'$ our refinement proof obligations will be phrased in these terms. Therefore, although expressions such as $account'.money'$ do not occur in an Object-Z specification, they will occur in a retrieve relation or as a primed state in the simulation conditions.

For any class A, we define $\theta A.\mathit{STATE}$ to be the normal Z binding, i.e., one giving values of all variables defined in A including any object references. We also define, for any object reference a, $\theta a.\mathit{STATE}$ as the binding of the variables of the instance a.

To verify refinements involving references to objects we need to define the relation corresponding to an operation in the presence of object references. In Z the relation corresponding to an operation AOp is given by

$$\mathsf{AOp} == \{AOp \bullet \theta A.\mathit{STATE} \mapsto \theta A.\mathit{STATE}'\}$$

In Object-Z we also need to take into account any changes of state in objects referenced in the class where AOp is defined. This can be formalised by giving Object-Z a history semantics (see, for example, [7]) which records the state of a class together with the state of any objects referenced in that class. An operation AOp changes this collective state, and this defines a relation corresponding to that operation.

For example, if we define

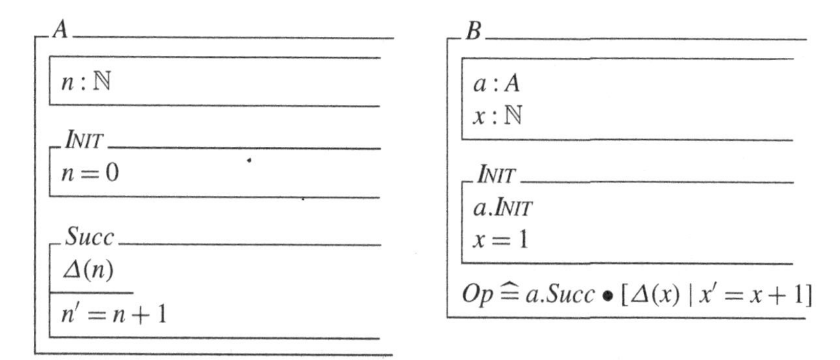

we can represent the state of B as a pair giving the state of B and the state of objects referenced from B. One such value might be, for some $\iota \in \mathbb{O}_A$, $(\langle\!| x == 1, a == \iota |\!\rangle, \langle\!| n == 0 |\!\rangle_a)$.

In this pair the first element is the state of B, i.e., $\theta B.\mathit{STATE}$, and the second element the state of the object referenced by a (hence the subscript). Application of the operation Op defines a relation on this state space, and, for example, maps $(\langle\!| x == 1, a == \iota |\!\rangle, \langle\!| n == 0 |\!\rangle_a)$ to $(\langle\!| x == 2, a == \iota |\!\rangle, \langle\!| n == 1 |\!\rangle_a)$.

17.1.2 Verifying the Bank Refinement

To see these ideas in practice, let us look at the operations in the *Bank* classes.

The operations in $Bank_0$ define relations which change the values of the object references *accts* and *account* and also change values in the state of the referenced objects. Operations in $Bank_1$ will define similar relations. For example, $ChangeName_0$ alters the *account* component in the state, and thus the relation corresponding to this operation is

$$\begin{array}{l}\{(\langle\!| accts == \Sigma, account == \iota |\!\rangle, \langle\!| money == n |\!\rangle_{account}, \ldots) \mapsto \\ \quad (\langle\!| accts == \Sigma, account == \kappa |\!\rangle, \langle\!| money == m |\!\rangle_{account}, \ldots) \\ \qquad \mid \iota, \kappa \in \mathbb{O}_{BankAcct} \wedge \kappa \neq \iota \wedge \Sigma \subseteq \mathbb{O}_{BankAcct} \wedge n, m \in \mathbb{N}\}\end{array}$$

To verify the refinement we use a retrieve relation R between the two specifications.

R

$Bank_0.\mathit{STATE}$
$Bank_1.\mathit{STATE}$

$Bank_0.accts = \operatorname{ran} Bank_1.accts$
$Bank_0.account = Bank_1.account$

To simplify the presentation we will write $account_0$ in place of $Bank_0.account$ etc. Then R' will be

$$\begin{array}{|l}\hline R' \\ \hline Bank_0.\mathit{STATE}' \\ Bank_1.\mathit{STATE}' \\ \hline accts'_0 = \operatorname{ran} accts'_1 \\ account'_0 = account'_1 \\ \hline \end{array}$$

Then to verify the refinement of the *Pick* operation we have to show (R is functional):

$$\forall Bank_0.\mathit{STATE};\ Bank_1.\mathit{STATE} \bullet R \Rightarrow (\operatorname{pre} Pick_0 \Leftrightarrow \operatorname{pre} Pick_1)$$
$$\forall Bank_0.\mathit{STATE};\ Bank_0.\mathit{STATE}';\ Bank_1.\mathit{STATE};\ Bank_1.\mathit{STATE}' \bullet$$
$$R \wedge Pick_1 \wedge R' \Rightarrow Pick_0$$

Applicability is straightforward, and expanding correctness the relevant part becomes the following which is easily verified.

$$\begin{array}{l} \forall accts_0, accts_1, accts'_0, accts'_1 \bullet \\ \quad (accts_0 = \operatorname{ran} accts_1) \wedge \\ \quad (o! \in \operatorname{ran} accts_1 \wedge \forall o : \operatorname{ran} accts_1 \bullet o!.money \leq o.money) \wedge \\ \quad (accts'_1 = accts_1 \wedge accts'_0 = \operatorname{ran} accts'_1) \\ \qquad \Rightarrow (o! \in accts_0 \wedge accts'_0 = accts_0) \end{array}$$

Notice again the use of the information from the Δ-list of *Pick*.

In a similar way the refinement of *Open* can be verified, where the inclusion of the predicate $o?.\mathit{INIT}$ merely adds the deduction $o?.\mathit{INIT} \Rightarrow o?.\mathit{INIT}$ to the conditions.

For *ChangeName* the applicability condition is (R is the identity on *account* so we drop the subscripts):

$$\begin{array}{l} \exists account' \bullet account' \neq account \wedge account'.name \neq account.name \\ \quad \Leftrightarrow \exists account' \bullet account' \neq account \end{array}$$

Since there is an object reference for each instantiation of *BankAcct*, applicability is obvious assuming that *NAME* contains more than one element.

Now let us consider the two operations which involve promotions of operations on objects. In the case of *Balance* applicability is:

$$\begin{array}{l} \forall accts_0, accts_1 \bullet R \Rightarrow \\ \quad (\operatorname{pre}(Select_0 \bullet o?.Balance) \Leftrightarrow \operatorname{pre}(Select_1 \bullet o?.Balance)) \end{array}$$

Now since $\operatorname{pre}(Select_0 \bullet o?.Balance) = \operatorname{pre} Select_0 \bullet \operatorname{pre}(o?.Balance)$ applicability follows. Correctness is similar, also requiring the obvious $\Xi(accts_0) \Rightarrow \Xi(\operatorname{ran} accts_1)$.

Finally for *OpenDeposit*, applicability requires calculation of the precondition of $(OD \bullet account'.Deposit \wedge \cdots)$ in each bank class. Upon doing this we find that both evaluate to the conjunction of $acct \neq account$ and $\mathit{INIT} \Rightarrow \operatorname{pre} Deposit$ in *BankAcct*.

17.2 Class Simulations

Having seen how to verify refinements which involve objects as variables, we will now look at refinements where the structure of the classes in the specifications changes. We call such refinements *class simulations*. To motivate the discussion recall the *Count* example, where the two classes are given by:

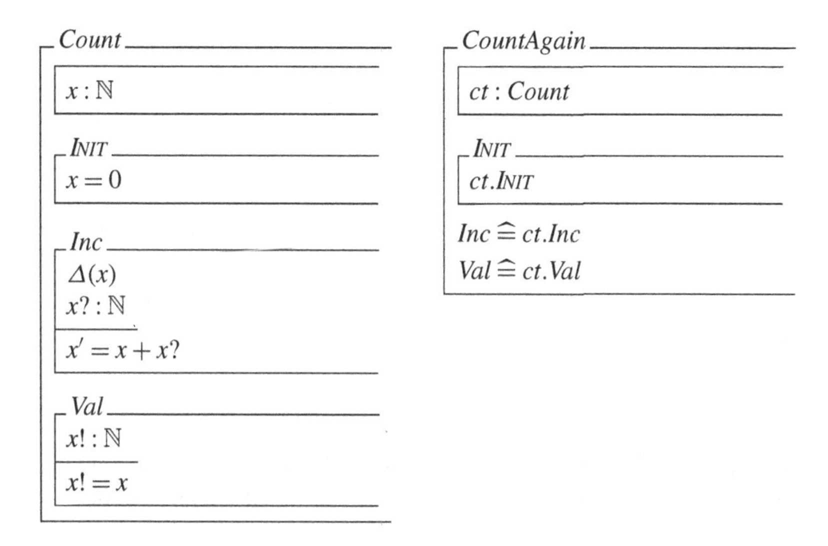

Clearly the behaviours of *Count* and *CountAgain* are identical, and our simulation rules should be able to verify the refinement. Before we do so we introduce some notation to make clear which classes are included in a specification and which is the system class.

Definition 17.1 If a specification includes classes $C_1, \ldots, C_n$ together with *Main* as the system class, then we refer to the complete specification by $(Main \bullet C_1, \ldots, C_n)$. □

Example 17.1 We now show that *Count* is refined by $(CountAgain \bullet Count)$. The retrieve relation R is:

```
┌─R─────────────────────────
│ Count.STATE
│ CountAgain.STATE
├───────────────
│ x = ct.x
└───────────────────────────
```

Thus the predicate in R' will be $x' = ct'.x'$. For the operation *Inc*, applicability requires

$$\forall x, ct \bullet R \Rightarrow (\text{pre}\, Inc \Leftrightarrow \text{pre}(ct.Inc))$$

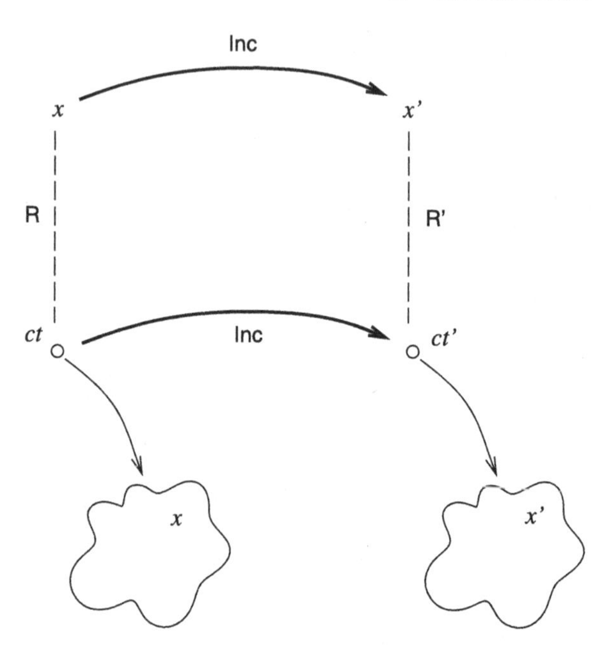

Fig. 17.2 The effect of the operation *Inc*

The Object-Z reference manual [8] gives rules for defining expansions of Object-Z operation schemas, and if we expand *ct.Inc* to its equivalent semantic form

Inc

$\Delta(ct.x, ct)$
$x? : \mathbb{N}$

$ct'.x' = ct.x + x?$
$ct' = ct$

where $\Delta(ct.x)$ allows the variable x to change in the object referenced by ct. Applicability then reduces to

$$\forall x, ct \bullet x = ct.x \Rightarrow (x + x? \in \mathbb{N} \Leftrightarrow ct.x + x? \in \mathbb{N})$$

In a similar fashion correctness requires:

$$\forall x, ct, x', ct' \bullet R \wedge ct.Inc \wedge R' \Rightarrow Inc$$

To show this, we again use the expanded form of *ct.Inc*, but, in addition, since ct is not in the Δ-list of the operation, the predicate $ct = ct'$ is available to us making the deduction straightforward. Figure 17.2 illustrates the changes of state, objects and object references that take place when the operation *Inc* is applied. □

As this example demonstrates, the technique for verifying class simulations involves:

- expanding the operations to their equivalent semantic definition;
- using information from the Δ-list about which object references are unaltered; and
- decorating all variables in the predicate of a retrieve relation R with $'$ in deriving R' as described above.

Example 17.2 Consider the teller machine ATM_0 in Example 16.1 (p. 387). In an implementation the branch of a user is not necessarily co-located with the cashpoint being used, and we therefore want to split this description into two separate classes: *CashPoint* and *Bank*. These are given as follows.

Bank

$\upharpoonright(accts, \mathrm{INIT}, OpenAcct, Withdraw)$

$$accts : \mathbb{N} \twoheadrightarrow \mathbb{N}$$

INIT

$$accts = \varnothing$$

OpenAcct

$$\Delta(accts)$$
$$account! : \mathbb{N}$$

$$account! = (max\,\mathrm{dom}\,accts) + 1$$
$$accts' = accts \cup \{(account!, 0)\}$$

Withdraw

$$\Delta(accts)$$
$$amt! : \mathbb{B}$$
$$amount?, money!, acct? : \mathbb{N}$$

$$accts' = accts \oplus \{acct? \mapsto accts(acct?) - money!\}$$
$$((money! = 0 \land \neg amt!) \lor$$
$$(0 \leq money! \leq amount? \land amt!))$$

CashPoint

$\upharpoonright(pins, \mathrm{INIT}, Eject, SupplyPin, Insert, Passwd, Withdraw)$

$$pins : \mathbb{N} \twoheadrightarrow \mathbb{N} \times \mathbb{N} \times \mathbb{N} \times \mathbb{N}$$
$$card : \mathbb{N}$$
$$trans : \mathbb{B}$$

INIT

$$pins = \varnothing \land card = 0 \land \neg trans$$

Eject

$$\Delta(card, trans)$$

$$card' = 0 \land \neg trans'$$

SupplyPin

$$\Delta(pins)$$
$$account? : \mathbb{N}$$
$$pin? : \mathbb{N} \times \mathbb{N} \times \mathbb{N} \times \mathbb{N}$$

$$pins' = pins \cup \{(account?, pin?)\}$$

Insert

$$\Delta(card)$$
$$card? : \mathbb{N}$$

$$card = 0 \land account? \in \operatorname{dom} pins$$
$$card' = card?$$

Passwd

$$\Delta(trans)$$
$$pin? : \mathbb{N} \times \mathbb{N} \times \mathbb{N} \times \mathbb{N}$$

$$card \neq 0 \land \neg trans$$
$$pins(card) = pin? \Rightarrow trans'$$
$$pins(card) \neq pin? \Rightarrow \neg trans'$$

Withdraw

$$\Delta(card, trans)$$
$$amt? : \mathbb{B}$$
$$acct! : \mathbb{N}$$

$$trans$$
$$acct! = card$$
$$((card' = 0 \land \neg amt? \land \neg trans') \lor$$
$$(card' = card \land amt? \land trans'))$$

The complete behaviour, including communication between the two components, is given by the main class ATM_5. This includes a bank and a cashpoint, and it promotes operations from these objects to the overall class. A customer can then *Withdraw* from the bank when the cashpoint gives permission to proceed and has communicated for which *acct*! permission is being granted. The communication $\|$ then identifies *acct*! with *acct*? in *Withdraw* and the correct account is debited.

ATM_5

$$c : CashPoint$$
$$b : Bank$$

$$\operatorname{dom} b.accts = \operatorname{dom} c.pins$$

INIT

$$b.\mathrm{INIT} \land c.\mathrm{INIT}$$

$$
\begin{array}{|l}
Insert \mathrel{\widehat=} c.Insert \land b.\Xi \\
Eject \mathrel{\widehat=} c.Eject \land b.\Xi \\
Passwd \mathrel{\widehat=} c.Passwd \land b.\Xi \\
OpenAcct \mathrel{\widehat=} (c.SupplyPin \parallel_! b.OpenAcct) \\
Withdraw \mathrel{\widehat=} (b.Withdraw \parallel c.Withdraw)
\end{array}
$$

Then ATM_5 is a refinement of the class ATM_0. To show this formally we need to define a retrieve relation between the two main classes:

$$
\begin{array}{|l}
\multicolumn{1}{l}{R} \\
\hline
ATM_0.\mathit{STATE} \\
ATM_5.\mathit{STATE} \\
\hline
ATM_0.accts = ATM_5.b.accts \\
ATM_0.pins = ATM_5.c.pins \\
ATM_0.trans = ATM_5.c.trans \\
ATM_0.card = ATM_5.c.card \\
\hline
\end{array}
$$

Then, for example, for the *Insert* operation we need to verify that

$$
\begin{array}{l}
R \Rightarrow (\mathrm{pre}\, ATM_0.Insert \Leftrightarrow \mathrm{pre}\, ATM_5.Insert) \\
R \land ATM_5.Insert \land R' \Rightarrow ATM_0.Insert
\end{array}
$$

For example, applicability for *Insert* now requires that

$$
\begin{array}{l}
(pins = c.pins \land card = c.card) \Rightarrow \\
\quad ((card = 0 \land card? \in \mathrm{dom}\, pins) \Leftrightarrow c.card = 0 \land card? \in \mathrm{dom}\, c.pins)
\end{array}
$$

Similarly correctness requires that

$$
\begin{array}{l}
(pins = c.pins \land card = c.card \land \cdots) \land \\
(c.card = 0 \land card? \in \mathrm{dom}\, c.pins \ \land c' = c \ \land c.card' = card?) \\
\Rightarrow \exists ATM_0.\mathit{STATE}' \bullet (pins' = c'.pins' \land card' = c'.card' \land \cdots) \land \\
\qquad\qquad (card = 0 \land card? \in \mathrm{dom}\, pins \ \land card' = card?)
\end{array}
$$

and it uses $b.\Xi$ to re-establish the equality between $ATM_0.accts$ and $ATM_5.b.accts$ in the after-state. □

Example 17.3 A further illustration of class refinement is given by the following example where we refine a single class $Move_0$ into one containing three further objects. The class $Move_0$ contains a single operation *Transfer* which transfers a single element between the sequences *src* and *dst*.

$$
\begin{array}{l}
[T] \\
ValT ::= empty \mid val\langle\langle T\rangle\rangle
\end{array}
$$

$Move_0$

$src, dst : \mathrm{seq}\, T$

INIT
$dst = \langle\rangle$

Transfer
$\Delta(src, dst)$

$src \neq \langle\rangle \wedge src' = tail\, src$
$dst' = dst \frown \langle head\, src \rangle$

In our refinement the system class $Move_1$ contains a number of interacting components: a producer, a consumer and a one-place buffer. The producer reads the contents of *src* and passes the contents onto the buffer, from which the consumer builds up *dst*. In $Move_1$, *Transfer* consists of a *Put* followed by a *Get* and these operations involve a communication by the buffer with either the producer of the consumer. The absence of INIT in *Prod* means that *src* is chosen non-deterministically (as in $Move_0$).

$Move_1$

$b : Buffer;\ p : Prod;\ c : Cons$

$\mathrm{INIT} \mathrel{\widehat{=}} [b.\mathrm{INIT} \wedge p.\mathrm{INIT} \wedge c.\mathrm{INIT}]$
$Put \mathrel{\widehat{=}} p.Call \| b.Put \wedge c.\Xi$
$Get \mathrel{\widehat{=}} b.Get \| c.Shift \wedge p.\Xi$
$Transfer \mathrel{\widehat{=}} Put \mathbin{⨾} Get$

Prod

$src : \mathrm{seq}\, T$

Call
$\Delta(src)$
$x! : T$

$src \neq \langle\rangle \wedge src' = tail\, src$
$x! = head\, src$

Cons

$dst : \mathrm{seq}\, T$

INIT
$dst = \langle\rangle$

Shift
$\Delta(dst)$
$x? : T$

$dst' = dst \frown \langle x? \rangle$

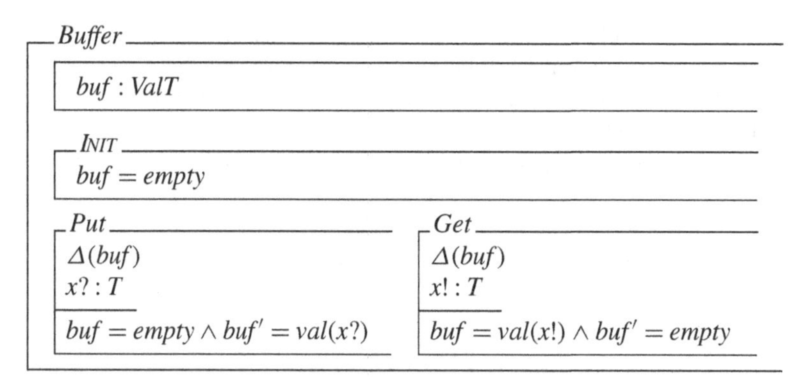

Then ($Move_1 \bullet Prod, Cons, Buffer$) refines $Move_0$. To show this we document the retrieve relation between the state spaces of the system classes:

R

$Move_0.\mathrm{STATE}$
$Move_1.\mathrm{STATE}$

$Move_0.src = Move_1.p.src$
$Move_0.dst = Move_1.c.dst$
$Move_1.b.buf = empty$

With this in place we can verify the three necessary refinement conditions. The initialisation condition is:

$$\forall Move_1.\mathrm{STATE} \bullet Move_1.\mathrm{INIT} \Rightarrow (\exists Move_0.\mathrm{STATE} \bullet Move_0.\mathrm{INIT} \wedge R)$$

This reduces to checking the following which is easily done.

$$\begin{aligned} &Move_1.c.dst = \langle\rangle \wedge Move_1.b.buf = empty \\ &\qquad \Rightarrow (\exists Move_0.src, Move_0.dst \bullet Move_0.dst = \langle\rangle \wedge R) \end{aligned}$$

To verify applicability we need to calculate the precondition of *Transfer* in $Move_1$, i.e., calculate $\mathrm{pre}(Put \mathbin{⨾} Get) = \mathrm{pre}(Put) = \mathrm{pre}(p.Call \| b.Put)$. This evaluates to $\mathrm{pre}(p.Call) \wedge \mathrm{pre}(b.Put) = [p.src \neq \langle\rangle] \wedge [b.buf = empty]$, and this is sufficient to verify applicability for *Transfer*.

Correctness is similarly verified. □

In a class simulation internal operations are sometimes introduced in order to communicate values between the classes in a refinement. In such cases weak simulations can be used to verify the refinement.

Example 17.4 A lift controller is required to control a lift in a building. Requests for the lift can be made by users in the lift or on one of the floors of the building, and the purpose of the controller is to service these requests. Initially we describe the controller as a single class $LiftCon_0$ with five operations: *RequestD* and *RequestE* model requests by users in the lift and on the floors respectively; *Close* and *Open* represent the closing and opening of the lift doors, and *Arrive* controls the movement of the lift between floors.

Floors are given as follows, and we also need a lift *STATUS*.

$$\begin{array}{|l} maxFloor : \mathbb{N} \\ \hline 1 < maxFloor \end{array}$$

$$FLOOR == 1..maxFloor$$
$$STATUS ::= open \mid closed \mid stop$$

$LiftCon_0$

$$\begin{array}{|l} req : \mathbb{P}\, FLOOR \\ pos : FLOOR \\ state : STATUS \end{array}$$

INIT

$$req = \varnothing \land pos = 1 \land state = open$$

RequestD

$$\begin{array}{|l} \Delta(req) \\ f? : FLOOR \\ \hline req' = req \cup \{f?\} \end{array}$$

RequestE

$$\begin{array}{|l} \Delta(req) \\ f? : FLOOR \\ \hline req' = req \cup \{f?\} \end{array}$$

Close

$$\begin{array}{|l} \Delta(state) \\ \hline req \neq \varnothing \\ state = open \land state' = closed \end{array}$$

Open

$$\begin{array}{|l} \Delta(state) \\ \hline state = stop \land state' = open \end{array}$$

Arrive

$$\begin{array}{|l} \Delta(req, pos, state) \\ f! : FLOOR \\ \hline state = closed \land state' = stop \\ pos' \in req \\ req' = req \setminus \{pos'\} \\ pos' = f! \end{array}$$

In this description no difference is made between requests emanating from a customer inside the lift and from requests from those on floors in the building. Furthermore, the movement of the lift is non-deterministic; any valid request can be taken as the next floor to move to.

As a first step towards implementation we will refine the lift system into two separate components, namely the lift itself and a controller. The lift component keeps track of its position *pos*, its next destination (*target*) and the requests made by those inside the lift. The controller *Con* has a similar set of requests from users on the floors of the building.

Lift

$pos, target : FLOOR$
$req : \mathbb{P}\, FLOOR$
$state : STATUS$
$target\_received : \mathbb{B}$

INIT

$pos = 1 \wedge state = open \wedge req = \varnothing \wedge \neg\, target\_received$

Request

$\Delta(req)$
$f? : FLOOR$

$req' = req \cup \{f?\}$

Close

$\Delta(state)$

$target\_received$
$state = open \wedge state' = closed$

Open

$\Delta(state)$

$state = stop \wedge state' = open$

GetTarget

$\Delta(target, target\_received)$
$f? : FLOOR$
$target\_sent? : \mathbb{B}$

$\neg\, target\_received \wedge target\_received'$
$state = open$
$\neg\, target\_sent? \Rightarrow target' \in req$
$target\_sent? \Rightarrow target' \in req \vee target' = f?$

MoveOneFloor

$\Delta(pos)$

$state = closed \wedge pos \neq target$
$pos > target \Rightarrow pos' = pos - 1$
$pos < target \Rightarrow pos' = pos + 1$

Arrive

$\Delta(state, target\_received)$
$f! : FLOOR$

$state = closed \wedge state' = stop$
$\neg\, target\_received'$
$pos = target \wedge f! = pos$
$req' = req \setminus \{pos\}$

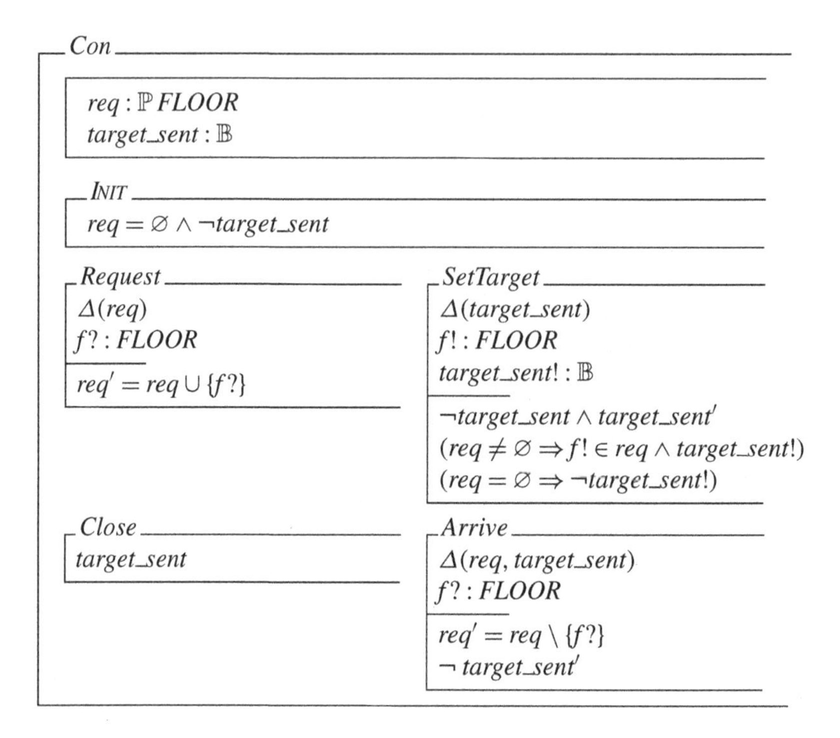

These classes contain additional operations to control the movement and scheduling of the lift. *MoveOneFloor* moves the lift one floor at a time. *SetTarget* sets a potential target for the lift to move to by choosing a request from someone in the building. This will be communicated to the *Lift* through its *GetTarget* operation which either accepts this target or uses one taken from the requests made inside the lift.

The specification is completed by describing how the two components communicate.

$LiftCon_1$

$\upharpoonright(Open, Close, RequestD, RequestE, Arrive)$

$Internals(Target, MoveOneFloor)$

$l : Lift$
$c : Con$

INIT

$l.\mathrm{INIT} \wedge c.\mathrm{INIT}$

$Open \mathrel{\widehat{=}} l.Open \wedge c.\Xi$

$Close \mathrel{\widehat{=}} l.Close \wedge c.Close$

$$
\begin{array}{|l}
RequestD \mathrel{\widehat{=}} l.Request \wedge c.\Xi \\
RequestE \mathrel{\widehat{=}} c.Request \wedge l.\Xi \\
Arrive \mathrel{\widehat{=}} l.Arrive \parallel_! c.Arrive \\
Target \mathrel{\widehat{=}} l.GetTarget \parallel_! c.SetTarget \\
MoveOneFloor \mathrel{\widehat{=}} l.MoveOneFloor \wedge c.\Xi \\
\hline
\end{array}
$$

We can now show that $(LiftCon_1 \bullet Lift, Con)$ is a weak class refinement of $LiftCon_0$. The retrieve relation is:

$$
\begin{array}{|l}
\multicolumn{1}{l}{R} \\
\hline
LiftCon_0.\mathit{STATE} \\
LiftCon_1.\mathit{STATE} \\
\hline
LiftCon_0.req = LiftCon_1.l.req \cup LiftCon_1.c.req \\
LiftCon_0.pos = LiftCon_1.l.pos \\
LiftCon_0.state = LiftCon_1.l.state \\
\hline
\end{array}
$$

Then, first note that there are no internal operations in the abstract $LiftCon_0$ class, and therefore this simplifies the verification. Second, by examining the state and effect of the concrete operations, it can be seen that $l.target\_received$ is true if, and only if, $l.req \cup c.req \neq \varnothing$.

Next we note that neither of the object references l and c are changed by any of the operations in the concrete specification, and thus their *rôle* in the refinement proof obligations is the same as in Example 17.2.

For applicability of *Close* we note that pre *Close* in $LiftCon_0$ is $[req \neq \varnothing \wedge state = open]$ whereas $\text{pre}\, l.Close \wedge \text{pre}\, c.Close$ is $[l.target\_received \wedge c.target\_sent \wedge l.state = open]$. As we have seen $l.target\_received$ is equivalent to $l.req \cup c.req \neq \varnothing$, i.e., to $req \neq \varnothing$, and adding in the effect of the internal *Target* operation we find that

$$R \Rightarrow (\text{pre}\, Close \Leftrightarrow \text{pre}(Target \mathbin{⨟} (l.Close \wedge c.Close)))$$

In a similar way it is possible to show that the effect of *Arrive* in $LiftCon_0$ is simulated by $MoveOneFloor^n \mathbin{⨟} Arrive$ (for some n) in $LiftCon_1$.

Finally to deal with potential divergence of the internal operations, we note that *SetTarget* and *GetTarget* can only occur once between visible operations, and *MoveOneFloor* can only occur a maximum of *maxFloor* number of times before an *Arrive* operation has to occur. This prevents any divergence being introduced. □

Class refinements can be quite complex as the next example shows. Although we have restricted ourselves so far to non-recursive object instantiations there is scope to extend the refinement methodology to cope with recursive definitions.

Example 17.5 A prime number sieve generates an increasing sequence of prime numbers. Our initial specification is given by the class *Primes*.

Primes

$$number : \mathbb{N}$$
$$primes : \operatorname{seq} \mathbb{N}$$

INIT

$$number = 1$$
$$primes = \langle\, \rangle$$

Sieve

$$\Delta(number, primes)$$

$$number' = number + 1$$
$$(\forall p : \operatorname{ran} primes \bullet number' \bmod p \neq 0) \Rightarrow$$
$$\qquad primes' = primes \frown \langle number' \rangle$$
$$\neg(\forall p : \operatorname{ran} primes \bullet number' \bmod p \neq 0) \Rightarrow primes' = primes$$

The operation *Sieve* simply checks the next possible number against all the primes generated so far, updating *primes* if a new prime is found. We refine this description to a specification ($Main \bullet Store$) where *Main* generates the increasing sequence of natural numbers, which it passes into a chain of objects of type *Store*. Each object in the chain then stores the first number that is passed to it, and of the rest it passes on only those numbers which are not divisible by its stored prime.

Main

$$number : \mathbb{N}$$
$$next : Store$$

INIT

$$number = 1$$
$$next.\mathrm{INIT}$$

Pass

$$\Delta(number)$$
$$n! : \mathbb{N}$$

$$number' = number + 1$$
$$n! = number'$$

$$Sieve \mathrel{\widehat{=}} Pass \parallel next.Sieve$$

Store

$n : \mathbb{N}$
$created : \mathbb{B}$
$next : Store$

INIT

$n = 0$
$\neg created$

Sieve1

$\Delta(created, n, next)$
$n? : \mathbb{N}$

$\neg created \wedge created'$
$n' = n?$
$next'.\mathrm{INIT}$

Sieve2

$n? : \mathbb{N}$

$created$
$n? \bmod n = 0$

Sieve3

$n?, n! : \mathbb{N}$

$created$
$n? \bmod n \neq 0$
$n! = n?$

$Sieve \mathrel{\widehat=} Sieve1 \;[]\; Sieve2 \;[]\; (Sieve3 \parallel next.Sieve)$

In the *Sieve* operation, *Sieve*1 initialises the *next* object in the chain and reads in the number passed to it by the previous one. *Sieve*2 and *Sieve*3 determine what to do according as to whether it is still a candidate new prime, and, if it is, it is then communicated to *next* via *Sieve*3.

The relationship between (*Main* • *Store*) and *Primes* is then given by

$$Primes.primes \subseteq \langle Main.number, Main.next.n, Main.next.next.n, \ldots \rangle$$

(recall that $\subseteq$ on sequences is prefix). □

17.3 Issues of Compositionality

In Chap. 6 we discussed compositionality in the context of promotions of local state to global state in Z. In this section we look at the issue in Object-Z.

In Z we promote a local state and operations to a global state with an indexed collection of components, and promote the local operations to this global state. Under certain conditions refinement distributes through the promotion.

In Object-Z we are faced with a similar issue: if a class A is refined by a class B then will the specification $(C \bullet A)$ be refined by the specification $(C \bullet B)$, i.e., can we promote the refinement to any class C that uses objects of type A? As in Z, a positive answer depends on certain conditions being satisfied. One requirement is that there must, in general, be no direct access to state variables in the class C (and for a similar reason no inheritance). The next example illustrates why.

Example 17.6 Consider the *Count* example again. An alternative description is

$Count_1$

$x : \mathbb{Z}$

INIT
$x = -1$

Inc
$\Delta(x)$
$x? : \mathbb{N}$
$x' = x + x?$

Val
$x! : \mathbb{N}$
$x! = x + 1$

$Count_1$ and *Count* are refinements of each other. However, if we could access the variables directly then using the following context

$AbuseCount[X]$

$ct : X$

INIT
$ct.\mathrm{INIT}$

$Inc \mathrel{\widehat{=}} ct.Inc$

Val
$x! : \mathbb{N}$
$x! = ct.x$

we would find that $(AbuseCount[Count] \bullet Count)$ and $(AbuseCount[Count_1] \bullet Count_1)$ are now *not* refinements of each other. We have lost compositionality because we have broken the object encapsulation and accessed the variables directly. □

In addition to accessing the variables directly, the use of schema composition or preconditions can also break compositionality. The next example illustrates the problem with schema composition and the subsequent example preconditions.

Example 17.7 Consider the classes A and B.

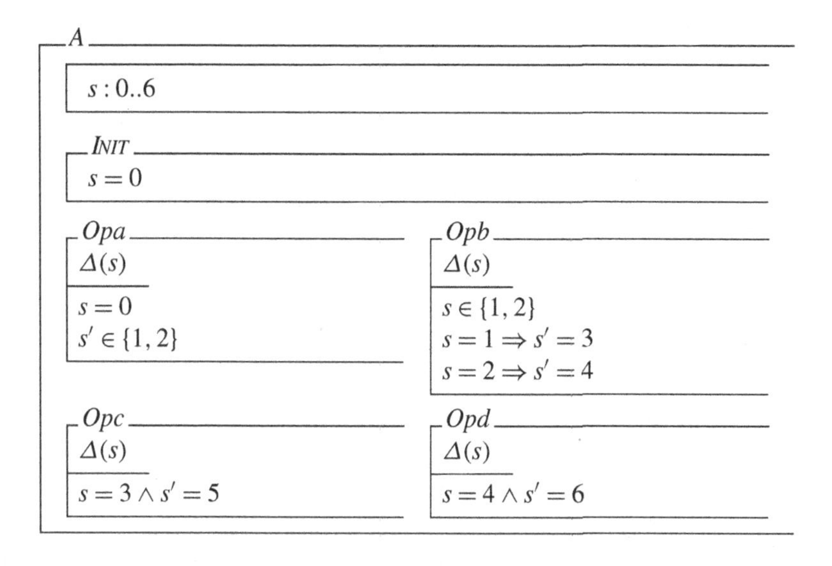

B

$s : 7..12$

INIT

$s = 7$

Opa

$\Delta(s)$

$s = 7 \wedge s' = 8$

Opb

$\Delta(s)$

$s = 8$

$s' \in \{9, 10\}$

Opc

$\Delta(s)$

$s = 9 \wedge s' = 11$

Opd

$\Delta(s)$

$s = 10 \wedge s' = 12$

which have behaviours represented in Fig. 17.3.

Classes A and B are equivalent, however, using schema composition it is possible to construct a context that distinguishes them. Consider the following class.

$C[X]$

$x : X$

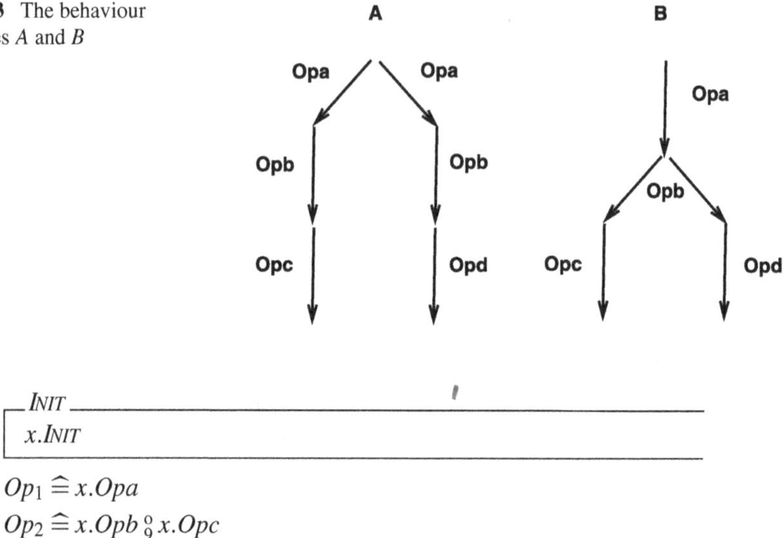

Fig. 17.3 The behaviour of classes A and B

INIT

$x.\mathit{INIT}$

$Op_1 \mathrel{\widehat{=}} x.Opa$

$Op_2 \mathrel{\widehat{=}} x.Opb \mathbin{;} x.Opc$

Then $C[A]$ and $C[B]$ are not equivalent, since in $C[B]$ after Op_1 it is always possible to perform Op_2 whereas in $C[A]$ after Op_1, Op_2 may be blocked. □

Example 17.8 Consider the following classes which are equivalent to one another.

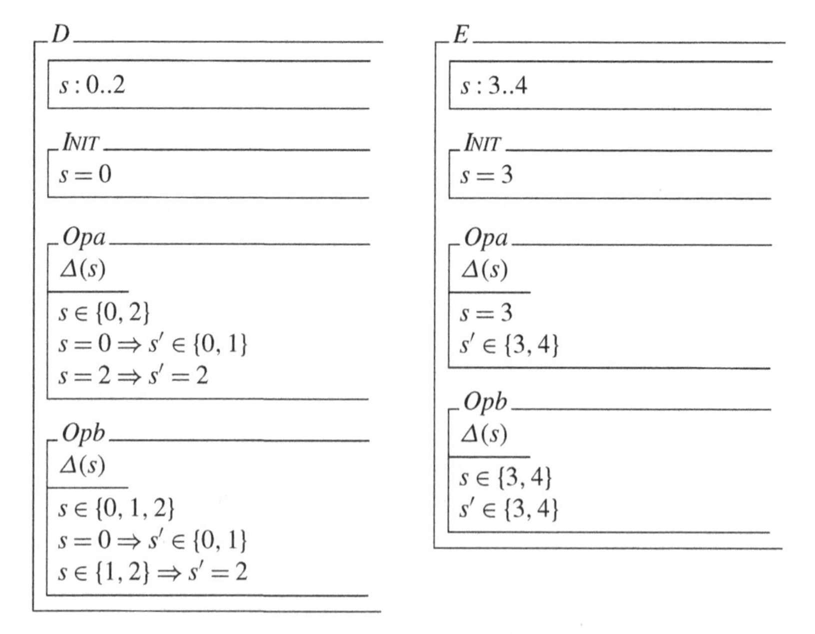

D

$s : 0..2$

INIT

$s = 0$

Opa

$\Delta(s)$

$s \in \{0, 2\}$
$s = 0 \Rightarrow s' \in \{0, 1\}$
$s = 2 \Rightarrow s' = 2$

Opb

$\Delta(s)$

$s \in \{0, 1, 2\}$
$s = 0 \Rightarrow s' \in \{0, 1\}$
$s \in \{1, 2\} \Rightarrow s' = 2$

E

$s : 3..4$

INIT

$s = 3$

Opa

$\Delta(s)$

$s = 3$
$s' \in \{3, 4\}$

Opb

$\Delta(s)$

$s \in \{3, 4\}$
$s' \in \{3, 4\}$

If the operator pre was included in the Object-Z schema calculus then the following context could be constructed.

$F[X]$

$x : X$

INIT
$x.\mathrm{INIT}$

$Op_1 \mathrel{\widehat{=}} x.Opa$

Op_2
$x.Opb$
$\neg \operatorname{pre} Op_1$

Then $F[D]$ and $F[E]$ are not equivalent since they have different sets of traces (i.e., sequences of permissible events). For example, in $F[D]$ after Op_1 there can be at most one occurrence of Op_2, whereas in $F[E]$ after Op_1 there can be an arbitrary number of occurrences of Op_2. The two classes therefore cannot be equivalent. □

Clearly if a class B is not a refinement of a class A then we can find a context such that the promotion is not a refinement, by simply promoting all the operations in A that have a different behaviour from B. However, it is easy to construct examples with two classes A and B which are not related by refinement but which have contexts in which they are equivalent (by not promoting the operations which differ).

17.3.1 Promoting Refinements Between Object-Z Classes

The general result we state will guarantee compositionality when our promotions conform to a particular form (essentially one of the commonly occurring promotions of Sect. 6.5).

In particular, we allow declarations which are arbitrary partial functions, indexed over some set I, of the form $f_D : I \rightarrow\!\!\!\!\!\!+ \; D$, where D is a class. Thus $f_D(d)$ will be an object of type D whenever $d \in \operatorname{dom} f_D$. This allows objects to be used in fairly general ways within a class including single instances, sets, sequences and partial or total functions.

The operation promotions we shall consider will be of the form

$$Op \mathrel{\widehat{=}} [i? : I \mid i? \in \operatorname{dom} f_D] \bullet f_D(i?).Op$$

This promotes an operation in D to an operation in the main class by applying it to a selected component.

Theorem 17.1 *Let $A = (A.\mathrm{STATE}, A.\mathrm{INIT}, \{Op\})$ and $C = (C.\mathrm{STATE}, C.\mathrm{INIT}, \{Op\})$ be Object-Z classes. Suppose that C is a downward simulation of A, and let the context $\Phi[D]$ be defined by*

$$
\begin{array}{|l}
\hline
\Phi[D] \\
\quad \begin{array}{|l} f_D : I \twoheadrightarrow D \end{array} \\
\quad \begin{array}{|l} \textit{INIT} \\ \hline \forall i : \operatorname{dom} f_D \bullet f_D(i).\textit{INIT} \end{array} \\
\quad Op \mathrel{\widehat=} [i? : I \mid i? \in \operatorname{dom} f_D] \bullet f_D(i?).Op \\
\quad \vdots \\
\hline
\end{array}
$$

Then $\Phi[C]$ *is a downward simulation of* $\Phi[A]$.

In order to verify this result we will need to use the retrieve relation used in the refinement of A. Let us call this retrieve relation R, and thus we know that for all operations Op:

$$
\begin{array}{l}
\forall C.\textit{STATE} \bullet C.\textit{INIT} \Rightarrow (\exists A.\textit{STATE} \bullet A.\textit{INIT} \land R) \\
\forall A.\textit{STATE};\ C.\textit{STATE} \bullet R \Rightarrow (\operatorname{pre} A.Op \Leftrightarrow \operatorname{pre} C.Op) \\
\forall A.\textit{STATE};\ C.\textit{STATE};\ C.\textit{STATE}' \bullet \\
\qquad R \land C.Op \Rightarrow \exists A.\textit{STATE}' \bullet R' \land A.Op
\end{array}
$$

Using this we define a retrieve relation P_R by

$$
\begin{array}{|l}
P_R \\
\hline
f_A : I \twoheadrightarrow A \\
f_C : I \twoheadrightarrow C \\
\hline
\operatorname{dom} f_A = \operatorname{dom} f_C \\
\forall i : \operatorname{dom} f_A \bullet \exists R \bullet \theta f_A(i).\textit{STATE} = \theta A.\textit{STATE} \land \\
\qquad\qquad\qquad\qquad \theta f_C(i).\textit{STATE} = \theta C.\textit{STATE} \\
\hline
\end{array}
$$

and show that

$$
\begin{array}{l}
\forall \Phi[C].\textit{STATE} \bullet \Phi[C].\textit{INIT} \Rightarrow (\exists \Phi[A].\textit{STATE} \bullet \Phi[A].\textit{INIT} \land P_R) \\
\forall \Phi[A].\textit{STATE};\ \Phi[C].\textit{STATE} \bullet P_R \Rightarrow (\operatorname{pre} \Phi[A].Op \Leftrightarrow \operatorname{pre} \Phi[C].Op) \\
\forall \Phi[A].\textit{STATE};\ \Phi[C].\textit{STATE};\ \Phi[C].\textit{STATE}' \bullet \\
\qquad P_R \land \Phi[C].Op \Rightarrow \exists (\Phi[A].\textit{STATE})' \bullet P_R' \land \Phi[A].Op
\end{array}
$$

For initialisation we need to show that given

$$\forall i : \operatorname{dom} f_C \bullet f_C(i).\textit{INIT}$$

does there exist an f_A with $\forall i : \operatorname{dom} f_A \bullet f_A(i).\textit{INIT} \land P_R$. In fact such an f_A can be found by letting $\operatorname{dom} f_A = \operatorname{dom} f_C$ and $\forall i \in \operatorname{dom} f_C \bullet \theta f_A(i).\textit{STATE} = \theta A.\textit{INIT}$.

For applicability, we need to show that

$$
\begin{array}{l}
P_R \Rightarrow ([i? : I \mid i? \in \operatorname{dom} f_A] \bullet f_A(i?).\operatorname{pre} Op \Leftrightarrow \\
\qquad\qquad [i? : I \mid i? \in \operatorname{dom} f_C] \bullet f_C(i?).\operatorname{pre} Op)
\end{array}
$$

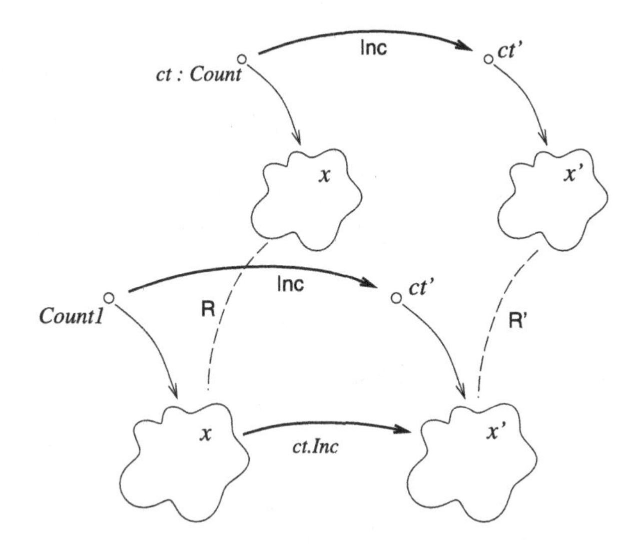

Fig. 17.4 The effect of the operation *Inc*

Now $\operatorname{dom} f_A = \operatorname{dom} f_C$ ensures one part of the deduction, and the second follows because P_R ensures that the states of the objects $f_A(i?)$ and $f_C(i?)$ are related by the retrieve relation R. The refinement of A to C then ensures that the preconditions of Op in these objects are equivalent. Correctness can be informally verified in a similar manner. □

Example 17.9 Consider the following class.

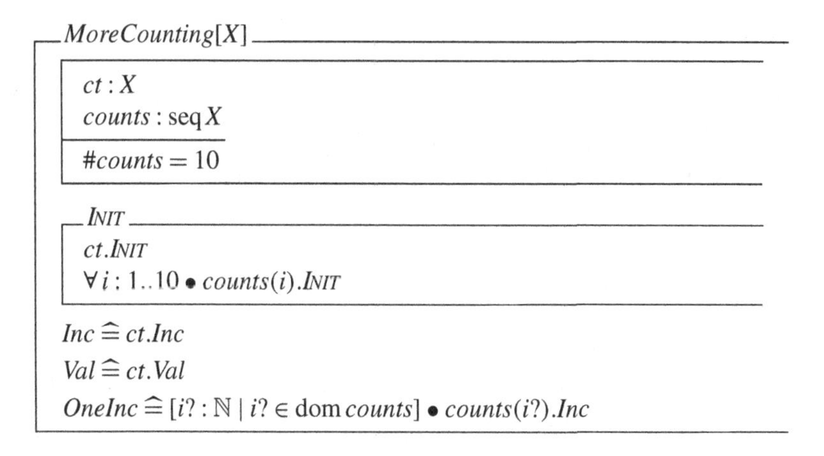

MoreCounting[*X*]

$ct : X$
$counts : \operatorname{seq} X$

$\#counts = 10$

INIT
$ct.\mathrm{INIT}$
$\forall i : 1..10 \bullet counts(i).\mathrm{INIT}$

$Inc \mathrel{\widehat=} ct.Inc$
$Val \mathrel{\widehat=} ct.Val$
$OneInc \mathrel{\widehat=} [i? : \mathbb{N} \mid i? \in \operatorname{dom} counts] \bullet counts(i?).Inc$

Now since $Count_1$ refines $Count$, application of Theorem 17.1 ensures that $MoreCounting[Count_1]$ refines $MoreCounting[Count]$. Figure 17.4 illustrates the retrieve relation. □

Although this result allows some limited promotion of refinement, it does not cover the cases when a selection operation of the form

Select
$c? : Count$
$c? \in counts$

is used. Here we are using the object reference $c?$ as input, and the reference semantics of Object-Z complicates the verification of refinement in cases where the class of $c?$ is refined.

17.4 Bibliographical Notes

The idea of a class simulation has been discussed by Goldsack and Lano in [4, 5] in the context of VDM^{++}, an object-oriented extension of VDM. The technique there has been called annealing, and VDM^{++} class structures are refined into a structure composed of objects of simpler classes in a similar way to class simulation.

Refinements which involve changes of class structure have also been considered in the context of object-oriented action systems by Bonsangue, Kok and Sere [1, 2].

Early work on compositionality in Object-Z includes [7], and Examples 17.7 and 17.8 are due to Smith. The omission of the operator pre from the Object-Z schema calculus is partly due to compositionality failing in its presence. Schema composition was also omitted in [7] for similar reasons.

References

1. Bonsangue, M., Kok, J., & Sere, K. (1998). An approach to object-orientation in action systems. In J. Jeuring (Ed.), *Mathematics of Program Construction (MPC'98). Lecture Notes in Computer Science: Vol. 1422* (pp. 68–95). Berlin: Springer.
2. Bonsangue, M., Kok, J., & Sere, K. (1999) Developing object-based distributed systems. In Ciancarini et al. [3] (pp. 19–34).
3. Ciancarini, P., Fantechi, A., & Gorrieri, R. (Eds.) (1999). *Formal Methods for Open Object-Based Distributed Systems*. Dordrecht: Kluwer Academic.
4. Goldsack, S. J., Lano, K., & Dürr, E. H. (1996). Annealing and data decomposition in VDM^{++}. *ACM SIGPLAN Notices*, *31*(4), 32–38.
5. Lano, K., & Goldsack, S. Refinement of distributed object systems. In Najm and Stefani [6] (pp. 99–114).
6. Najm, E. & Stefani, J. B. (Eds.) (1996). *First IFIP International Workshop on Formal Methods for Open Object-Based Distributed Systems*, Paris, March 1996. London: Chapman & Hall.
7. Smith, G. (1995). A fully abstract semantics of classes for Object-Z. *Formal Aspects of Computing*, *7*(3), 289–313.
8. Smith, G. (2000). *The Object-Z Specification Language*. Dordrecht: Kluwer Academic.

Part IV
Modelling State and Behaviour

In this part we turn our attention to combining notions of state and behaviour when specifying and refining systems. The use of Object-Z provides a notion of state, which we combine with the notion of behaviour that is provided by the process algebra CSP.

Chapter 18 introduces a particular style of using CSP in combination with Object-Z. First, we give a short introduction to CSP. Then we show how component specifications can be given in Object-Z, and then combined using CSP to specify their interactions. This integration is possible due to the existence of a common semantic model.

Chapter 19 discusses some of the possibilities for refining systems defined in this way. In particular, it turns out that they can be refined component by component, due to a correlation between refinement in Object-Z and refinement in CSP.

Chapter 18
Combining CSP and Object-Z

CSP is a language for describing components which run concurrently and communicate over unbuffered channels by synchronising on events. The purpose of using CSP together with Object-Z is that these two languages provide complementary advantages. In particular, CSP provides primitives to express the temporal ordering of events and their communication and synchronisation. All of this is possible in Object-Z, but can sometimes be cumbersome. For example, we have to use Boolean variables, or similar, to control the desired sequencing of operations, and such a specification can take some unravelling to understand the intention. On the other hand, in CSP such temporal ordering is specified in an easy and readable fashion.

What CSP lacks, however, is a rich language for describing data aspects, which is precisely what is provided in Object-Z. Linking the two notations together is a common view of their semantics which enables Object-Z classes to be interpreted as processes and operations as events. This gives rise to a smooth specification style and allows classes specified in Object-Z to be used directly within the CSP part of the specification.

In this chapter we provide an overview of this type of specification style, including an introduction to CSP. Then, in Chap. 19, we look at refinement in combinations of Object-Z and CSP where we show how Object-Z refinement techniques can be used in the combined notation to refine the specification component by component.

18.1 An Introduction to CSP

CSP (for Communicating Sequential Processes) describes a system as a collection of communicating processes running concurrently and synchronising on events. A process is a self-contained component with an *interface*, through which it interacts with

Excerpts from p. 110 of [43] are reprinted within Sect. 18.4.1 with kind permission of Springer Science and Business Media.

J. Derrick, E.A. Boiten, *Refinement in Z and Object-Z*,
DOI 10.1007/978-1-4471-5355-9_18, © Springer-Verlag London 2014

the environment, and the interface is described as a set of *events*. Events are instantaneous and atomic, as are operations in Z or Object-Z (unless dealing with non-atomic refinement). Σ is usually used to denote the set of all possible events.

Processes are described by guarded equations such as, for example,

$$P = a \to b \to P$$

Stop *stop* is the process which is never prepared to engage in any activity, nor does it successfully terminate. A process is said to *deadlock* if it is not able to perform any activity or terminate.

Skip *skip* similarly is never prepared to engage in any activity, but instead it terminates successfully, and thus allows control to pass to another process.

Event Prefix Event prefix, as in the process specification

$$a \to P$$

describes a process which is initially able to perform only a, and after performing a it behaves like P. For example,

$$P = \mathit{Timeout} \to \mathit{Halt} \to \mathit{skip}$$

describes a process which can initially perform a *Timeout*, then subsequently a *Halt*, and then terminate.

Input and Output Events can be structured and, in particular, can consist of a *channel* which can carry a value. If values v of type T are being communicated along channel c, the set of events associated with this communication is $\{c.v \mid v \in T\}$.

If c is a channel and v a value, then output is modelled by the CSP expression

$$c!v \to P$$

which describes a process that is initially willing to output v along channel c and subsequently behaves like P. In particular, the only event that this process is able to perform initially is $c.v$. On the other hand we can describe input by the CSP expression

$$c?v : T \to P$$

which describes a process that is initially willing to accept any value of type T along c, which it will then bind to the name v, and it subsequently behaves like P, possibly using the value of v received. Thus this process can perform a number of initial events, i.e., any event of the form $c.v$ for $v \in T$.

We will sometimes omit the typing information on the inputs where this is clear from the context, and simply write $c?v \to P$.

Choice *External* choice, in which control over the choice is external to the process, is described by

$$P_1 \Box P_2$$

This process is initially ready to perform the events that either process can engage in, and the choice is resolved by the first event in favour of the process that performs it. The choice is external because events of both P_1 and P_2 are initially available. For example, in

$$\begin{aligned} P = {} & Timeout \rightarrow Halt \rightarrow skip \\ & \Box \\ & Start \rightarrow stop \end{aligned}$$

both events *Timeout* and *Start* are enabled, however once the environment offers one of these, say *Timeout*, then subsequently the process behaves like $Halt \rightarrow skip$.

Internal choice, on the other hand, describes a process in which the choice is resolved by the process itself, and is written

$$P_1 \sqcap P_2$$

Thus, for example, given

$$\begin{aligned} Q = {} & Timeout \rightarrow Halt \rightarrow skip \\ & \sqcap \\ & Start \rightarrow stop \end{aligned}$$

the process will non-deterministically choose which branch of the choice to offer to the environment.

To understand the difference between internal and external choice, consider an environment which is initially only willing to perform a *Timeout*. In such an environment process P can always proceed since it is willing to perform either a *Timeout* or a *Start* initially. However, process Q may deadlock since the process itself makes the non-deterministic choice as to which event is initially enabled.

Recursion Process definitions can be recursive or mutually recursive. For example,

$$WDH = Start \rightarrow Enable \rightarrow WDH$$

describes a process that can repeatedly perform a *Start* followed by an *Enable*. Mutual recursion, as in

$$\begin{aligned} Lift &= GetTarget \rightarrow Close \rightarrow Q \\ Q &= MoveOneFloor \rightarrow Q \\ & \quad \Box \\ & \quad Arrive \rightarrow Open \rightarrow Lift \end{aligned}$$

is frequently used in process definitions. In recursive definitions, processes are often parameterised by values such as sets or sequences. For example, the following specifies a one place buffer holding values of type M.

$$Buf(\langle\rangle) = Put?x : M \rightarrow Buf(\langle x \rangle)$$
$$Buf(\langle x \rangle) = Get!x \rightarrow Buf(\langle\rangle)$$

Renaming Events (or channels) can be renamed using a substitution notation $P[\![a/b]\!]$ which means that the event or channel b has been replaced by a throughout the process definition. For example, $Buf[\![Mid/Put]\!]$ is a buffer with channel name Put renamed to Mid.

Interleaving So far the process definitions we have described only allow for sequential execution of events, however, the power of CSP really comes from its description of how processes can be combined together in forms of parallel execution.

The simplest of these is the notion of interleaving. Interleaving is the concurrent execution of two processes where no synchronisation is required between the components. It is described by the process

$$P_1 \mid\mid\mid P_2$$

in which processes P_1 and P_2 execute completely independently of each other (and do not synchronise even on events in common). For example, given

$$P = Coin \rightarrow P$$
$$Q = Ticket \rightarrow Q$$

the process $P \mid\mid\mid Q$ can initially perform either a $Coin$ or $Ticket$ and also subsequently perform either a $Coin$ or $Ticket$. Interleaving is sometimes used to describe replication of a resource or process, e.g., two independent lifts can be specified as

$$Lift \mid\mid\mid Lift$$

Parallel Composition CSP has a number of operators to describe parallel composition which allow processes to selectively synchronise on events. Here we shall consider just the *interface parallel* operator $\underset{A}{\|}$. The process

$$P_1 \underset{A}{\|} P_2$$

synchronises on events in the set A but all other events are interleaved. Thus P_1 and P_2 evolve separately, but events in A are enabled only when they are enabled in *both* P_1 and P_2.

For example, given

$$P = Coin \rightarrow Dest \rightarrow P$$
$$Q = Coin \rightarrow Ticket \rightarrow Q$$

the process $P \underset{\{Coin\}}{\parallel} Q$ can perform the *Coin* event, after which it may perform either *Dest* or *Ticket* next, and when it has performed both of these it can again synchronise on the *Coin* event.

When the required synchronisation set is the intersection of the interfaces of the two processes we omit explicit reference to the synchronisation set and simply write $P \parallel Q$.

The synchronisation allows communication between input and output values in events. So, given

$$\begin{aligned} Lift &= Target?x : \mathbb{N} \rightarrow Floor!x \rightarrow Lift \\ Con &= Target!7 \rightarrow Request \rightarrow Con \end{aligned}$$

the process $Lift \parallel Con$ synchronises on the *Target* channel. The effect of this process is to first communicate the value 7 to *Lift* and then to interleave the *Floor.7* and *Request* events.

Hiding Hiding encapsulates an event within a process and removes it from its interface. In

$$P \setminus L$$

the process behaves as P except that events in L are hidden from the environment and the process P has complete control over its hidden (or internal) events.

We can use hiding (together with composition and renaming) to specify a two place buffer by using two copies of the one place buffer.

$$Buf_2 = (Buf[\![Mid/Get]\!] \parallel Buf[\![Mid/Put]\!]) \setminus \{Mid.x \mid x \in M\}$$

There are further CSP operators, such as chaining $\gg$, sequential composition ; and interrupt Δ which we do not detail here since the subset of CSP that we use excludes them. Full details of their definition and use can be found in [30] or [32].

18.1.1 The Semantics of CSP

There are a number of different semantic models for CSP which are used to capture various aspects of a process and its interaction with its environment. However, the semantics we shall use here for the combination of CSP and Object-Z is the standard failures-divergences semantics.

To begin, we consider the *traces* of a process. These are the finite sequences of visible events that the process may undergo, and are straightforward to calculate. In what follows, *traces*(P) will denote the set of all traces associated with the process P. We will illustrate the calculation of traces here by a few examples; a full definition of their calculation appears in the text by Schneider [32].

Example 18.1 The set of all traces associated with the process $P = Start \rightarrow stop$ is $\{\langle\,\rangle, \langle Start\rangle\}$. Similarly, the set of all traces associated with the process

$$
\begin{aligned}
Q = {} & Timeout \rightarrow Halt \rightarrow skip \\
& \Box \\
& Start \rightarrow stop
\end{aligned}
$$

is $\{\langle\,\rangle, \langle Start\rangle, \langle Timeout\rangle, \langle Timeout, Halt\rangle, \langle Timeout, Halt, \checkmark\rangle\}$. The presence of $\checkmark$ in the last trace in this set is due to the *skip* action, and denotes successful termination at the semantic level. It is needed to distinguish *skip* from *stop*.

Finally consider

$$
\begin{aligned}
R = {} & Timeout \rightarrow Halt \rightarrow skip \\
& \sqcap \\
& Start \rightarrow stop
\end{aligned}
$$

Although we have replaced external choice with internal choice, the set of traces of R is exactly the same as the set of traces of P. $\Box$

The traces associated with input and output constructors include components associated with the values communicated.

Example 18.2 The traces of the process $Lift = Target?x : \mathbb{N} \rightarrow Floor!x \rightarrow stop$ are described by the following set: $\{\langle\,\rangle\} \cup \{\langle Target.v\rangle \mid v \in \mathbb{N}\} \cup \{\langle Target.v, Floor.v\rangle \mid v \in \mathbb{N}\}$. $\Box$

Because traces are only concerned with the visible events of a process, events which are hidden do not appear in the traces of a process.

Example 18.3 The traces of the process $(a \rightarrow b \rightarrow a \rightarrow c \rightarrow stop) \setminus \{a\}$ are described by the following set: $\{\langle\,\rangle, \langle b\rangle, \langle b, c\rangle\}$. $\Box$

The set of traces is not always finite, for example, recursion typically produces an infinite set of (finite) traces.

Example 18.4 Given

$$
\begin{aligned}
P &= Coin \rightarrow Dest \rightarrow P \\
Q &= Coin \rightarrow Ticket \rightarrow Q
\end{aligned}
$$

we can calculate the traces of $P \underset{\{Coin\}}{\parallel} Q$. This will consist of the collection: $\langle\,\rangle$, $\langle Coin\rangle$, $\langle Coin, Dest\rangle$, $\langle Coin, Ticket\rangle$, $\langle Coin, Dest, Ticket\rangle$, $\langle Coin, Ticket, Dest\rangle$, $\langle Coin, Dest, Ticket, Coin\rangle$, $\langle Coin, Ticket, Dest, Coin\rangle, \ldots$. $\Box$

However, note that although the *set* is infinite, the traces themselves consist of finite sequences of events, and the semantics assumes that the infinite traces can be extrapolated from the finite ones. This, in general, is true except in the presence of unbounded non-determinism, and leads us to make a restriction on the use of hiding when combining Object-Z with CSP operators.

Aside Unbounded non-determinism refers to the situation when a process can choose from an infinite set of options. For example, a process which non-deterministically selects *any* natural number n and performs an event n times contains unbounded non-determinism. □

The trace model is not very discriminating, as we have seen it failed to distinguish between internal and external choice. This is because traces tell us what a process *can* do but nothing about what it *must* do. For this reason we also consider refusals.

Definition 18.1 (Refusal set) A set of events X is a refusal set for a process P if there is no $a \in X$ such that P can perform a. □

The failures-divergences semantics uses this idea of a refusal and considers, in addition to the traces, the refusal set after each trace. Each process is now modelled by a triple (A, F, D) where A is its alphabet, F its failures and D its divergences. The failures of a process are pairs (t, X) where t is a trace of the process, and X is a set of events the process may refuse to perform after undergoing t.

The failures of a process with alphabet A are defined to be a set

$$F \subseteq A^* \times \mathbb{P} A$$

which satisfies the following.

$$
\begin{array}{l}
(\langle\rangle, \varnothing) \in F \\
(t_1 \frown t_2, \varnothing) \in F \Rightarrow (t_1, \varnothing) \in F \\
(t, X) \in F \wedge Y \subseteq X \Rightarrow (t, Y) \in F \\
(t, X) \in F \wedge (\forall e \in Y \bullet (t \frown \langle e \rangle, \varnothing) \notin F) \Rightarrow (t, X \cup Y) \in F
\end{array}
$$

The first two properties capture the requirement that the sequences of events a process can undergo form a non-empty, prefix-closed set. The next states that if a process can refuse all events in a set X then it can refuse all events in any subset of X. Finally we require that a process can refuse any event which cannot occur as the next event.

Example 18.5 In this example we assume that the alphabet of each process is restricted to consist of only those events mentioned in their definition. The process $P = \mathit{Start} \rightarrow \mathit{stop}$ has failures: $(\langle\rangle, \varnothing)$, $(\langle \mathit{Start} \rangle, \varnothing)$ and $(\langle \mathit{Start} \rangle, \{\mathit{Start}\})$.

Similarly, the failures of

$$
\begin{array}{l}
Q = a \rightarrow b \rightarrow \mathit{stop} \\
\qquad \Box \\
\qquad b \rightarrow a \rightarrow \mathit{stop}
\end{array}
$$

include $(\langle\rangle, \varnothing)$, $(\langle a\rangle, \{a\})$, $(\langle b\rangle, \{b\})$, $(\langle a, b\rangle, \{a, b\})$, and $(\langle b, a\rangle, \{a, b\})$.

In a similar way, the failures of

$$\begin{aligned} R = ((b \to c \to stop)\\ \Box\\ (c \to a \to stop)) \setminus \{c\} \end{aligned}$$

are

$$\{(\langle\rangle, X) \mid X \subseteq \{b, c\}\} \cup$$
$$\{(\langle a\rangle, X), (\langle b\rangle, X) \mid X \subseteq \{a, b, c\}\}$$

Finally consider

$$\begin{aligned} S = (a \to stop)\\ \sqcap\\ (b \to stop) \end{aligned}$$

which has as failures

$$\{(\langle\rangle, X) \mid \{a, b\} \not\subseteq X\} \cup$$
$$\{(\langle a\rangle, X), (\langle b\rangle, X) \mid X \subseteq \{a, b\}\}$$

Notice that, for a larger alphabet, the failures of each process would include additional elements due to the refusal of events from the enlarged alphabet. □

The divergences of a process are the sequences of events after which the process may undergo an infinite sequence of internal events, i.e., livelock. Divergences also result from unguarded recursion (e.g., processes such as $P = P \Box (a \to stop)$). We define divergences of a process with alphabet A and failures F to be a set $D \subseteq A^*$ such that

$$D \subseteq \text{dom}\, F$$
$$t_1 \in D \wedge t_2 \in A^* \Rightarrow t_1 \frown t_2 \in D$$
$$t \in D \wedge X \subseteq A \Rightarrow (t, X) \in F$$

The first property simply states that a divergence is a possible sequence of events of the process. The last two properties model the requirement that it is impossible to determine anything about a divergent process in a finite time. Therefore, the possibility that it might undergo further events cannot be ruled out. In other words, a divergent process behaves *chaotically* (see also Sect. 11.6).

18.2 Integrating CSP and Object-Z

There are a number of possible approaches to combining a state-based language such as Object-Z with a process algebra such as CSP, but the most natural way is to relate Object-Z classes to CSP processes. This allows Object-Z classes to be used in expressions involving CSP operators in an intuitive way, and the correspondence between the two languages is enhanced by relating Object-Z operations to CSP events.

With this type of common semantic model we will be able to use CSP and Object-Z together in two ways:

- by combining CSP processes and Object-Z classes in parallel composition;
- by combining a collection of Object-Z classes together using CSP parallel and interleaved composition.

The first allows the required temporal ordering in a specification to be expressed in CSP and the parallel composition acts as a conjunction between the two sets of requirements. The second approach enables CSP operators to be used to express communication and synchronisation requirements between the components.

We will restrict ourselves in both approaches to Object-Z classes without internal operations, and in the second approach we will make further restrictions on both the Object-Z and CSP parts.

The use of the CSP parallel composition operator requires that synchronising events communicate an identical number of parameters. This means that in both approaches it is sometimes necessary to introduce dummy parameters to allow operations and events with different parameters to synchronise.

However, before we give some examples we need to explain the basis of the integration between the two languages.

18.2.1 A Common Semantic Model

The semantic model of CSP we use here is the standard failures-divergences semantics as discussed above. Then, combined Object-Z and CSP specifications are given a well-defined meaning by giving the Object-Z classes a failures-divergences semantics identical to that of a CSP process.

The failures of a class are derived from its history semantic model (defined in [39]), and by restricting ourselves to Object-Z classes without internal operations, the divergences of a class will be empty. (Note that this is a subtly different semantic model from the relational one, see Chap. 19.)

The key to making this link work is the treatment of operations and events. These are related by mapping each operation invocation (i.e., operation plus values for its parameters) to the appropriate event. To do this we use the following function which turns an operation op with assignment of values to its parameter p to an event:

$$event((op, p)) = op.\beta(p)$$

The meta-function β replaces each parameter name in p by its base name, i.e., it removes the ? or !. The net effect of this is that the event corresponding to an operation (op, p) is a communication event with the operation name op as the channel and an assignment of values to the base names of the operation's parameters as the value passed on that channel.

For example, given an operation

$$\begin{array}{|l}
\hline \textit{Inc} \\
\Delta(x) \\
y? : \mathbb{N} \\
\hline
x' = x + y? \\
\hline
\end{array}$$

the event corresponding to this operation with value 5 for input is $Inc.\{(y, 5)\}$. Similarly, the event corresponding to the operation

$$\begin{array}{|l}
\hline \textit{Val} \\
z! : \mathbb{N} \\
\hline
z! = x \\
\hline
\end{array}$$

will be $Val.\{(z, x)\}$.

The mapping of operations to events where β is used to return the base name of a parameter means that events can synchronise in a combined specification in a number of ways. Usually synchronisation will have been used to communicate between input and output parameters either between two Object-Z classes or between an Object-Z class and a CSP process. However, synchronisation will also occur between two inputs and between two outputs. The former represents the sharing of an input value, while the latter represents cooperation of two components to produce an output.

18.3 Combining CSP Processes and Object-Z Classes

With the relationship between classes and processes in place we can give meaning to specifications involving Object-Z classes and CSP processes. By putting a CSP process in parallel with an Object-Z class we can use the CSP part to define the temporal ordering of the combined specification. The approach is best illustrated by some examples.

Example 18.6 Consider the specification of the lift controller given by $LiftCon_0$ in Example 17.4 (p. 415). The purpose of the variable *state* was merely to control the correct sequencing of operations, and we can specify this required ordering via a CSP process *LTO*.

$$LTO = Close \rightarrow Arrive?f \rightarrow Open \rightarrow LTO$$

The complete specification will be given by

$$LiftCon = LTO \parallel LiftCon_2$$

where $LiftCon_2$ is the following Object-Z class.

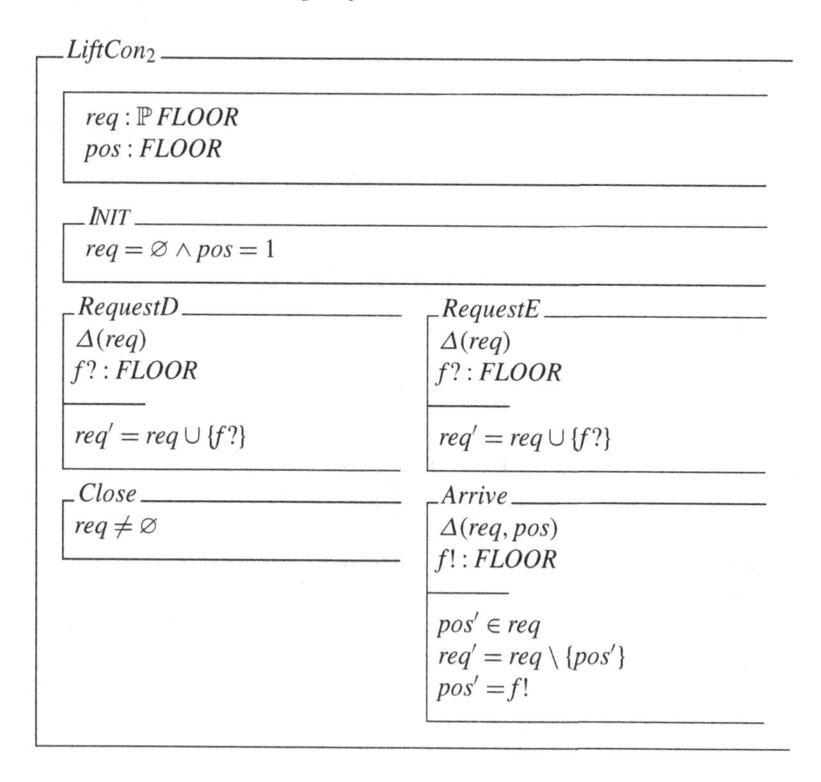

The effect of *LiftCon* is as follows. The CSP parallel operator $\parallel$ causes *LTO* and $LiftCon_2$ to run concurrently synchronising on common events, with all other events being interleaved. Thus *Close*, *Arrive* and *Open* occur in sequence with any number of *RequestD* and *RequestE* happening at any point in time (i.e., the requests can be interleaved with the movement of lift and doors). This gives us the required temporal ordering that was in the original $LiftCon_0$ class without the use of the variable *state*.

Notice that since *Open* had no behaviour other than to occur in sequence it need not occur in the Object-Z class $LiftCon_2$, however, the overall specification still has an event called *Open* due to its presence in the CSP part of the specification.

The remaining point to note is the use of the input parameter on *Arrive* in the CSP expression. This is necessary because the CSP parallel composition operator $\parallel$ requires synchronisation on identical events, and in the case of events with communication the event comprises the channel name together with the communicated value. Thus if we wrote

$$LTO = Close \to Arrive \to Open \to LTO$$

for the CSP part of the specification it would deadlock due to this *Arrive* not being compatible with the communicated value in the Object-Z *Arrive* operation.

Since we do not want to constrain the value of the output $f!$ in $LiftCon_2$ we simply accept it as an input in the CSP part, this value is not hidden in the complete specification and thus *LiftCon* achieves the same effect as the original $LiftCon_0$. □

The key aspect to note from this is the use of input in the *Arrive* event. This approach works in general: if the CSP part is just used to define the required temporal ordering, then the events corresponding to Object-Z operations have associated input parameters. This allows synchronisation without restricting the values used for input or output further than the values accepted in the Object-Z part.

Of course the CSP part could, in addition to defining the temporal ordering of events, also constrain the inputs and outputs further if desired. The CSP part could also introduce events not mentioned in the Object-Z components, their total effect would then be defined just within the CSP defining process. For example, the CSP part might introduce an event *EmergencyStop* which did not have a corresponding operation in the Object-Z components.

The Object-Z part can also define requirements that affect the allowable temporal ordering of events as the following example illustrates.

Example 18.7 A ticket machine in an underground railway station allows either single or return tickets to be dispensed to a number of destinations. We can specify its behaviour using a CSP process together with an Object-Z class. The class has a *Coin* operation to accept money for the trip up to the *limit* of the machine, the *Ticket* operation allows a choice between single and return and *Dest* dispenses the ticket and change according to the chosen destination. We use a function *price* to determine the correct fare given state and ticket type.

$$[STATION]$$
$$PRINT ::= print\langle\langle STATION\rangle\rangle$$
$$TICKET == PRINT \times \mathbb{B}$$

$$\begin{array}{|l} price : STATION \times \mathbb{B} \to \mathbb{N} \\ limit : \mathbb{N} \end{array}$$

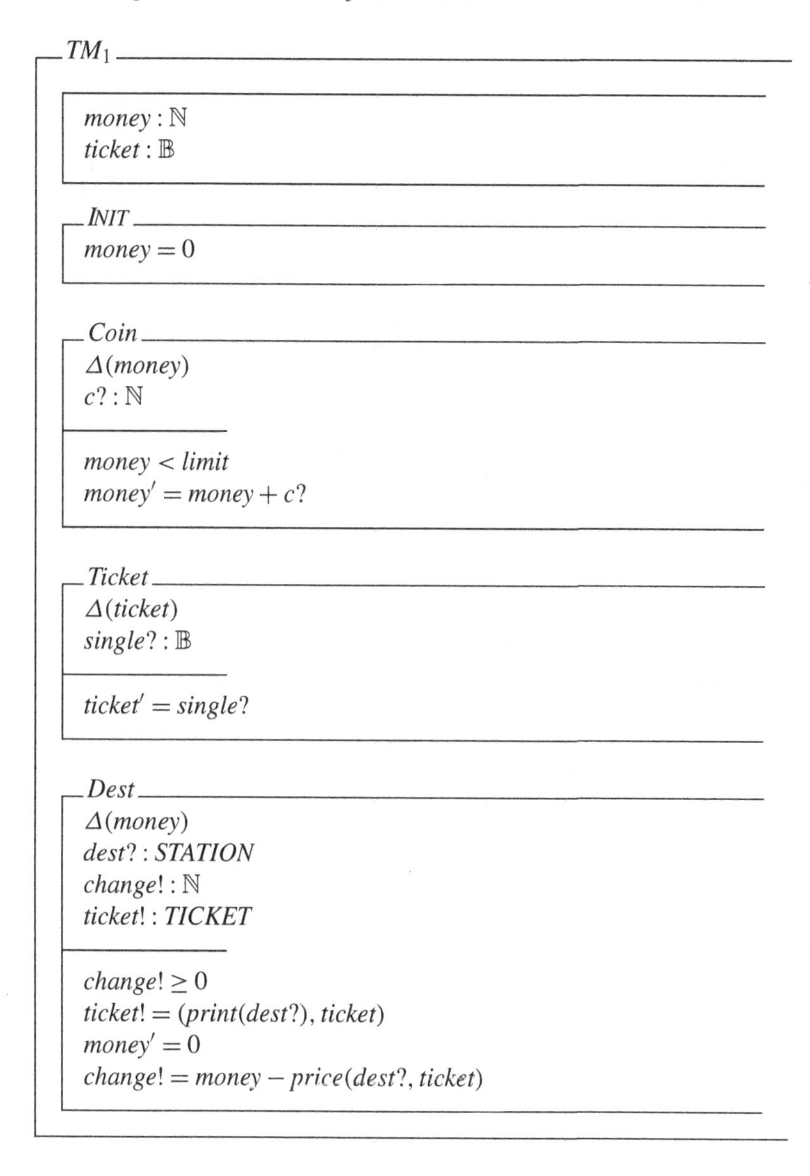

The CSP part then defines the acceptable temporal ordering which allows any number of coins to be inserted followed by *Ticket* then *Dest*. As explained in the previous example, we need to ensure that identical communication takes place on these events. The CSP part is thus:

$$
\begin{aligned}
TM_2 = Coin?c \rightarrow (&TM_2\\
&[]\\
&Ticket?single \rightarrow Dest?dest?change?ticket \rightarrow TM_2)
\end{aligned}
$$

The complete specification being

$$TM = TM_1 \parallel TM_2$$

Although the CSP process TM_2 restricts the temporal ordering of events in TM to that allowed by TM_2, further constraints can be embedded as part of the Object-Z class. This is the case here. TM_2 allows an indefinite number of coins to be inserted, but this is restricted in TM_1 which makes a requirement that $money < limit$ in the precondition of *Coin*.

Similarly, the *Ticket* event is enabled after the first coin event in the CSP process, but the Object-Z class refuses this event unless there is enough money in the machine to buy the requested ticket.

We can amend the ticket machine to provide a cancellation facility, which allows a user to change a ticket type if they change their mind. We can do this by changing just the CSP part of the specification to:

$$\begin{array}{l} TM_3 = Coin?c \rightarrow (TM_3 \; [\,] \; Q) \\ Q = Ticket?single \rightarrow (Cancel \rightarrow Q \\ \qquad\qquad\qquad\qquad\qquad [\,] \\ \qquad\qquad\qquad\qquad\qquad Dest?dest?change?ticket \rightarrow TM_3) \end{array}$$

Alternatively we could provide a cancellation facility which could be invoked at any time, and which returns any money currently in the machine. We can specify this by amending just the Object-Z part of TM to

$$\begin{array}{|l} \hline TM_4 \\ \hline TM_2 \\ \\ \quad \begin{array}{|l} \hline Cancel \\ \hline \Delta(money) \\ change! : \mathbb{N} \\ \hline change! = money \\ money' = 0 \\ \hline \end{array} \\ \hline \end{array}$$

□

18.4 Combining Object-Z Classes Using CSP Operators

An alternative way to use Object-Z and CSP together is to combine a collection of Object-Z classes together using the CSP parallel, interleaving and hiding operators. The benefits of this approach are that synchronisation and concurrency can be defined at the class level instead of on an operation by operation basis as is necessary if using just Object-Z.

The general approach in this method is to define the component processes as Object-Z classes, and these components are then combined using composition and hiding. This produces specifications of the form $((C_1 \parallel C_2) \interleave C_3) \setminus L$, where C_i are Object-Z classes and L a set of operations which are being hidden (i.e., made internal).

When combining Object-Z classes with CSP we will use subscripts on a class name to denote any restrictions of initial value and constants. For example, $BankAcct_{\{(name,n)\}}$ will denote the *BankAcct* class with constant *name* instantiated to n. The absence of a subscript means there are no restrictions on the initial state.

Classes with generic parameters are referenced with the actual parameters in brackets following the class name.

In using this specification style we make some syntactic restrictions on both Object-Z and CSP as follows. Since process interaction is defined using CSP operators we can constrain ourselves to using a subset of Object-Z. In particular, we can dispense with object instantiation and therefore also polymorphism, class union, object containment and the Object-Z operators $\parallel$, $\parallel_!$ and enrichment.

Similarly we only use a subset of CSP operators consisting of parallel, interleaving and hiding, so, for example, piping and sequential composition are not needed.

For semantic reasons we place a restriction on the use of CSP hiding to prevent problems due to unbounded non-determinism. The problem is caused by the fact that the standard failures-divergences semantics of CSP does not support the specification of unbounded non-determinism, since this semantic model assumes that the infinite sequences of events of a process can be extrapolated from its finite sequences.

Therefore we have to place a restriction on hiding to make sure that it does not introduce unbounded non-determinism. There are a number of ways in which this can be done, the simplest is the following restriction which says that no trace of P can be extended by an unlimited number of actions in C and still be a trace of P. Formally given a process $P \setminus C$, the failures F of P must satisfy:

$$\forall s \in \operatorname{dom} F \bullet \neg(\forall n \in \mathbb{N} \bullet \exists t \in C^* \bullet \#t > n \wedge s \frown t \in \operatorname{dom} F)$$

Application of this condition (which, although it looks complex, is quite natural and our examples below conform to it) prevents unbounded sequences of events from being hidden.

Example 18.8 CSP interleaving can be used to define multiple copies of a component in a natural fashion. Consider the *BankAcct* class from Sect. 15.2.1 (p. 369) which defines a single account identified by the constant *name*. A collection of independent accounts is specified by first defining a process $BankAcct_n$ corresponding to the account with name n:

$$BankAcct_n = BankAcct_{\{(name,n)\}}$$

then using the set *NAME* we can define the collection as:

$$AcctCollection = \interleave_{n:NAME} BankAcct_n$$

Alternatively a set of users might interact with a single account. For example, if we define

$[ID]$

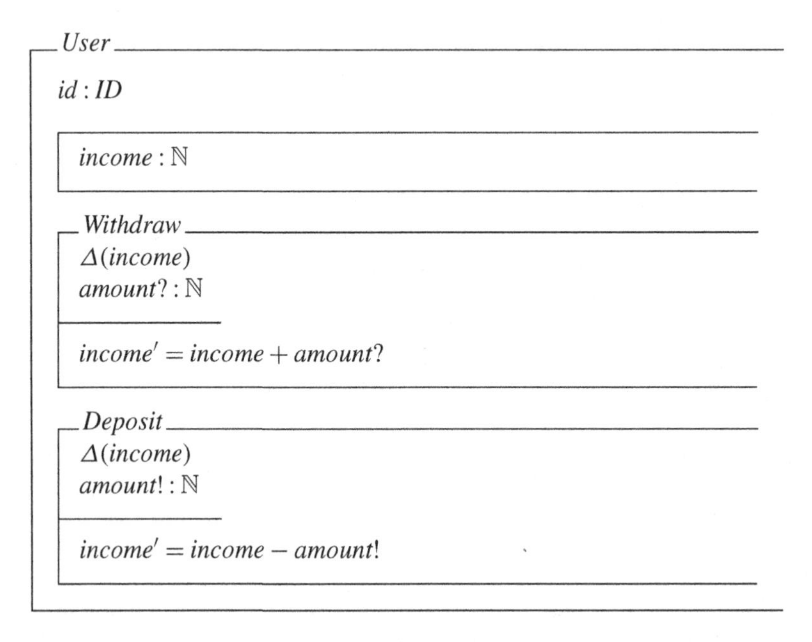

then the interaction can be described by

$$Banking = (|||_{n:ID} User_n) \parallel BankAcct$$

where $User_n$ is defined in a similar way to $BankAcct_n$. □

Renaming can be used to amend the interface of a component prior to synchronisation as the following example illustrates.

Example 18.9 We build a $2n$ place buffer out of two n place buffers in the standard way. If the n place buffer is specified as (for some type T):

Buf

$s : \mathrm{seq}\, T$

$\#s \leq n$

INIT

$s = \langle\, \rangle$

Put	*Get*
$\Delta(s)$ $x? : T$	$\Delta(s)$ $x! : T$
$s' = \langle x? \rangle \frown s$	$s = s' \frown \langle x! \rangle$

Then the $2n$ place buffer can be expressed as

$$(Buf[Mid/Get] \parallel Buf[Mid/Put]) \setminus \{Mid.\{(x, y)\} \mid y \in T\}$$

We have used renaming here to rename operations *Put* and *Get* to *Mid* in each of the two buffers, thus when synchronised the value from the first will be passed to the second. We have then hidden this communication since we do not wish it to be observable. In hiding we have to hide each event explicitly, i.e., we use the name of the channel together with a value corresponding to its parameter. It is preferable to use a shorthand for this, for example, we could write specifications like this as:

$$(Buf[Mid/Get] \parallel Buf[Mid/Put]) \setminus \{Mid\} \qquad \square$$

To compare the use of CSP and Object-Z together against the composition facilities offered by Object-Z alone, we re-specify the cashpoint machine given in Example 17.2 (p. 411). As mentioned above, it is sometimes necessary to introduce dummy parameters and this example illustrates how this is done.

Example 18.10 Instead of using the class ATM_5 to define the interaction between the components we can use parallel composition together with hiding to achieve the same effect.

Although we can use synchronisation to good effect, when using the CSP parallel operator we need to introduce dummy input and output parameters for the synchronisation to work. This is necessary because CSP and Object-Z are subtly different in how their $\parallel$ operator works. In Object-Z, operations involved in a composition using $\parallel$ or $\parallel_!$ can have different input and output parameters, all the operator does is to synchronise on any commonly named parameters if there are any. Any amount of additional parameters can be contained in either operation, compare, for example, the *Withdraw* operation in the classes *CashPoint* and *Bank*.

However, in CSP, in order for a correct synchronisation to occur the channels have to communicate identical parameters. It was for this same reason that we needed a redundant input parameter on *Arrive* in Example 18.6, and we have to do the same here in the Object-Z components. With Object-Z we can achieve the desired result with the reuse provided by inheritance. Thus we define two new classes:

$CashPoint_1$

$CashPoint$

$Withdraw$

$amount?, money? : \mathbb{N}$

$Bank_1$

$Bank[SupplyPin/OpenAcct]$

$SupplyPin$

$pin? : \mathbb{N} \times \mathbb{N} \times \mathbb{N} \times \mathbb{N}$

Then we can define the interaction between the two components by writing:

$$ATM_6 = CashPoint_1 \parallel Bank_1$$

However, even this does not quite have the same observable effect as ATM_5. The problem is the *Withdraw* operation. In the Object-Z version parameters *amt*? and *acct*? were introduced to communicate between the classes. In ATM_5 the operation definition

$$Withdraw \mathrel{\widehat=} (b.Withdraw \| c.Withdraw)$$

these will be hidden, however, *amount*? and *money*! in the same operation are not hidden, and indeed we wish them to be visible to the environment. But in the CSP specification we cannot hide just some of the parameters and not others, thus in ATM_6 all the values communicated on the parameters (including *amt*? and *acct*?) are visible to the environment. One solution to this (which we do not detail here) is to separate out the communication between the two classes into another operation separate from *Withdraw*. This new event can be made internal by hiding it, leaving *Withdraw* visible together with just the values due to *amount*? and *money*!.

18.4.1 Semantic Considerations—Models of Operations and Outputs

There are a couple of subtle choices about the semantics used for Object-Z when Object-Z classes are combined using CSP operators. Specifically one can make a choice of how preconditions and outputs are dealt with, and this choice affects the design approach one takes in the language. Although Object-Z normally takes a blocking approach to the domain of an operation, it is possible to use the integration with the non-blocking interpretation. This gives one design choice to make. The other concerns outputs, specifically whether to adopt a *demonic* or *angelic* model of outputs—whereby we mean that an operation's output is angelic if it satisfies the environment's constraints on them and demonic if outputs are not influenced by the environment at all.

To understand the latter point consider the following two simple classes:

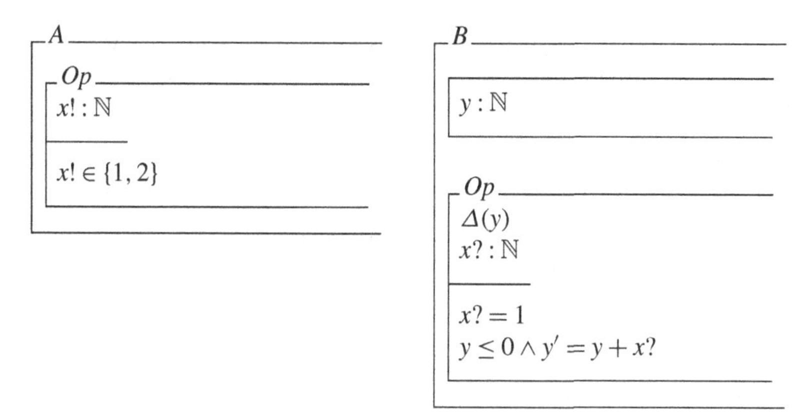

with the overall specification

$$Spec = A \parallel B$$

A has a non-deterministic output, matched by an input in the corresponding operation in *B*, but this input has some constraints on the possible values it can take. There is a choice about whether component *B* influences the choice of output chosen in *A*. The demonic model of outputs views that *B* has no influence, thus the non-determinism of the value output by *Op* in *A* is entirely internal to that component. Thus, if the chosen output is incompatible with the expected input in B, the combined operation's precondition is not satisfied (resulting in the operation being blocked). However, using an angelic model of outputs allows *B* to affect the non-determinism in *A* by choosing a value to synchronise on, if one can be found, thus allowing the operation to be enabled if at all possible.

In this particular example then, using the standard blocking model the operation *Op* in *A* and *B* must synchronise, with agreement on the value communicated, so if this event occurs the value 1 will be communicated from *A* to *B*. Whether the event occurs depends on the model of outputs. In the demonic model *A* can non-deterministically output 2, thus the operation will not be enabled, whereas in the angelic model the non-determinism in *A* is restricted by what the environment is prepared to accept, thus the operation in *Spec* will be enabled, and the value 1 will be communicated from *A* to *B*.

Use of the blocking model with the angelic model of outputs can lead to a nice abstract approach to specification using the combined Object-Z and CSP notation. The blocking model, appropriate anyway for the specification of concurrent systems, abstracts away from a certain amount of detail in that one does not need to specify what happens outside an operation's precondition. The use of the angelic model of outputs can further add to this abstraction, since it allows one to model cooperation between processes in a simple fashion. This can allow one to abstract away from the actual cooperation mechanism which in general would require additional message passing. The following example illustrates this.

Example 18.11 Consider the *BankAcct* class from Example 15.2.1 (p. 369) which defines a number of Users of a Bank Account. We can enhance our Bank Account and Users with the ability to set a password as follows. We define a new given type for the collection of passwords, and a user can now set their own password in a specific operation.

$[Passwords]$

User1

$User$

SetPasswd

$id! : ID$
$passwd! : Passwords$

$id! = id$

The Bank Account contains the restriction that the new password (as a result of a *SetPasswd* operation) must be different from the old one. This is achieved by recording passwords for users in a *passwordfile*.

BankAcct1

$BankAcct$

$passwordfile : ID \nrightarrow Passwords$

INIT

$passwordfile' = \varnothing$

SetPasswd

$\Delta(passwordfile)$
$id? : ID$
$passwd? : Passwords$

$(id?, passwd?) \notin passwordfile$
$passwordfile' = passwordfile \oplus \{(id?, passwd?)\}$

Then as before we can specify the interaction between users and the bank by

$$Banking = (|||_{n:ID} User1_n) \parallel BankAcct1$$

In particular, the parallel composition ensures any user setting their password will synchronise with the bank performing the same event. Under the blocking model of operations and the angelic model of outputs, the synchronisation corresponding to setting the password ensures that a valid (i.e., new) password is chosen—and this is abstractly modelling a sequence of communications. One possible implementation is that the user provides a preference for their password, and the bank either acknowledges this new setting or indicates that it is already taken and asks for a second attempt. Using the blocking model and angelic outputs, the actual implementation need not be specified at this stage.

This is not the case, however, if either the non-blocking model of operations or demonic model of outputs is used. Both of these require additional information to be passed between the user and bank before a new password can be set.

Adopting the non-blocking model, the bank would accept any id and password (since preconditions just specify the correct behaviour) as the outcome of the operation is unspecified outside the precondition $(id?, passwd?) \notin passwordfile$.

In this case it would be necessary to specify an exception which modelled the bank indicating that the suggested combination could not be used. With demonic outputs, additional information would need to be communicated to the user (for example using the implementation strategy above) before the password could be successfully set. Otherwise, there is the possibility of user processes deadlocking when their booking is unavailable. Once again, this is not the correct behaviour.

18.5 Bibliographical Notes

CSP is one of a family of process algebras, other similar languages include CCS [24] and the ISO standard LOTOS [3]. Introductions to CSP include the texts by Hoare [20], Roscoe [30, 31] and Schneider [32].

There are a number of possible approaches to combining state-based languages with process algebras. The composition of Object-Z classes with CSP operators was first detailed by Smith in [40], see also the work of Smith and Derrick [41, 42] which consider, in addition, refinement and verification in this notation. Work on the different semantic models of operations and outputs was considered in [43]. More details of this and other work on this combination of CSP and Object-Z are given in Sect. 19.6 below. In [14] Fischer and Smith consider the infinite trace model as an alternative semantics for combinations of CSP and Object-Z.

Alternative approaches to combining Z or Object-Z with CSP or CCS include the work of Fischer [12], Mahony and Dong [22, 23], Galloway and Stoddart [18] and Taguchi and Araki [44]. A survey of some of these approaches is given in [13].

The work of Fischer, for example, also combines Object-Z with CSP by using a failure-divergence semantics as the basis for the integration. However, Object-Z classes are extended with channel definitions and a CSP process. In addition, both the precondition and guard of an operation are defined and events can either be atomic or have duration (and therefore have a start and end). The resultant language, called CSP-OZ, is complex but very expressive and benefits from a theory of refinement and tool support (see, for example, [15]).

Although Object-Z adopts a blocking model of operations, not all integrations of Object-Z and CSP have used that interpretation. For example, although the blocking model is used in [40, 42], in [12, 22] a non-blocking model was adopted. The majority of work on integrating Object-Z and CSP has used the demonic model of outputs, including [12, 22, 40, 42]. This is also the main interpretation used in other work on the semantics of value-passing communication. For example, [7] places conditions on composed value-passing action systems which ensure that outputs are always accepted by the environment. More details on how the differing models of operations and outputs affect the semantics of the integration between Object-Z and CSP are to be found in [43].

Mahony and Dong [23] use a combination of Object-Z and timed CSP in their TCOZ language. However, instead of identifying operations and events, operation parameters are mapped to events. See also [29] and [10].

Similar approaches, where the granularity of operations and events is different, but using Z and CCS include the work of Galloway [18] and Taguchi and Araki [44].

There has also been similar work on combining CSP with the B notation. Butler [6, 8] uses CSP to control the temporal ordering of B operations in a manner similar to the approach discussed in this chapter. However, to define the meaning of the specification he defines a translation of CSP into B, and therefore uses a restricted subset of CSP.

Treharne and Schneider [33, 45, 46] also consider a combination of CSP and B, which they denote $CSP \parallel B$. However, in addition to controlling the temporal ordering of events, the CSP acts as a control executive for the B machine. That is, in addition to inputs communicated from the environment, internal inputs and outputs are also passed between the CSP process and the B machine parts of the combined specification. An approach to modular verification in the notation is discussed in [35], and [34] defines interface refinement for $CSP \parallel B$. The latter defines non-atomic refinement in the context of $CSP \parallel B$, so that abstract events can be refined to a sequence of concrete events, and it also uses a notion of IO refinement in the framework. Tool support is discussed in [11].

This work has been extended to define combinations of CSP and Event-B, the definition of this combination is given in [36], and [38] defines a CSP semantics for Event-B refinement. For a simple case study see [37].

Circus [26] is a notation that also combines Z and CSP as well as Morgan's specification statements and Dijkstra's guarded commands. Its semantics is based on Hoare and He's Unifying Theories of Programming (UTP) [21]. Circus specifications define both data and behavioural aspects of systems using a combination of Z and CSP constructs, where the Z notation is used to define most of the data aspects, and CSP is used to define behaviour. Again the failures-divergences model is used as the basis for refinement in Circus, but this time in UTP, as this provides a framework that unifies programming discipline across many different computational paradigms. Publications include extensions to incorporate time [47, 49], tool support for refinement [50], tactics for refinement [27, 28] and many others, see http://www.cs.york.ac.uk/circus/index.html.

References

1. Araki, K., Galloway, A., & Taguchi, K. (Eds.) (1999). *International Conference on Integrated Formal Methods 1999 (IFM'99)*. York: Springer.
2. Araki, K., Gnesi, S., & Mandrioli, D. (Eds.) (2003). *Formal Methods Europe (FME 2003). Lecture Notes in Computer Science: Vol. 2805*. Berlin: Springer.
3. Bolognesi, T., & Brinksma, E. (1988). Introduction to the ISO Specification Language LOTOS. *Computer Networks and ISDN Systems*, *14*(1), 25–59.
4. Bowen, J. P., Fett, A., & Hinchey, M. G. (Eds.) (1998). *ZUM'98: the Z Formal Specification Notation. Lecture Notes in Computer Science: Vol. 1493*. Berlin: Springer.
5. Bowen, J. P., Dunne, S., Galloway, A., & King, S. (Eds.) (2000). *ZB2000: Formal Specification and Development in Z and B. Lecture Notes in Computer Science: Vol. 1878*. Berlin: Springer.
6. Butler, M. (1999) csp2B: A practical approach to combining CSP and B. In Wing et al. [48] (pp. 490–508).
7. Butler, M. J. (1993). Refinement and decomposition of value-passing action systems. In *Proceedings of the 4th International Conference on Concurrency Theory, CONCUR '93*, London, UK (pp. 217–232). Berlin: Springer.
8. Butler, M. J., & Leuschel, M. (2005) Combining CSP and B for specification and property verification. In Fitzgerald et al. [16] (pp. 221–236).
9. Derrick, J., Boiten, E. A., & Reeves, S. (Eds.) (2011). *Proceedings of the 15th International Refinement Workshop. Electronic Proceedings in Theoretical Computer Science: Vol. 55*. Open Publishing Association.
10. Dong, J. S., Hao, P., Qin, S., Sun, J., & Yi, W. (2004). Timed patterns: TCOZ to timed automata. In J. Davies, W. Schulte, & M. Barnett (Eds.), *ICFEM'04—Formal Methods and Software Engineering, 6th International Conference on Formal Engineering Methods. Lecture Notes in Computer Science: Vol. 3308* (pp. 483–498). Berlin: Springer.
11. Evans, N., & Treharne, H. (2007). Interactive tool support for CSP||B consistency checking. *Formal Aspects of Computing*, *19*(3), 277–302.
12. Fischer, C. (1997). CSP-OZ—a combination of CSP and Object-Z. In H. Bowman & J. Derrick (Eds.), *Second IFIP International Conference on Formal Methods for Open Object-Based Distributed Systems* (pp. 423–438). London: Chapman & Hall.
13. Fischer, C. (1998) How to combine Z with a process algebra. In Bowen et al. [4] (pp. 5–23).
14. Fischer, C., & Smith, G. (1997) Combining CSP and Object-Z: finite or infinite trace semantics. In Mizuno et al. [25] (pp. 503–518).
15. Fischer, C., & Wehrheim, H. (1999) Model checking CSP-OZ specifications with FDR. In Araki et al. [1] (pp. 315–334).
16. Fitzgerald, J. A., Hayes, I. J., & Tarlecki, A. (Eds.) (2005). *FM 2005: Formal Methods, International Symposium of Formal Methods Europe. Lecture Notes in Computer Science: Vol. 3582*. Berlin: Springer.
17. Fitzgerald, J. A., Jones, C. B., & Lucas, P. (Eds.) (1997). *FME'97: Industrial Application and Strengthened Foundations of Formal Methods. Lecture Notes in Computer Science: Vol. 1313*. Berlin: Springer.
18. Galloway, A., & Stoddart, W. (1997) An operational semantics for ZCCS. In Hinchey and Liu [19] (pp. 272–282).

19. Hinchey, M. G. & Liu, S. (Eds.) (1997). *First International Conference on Formal Engineering Methods (ICFEM'97)*. Hiroshima, Japan, November 1997. Los Alamitos: IEEE Comput. Soc.
20. Hoare, C. A. R. (1985). *Communicating Sequential Processes*. New York: Prentice Hall.
21. Hoare, C. A. R., & He, J. (1998) *Unifying Theories of Programming*. *International Series in Computer Science*. New York: Prentice Hall.
22. Mahony, B. P., & Dong, J. S. (1998). Blending Object-Z and timed CSP: an introduction to TCOZ. In K. Futatsugi, R. Kemmerer, & K. Torii (Eds.), *20th International Conference on Software Engineering (ICSE'98)*. New York: IEEE Press.
23. Mahony, B. P., & Dong, J. S. (2000). Timed communicating Object-Z. *IEEE Transactions on Software Engineering*, *26*(2), 150–177.
24. Milner, R. (1989). *Communication and Concurrency*. New York: Prentice Hall.
25. Mizuno, T., Shiratori, N., Higashino, T., & Togashi, A. (Eds.) (1997) *FORTE/PSTV'97*, Osaka, Japan, November 1997. London: Chapman & Hall.
26. Oliveira, M. V. M., Cavalcanti, A. L. C., & Woodcock, J. C. P. (2007). A UTP semantics for *Circus*. *Formal Aspects of Computing*, *21*(1), 3–32.
27. Oliveira, M. V. M., Gurgel, A. C., & de Castro, C. G. (2008). CRefine: support for the *Circus* refinement calculus. In A. Cerone & S. Gruner (Eds.), *6th IEEE International Conference on Software Engineering and Formal Methods* (pp. 281–290). Los Alamitos: IEEE Comput. Soc.
28. Oliveira, M. V. M., Zeyda, F., & Cavalcanti, A. L. C. (2011). A tactic language for refinement of state-rich concurrent specifications. *Science of Computer Programming*, *76*(9), 792–833.
29. Qin, S., Dong, J. S., & Chin, W.-N. (2003) A semantic foundation for TCOZ in Unifying Theories of Programming. In Araki et al. [2] (pp. 321–340).
30. Roscoe, A. W. (1998). *The Theory and Practice of Concurrency*. *International Series in Computer Science*. New York: Prentice Hall.
31. Roscoe, A. W. (2010). *Understanding Concurrent Systems*. Berlin: Springer.
32. Schneider, S. (2000). *Concurrent and Real-Time Systems: the CSP Approach*. New York: Wiley.
33. Schneider, S. A., & Treharne, H. (2005). CSP theorems for communicating B machines. *Formal Aspects of Computing*, *17*(4), 390–422.
34. Schneider, S. A., & Treharne, H. (2011). Changing system interfaces consistently: a new refinement strategy for CSP||B. *Science of Computer Programming*, *76*(10), 837–860.
35. Schneider, S. A., Treharne, H., & Evans, N. (2005). Chunks: component verification in CSP||B. In J. Romijn, G. Smith, & J. van de Pol (Eds.), *IFM 2005—Integrated Formal Methods, 5th International Conference*. *Lecture Notes in Computer Science: Vol. 3771* (pp. 89–108). Berlin: Springer.
36. Schneider, S. A., Treharne, H., & Wehrheim, H. (2010). A CSP approach to control in Event-B. In D. Méry & S. Merz (Eds.), *Proceedings of 8th International Conference on the Integrated Formal Methods—IFM 2010*, Nancy, France, October 11–14, 2010. *Lecture Notes in Computer Science: Vol. 6396* (pp. 260–274). Berlin: Springer.
37. Schneider, S. A., Treharne, H., & Wehrheim, H. (2011). Bounded retransmission in Event-B||CSP: a case study. *Electronic Notes in Theoretical Computer Science*, *280*, 69–80.
38. Schneider, S. A., Treharne, H., & Wehrheim, H. (2011) A CSP account of Event-B refinement. In Derrick et al. [9] (pp. 139–154).
39. Smith, G. (1995). A fully abstract semantics of classes for Object-Z. *Formal Aspects of Computing*, *7*(3), 289–313.
40. Smith, G. (1997) A semantic integration of Object-Z and CSP for the specification of concurrent systems. In Fitzgerald et al. [17] (pp. 62–81).
41. Smith, G., & Derrick, J. (1997) Refinement and verification of concurrent systems specified in Object-Z and CSP. In Hinchey and Liu [19] (pp. 293–302).
42. Smith, G., & Derrick, J. (2001). Specification, refinement and verification of concurrent systems—an integration of Object-Z and CSP. *Formal Methods in Systems Design*, *18*, 249–284.

43. Smith, G., & Derrick, J. (2002). Abstract specification in Object-Z and CSP. In C. George & H. Miao (Eds.), *ICFEM. Lecture Notes in Computer Science: Vol. 2495* (pp. 108–119). Berlin: Springer.
44. Taguchi, K., & Araki, K. (1997) The state-based CCS semantics for concurrent Z specification. In Hinchey and Liu [19] (pp. 283–292).
45. Treharne, H., & Schneider, S. (1999) Using a process algebra to control B operations. In Araki et al. [1] (pp. 437–456).
46. Treharne, H., & Schneider, S. A. (2000) How to drive a B machine. In Bowen et al. [5] (pp. 188–208).
47. Wei, K., Woodcock, J. C. P., & Cavalcanti, A. L. C. (2012). *Circus time* with reactive designs. In B. Wolff, M.-C. Gaudel, & A. Feliachi (Eds.), *4th International Symposium on Unifying Theories of Programming. Lecture Notes in Computer Science: Vol. 7681* (pp. 68–87). Berlin: Springer.
48. Wing, J. M., Woodcock, J. C. P., & Davies, J. (Eds.) (1999). *FM'99 World Congress on Formal Methods in the Development of Computing Systems. Lecture Notes in Computer Science: Vol. 1708*. Berlin: Springer.
49. Woodcock, J. C. P., Oliveira, M. V. M., Burns, A., & Wei, K. (2010). Modelling and implementing complex systems with timebands. In *4th IEEE International Conference on Secure System Integration and Reliability Improvement, SSIRI 2010* (pp. 1–13). Los Alamitos: IEEE Comput. Soc.
50. Zeyda, F., Oliveira, M. V. M., & Cavalcanti, A. L. C. (2012). Mechanised support for sound refinement tactics. *Formal Aspects of Computing*, *24*(1), 127–160.

Chapter 19
Refining CSP and Object-Z Specifications

The previous chapter has defined ways of integrating CSP and Object-Z, and we now turn our attention to issues of refinement in these notations. In particular what we do is to discuss how we can refine these specifications component by component. That is, we discuss how a single Object-Z class or a single CSP process can be refined in such a way that the overall specification is refined. This has echoes of the question of compositionality that we looked at in Sect. 17.3, however, because our specifications have a particular set form, compositionality is easier to prove.

In Sect. 19.1 we discuss definitions of refinement in CSP and the relationship to Object-Z refinement. At the heart of the integration of the two languages is a correlation between refinement in Object-Z and refinement in CSP. More specifically, we give versions of the downward and upward simulation conditions which are sound and jointly complete with respect to failures-divergences refinement in CSP. We apply this result in Sects. 19.3 and 19.4, showing how CSP and Object-Z components can be refined in a compositional manner.

19.1 Refinement in CSP

There are a number of different refinement relations in CSP based upon the strength of the underlying semantic model. The simplest notion of refinement is *trace refinement*.

Definition 19.1 (Trace refinement) A process Q is a trace refinement of a process P, denoted $P \sqsubseteq_T Q$, if

$$traces(Q) \subseteq traces(P) \qquad \Box$$

Excerpts from pp. 66, 73 of [4] are reprinted within this chapter with kind permission of Springer Science and Business Media.
Excerpts from pp. 200–201 of [7] are reprinted within Sect. 19.2.1 with kind permission of Springer Science and Business Media.

J. Derrick, E.A. Boiten, *Refinement in Z and Object-Z*,
DOI 10.1007/978-1-4471-5355-9_19,

Thus, for example, $Q = Start \to stop$ is a trace refinement of

$$P = Timeout \to Halt \to skip \\ \Box \\ Start \to stop$$

and, furthermore, P is a trace refinement of

$$R = Timeout \to Halt \to skip \\ \sqcap \\ Start \to stop$$

As this last example shows, trace refinement is rather weak, even more so once we realise that *stop* is a trace refinement of any process. This is hardly a good basis for implementation! (See also Sect. 3.3.3 where we discuss trace refinement in the relational setting.)

More discriminating is failures-divergences refinement, which is based upon the failures-divergences semantics discussed above.

Definition 19.2 (Failures-divergences refinement) A process Q is a failures-divergences refinement of a process P, denoted $P \sqsubseteq_F Q$, if

$$failures(Q) \subseteq failures(P) \quad \text{and} \quad divergences(Q) \subseteq divergences(P)$$

Two processes are failures-divergences equivalent if they are both failures-divergences refinements of each other. □

Example 19.1 As an example, consider $Q = Start \to stop$ and

$$P = Timeout \to Halt \to skip \\ \Box \\ Start \to stop$$

The failures of Q with respect to the alphabet $\{Timeout, Halt, Start\}$ include $(\langle\rangle, \{Timeout\})$, which is not in the failures of P. On the other hand, the failures of P include $(\langle Timeout\rangle, \{Start, Timeout\})$ which is not in the failures of Q. Thus the two processes are not comparable in terms of failures-divergences refinement.

Now consider

$$R = Timeout \to Halt \to skip \\ \sqcap \\ Start \to stop$$

In terms of traces, P and R are equivalent, i.e., each is a trace refinement of the other. However, this is not the case when measured against failures-divergences refinement. In fact P is a failures-divergences refinement of R but R is not a failures-divergences refinement of P. To see the latter, note that one possible failure of R is $(\langle\rangle, \{Start\})$ which is not included in the failures of P. □

Failures-divergences refinement only requires inclusion of failures and divergences, not equality, and this allows the reduction of non-determinism in a fashion which should be familiar by now.

Example 19.2 Consider the processes P_1, P_2 and P_3:

$$P_1 = (a \to stop)\Box(b \to stop)$$
$$P_2 = (a \to stop) \sqcap (b \to stop)$$
$$P_3 = stop \sqcap ((a \to stop)\Box(b \to stop))$$

Then $P_3 \sqsubseteq_F P_2 \sqsubseteq_F P_1$. To see this, note that the key trace to consider is the empty trace, since the refusals after $\langle a \rangle$ or $\langle b \rangle$ are identical for all three processes.

However, after $\langle\rangle$ the refusals of P_1 are $\varnothing$, those of P_2 are $\varnothing$, $\{a\}$, $\{b\}$ but crucially not $\{a, b\}$, whereas the refusals of P_3 are any $X \subseteq \{a, b\}$. Hence $failures(P_3) \subseteq failures(P_2) \subseteq failures(P_1)$. The divergences of all three processes are empty, and hence $P_3 \sqsubseteq_F P_2 \sqsubseteq_F P_1$. □

19.2 The Correspondence Between Data Refinement and Refinement in CSP

We have given combined Object-Z and CSP specifications a CSP semantics (see Sect. 18.2.1), and therefore can use CSP refinement relations to refine integrated specifications. For example, we could show that the failures and divergences of one specification were contained in the failures and divergences of another and hence verify a failures-divergences refinement.

However, we can do better than this. In particular, in order to refine specifications which combine CSP and Object-Z we can use simulations between the Object-Z components, in addition to techniques based around calculating failures and divergences.

This is because there are versions of the downward and upward simulation conditions that are sound and jointly complete with respect to failures-divergences refinement. This result goes back to Josephs [15], set in the context of refinement methods for state-based concurrent systems defined by transition systems where processes do not diverge. Josephs defines downward and upward simulations between CSP processes, and shows that, if we do not have divergence, these are sound and jointly complete with respect to failures-divergences refinement. (Since we have restricted ourselves in our language integration to Object-Z classes without internal operations, as long as we stay with the blocking model the Object-Z components we use represent processes which do not diverge as required by [15].)

These simulation methods can be adapted for use on Object-Z classes (see Smith and Derrick [19, 20]), and provide a subtle strengthening of the simulation rules as presented in Definitions 16.1 and 16.2.

The (mildly) surprising point to note is that augmented simulation rules need to be used in order to derive failures-divergences refinement. The issue is that one needs to be able to explicitly observe refusals to construct a correspondence with the failures-divergences refinement relation. The traces are contained in the standard relational semantics—through the notion of programs, however, the refusals are not present in the standard embedding we have used. Thus to fully encode failures it is necessary to enhance the standard relational theory by adding the observation of refusals at the end of every program. This results in a change to the global state, and more importantly the finalisation, and [7] shows how to do this via the definition of a refusal embedding, which observes the set of refused actions in the final state of a program.

To define the correct simulation rules, one needs to be able to calculate the traces, failures and divergences in the Object-Z model used. With no inputs or outputs they can be defined easily in each of the two models along the following lines.

Blocking Model Traces arise from sequences of operations which are defined within their guards. Refusals indicate the impossibility of applying an operation outside its precondition. Furthermore, there are no divergences since each operation is either blocked or gives a well-defined result.

Non-blocking Model As no operation is blocked, every trace is possible: those that arise in the blocking model, and any other ones following divergence. There are no refusals beyond those after a divergence, since before the ADT diverges, no operation is blocked, it either gives a well-defined result or causes divergence. There are now, however, divergences, which arise from applying an operation outside its precondition.

Since refusals are not normally observed in data refinement, it is necessary to observe the refusals directly via the simulation rules, and for this reason the refusal embedding is used. For data types without output the consequences are relatively straightforward, and result in a change to the upward simulation rule only.

The Simulation Conditions The downward and upward simulation rules as expressed for partial relations contain conditions on the finalisations. When we use the refusal embedding, and thus alter the finalisation, these conditions impose additional constraints.

Here, even with a refusal embedding a downward simulation places no further constraints than already present in the standard definition. For an upward simulation the amended finalisation conditions are:

$$\mathsf{CFin} \subseteq \mathsf{T} \mathbin{⨾} \mathsf{AFin}$$
$$\forall \mathsf{c} : \mathsf{CState} \bullet \mathsf{T}(\!|\, \{\mathsf{c}\} \,|\!) \subseteq \mathrm{dom}\, \mathsf{AFin} \Rightarrow \mathsf{c} \in \mathrm{dom}\, \mathsf{CFin}$$

With the refusals embedding the second is always satisfied, however, the first leads to a strengthening of the standard applicability condition from $\forall i : I \bullet \forall CState \bullet \exists AState \bullet T \wedge (\mathrm{pre}\, AOp_i \Rightarrow \mathrm{pre}\, COp_i)$ to the condition (as appears in [15]):

$$\forall CState \bullet \exists AState \bullet \forall i : I \bullet T \wedge (\mathrm{pre}\, AOp_i \Rightarrow \mathrm{pre}\, COp_i) \tag{19.1}$$

As noted in [7]: "The standard upward simulation applicability condition requires that we have to consider pairs of abstract and concrete states for each operation. The finalisation condition, on the other hand, requires that for every abstract state we can find a *single* concrete state such that all the preconditions of the abstract operations imply the preconditions of their concrete counterparts."

19.2.1 Adding Outputs to the Model

The presence of outputs (which were not considered in the model in [15]) has an effect on the refusals observed. In particular, when a system includes a non-deterministic choice of different output values, is the environment able to 'select' its choice among these, or is the choice entirely inside the system? The refusals in each model differ slightly, and only the latter option causes refusals through choice of output. These options are called the angelic and demonic models of outputs (see Sect. 18.4.1), respectively. In the *angelic* model the only refusals are the ones arising when an operation is not applicable, there are no refusals due to outputs. In the *demonic* model a process is, in addition, allowed to refuse all but one of the possible outputs.

Demonic Model The consequences again lie with the upward simulation conditions, since the downward simulation condition imposed by a refusal embedding is subsumed by the normal applicability and correctness rules.

For an upward simulation the finalisation condition again leads to an extra condition, which is somewhat complicated involving the need to look at combinations of different operations, whilst considering possible output values individually. This complexity arises from the presence of non-deterministic outputs, and their interaction with the refusal sets. With only deterministic outputs the upward simulation condition is as above (i.e., just involving a strengthened applicability condition).

In a model with (potentially non-deterministic) outputs, the requirement that $\mathsf{CFin} \subseteq \mathsf{T} \mathbin{;} \mathsf{AFin}$, forces one to consider different linked abstract states for different maximal concrete refusal sets. In particular, even with just a single operation, it is, in general, necessary to look at different linked states for different output values. Details are not simple, and the conditions are given in full in [7].

Angelic Model The effects of the finalisation on the simulation rules are less drastic in the angelic model. Whereas in the demonic model one can reduce non-determinism in the outputs, in the angelic model one cannot. This difference is easily expressed in the simulation rules by using a different precondition operator, defined as $\mathrm{Pre}\, Op \mathrel{\widehat=} \exists\, State' \bullet Op$, and one then uses Pre in place of pre in the strengthened applicability conditions.

A characterisation of the simulation rules is as follows. For downward simulation, we have the following potential collection of conditions (again see [7]).

DS.Init $\forall\, CState' \bullet CInit \Rightarrow \exists\, AState' \bullet AInit \wedge R'$

DS.App $\forall\, CState;\, AState;\, i : I \bullet R \wedge \text{pre}\, AOp_i \Rightarrow \text{pre}\, COp_i$

DS.CorrNonBlock $\forall\, i : I;\, Output;\, CState';\, CState;\, AState \bullet \text{pre}\, AOp_i \wedge R \wedge COp_i \Rightarrow \exists\, AState' \bullet R' \wedge AOp_i$

DS.CorrBlock $\forall\, i : I;\, Output;\, CState';\, CState;\, AState \bullet R \wedge COp_i \Rightarrow \exists\, AState' \bullet R' \wedge AOp_i$

DS.FinAng $\forall\, CState;\, AState;\, i : I;\, Output \bullet R \wedge \text{Pre}\ AOp_i \Rightarrow \text{Pre}\ COp_i$

For a basic embedding, non-blocking data refinement requires: **DS.Init**, **DS.App** and **DS.CorrNonBlock**, and blocking data refinement requires: **DS.Init**, **DS.App** and **DS.CorrBlock**.

For a refusals embedding in the blocking model the table below gives the rules that are required in the various situations. Each column represents a particular model for outputs; a missing entry indicates a condition dominated by the other conditions in the same column.

Outputs:	none	demonic	angelic
Init	**DS.Init**		
App	**DS.App**		-
Corr	**DS.CorrBlock**		
Fin	-		**DS.FinAng**

For upward simulation in the blocking model, we have the totality of T on $CState$ plus the following set:

US.Init $\forall\, CState';\, AState' \bullet T' \wedge CInit \Rightarrow AInit$

US.AppBlock $\forall\, i : I;\, Output \bullet \forall\, CState \bullet \exists\, AState \bullet T \wedge \text{pre}\, AOp_i \Rightarrow \text{pre}\, COp_i$

US.CorrBlock $\forall\, i : I;\, Output;\, AState';\, CState';\, CState \bullet T' \wedge COp_i \Rightarrow \exists\, AState \bullet T \wedge AOp_i$

US.FinRef $\forall\, CState \bullet \exists\, AState \bullet T \wedge \forall\, i : I \bullet \text{pre}\, AOp_i \Rightarrow \text{pre}\, COp_i$

US.FinAng $\forall\, CState \bullet \exists\, AState \bullet T \wedge \forall\, i : I;\, Output \bullet \text{Pre}\, AOp_i \Rightarrow \text{Pre}\, COp_i$

The basic blocking data refinement then requires: **US.Init**, **US.App** and **US.CorrBlock**. For a refusals embedding in the blocking model the following rules are required in the various situations.

Outputs:	none	demonic	angelic
Init	**US.Init**		
App	-		
Corr	**US.CorrBlock**		
Fin	**US.FinRef**	**US.FinDem**	**US.FinAng**

For the complex definition of property **US.FinDem**, see [7]

In summary, the result of all this is that because we have set up a correspondence between operations and events, we can compare a state-based refinement between two Object-Z classes with a CSP refinement between their corresponding interpretation as processes, with two degrees of freedom in this interpretation.

To verify a refinement in an integrated CSP/Object-Z specification we can therefore use either the CSP failures-divergences approach or the state-based simulation methods as defined above. If we stick to downward simulations we can even use the conditions as presented for Object-Z. The failures approach involves direct calculation of the failures and divergences of a specification, or components in a specification, and this can be done for either CSP or Object-Z. However, calculation of the failures of an Object-Z class is complex (see [19] for an example) and not recommended. Preferable, therefore, is the use of the simulation methods for verifying a refinement between Object-Z components.

Such refinements are also compositional. That is, if we construct a specification which combines components (whether specified in CSP or Object-Z) using CSP operators, then a failures-divergences refinement of any individual component induces a refinement of the whole specification due to the compositional nature of CSP failures-divergences refinement. The correlation between the simulation methods and CSP refinement then gives us the result we want.

The next sections illustrate this in practice, firstly for specifications where we refine the CSP processes, and secondly where we refine the Object-Z components.

19.3 Refinement of CSP Components

Where Object-Z classes have been composed with CSP processes, both the Object-Z and the CSP components can be refined separately, and the next example illustrates the latter case.

Example 19.3 The ticket machines in the underground railway station are being upgraded to a new "improved" version. Unfortunately this new version is faulty and has behaviour described by the following CSP/Object-Z specification.

$$FTM = FTM_1 \parallel TM_1$$

where TM_1 is the Object-Z component given in Example 18.7 and FTM_1 is the CSP specification:

$$\begin{aligned} FTM_1 = Coin?c \rightarrow (&FTM_1 \\ &\sqcap \\ &Ticket?single \rightarrow Dest?dest?change?ticket \rightarrow FTM_1 \\ &\sqcap \\ &Dest?dest?change?ticket \rightarrow FTM_1) \end{aligned}$$

As we can see the manufacturers have introduced two faults in the new machine. Firstly, a ticket might be dispensed without the user having specified whether a single or return is desired (the option will be non-deterministically chosen), and secondly the machine sometimes only accepts coins irrespective of whether the correct amount has already been inserted.

Fortunately we can implement this specification by something more sensible since TM_2 is a failures-divergences refinement of FTM_1, and hence TM is a failures-divergences refinement of FTM. Unfortunately so too are the machines with CSP components being either of the following:

$$\begin{aligned} FTM_2 &= Coin?c \rightarrow FTM_2 \\ FTM_3 &= Coin?c \rightarrow (FTM_3 \\ &\qquad\qquad\qquad \sqcap \\ &\qquad\qquad Dest?dest?change?ticket \rightarrow FTM_3) \end{aligned}$$

since these are also both failures-divergences refinement of FTM_1. □

19.4 Refinement of Object-Z Components

As well as refining the CSP component we can refine the Object-Z component independently of its use in the overall specification.

Example 19.4 In Example 18.6 (p. 440) a lift was described as having two components, *LTO* and $LiftCon_2$, described in CSP and Object-Z respectively. One possible refinement of the $LiftCon_2$ class would be to implement a strategy for choosing which floor to go to next, since the original component was completely underspecified in this respect.

For example, we could choose to service the requests in the order that they were made, although the movement would be rather erratic if we did so. This might be described by the following refinement of $LiftCon_2$:

$LiftCon_3$

$req : \mathrm{seq}\, FLOOR$
$pos : FLOOR$

INIT

$req = \langle\rangle \land pos = 1$

RequestD

$\Delta(req)$
$f? : FLOOR$

$req' = req \frown \langle f? \rangle$

RequestE

$\Delta(req)$
$f? : FLOOR$

$req' = req \frown \langle f? \rangle$

$$\begin{array}{l}
\underline{\mathit{Close}} \\
req \neq \langle \rangle
\end{array}
\qquad
\begin{array}{l}
\underline{\mathit{Arrive}} \\
\Delta(req, pos) \\
f! : FLOOR \\
\hline
pos' = head\ req \\
req' = tail\ req \\
pos' = f!
\end{array}$$

We can then conclude that $LTO \parallel LiftCon_3$ is a refinement of $LTO \parallel LiftCon_2$. □

Similarly, we can refine the Object-Z components in a specification which consists of Object-Z classes combined using CSP operators.

Example 19.5 We might choose to perform a data refinement of one of the buffer components from Example 18.9 (p. 446), producing the following:

$$\begin{array}{l}
\underline{Buf_1} \\
\quad r : \mathrm{seq}\, T \\
\quad \hline
\quad \#r \leq n \\
\\
\quad \underline{\mathit{INIT}} \\
\quad r = \langle \rangle \\
\\
\quad \underline{\mathit{Mid}} \\
\quad \Delta(r) \\
\quad x? : T \\
\quad \hline
\quad r' = r \frown \langle x? \rangle \\
\\
\quad \underline{\mathit{Get}} \\
\quad \Delta(r) \\
\quad x! : T \\
\quad \hline
\quad r = \langle x! \rangle \frown r'
\end{array}$$

Then Buf_1 refines $Buf[Mid/Put]$ and thus, for example,

$$(Buf[Mid/Get] \parallel Buf_1) \setminus \{Mid.\{(x, y)\} \mid y \in T\}$$

refines the $2n$ place buffer described originally. □

Example 19.6 In the specification

$$ATM_6 = CashPoint_1 \parallel Bank_1$$

from Example 18.10 (p. 447), the *Withdraw* operation in $Bank_1$ was non-deterministic in the amount of money it would deliver. We could thus refine $Bank_1$ to $Bank_2$:

$Bank_2$

$Bank_1$

Withdraw

$amt! : \mathbb{B}$
$amount?, money!, acct? : \mathbb{N}$

$((\neg amt! \wedge amount? \geq accts(acct?)) \vee$
$(money! = amount? \wedge amt! \wedge amount? < accts(acct?)))$

in the knowledge that $CashPoint_1 \parallel Bank_2$ then refines ATM_6. □

19.4.1 Example: The Lift Specification

In Example 17.4 (p. 415) we described a lift and its controller using two components, *Lift* and *Con*, together with a class $LiftCon_1$ which described how they interact, and this specification was a weak refinement of the original $LiftCon_0$ class. We can give an alternative specification of $LiftCon_1$ as a CSP synchronisation of the *Lift* and *Con* components as $LiftCon_4 \setminus \{SetTarget, MoveOneFLoor\}$, where $LiftCon_4$ is given as:

$$LiftCon_4 = Lift[SetTarget/GetTarget, RequestD/Request] \parallel Con[RequestE/Request]$$

This specification can be refined by refining the individual components, for example, we might consider alternative strategies for scheduling the movement of the lift.

In $LiftCon_4$ the next floor the lift moves to is determined by the *GetTarget* and *SetTarget* operations. If the controller *Con* has unsatisfied requests from customers waiting on the floors of the building it sends a potential next *target* to the lift. *GetTarget* will then either use this target or pick one from its own list of requests derived from those customers inside the lift.

One obvious implementation is to give priority to the customers actually inside the lift, as currently they may never get to the floor they want! Thus we refine *GetTarget* in *Lift* to

$GetTarget_1$

$\Delta(target, target\_received)$
$f? : FLOOR$
$target\_sent? : \mathbb{B}$

$\neg\, target\_received \wedge target\_received'$
$state = open$
$\neg\, target\_sent? \Rightarrow target' \in req$
$target\_sent? \wedge req = \varnothing \Rightarrow target' = f?$
$target\_sent? \wedge req \neq \varnothing \Rightarrow target' \in req$

which only services a request from the controller if it has no requests of its own.

Having done this we can tackle how a request is actually chosen from the set of requests in the lift. One option is, of course, to service requests in the order they were made, as we did in Example 19.4 with a data refinement of *Lift* where *req* becomes a sequence.

An alternative option would be to move to the nearest internally requested floor to the current position, and thus we refine $GetTarget_1$ to $GetTarget_2$.

$$
\begin{array}{|l}
\hline
GetTarget_2 \\
\Delta(target, target\_received) \\
f? : FLOOR \\
target\_sent? : \mathbb{B} \\
\hline
\neg\, target\_received \wedge target\_received' \\
state = open \\
(\neg\, target\_sent? \Rightarrow \\
\qquad target' \in req \wedge (\neg \exists p : req \bullet | pos - p | < | pos - target' |)) \\
(target\_sent? \wedge req = \varnothing \Rightarrow target' = f?) \\
(target\_sent? \wedge req \neq \varnothing \Rightarrow \\
\qquad target' \in req \wedge (\neg \exists p : req \bullet | pos - p | < | pos - target' |)) \\
\hline
\end{array}
$$

The same effect could, of course, be achieved in *SetTarget* if we amended the specifications to communicate the current position of the lift to the controller.

In our next iteration, instead of hopping all over the place we might think it sensible to take the direction of the lift into account, and to service the closest floor given the current direction of the lift. The direction of movement of the lift is in fact already available implicitly in the specification, but we can make it explicit in the following data refinement, which now takes this into account when determining the next target.

$$DIRECTION ::= up \mid down$$

$$
\begin{array}{|l}
\hline
Lift_1 \\
Lift \\
\quad \begin{array}{|l}
\hline
dir : DIRECTION \\
\hline
\end{array} \\
\quad \begin{array}{|l}
\hline
INIT \\
dir = up \\
\hline
\end{array}
\end{array}
$$

$GetTarget_3$

$\Delta(target, target\_received)$
$f? : FLOOR$
$target\_sent? : \mathbb{B}$

$\neg\ target\_received \wedge target\_received'$
$state = open$
$\neg\ target\_sent? \Rightarrow target' \in req$
$target\_sent? \wedge req = \varnothing \Rightarrow target' = f?$
$target\_sent? \wedge req \neq \varnothing \Rightarrow target' \in req \wedge$
$((dir = down \wedge target' < pos \wedge$
$(\neg \exists p : req \bullet p < pos \wedge (pos - p) < (pos - target'))) \vee$
$(dir = up \wedge target' > pos \wedge$
$(\neg \exists p : req \bullet p > pos \wedge (p - pos) < (target' - pos))) \vee$
$(dir = down \wedge \neg \exists t \in req \bullet t < pos \wedge dir' = up \wedge$
$(\neg \exists p : req \bullet | pos - p | < | pos - target' |)) \vee$
$(dir = up \wedge \neg \exists t \in req \bullet t > pos \wedge dir' = down \wedge$
$(\neg \exists p : req \bullet | pos - p | < | pos - target' |)))$

$MoveOneFloor$

$\Delta(pos)$

$state = closed$
$pos > target \Rightarrow dir' = down$
$pos < target \Rightarrow dir' = up$

Then in this case, as with the other lift refinements, $Lift_1$ refines $Lift$, and thus

$$Lift_1[SetTarget/GetTarget, RequestD/Request] \parallel Con[RequestE/Request]$$

refines $LiftCon_4$.

Finally we note that $Lift$ (and $Lift_1$ etc.) itself contains several variables whose sole purpose is to determine the correct temporal ordering, and this could be extracted into a CSP process. Hence we rewrite $Lift$ as

$$LTO_1 \parallel Lift_2$$

where the components LTO_1 and $Lift_2$ are given by:

$$LTO_1 = SetTarget?f?target\_sent \rightarrow Close \rightarrow Q$$
$$Q = MoveOneFloor \rightarrow Q$$
$$\Box$$
$$Arrive!f \rightarrow Open \rightarrow LTO_1$$

$Lift_2$

$pos, target : FLOOR$
$req : \mathbb{P}\, FLOOR$

$INIT$
$pos = 1 \wedge req = \varnothing$

$Request$
$\Delta(req)$
$f? : FLOOR$
$req' = req \cup \{f?\}$

$GetTarget$
$\Delta(target)$
$f? : FLOOR$
$target\_sent? : \mathbb{B}$
$\neg\ target\_sent? \Rightarrow target' \in req$
$target\_sent? \Rightarrow target' \in req \vee target' = f?$

$MoveOneFloor$
$\Delta(pos)$
$pos \neq target$
$pos > target \Rightarrow pos' = pos - 1$
$pos < target \Rightarrow pos' = pos + 1$

$Arrive$
$f! : FLOOR$
$pos = target \wedge f! = pos$
$req' = req \setminus \{pos\}$

The final specification being

$$LTO_1 \parallel Lift_2[SetTarget/GetTarget, RequestD/Request] \parallel Con[RequestE/Request]$$

19.5 Structural Refinements

Not all refinements in integrated CSP/Object-Z specifications can be verified using the simulation rules on individual components, and in these circumstances more complicated simulation conditions need to be employed.

Considering yet again our lift specification, we notice that we have described the following refinements.

$$\begin{aligned} \mathit{LiftCon}_0 \sqsubseteq_w\ & \mathit{LiftCon}_1 \\ & \equiv \mathit{LiftCon}_4 \setminus \{\mathit{SetTarget}, \mathit{MoveOneFloor}\} \\ & \sqsubseteq_F (\mathit{Lift}_1[\cdots] \parallel \mathit{Con}[\cdots]) \setminus \{\mathit{SetTarget}, \mathit{MoveOneFloor}\} \end{aligned}$$

where $\sqsubseteq_w$ denotes weak refinement in Object-Z and $\sqsubseteq_F$ CSP failures-divergences refinement.

The simulation rules we have described here work on the individual components in a combined CSP and Object-Z specification, however, they are not sufficient to be able to verify a refinement step from a single Object-Z component to a specification involving components combined via CSP operators.

For example, we cannot use the simulation rules to verify directly that $(\mathit{Lift}_1[\cdots] \parallel \mathit{Con}[\cdots]) \setminus \{\mathit{SetTarget}, \mathit{MoveOneFloor}\}$ is a failures-divergences refinement of $\mathit{LiftCon}_0$. To tackle situations such as these, structured simulation rules have been developed by Derrick and Smith [9]. These rules allow a single Object-Z component to be refined to a number of communicating or interleaved classes as in the lift example where the CSP parallel composition has been introduced in the refinement step from $\mathit{LiftCon}_0$ to $(\mathit{Lift}_1[\cdots] \parallel \mathit{Con}[\cdots]) \setminus \{\mathit{SetTarget}, \mathit{MoveOneFloor}\}$.

The structural simulation rules are also useful when using the blocking model with an angelic model of outputs. As discussed in Sect. 18.4.1 the motivation for use of that interpretation was to model cooperative communication between components without detailing the method of cooperation. It achieves this by allowing components to agree on values without having to describe explicitly how this agreement is achieved. However, in an implementation the actual agreement mechanism used will need to be made explicit, and the structural simulation rules allow one to introduce the exact mechanism as a series of refinement steps, allowing the abstract description of the communicating components to be refined to an implementation-oriented view.

Thus the blocking model with an angelic model of outputs does not constrain the design to necessarily adopt the abstract communication mechanism that motivated it. By using structural refinements, it is possible to introduce explicit mechanisms to communicate and negotiate between the components. Details of the rules and examples for these scenarios can be found in [21].

19.6 Bibliographical Notes

Work on state-based refinement relations for concurrent systems goes back to the end of the 1980s, and includes Josephs [15], He [13], Woodcock and Morgan [22], with renewed interest in the early 2000s, including work by Bolton and Davies [5, 6], and Derrick and Boiten [3, 4, 7]. That due to Josephs [15], He [13], Woodcock and Morgan [22] defines a basic correspondence between simulation rules and failures-divergences refinement. The more recent work of Bolton and Davies [5, 6], and Derrick and Boiten [3, 4, 7] investigates a direct correspondence between the relational model and process semantics, and includes specific consideration of input and output which introduces some subtleties.

The aim is to work out the correspondence between relational data refinement and the refinement relation induced by a given process semantics. Varying the type of relational model and the relational refinement relation gives rise to different refinement relations in the process semantics. Of course there is a whole variety of process semantics one could look at, but in integrations of Object-Z and CSP we are particularly interested in the semantic models of CSP, for example, traces-divergences or failures-divergences and how they related to data refinement in a relational model.

The variations in the relational model include whether to use the blocking or non-blocking model, but also the observations made. Apart from variations such as grey box refinement, the observations made are usually the input/output of the ADT, however, these can be extended to include, for example, the refusals in a given state. Understanding the consequence of these variations gives rise to the following type of result (this one proposed by Schneider):

> In the non-blocking relational model with standard observations, data refinement corresponds to traces-divergences refinement.

There is a version of this result in the blocking model that is due to Bolton and Davies, where the process semantics induced is a singleton-failures model. Subsequently, Reeves and Streader [18] showed that this equivalence requires a non-standard assumption on the blocking model, namely that the exact point of blocking is recorded in observations. In [7] it is shown what additional observations are needed to induce failures-divergences refinement. The latter derive simulation rules for failures-divergences refinement in the blocking model, and this was extended in [4] to include the integration of internal operations into this model. The results for refinement relations that do not consider internal operations are summarised in the following table:

Relational refinement	Process model	Citations
Non-blocking data refinement	Traces-divergences	Schneider (unpublished)
Blocking data refinement with deterministic outputs	Singleton failures	Bolton and Davies [5, 6]
Blocking data refinement	Singleton failures of process and input process	Bolton and Davies [5, 6]
Blocking data refinement with strengthened applicability but no input/output	Failures	Josephs [15]
Blocking data refinement with extended finalisations	Failures-divergences	Derrick and Boiten [3, 7]
Non-blocking data refinement with extended finalisations but no input/output	Failures-divergences	Derrick and Boiten [3, 7]

In [4] the work of [7] is extended by adding in the consideration of the two main 'erroneous' concurrent behaviours: deadlock and divergence, into relational refinement. It shows how to define simulation rules that correspond to failures-divergences refinement in the presence of internal operations (in the various output models). It also mechanises the theory and verifies the main results using the interactive theorem prover KIV.

In addition to the correspondence with failures-divergences, Derrick and Boiten [7] showed how to define relational refinement rules that correspond to process algebraic *readiness refinement* relation [17].

This program of work is continued in [8] where simulation conditions for process algebraic refinement are derived by defining further embeddings into the relational model: traces, completed traces, failure traces and extension. The framework is also extended to include various notions of automata based refinement, as automata offer another perspective on refinement to those given by a process algebra or state-based context. In [16] Lynch and Vaandrager provide a comprehensive treatment of refinement for automata, defining a number of simulation definitions and results relating them. In [8] the relationship between automata based refinement and the relational model is derived, hence answering the question raised in [16] concerning their connection. Derrick and Boiten [8] also consider IO-automata and provide a relational characterisation for IO-automata refinement and a set of simulation rules.

The first use of the results of Josephs [15] in the context of Object-Z components in an integrated CSP/Object-Z specification was due to Smith and Derrick [19, 20]. As stated earlier these results have been generalised in [9] to a set of structural refinement rules, where a development step can now change the granularity of components and introduce additional Object-Z classes communicating and synchronising via CSP operators. [21] also considered structural refinement rules when the differing models of output are used.

Simulation methods have also been developed for the CSP-OZ language. In [10] Fischer derives downward and upward simulations for use on the components in a CSP-OZ specification. Instead of structural simulation rules, model checking techniques based upon the CSP model checking tool FDR [12] have been developed [11] to verify simulations where the structure of the specification changes.

References

1. Araki, K., Galloway, A., & Taguchi, K. (Eds.) (1999). *International Conference on Integrated Formal Methods 1999 (IFM'99)*. York: Springer.
2. Bjørner, D., Hoare, C. A. R., & Langmaack, H. (Eds.) (1990). *VDM'90: VDM and Z!—Formal Methods in Software Development. Lecture Notes in Computer Science: Vol. 428*. Berlin: Springer.
3. Boiten, E. A., & Derrick, J. (2002). Unifying concurrent and relational refinement. *Electronic Notes in Theoretical Computer Science*, *70*(3), 94–131.
4. Boiten, E. A., Derrick, J., & Schellhorn, G. (2009). Relational concurrent refinement II: internal operations and outputs. *Formal Aspects of Computing*, *21*(1–2), 65–102.

5. Bolton, C., & Davies, J. (2002). Refinement in Object-Z and CSP. In M. Butler, L. Petre, & K. Sere (Eds.), *Integrated Formal Methods (IFM 2002). Lecture Notes in Computer Science: Vol. 2335* (pp. 225–244). Berlin: Springer.
6. Bolton, C., & Davies, J. (2006). A singleton failures semantics for communicating sequential processes. *Formal Aspects of Computing, 18*, 181–210.
7. Derrick, J., & Boiten, E. A. (2003). Relational concurrent refinement. *Formal Aspects of Computing, 15*(2–3), 182–214.
8. Derrick, J., & Boiten, E. A. (2012). Relational concurrent refinement part III: traces, partial relations and automata. *Formal Aspects of Computing*. doi:10.1007/s00165-012-0262-3
9. Derrick, J., & Smith, G. (2000). Structural refinement in Object-Z/CSP. In W. Grieskamp, T. Santen, & B. Stoddart (Eds.), *International Conference on Integrated Formal Methods 2000 (IFM'00). Lecture Notes in Computer Science: Vol. 1945* (pp. 194–213). Berlin: Springer.
10. Fischer, C. (2000). *Combination and implementation of processes and data: from CSP-OZ to Java*. PhD thesis, University of Oldenburg.
11. Fischer, C., & Wehrheim, H. (1999) Model checking CSP-OZ specifications with FDR. In Araki et al. [1] (pp. 315–334).
12. Formal Systems (Europe) Ltd. (1997). *Failures-Divergences Refinement: FDR 2*. FDR2 User Manual.
13. He, J. (1989). Process refinement. In J. McDermid (Ed.), *The Theory and Practice of Refinement*. Oxford: Butterworth-Heinemann.
14. Hinchey, M. G. & Liu, S. (Eds.) (1997). *First International Conference on Formal Engineering Methods (ICFEM'97)*. Hiroshima, Japan, November 1997. Los Alamitos: IEEE Comput. Soc.
15. Josephs, M. B. (1988). A state-based approach to communicating processes. *Distributed Computing, 3*, 9–18.
16. Lynch, N., & Vaandrager, F. (1995). Forward and backward simulations—Part I: Untimed systems. *Information and Computation, 121*(2), 214–233.
17. Olderog, E.-R., & Hoare, C. A. R. (1986). Specification-oriented semantics for communicating processes. *Acta Informatica, 23*, 9–66.
18. Reeves, S., & Streader, D. (2008). Data refinement and singleton failures refinement are not equivalent. *Formal Aspects of Computing, 20*(3), 295–301.
19. Smith, G., & Derrick, J. (1997) Refinement and verification of concurrent systems specified in Object-Z and CSP. In Hinchey and Liu [14] (pp. 293–302).
20. Smith, G., & Derrick, J. (2001). Specification, refinement and verification of concurrent systems—an integration of Object-Z and CSP. *Formal Methods in Systems Design, 18*, 249–284.
21. Smith, G., & Derrick, J. (2002). Abstract specification in Object-Z and CSP. In C. George & H. Miao (Eds.), *ICFEM. Lecture Notes in Computer Science: Vol. 2495* (pp. 108–119). Berlin: Springer.
22. Woodcock, J. C. P., & Morgan, C. C. (1990) Refinement of state-based concurrent systems. In Bjørner et al. [2] (pp. 340–351).

Chapter 20
Conclusions

We have come a long way since our initial definition of refinement in Chap. 2. Indeed the sign of a good theory is the richness that is produced from simple foundations, and the theory of refinement clearly falls in this category.

In part this stems from the use of simulations as the means to verify refinements, and they have become the *de facto* methodology in this respect. Their advantage is that they reduce verification of a refinement from substitutability of complete programs to a step-by-step comparison of individual operations. The presence of non-determinism gives rise to the necessity of having two different simulation rules if we are to provide a complete methodology.

Additional complexity then arises from the consideration of partial relations and operations, and the differing options for totalisation (including avoiding it). One totalisation, the contractual approach, gives us the simulation rules for use with Z, while the other, a behavioural interpretation, allows application of the rules to Object-Z.

In Part I of this book we have looked at some of this theory from the basic definition of refinement based upon substitutability to issues such as calculating refinements and the relationship between testing and refinement. All of this discussion took place within the standard theory of refinement with its usual presentation in Z.

Part II then went on to look at generalisations of simulations that arise from relaxing some of the assumptions made in Part I. The notion of what is observable is the crucial idea that underlies the chapters in this part. This idea of observability impacts on the input and output (IO and grey box refinement) as well as the operations themselves (weak and non-atomic refinement). This gives rise to a succession of generalisations that can be applied in a number of circumstances.

Part III took a different tack, and looked at how simulations can be adapted for use in Object-Z. The basics of this were simple, as were adapting the generalisations from Part II. The interesting aspects came, however, from the extra structuring provided in Object-Z, namely the use of objects and classes. This gave rise to the idea of a class simulation, which allowed the granularity of objects to be altered in a refinement.

J. Derrick, E.A. Boiten, *Refinement in Z and Object-Z*,
DOI 10.1007/978-1-4471-5355-9_20, © Springer-Verlag London 2014

Finally, in Part IV, we looked at how these ideas led us to combine state and behaviour in the form of Object-Z classes combined with CSP. This was possible due to a common semantic model of Object-Z and CSP which identifies objects with processes. The use of simulations then allowed Object-Z classes to be refined in a compositional manner.

There is much, however, we have not touched upon. There is, for example, a wealth of relational theory which we have not discussed and which is covered in depth in the monograph by de Roever and Engelhardt [16]. There has also been extensive work on refinement in other state-based notations, for example in VDM, B etc.

VDM standing for Vienna Development Method [24], bears many similarities to Z (for a discussion of the differences see [20]) and was developed around the same time. Being a methodology, as opposed to a notation, there is a well-defined notion of refinement, which for VDM is the use of downward simulations restricted to total surjective functions (from concrete to abstract). The use of total surjective functions, along with the omission of upward simulations, of course, makes the method incomplete.

Examples of refinement in VDM are given in Jones and Shaw [25], Jones [23, 24], and Bicarregui [5]. Clement discusses the VDM approach to refinement in [15] and extensions to refinement in VDM are considered in [17].

An alternative semantic basis to the relational framework that underpins Z and VDM is that of the weakest precondition semantics. Many of the results found in the relational setting have direct counterparts in the weakest precondition framework. For example, a single complete simulation rule can be derived similarly to the derivation in Chap. 8. This semantic basis has been used for a number of notations such as B and action systems.

B [1] like VDM, is a methodology and, again, refinement is restricted to downward simulations (but without the additional restrictions of VDM). Since a weakest precondition semantics has been adopted, the simulation rules are expressed as predicate transformers, but can be seen to directly correspond to the relational formalisation.

A collection of case studies in B refinement is presented in [36]. Further examples of refinement in B are given in [4, 27].

A correspondence between B and action systems has been used as a vehicle to specify and refine concurrent systems, see, for example, [12], and work which integrated B and CSP was described in Chap. 18.

Event-B [2, 3] is a trimmed down descendant of B, aiming at the specification, formal development, and analysis of event-based systems. This method has gained significant acclaim, particularly with the impressive tool support and industrial experience generated through EU projects Rodin (www.event-b.org, 2004–2007) and Deploy (www.deploy-project.eu, 2008–2012). Semantically, it concentrates on the link with action systems, moving away somewhat from the predicate transformer semantics used for B. Refinement in Event-B is mostly inherited from action systems, i.e., it is trace refinement with additional conditions preventing the introduction of global deadlock and divergence, see also [13].

The notion of refinement in Event-B is related to CSP refinement relations in [35]. The Event-B refinement design decisions, particularly with respect to non-atomic and weak refinement, are analysed in detail in [6].

The basic toolchain for Event-B supports the verification of individual refinement steps. A multi-level algorithm which executes multiple levels of an Event-B refinement hierarchy at once, checking the gluing invariant and the other refinement proof obligations on-the-fly while exploring the state space of the model is described in [19].

All of these notations are means to specify systems abstractly and then refine the abstract model. One aspect which we have not touched upon is the refinement of abstract models into code, and this is dealt with by the refinement calculus.

The refinement calculus [30] is a notation that allows incremental refinement of specification statements into code. The semantics of the refinement calculus is, like B, based upon a weakest precondition semantics.

In the refinement calculus a specification statement is an abstract program which consists of a precondition, postcondition and a list of variables known as a frame. This is interpreted as describing an operation which begins in a state satisfying the precondition, ends in a state satisfying the postcondition, and which only changes variables in the frame.

The refinement calculus provides laws to introduce programming constructs into specification statements, allowing computational detail to be included in the description of the operation.

Texts which consider how Z specifications can be integrated with the refinement calculus include those by Potter, Sinclair and Till [32] and Woodcock and Davies [38]. Cavalcanti and Woodcock have defined ZRC [14], a refinement calculus for Z. With Sampaio, they defined a language called Circus which includes the main elements of Z, CSP, and refinement calculus [33], grounded in Hoare and He's Unifying Theories of Programming [22].

A well-developed refinement theory also exists for the **ASM** (Abstract State Machine) [10] notation. In its essence an ASM defines a finite set of transition rules, which is enriched with a variety of function definitions in order for it to serve as a specification language. Although its theory of refinement is defined in terms of (generalised) simulations, its foundation is not one of observations and substitutivity, but rather it takes a generalised simulation diagram as its *starting point* rather than as an implementation of some underlying notion of refinement [8]. Such a generalised simulation diagram where m steps in the abstract machine are refined into n steps in the concrete is called an (m, n)-refinement. Given such diagrams, one can define that C is a correct refinement of A if and only if every (infinite) refined run simulates an (infinite) abstract run with equivalent corresponding states, and correctness of refinement implies for the special case of terminating runs the inclusion of the input/output behaviour of the abstract and the refined machine. Work by Schellhorn [34] has exposed these ideas in some more detail and related them to more general ideas of state-based systems refinement.

One aspect to note, and this relates to the definition of non-atomic refinement used in this book, is that the values of m and n in the generalised simulation rule are allowed to dynamically depend on the state. There are numerous examples of this, which are useful reminders that in practice refinements do not always take on simple forms. For example, in [9] a refinement step of Prolog to WAM code is described where n has no predetermined fixed bound and is determined only dynamically. In a specification of Lamport's "bakery" mutual exclusion algorithm, see [8], n is fixed but grows with the number of protocol members, and sometimes n is finite but without a priori bound, depending on the execution time of the participating processes. Another example is the correctness proof of a Java-to-JVM compiler in [7] which uses $(1, n)$-refinements with $0 \leq n \leq 3$ depending on the length of the computation which leads the JVM machine from one to its next state. Schellhorn in [34] showed that every (m, n)-refinement with $n > 1$ can be reduced to $(m, 1)$-refinements; this was used as the basis for a verification methodology implemented in the KIV theorem prover.

Finally, one should also mention work on **TLA+** [26] (TLA standing for Temporal Logic of Actions), for example, Merz [29] discusses the refinement concepts underlying **TLA+**, and Hesselink [21] has studied the foundations of refinement in the TLA context.

References

1. Abrial, J.-R. (1996). *The B-Book: Assigning Programs to Meanings*. Cambridge: Cambridge University Press.
2. Abrial, J.-R. (2010). *Modelling in Event-B*. Cambridge: Cambridge University Press.
3. Abrial, J.-R., Cansell, D., & Méry, D. (2005) Refinement and reachability in Event-B. In Treharne et al. [37] (pp. 222–241).
4. Abrial, J.-R., & Mussat, L. (1997). Specification and design of a transmission protocol by successive refinements using B. In M. Broy & B. Schieder (Eds.), *Mathematical Methods in Program Development. NATO ASI Series F: Computer and Systems Sciences: Vol. 158* (pp. 129–200). Berlin: Springer.
5. Bicarregui, J. C. (Ed.) (1998). *Proof in VDM: Case Studies. FACIT*. London: Springer.
6. Boiten, E. A. (2012). Introducing extra operations in refinement. *Formal Aspects of Computing*. doi:10.1007/s00165-012-0266-z.
7. Börger, E. (2002). The origins and the development of the ASM method for high level system design and analysis. *Journal of Universal Computer Science*, *8*(1), 2–74.
8. Börger, E. (2003). The ASM refinement method. *Formal Aspects of Computing*, *15*(2–3), 237–257.
9. Börger, E., & Rosenzweig, D. (1995). The WAM—definition and compiler correctness. In C. Beierle & L. Plümer (Eds.), *Logic Programming: Formal Methods and Practical Applications. Studies in Computer Science and Artificial Intelligence* (pp. 20–90). Amsterdam: North-Holland.
10. Börger, E., & Stark, R. F. (2003). *Abstract State Machines: A Method for High-Level System Design and Analysis*. New York: Springer.
11. Bowen, J. P., Hinchey, M. G., & Till, D. (Eds.) (1997). *ZUM'97: the Z Formal Specification Notation. Lecture Notes in Computer Science: Vol. 1212*. Berlin: Springer.
12. Butler, M. (1997) An approach to the design of distributed systems with B AMN. In Bowen et al. [11] (pp. 223–241).

13. Butler, M. J. (2009) Decomposition structures for Event-B. In Leuschel and Wehrheim [28] (pp. 20–38).
14. Cavalcanti, A., & Woodcock, J. C. P. (1998). ZRC—a refinement calculus for Z. *Formal Aspects of Computing*, *10*(3), 267–289.
15. Clement, T. (1994) Comparing approaches to data reification. In Naftalin et al. [31] (pp. 118–133).
16. de Roever, W.-P., & Engelhardt, K. (1998). *Data Refinement: Model-Oriented Proof Methods and Their Comparison*. Cambridge: Cambridge University Press.
17. Elvang-Goransson, M., & Fields, R. E. (1994) An extended VDM refinement relation. In Naftalin et al. [31] (pp. 175–189).
18. Frappier, M., Glässer, U., Khurshid, S., Laleau, R., & Reeves, S. (Eds.) (2010). *ABZ 2010. Lecture Notes in Computer Science: Vol. 5977*. Berlin: Springer.
19. Hallerstede, S., Leuschel, M., & Plagge, D. (2010) Refinement-animation for Event-B—towards a method of validation. In Frappier et al. [18] (pp. 287–301).
20. Hayes, I. J. (1992). VDM and Z: a comparative case study. *Formal Aspects of Computing*, *4*(1), 76–99.
21. Hesselink, W. H. (2005). Eternity variables to prove simulation of specifications. *ACM Transactions on Computational Logic*, *6*(1), 175–201.
22. Hoare, C. A. R., & He, J. (1998) *Unifying Theories of Programming. International Series in Computer Science*. New York: Prentice Hall.
23. Jones, C. B. (1980). *Software Development: A Rigorous Approach*. New York: Prentice Hall.
24. Jones, C. B. (1989). *Systematic Software Development Using VDM*. New York: Prentice Hall.
25. Jones, C. B. & Shaw, R. C. F. (Eds.) (1990). *Case Studies in Systematic Software Development*. New York: Prentice Hall.
26. Lamport, L. (2002). *Specifying Systems*. Boston: Addison-Wesley.
27. Lano, K. (1996). *The B Language and Method. FACIT*. Berlin: Springer.
28. Leuschel, M. & Wehrheim, H. (Eds.) (2009). *Integrated Formal Methods, 7th International Conference, IFM 2009. Lecture Notes in Computer Science: Vol. 5423*. Berlin: Springer.
29. Merz, S. (2008). The specification language TLA$^+$. In D. Bjørner & M. C. Henson (Eds.), *Logics of Specification Languages. Monographs in Theoretical Computer Science* (pp. 401–451). Berlin: Springer.
30. Morgan, C. C. (1994). *Programming from Specifications* (2nd ed.). *International Series in Computer Science*. New York: Prentice Hall.
31. Naftalin, M., Denvir, T., & Bertran, M. (Eds.) (1994). *FME'94: Industrial Benefit of Formal Methods, Second International Symposium of Formal Methods Europe. Lecture Notes in Computer Science: Vol. 873*. Berlin: Springer.
32. Potter, B., Sinclair, J., & Till, D. (1991). *An Introduction to Formal Specification and Z. International Series in Computer Science*. New York: Prentice Hall. 2nd ed., 1996.
33. Sampaio, A., Woodcock, J. C. P., & Cavalcanti, A. (2002). Refinement in Circus. In L.-H. Eriksson & P. A. Lindsay (Eds.), *FME. Lecture Notes in Computer Science: Vol. 2391* (pp. 451–470). Berlin: Springer.
34. Schellhorn, G. (2001). Verification of ASM refinements using generalized forward simulation. *Journal of Universal Computer Science*, *7*(11), 952–979.
35. Schneider, S. A., Treharne, H., & Wehrheim, H. (2012, in press). The behavioural semantics of Event-B refinement. *Formal Aspects of Computing*.
36. Sekerinski, E. & Sere, K. (Eds.) (1999). *Program Development by Refinement—Case Studies Using the B Method. FACIT*. Berlin: Springer.
37. Treharne, H., King, S., Henson, M. C., & Schneider, S. A. (Eds.) (2005). *ZB 2005: Formal Specification and Development in Z and B, 4th International Conference of B and Z Users. Lecture Notes in Computer Science: Vol. 3455*. Berlin: Springer.
38. Woodcock, J. C. P., & Davies, J. (1996). *Using Z: Specification, Refinement, and Proof*. New York: Prentice Hall.

Glossary of Notation

This glossary lists the notation used in this book. Further explanation of notation and the Z concrete syntax can be found in the Z standard. The Object-Z concrete syntax is included in the Object-Z reference manual [1].

Logic

Let P, Q be predicates and D a declaration. Let i be a name and x, y and e be expressions.

Notation		**Page**
$true, false$	logical constants	8
$\mathbb{B}$	Booleans: the set of logical constants	8
$\neg P$	negation: $true$ iff P is $false$	5
$P \wedge Q$	conjunction: $true$ iff both P and Q are $true$	5
$P \vee Q$	disjunction: $false$ iff both P and Q are $false$	5
$P \Rightarrow Q$	implication: $false$ iff P is $true$ and Q is $false$	5
$P \Leftrightarrow Q$	bi-implication: $(P \Rightarrow Q) \wedge (Q \Rightarrow P)$	5
$\forall D \bullet P$	universal quantification	5
$\exists D \bullet P$	existential quantification	5
$\exists_1 D \bullet P$	unique existential quantification	6
$\forall D \mid P \bullet Q$	$\forall D \bullet P \Rightarrow Q$	5
$\exists D \mid P \bullet Q$	$\exists D \bullet P \wedge Q$	5
if P then x else y	conditional expression	
let $i == x \bullet e(i)$	let expression: e with the value of x substituted for i	

J. Derrick, E.A. Boiten, *Refinement in Z and Object-Z*,
DOI 10.1007/978-1-4471-5355-9, © Springer-Verlag London 2014

Sets

Let S, T, S_i be sets, x, y, x_i, k terms, P a predicate, and p a tuple.

Notation		Page
$\varnothing$	empty set: the set with no members	6
$x \in S$	membership: *true* iff x is a member of S	6
$x \notin S$	non-membership: $\neg(x \in S)$	
$S \subseteq T$	subset: $\forall x : S \bullet x \in T$	6
$S = T$	equality: $S \subseteq T \wedge T \subseteq S$	6
$\{x\}$	singleton set: $y \in \{x\}$ iff $y = x$	
$S \cup T$	union: the set of x such that $x \in S \vee x \in T$	7
$S \cap T$	intersection: $\{x : S \mid x \in T\}$	7
$S \setminus T$	difference: $\{x : S \mid x \notin T\}$	7
$\{x_1, \ldots, x_n\}$	set extension: $\{x_1\} \cup \{x_2, \ldots, x_n\}$	6
$\{k : S \mid P\}$	comprehension: subset of elements k of S such that P	6
$\{k : S \mid P \bullet e\}$	comprehension	6
$\mathbb{P}\, S$	powerset: the set of all subsets of S	7
$\mathbb{F}\, S$	finite powerset: the set of all finite subsets of S	7
$\#S$	cardinality: the cardinality of a finite set S	7
(x, y)	the ordered pair	7
$S \times T$	Cartesian product	7
$x \mapsto y$	maplet: identical to (x, y)	11
first p	first: $first(x, y) = x$	12
second p	second: $second(x, y) = y$	12
$(x_1, \ldots, x_n)$	n-tuple: abbreviates $(x_1, (x_2, \ldots, x_n))$	
$S_1 \times \cdots \times S_n$	Cartesian product: abbreviates $S_1 \times (S_2 \times \cdots \times S_n)$	
$p.i$	component selection: returns ith element of the tuple	12

Numbers

Notation		Page
$\mathbb{Z}$	integers	8
$\mathbb{N}$	natural numbers	8
$\mathbb{N}_1$	positive natural numbers	8
$\mathbb{R}$	reals	8
div	integer division	
mod	remainder upon division	
$m..n$	set of integers between m and n inclusive	

Relations

Let X, Y and Z be sets; R and Q relations between X and Y; R_1 a relation between X and Y; R_2 a relation between Y and Z, and R_3 a relation between U and V. Finally, let S and T be sets with $S \subseteq X$ and $T \subseteq Y$.

Notation		**Page**
$X \leftrightarrow Y$	relations between X and Y: $\mathbb{P}(X \times Y)$	11
$id\,X$	identity: $\{x : X \bullet x \mapsto x\}$	11
$\mathrm{dom}\,R$	domain: $\{p : R \bullet \mathit{first}\ p\}$	12
$\mathrm{ran}\,R$	range: $\{p : R \bullet \mathit{second}\ p\}$	12
$S \lhd R$	domain restriction: $\{p : R \mid \mathit{first}\ p \in S\}$	12
$S \unlhd R$	domain subtraction: $\{p : R \mid \mathit{first}\ p \notin S\}$	12
$R \rhd T$	range restriction: $\{p : R \mid \mathit{second}\ p \in T\}$	12
$R \unrhd T$	range subtraction: $\{p : R \mid \mathit{second}\ p \notin T\}$	12
$R(\!\| S \|\!)$	relational image: $\{p : R \mid \mathit{first}\ p \in S \bullet \mathit{second}\ p\}$	13
$Q \oplus R$	relational overriding: $((\mathrm{dom}\,R) \unlhd Q) \cup R$	13
$R_1 \mathbin{;} R_2$	relational composition	14
$R^{\sim}$ or R^{-1}	relational inverse: $\{p : R \mid \mathit{second}\ p \mapsto \mathit{first}\ p\}$	13
$R_1 \parallel R_3$	relational parallel composition	14
S/R	weakest post-specification	14
$R \setminus T$	weakest pre-specification	14
$R*$	map	19

Functions

Let f be a function from X to Y and g a function from Y to Z; let X and Y be sets; and $t \in \mathrm{dom}\,f$.

Notation		**Page**
$f(t)$	function application	15
$f \circ g$	function composition: $g \mathbin{;} f$	15
$X \nrightarrow Y$	partial functions	15
$X \rightarrow Y$	total functions	15
$X \rightarrowtail Y$	total injective functions	16
$X \twoheadrightarrow Y$	total surjective functions	16
$X \rightarrowtail\!\!\!\!\!+ Y$	partial injective functions	16
$X \twoheadrightarrow\!\!\!\!\!+ Y$	partial surjective functions	16
$X \nrightarrow\!\!\!\!\!+ Y$	finite functions	16
$X \rightarrowtail\!\!\!\!\!+\!\!+ Y$	finite injective functions	16
$X \rightarrowtail\!\!\!\!\!\twoheadrightarrow Y$	total bijections	16

Sequences

Let X be a set; $x_i \in X$ and let s and t be sequences of type X.

Notation		Page
$\langle\,\rangle$	empty sequence	17
$\langle x_1, \ldots, x_n \rangle$	sequence listing	17
$\mathrm{seq}\, X$	sequences of type X	17
$\mathrm{seq}_1\, X$	non-empty sequences of type X	17
$\mathrm{iseq}\, X$	injective sequences of type X	17
$s \frown t$	concatenation	18
head s	head	18
tail s	tail	18
front s	front	18
last s	last	18
$\frown/$	distributed concatenation	18
$\fatsemi/$	distributed composition	18
$*$	map	18

Bags

Let X be a set; let B and C be bags of type X and $x, x_i \in X$. Let s be a sequence.

Notation		Page
$[\![\,]\!]$	empty bag	19
$[\![x_1, \ldots, x_n]\!]$	bag listing	19
$\mathrm{bag}\, X$	set of bags of type X	19
$x \mathrel{\mathrm{in}} B$	bag membership	19
items s	turns a sequence into a bag	19
count B x	bag count: the number of times x occurs in B	19
$B \uplus C$	bag union	19
$B \cup\!\!\!- C$	bag difference	19

Z Notation

Let *State*, S, S_1 and S_2 be schemas, let v, p and q be variables, and i a name and e an expression.

Notation		Page
$::=$	free types	8
$==$	abbreviation	10
$v?$	input	22
$v!$	output	22

v'	variable decoration	22
$State'$	schema decoration	23
$\Delta State$		23
$\Xi State$		24
$S[p/q]$	schema renaming	27
$S_1 \wedge S_2$	schema conjunction	27
$S_1 \vee S_2$	schema disjunction	28
$\neg S$	schema negation	30
$S_1 \Rightarrow S_2$	schema implication	31
ΣS	signature	31
$S \setminus (\cdots)$	hiding	33
$S_1 \gg S_2$	piping	35
$S_1 \mathbin{;} S_2$	schema composition	34
pre S	precondition	36
$?S$	input signature	36
$!S$	output signature	36
θ	theta	38
$\langle\!\vert i == e \vert\!\rangle$	binding extension	38
$\overline{S}$	IO decoration	244
$??S$	input view	247
$!!S$	output view	247

Object-Z Notation

Let f_i be operation names, variables or constants, and let v_i be variables. Let A, B and C be classes. Let o be an object expression and op, op_1, op_2 be operation expressions.

Notation		**Page**
$\upharpoonright(f_1, \ldots, f_n)$	visibility list	364
$\Delta(v_1, \ldots, v_n)$	Δ list	365
$\widehat{=}$	schema definition	365
INIT	initialisation schema	365
Δ	secondary variables indicator	367
$op_1 \wedge op_2$	conjunction	367
$op \setminus (\cdots)$	hiding	367
$op_1 \, [] \, op_2$	disjunction	367
$op_1 \parallel op_2$	parallel composition	368
$op_1 \parallel_! op_2$	associative parallel composition	368
$op_1 \bullet op_2$	environment enrichment	368
$op_1 \mathbin{;} op_2$	sequential composition	368
$op_1 \gg op_2$	piping	368
$o.op$	operation application	372

$o.\Xi$	implicit identity operation	373
$\Xi(S)$	implicit identity on set of objects	373
$v : A_{©}$	object containment	376
$\mathbb{O}$	universe of object identities	379
$\mathrm{pre}\, op$	precondition	380
$B \cup C$	class union	380
$v :\downarrow A$	polymorphic declaration	380
θ	theta	406
$(M \bullet C_1, \ldots, C_n)$	system class and auxiliary classes	409

References

1. Smith, G. (2000). *The Object-Z Specification Language*. Dordrecht: Kluwer Academic.

Index

J. Derrick, E.A. Boiten, *Refinement in Z and Object-Z*,
DOI 10.1007/978-1-4471-5355-9, © Springer-Verlag London 2014

Printed in the United States
By Bookmasters